ARCHAEOLINGUA

Edited by
ERZSÉBET JEREM and WOLFGANG MEID

Volume 21

FERENC GYULAI

ARCHAEOBOTANY IN HUNGARY

Seed, Fruit, Food and Beverage Remains
in the Carpathian Basin
from the Neolithic to the Late Middle Ages

BUDAPEST 2010

This book was published with the financial support by
the National Cultural Fund of Hungary,
the Ministry of National Cultural Heritage
and the Hungarian Academy of Sciences

Cover illustration
Seed remains from the Teleki mansion in the Buda castle, 15th century.
Photograph by
GÁBOR GYULAI

Hungarian text revised by
NÁNDOR KALICZ

Translated by
BÉLA BORSOS and PÉTER SZABÓ

Volume Editor
ERZSÉBET JEREM

ISBN 978 963 8046 93 2
HU-ISSN 1215-9239

2010

ARCHAEOLINGUA ALAPÍTVÁNY
H-1250 Budapest, Úri u. 49

Copy editing by Irwin Rovner, Péter Szabó and Réka Benczes
Desktop editing and layout by Rita Kovács, András Kardos

Printed by Prime Rate Kft.

Contents

1. Introduction

In recent decades, archaeology has benefitted substantially through interdisciplinary cooperation with many specialized areas of science. Scientific methods of analysis applied to archaeological research have increased the amount of information obtained at all levels of study; from a single artefact to whole assemblages of finds, from an excavation to a region of sites of all sizes and all ages. Archaeologists and historians now have greater ability to reconstruct ancient lifeways, art and culture, industries, architecture, settlement patterns, land use and more. Human impact on the environment through such achievements as the development of agriculture and the spread of cultivated plants can now be tracked using new and sophisticated types of scientific investigation. In Hungary, soil conditions in archaeological sites are often very favourable to the preservation of large quantities of seeds, cereal grains and weeds, an opportunity to assess the quality and quantity of plant usage, agricultural knowledge and technologies in individual cultures and regions of the Carpathian Basin.

In the last decade, the increasingly active relationship between archaeology and botany has lead to the establishment of several institutes of archaeobotany in Western Europe. Today, an archaeobotanist is a standard participant in significant archaeological projects. The archaeobotanist can aid the archaeologist in devising appropriate research and sampling strategies, apply specialized recovery techniques such as flotation, and conduct the all-important identification of recovered botanical remains. When archaeologists and archaeobotanists combine their experience and knowledge to interpret plant uses in earlier societies, a far more detailed and comprehensive understanding of history is achieved.

This active relationship is an increasing part of Hungarian archaeology. Archaeobotany has become a basic course taught in archaeology curricula at Hungarian universities. The objective of such training is to ensure that students gain a perspective of historical ecology, an acquaintance with the contributions of archaeobotany to history, a specialized knowledge of botanical systematics, and, especially, expert assistance in the recovery and identification of plant remains from archaeological sites. This is particularly important because even a single grain or a fragment of fruit found in an excavation can be a source of significant information. All recovered remains are "eco-facts," i.e. artefacts of the associated ecology. Remains of wild plants specifically collected for human use and especially cultivated plants are truly "human artefacts" as well. A great deal depends on the archaeologist leading the excavation, whether s/he recognizes the importance of such an interdisciplinary investigation or is content with the far more restricted information gleaned from traditional archaeological recovery limited to inorganic and non-biological objects.

Archaeobotanical investigation has demonstrated convincingly that the Carpathian Basin is one of the longest inhabited parts of Europe. Cultivated plants arrived in the Carpathian Basin with the first Neolithic agrarian culture some eight thousand years ago. Cultivation of vegetables has been pursued for four thousand years and fruit cultivation has been going on for at least two thousand years (GYULAI 2000a).

Archaeobotanical study of grain remains contributes not only to the history of plant cultivation but also contributes to our understanding of Hungary's historical ecology. Hungary devoted some 4.7 million hectares of land to agricultural cultivation at the end of the 20th century, With respect to agrobiodiversity, some 4000 recognised cultivated varieties of 350 arable crop and horticultural plant species were being produced (HESZKY – BÓDIS – KISS 1999). Hungarian gene banks store and study some 80,000 individual samples of genetic material (HOLLY 2000). In our rapidly changing world, the discovery and preservation of genetic reserves is exceedingly important. While it is obvious that continual developments in biotechnology allow new varieties to be produced much faster than using the traditional method of selection, the maintenance of plant species locally adapted to the Carpathian Basin after decades and centuries of cultivated evolution are irreplaceable. This, and only this, guarantees a sustainable food supply for the future as a storehouse of raw material for the breeding of plants. The special climatic and ecological conditions of the Carpathian Basin, the long period of time that cultivated plants have been grown and the accompanying processes of selection have resulted in a very high degree of diversity. Accordingly, these regional varieties should be regarded as a part of our cultural heritage, so that their preservation is a matter of national concern (V. MADARASSY 2000).

The plants of old – unlike today's often overbred, weakened, monoculturally grown varieties – were in harmony with their environment to such an extent that they virtually formed an organic whole with it. Individual regions and areas "bred" their own varieties. For want of a better term, we call such varieties "ancient landraces." These regional cereal grain and fruit varieties are genetically resistant, have high nutrient value for human consumption and produce significant yields. They were adapted successfully to their region of production prior to the development of modern, intensive – often expensive and ecologically adverse – agricultural practices. These are positive qualities for their reintroduction today into an ecologically benign, sustainable agriculture. The best hope for their preservation and maintenance in their original habitats is to produce them in "sensitive natural areas." A successful campaign is needed to convince people living and farming in these areas that it is in their interest to do natural farming including production of ancient landraces. This represents a new opportunity for the maintenance of the diversity and sustainability of cultivated species (TARDY 1994).

However, there seems to be no continuity of cultivated plants in successive archaeological eras up to the Middle Ages. The cultivation of several specific plant species is often associated only with a limited archaeological era or a specific culture. A review of the history of the Carpathian Basin over the last eight thousand years indicates no significant continued association between many cultivated plants (species and varieties) grown and individual human populations through time. New peoples moving into the area apparently introduced their own cultivated plants. The "disappearance" of certain archaeological cultures is associated with the almost complete disappearance of associated cultivated plant species that were previously grown extensively. Therefore, these earlier societies are often confidently identified by culture-association of their cultivated plants. However, these are not merely historical curiosities. Understanding how these introduced plants were adapted to their new region and why they ultimately disappeared (failed?) is relevant to any plan to reintroduce these ancient landraces in modern agriculture.

Although cultivated fields increasingly replaced areas of natural vegetation, plants often transferred from cultivation to wild contexts allowing them to persist within the limits of their ecological conditions, including climate change through time. (GYULAI 2000b). Many species, that had been cultivated plants characteristic of individual cultures, became weeds or disappeared almost without a trace. For example, bitter vetch (*Vicia ervilia*) appeared in Hungary only in the Middle and Late Bronze Age. This leguminous plant characterised vegetable gardens for over a thousand years. As the Bronze Age passed, it no longer occurred among plant remains, not even as a weed. On the other hand, some species that had been present for many millennia as weeds were transformed quite recently into cultivated plants. Rudolf Fleischmann, in the 1930s at Kompolt, experimenting with fodder vetch (*Vicia villosa*), created hairy vetch (*V. villosa* subsp. *pseudovillosa*), a plant with hairless leaves and no bitter component. The fodder species is still being grown today. There are ongoing experiments at Tápiószele to improve other vetch species (*V. sepium*, *V. articulata*, and *V. biennis*).

Many cultivated plants that had been forgotten for several thousand years are re-entering agriculture as important crops today. The highly nutritious grain, amaranth, was an important cultivated flour grain in Precolumbian cultures of South and Middle America before it was largely replaced by maize (*Zea mays* L.) and mostly forgotten for a very long time. Renewed interest in this species began in the United

States in the 1970s and in Hungary in the 1980s with the aim of producing the plant for food and fodder purposes (SZŐCS 1995). Today, the species has several officially recognised varieties. Further experiments are being conducted to improve amaranth and to introduce similarly "forgotten" crop plants of other cultures at the Agrobotanical Institute at Tápiószele (verbal communication from Béla Baji).

A major, critical unanswered question concerns the origins of cultivated plant varieties. Often we do not know when a landrace was first cultivated or who was responsible for its development. Based on written sources, the use of cultivated fruit varieties in Hungary can be traced back to the Middle Ages (SURÁNYI 1985). Metric measurements provide perhaps the best evidence in the case of grape varieties. Several varieties of black winegrapes were already grown in the Roman Period (DÁLNOKI 1998). A diversity of varieties grew in the Middle Ages (FACSAR 1970; but these data by no means establish the beginning of their use. The longest vegetable lineages go back only to the modern age (KAPÁS 1997). This is more difficult to do for cereal grains.

Our preliminary computer image analysis studies of recent cereal grain populations have indicated that the measured values (e.g. length, width, area, etc.) and non-dimensional ratios (e.g. ratio of length and width, roundness, formfactor, etc.) characteristic of grain seeds can be utilised to identify certain sub-species taxa (ROVNER – GYULAI 1999). But, such instances are rare and can only be observed in large grain populations. In any case, landrace varieties can be very difficult or even impossible to characterise using measurements or indices of form and shape.

In the case of preserved, subfossil archaeological seeds the lack of adequate reference data renders the problem of identification of certain varieties even more difficult. Yet several studies have reported that ecotypes/regional varieties of grain species may have existed even in early periods. On rare occasions, there are similarities between grains of a distinctive species found in only one archaeological era or in several excavations of one archaeological culture or in related cultures derived from the same place of origin. Deliberate breeding of distinctive cereal grain varieties is a modern biotechnological capability. Thus, the existence of distinctive cereal grain landraces in archaeological contexts are virtually certain to have been local varieties, i.e. types that adapted to specific conditions or which, in some instances, may have been "developed" through mass selection by members of a specific culture in a relatively narrow time and place (*Fig. 1*).

The earliest proof of the use of cereal grain landraces is from the Berettyóújfalu-Szilhalom site, dated to the first half of the 5th millennium BC Neolithic (excavation by Márta Sz. Máthé 1976). Emmer wheat grains found in various locations of the tell fall into two types (*Fig. 2*). The majority of the grains are similar to einkorn, but are wider and longer. Their base widens and the portion over the scutellum protrudes strongly. Their bottom side is flat but they are open at their curved bottom seam. This "elongated type" is very similar to the emmer grains found at the Dévaványa-Réhelyi gát (excavation by István Ecsedy 1970 – the site is a part of the so-called Szakálhát-Szilmeg group, an older set of Neolithic sites) (*Fig. 3*). A smaller number of emmer grains fell in the second, "wider type" (*Fig. 4*). Those grains are flatter and their portions over the scutellum protrude less. That type is identical to the emmer grains from the related and somewhat older Neolithic Pári-Altäcker site (excavation by István Torma 1968) (*Fig. 5*). Approximately 5 litres (!) of charred emmer grains found in Pári-Altäcker all fall into this landrace type (P. HARTYÁNYI – NOVÁKI 1975). The two sites are not close enough to each other to suppose a simple exchange of grains. These distinctive types of grains may be evidence for the existence of locally developed varieties produced by mass selection. A similar process of "selective improvement" is also observable in animal bones. For instance, animal bones found at Middle Bronze Age sites also showed a certain degree of differentiation: the bones of domesticated animals fell into large and small varieties (BARTOSIEWICZ 1999).

Despite numerous finds of cereal grains by our archaeobotanists, there are no evidence and no published reports suggesting that the distinctive Neolithic varieties were being used in the Bronze Age or the Roman Period. But, evidence from later eras shows a reappearance of earlier landraces. This may be inferred from the charred grains found in the Late Migration Period, 8th – 9th Century, lake dwelling settlement at Fonyód-Bélatelep (excavation by Béla Horváth 1964). Bread wheat grains found there are smaller than varieties cultivated today, but they are of a similar size to the contemporary Central European bread wheat finds (TEMPÍR 1961). The measurement data (length, width and height of grains) seem to indicate that there may have been several form classes (ecotypes or varieties?) of the bread and club wheats and rye recovered (GYULAI 1986). Two and a half thousand grains of common bread wheat (*Triticum aestivum* subsp. *vulgare*) fall into two types according to their shape (*Fig. 6*). The grains of the first type are oval and relatively flat. Their scutellum is less acute and tips are oval. Tufts at the tips were preserved in several cases. The bottom line is straight, in most cases a deep channel opened up as the grains were burned. The number of grains falling into this type was one and a half times the number of grains of the second type. The second type of grains are not entirely oval, but are "broad-shouldered", i.e. they have a curved widening towards the scutellum. This differentiating feature could not be the result of burning, but is much more likely a feature of a distinct variety.

In addition to Fonyód-Bélatelep, the latter variety of common wheat, which widens towards the scutellum, was also found at the 9th–10th-century site next to the Sopron City Hall (P. HARTYÁNYI – NOVÁKI – PATAY 1967–68) and the 10th-century comes' castle at Hont (P. HARTYÁNYI 1981–83) (*Figs. 7–9*).

A similar observation was made in the case of the club wheat (*Triticum aestivum* subsp. *compactum* (Host) MacKey) from Fonyód Bélatelep. Ninety percent of almost three thousand seeds are rounded on all sides and compressed – typical representatives of the "compactoid" type (*Figs. 10–11*). The tops are protuberant and the oval scutellum is strongly curved. Several seeds retain the tuft at the tip. As a result of charring, several seeds were inflated and the sutura ventralis opened up. The seeds of a second type differ from those of the first only in their basal portion which is rounded, not square. Actually, the latter type of club wheat shares similarities to the club wheat used in the variety maintenance and breeding program of the Georgicon Faculty of Veszprém University at Keszthely. With respect to size, seeds of modern club wheat are some 20% larger. The difference may be attributable to genetic, edaphic and/or climatic factors.

Two thousand seeds of rye (*Secale cereale*) found at the Late Migration Period (7th–9th centuries) settlement at Fonyód Bélatelep are mostly of a larger type, yet the shape of the seeds puts them in the "squat" class, while a smaller proportion of the seeds are smaller and of a "narrow" type. The two types imply two varieties (*Figs. 12–13*).

Charred rye seeds found in the 15th–16th century houses at Nagyvázsony-Csepely (excavation by Júlia Kovalovszky 1957–58) are also of two types, a longer squat and a shorter narrow type (HARTYÁNYI – NOVÁKI – PATAY 1967–68). They may furnish evidence for the use of varieties in the Middle Ages (*Fig. 14*).

The natural flora of any given period gives a more direct indication of the environmental context than animal remains. Evidence of animal husbandry (and in particular, any export-import activities) is really interesting only if it is reviewed in parallel with a reconstruction of the environment and the finds associated with plant cultivation (*Figs. 15–18*).

Before we start our summary of the history of the long, early periods of antiquity resulting from the last century of Hungarian archaeobotanical research, we believe it is important to discuss the history of archaeobotanical research itself.

Acknowledgements

I hereby thank my masters and professors, above all István Kárpáti, István Szabó, Miklós Füzes, Stefanie Jacomet, Max Währen and Benno Richter for setting me on my way; István Ecsedy for providing the financial resources for the English translation of the manuscript; the Hungarian Agricultural Museum for permission to use the visual documentation; Nándor Kalicz for reviewing the original manuscript; Béla Borsos, Péter Szabó and Réka Benczes for the English translation, Irwin Rovner for reviewing the translation; Gábor Gyulai for the cover illustration; Tünde Kovács and Endre Tóth for photos and illustrations; the Archaeolingua Publishing House: Erzsébet Jerem and her colleagues for their high-quality editorial work; Tamás Németh for providing support for the publication of the book; current and former colleagues, my students, friends, relatives and everyone who helped my work with material or encouragement.

Emmer (*Triticum turgidum* subsp. *dicoccum*)

Elongated or longshaped typ

Neolithikum (End of 5th TY.BC.)	*Length mm*	*Width mm*	*Height mm*	*Measured piece*
Berettyóújfalu-Szilhalom 9/1976 (average)	5.92	2.65	2.18	40
Minimalvalue – Maximalvalue (mm)	4.3–8.1	1.8–3.5	1.6–2.6	
Dévaványa-Réhelyi Flur 1970 (average)	6	2.6	2.3	50

Wider or wideshaped typ

Neolithikum (End of 5th TY.BC.)	*Length mm*	*Width mm*	*Height mm*	*Measured piece*
Berettyóújfalu-Szilhalom 6/1976 (average)	6.67	3.39	2.89	55
Minimalvalue – Maximalvalue (mm)	5.0–8.0	2.1–4.0	2.0–3.7	
Pári-Altäcker Flur 6/1970 (average)	6.56	3.19	2.9	50

Common bread wheat (*Triticum aestivum* subsp. *vulgare*)

Longshaped (oval and relatively flat) typ

Late migration period – Hungarian Conquest time	*Length mm*	*Width mm*	*Height mm*	*Measured piece*
Fonyód-Bélatelep (7–9th century AD) "A" group (average)	4.76	3.06	2.63	50
Minimum – Maximum	3.76–5.48	2.66–3.28	2.28–2.99	
Sopron-Városháza St. (9–11th century AD) (average)	4.71	2.54	2.13	50
Hont-Ispánsági vár (10th century AD) (average)	4.67	3.36	3.43	50
Minimum – Maximum	3.9–5.5	2.7–4.2	2.2–3.3	

Oboval (broad-shouldered) typ

Late migration period – Hungarian Conquest time	*Length mm*	*Width mm*	*Height mm*	*Measured piece*
Fonyód-Bélatelep (7–9th century AD) "B" group (average)	4.75	3.27	2.68	50
Minimum – Maximum	3.79 5.92	2.52–4.18	1.94–3.32	
Sopron-Városháza St. (9–11th century AD) (average)	4.89	2.86	2.57	50
Recent				
Minimum – Maximum	5.0–7.1	2.5–3.6	2.3–3.0	

Club wheat (*Triticum aestivum* subsp. *compactum*)

Strong compactoid typ

Late migration period	*Length mm*	*Width mm*	*Height mm*	*Measured piece*
Fonyód-Bélatelep (7–9th century AD) "A" group (average)	4.21	3.45	2.79	50
Minimum – Maximum	3.75–5.13	2.95–4.00	2.47–3.05	

Soft compactoid (round) typ

Late migration period	*Length mm*	*Width mm*	*Height mm*	*Measured piece*
Fonyód-Bélatelep (7–9th century AD) "B" group (average)	4.32	3.31	2.79	50
Minimum – Maximum	3.38–5.42	2.77–3.88	2.18–3.18	

Rye (*Secale cereale*)

Big squat typ

Late migration period	*Length mm*	*Width mm*	*Height mm*	*Measured piece*
Fonyód-Bélatelep (7–9th century AD) "A" group (average)	6.17	2.53	2.29	50
Minimum – Maximum	5.02–7.15	2.01–2.88	1.88–2.72	

Narrow small and thin typ

Late migration period – Hungarian Conquest time	*Length mm*	*Width mm*	*Height mm*	*Measured piece*
Fonyód-Bélatelep (7–9th century AD) "B" group (average)	5.58	2.21	1.74	50
Minimum – Maximum	4.85–6.28	1.86–2.80	1.58–1.96	
Hont-Hont ispánsági vár (10th century AD) (average)	5.86	2.34	2.14	50
Minimum – Maximum	4.6–7.5	1.7–2.9	1.5–2.6	

Fig. 1. Species types – Plantbreeding in the prehistoric and historic periods.

Fig. 2. Emmer (*Triticum turgidum* subsp. *dicoccum*) grains „longshaped form" from Berettyóújfalu-Szilhalom, first part of the 5th millenium BC. After P. HARTYÁNYI – SZ. MÁTHÉ 1980.

Fig. 3. Emmer (*Triticum turgidum* subsp. *dicoccum*) grains „longshaped form" from Dévaványa-Réhelyi gát, end of the 6th millennium BC. After P. HARTYÁNYI – SZ. MÁTHÉ 1980.

Fig. 4. Emmer (*Triticum turgidum* subsp. *dicoccum*) grains „wideshaped form" from Berettyóújfalu-Szilhalom, first part of the 5th millennium BC. After P. HARTYÁNYI – SZ. MÁTHÉ 1980.

Fig. 5. Emmer (*Triticum turgidum* subsp. *dicoccum*) grains „wideshaped form" from Pári-Altäcker dűlő, second part of the 6th millennium BC. After P. HARTYÁNYI – SZ. MÁTHÉ 1980.

Fig. 6. Common bread wheat (*Triticum aestivum* subsp. *vulgare*) grains „longshaped form" from Fonyód-Bélatelep Late Migration Period lake dwelling settlement. Photograph by the author.

Fig. 7. Common bread wheat (*Triticum aestivum* subsp. *vulgare*) grains „longshaped form" from Sopron –Városháza Str, 9th–10th centuries. After P. HARTYÁNYI *et al.* 1967–68.

Fig. 8. Common bread wheat (*Triticum aestivum* subsp. *vulgare*) grains „oboval" from Hont-Ispánsági vár, Hungarian, second half of the 10th century. After P. HARTYÁNYI 1981–83.

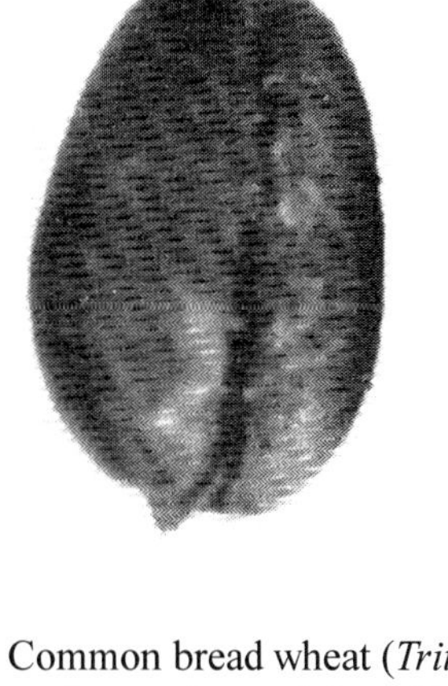

Fig. 9. Common bread wheat (*Triticum aestivum* subsp. *vulgare*) "oboval" grains (spreading towards the germ disk) from Fonyód-Bélatelep Late Migration Period lake dwelling settlement. Photograph by the author.

Fig. 10. Club wheat (*Triticum aestivum* subsp. *compactum*) "strongly compactoid " grains from Fonyód-Bélatelep Late Migration Period lake dwelling settlement. Photograph by the author.

Fig. 11. Club wheat (*Triticum aestivum* subsp. *compactum*) "oboval" grains (grains rounded towards the base) from Fonyód-Bélatelep Late Migration Period lake dwelling settlement. Photograph by the author.

Fig. 12. Rye (*Secale cereale*) short (wide) grains from Fonyód-Bélatelep Late Migration Period lake dwelling settlement. Photograph by the author.

Fig. 13. Rye (*Secale cereale*) thin and large grains from Fonyód-Bélatelep Late Migration Period lake dwelling settlement. Photograph by the author.

Fig. 14. Rye (*Secale cereale*) thin and large grains from Hont-Ispánsági vár, Hungarian, second half of the 10th century. After P. HARTYÁNYI 1981–83.

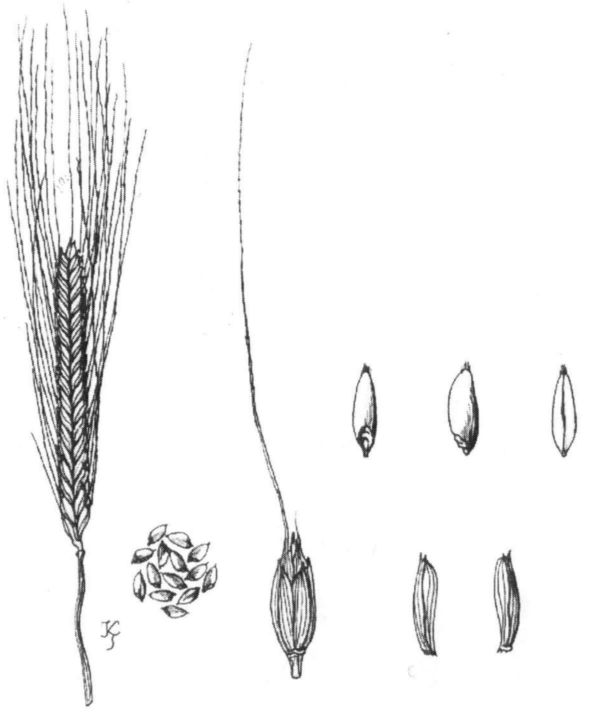

Fig. 15. Einkorn (*Triticum monococcum*) drawing by Tünde Kovács.

Fig. 16. Emmer (*Triticum turgidum* subsp. *dicoccum*) drawing by Tünde Kovács.

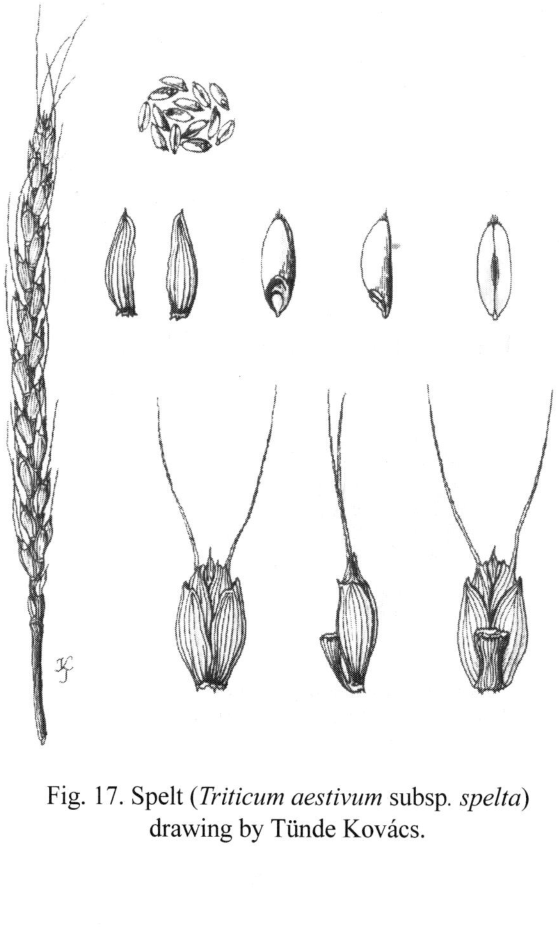

Fig. 17. Spelt (*Triticum aestivum* subsp. *spelta*) drawing by Tünde Kovács.

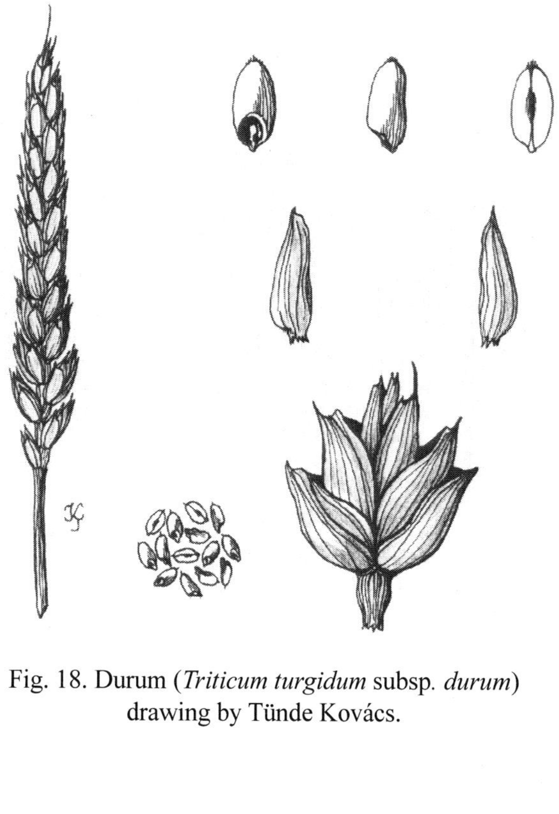

Fig. 18. Durum (*Triticum turgidum* subsp. *durum*) drawing by Tünde Kovács.

2. The history of Hungarian archaeobotanical research

The writings of 19th-century scientists working on the evolution of plants such as DARWIN (1859), DE CANDOLLE (1894) and the later genetic hearth theories of VAVILOV (1928, 1934) along with his collecting expeditions to genetic centres for bread wheat species in the Middle East (VAVILOV 1950) had a definitive effect on the development of archaeobotany as an independent branch of the natural sciences.

The first book on archaeobotany, written by Franz UNGER in 1851, was a study of Early Iron Age plant remains from the Salzkammergut. Excavation of lake dwelling settlements in Switzerland began in 1854. The resulting archaeobotanical study, HEER (1865), of the large quantities of recovered plant remains was soon published. Later NEUWEILER (1905) and BERTSCH (1932) also examined plant remains from pile dwellings. Plant remains were also found at excavations in Egypt (UNGER 1862; SCHWEINFURT 1884), and at Troy and Pompeii (WITTMACK 1890, 1903). Quickly developing botanical studies, together with linguistic investigations, resulted in works recounting the knowledge of ancient Greeks and Romans about plants that have become classics, such as LENZ (1859) and HEHN (1877). Further comprehensive studies were published on the development and spread of cultivated plants (NETOLITZKY 1931; SCHEIMANN 1932; BETSCH – BETSCH 1949; SCHWANITZ 1973).

By the middle of the 20th Century, excellent seed and fruit identification handbooks (e.g. BEIJERINK 1947) assisted archaeobotanical research. An additional breakthrough in the processing of finds was furnished by the floristic-geobotanical research results of FIRBAS (1949) and the ecological evaluation systems of LANDOLT (1977) and ELLENBERG (1979).

After World War II, ecological, economic, vegetation-historical and climate-ecological inferences that could be drawn from plant remains gained centre stage. The problems encountered grew beyond the framework of geobotany. Study of the ecological needs of wild plants and weeds began (ROTHMALER 1953). Archaeobotanical processing work was becoming uniform in its processes and equipment (WILLERDING 1970). Analytically, the trend towards the reconstruction of environments came to dominate. In Switzerland, a computer program was developed for archaeological registration and assessment (JACOMET – BROMBACHER – DICK 1989).

Large-scale environmental archaeology projects spanning several decades were launched. Outside Europe, such excavations were primarily conducted in the Middle East, in particular the "Fertile Crescent" area, considered to be the cradle of grain species domestication – until they were interrupted by wars in the area. Excellent material found made significant contributions to our knowledge of the domestication and development of cultivated plants (ZOHARY – HOPF 1988). Several comprehensive works were published on the history of plant cultivation and gardening, segetal weeds and vegetation changes (e.g. WILLERDING 1988).

At present, several institutes of archaeobotany, engaged in research and education, are operating in Europe: Basel, Stuttgart, Munich, Kiel, Vienna, and Nitra to mention only those closest to Hungary.

The history of archaeobotany in Hungary is no less interesting. According to historical records, grains from the Roman Period were found at Szombathely in the period 1836 to 1845. During excavations of prehistoric sites, grains were found at Felsődobsza between 1857 and 1877, at Szíhalom in 1870, and at the Pécs-Makárhegy site in 1871. The 1876 international archaeological conference in Budapest lent tremendous impetus to Hungarian archaeology. During excavations at the sites of Aggtelek cave, Tószeg, Nagyrév and Tiszafüred, the search for preserved cereal grains was included in the recovery of prehistoric remains.

Imre Deininger (1844–1918), considered the father of Hungarian archaeobotany, began his archaeobotanical work in 1876 when he studied plant remains found at the Aggtelek cave. He went on to study finds from the sites of Felsődobsza (*Figs. 19–20*), Tószeg and Torna-Szádelő Valley. His last research work in archaeobotany, on plant remains found at a prehistoric settlement in Poland, was published in 1892. Unfortunately, Deininger had no followers, so the work that he started with great promise virtually ended, although archaeological publications continued to record recovery of large quantities of grains. The next record we have of the botanical study of grains and fruits is from 1917, when Georg Lindau assessed the grains found at the 1906 excavation at Tószeg. At the beginning of the century, Árpád Csák also reported grain remains at his excavation in the Roman garrison at Keszthely-Fenékpuszta.

Although he himself did not do any archaeobotanical work, Raymund Rapaics (1885–1954) made a significant contribution to the history of cultivated plants in Hungary. He published the results of his investigations of historical records, iconography, linguistics and botany in several books in the 1940s, which are still important resources for people working in that field.

Between 1942 and 1964, albeit only in passing, Ádám Boros, Emil Gubányi, Zoltán Szák, János Maácz and Zdenek Tempír all worked on the study of Hungarian plant remains.

At the beginning of the 1960s, István Bogdán, György Mándy and József Mesch conducted special studies that were infused with an interest in archaeobotany. Imre Wellmann studied grain seeds that were found attached to archival materials from the 17th–18th centuries (WELLMANN – MÁNDY – MESCH 1963).

Regular archaeobotanical processing work only began at the beginning of 1960. The archaeobotanist Borbála P. Hartyányi worked together with garden historian Magdolna P. Erményi, for a brief period with Katalin M. Kassai, with the archaeologists Árpád Patay and Gyula Nováki of the Hungarian Agricultural Museum in Budapest, the archaeobotanist Miklós Füzes and the archaeologist Károly Sági of the Balaton Museum in Keszthely. These collaborating colleagues were leaders in the field during the decade of the 1960's (*Figs. 21–22*). At the Domokos Kunyi Museum in Tata, scientific museologist István Skoflek and teacher Veronika Árendás worked on processing grain remains found in the northwest of the country, primarily around Tata. To a great extent it is thanks to them that archaeobotany became an established and independent branch of science in this country and beyond, gaining international recognition. As a result, archaeologists have come to rely on archaeology as an applied discipline. Comparative collections of seeds and fruits at Budapest and Keszthely are significant reference resources for research throughout Europe.

The plant gene centre theories of György MÁNDY (1972) and the genetic theory of origin of wheat species developed by Adonisz BELEA (1986) provided a solid theoretical foundation for renewed archaeobotanical research in Hungary. The study of the nutrient content of ancient bread wheat species undertaken by Ernő BOCZ (1998) yielded important information for the discovery of valuable properties of old grain varieties. In addition, excellent seed catalogues by Szilárd SCHERMANN (1966) and Gyula BRECHER (1960) provide assistance with processing and identifying archaeobotanical material.

The seminal achievement of the modern age, the computer, has also received a role in processing. Initially, it was only used for maintaining records and for evaluation, but more recently its role has come to include morphometric studies of shape and size that use computer image analysis (ROVNER – GYULAI 2007). A computer seed identifier based on image analysis techniques is also in progress.

Experimental archaeology, a discipline in the process of being established in Hungary, also plays a significant role in answering questions associated with plant cultivation in the course of archaeobotanical processing work.

The publication of many books demonstrates that there is tremendous interest in plant history monographs that, among other things, take into account archaeobotanical finds. László GAÁL (1978),

Sándor KAPÁS (1997) and Dezső SURÁNYI (1985) summarised the history of the cultivation of grains, vegetables and fruit in Hungary in excellent monographs.

Research in archaeobotany continues and increasingly develops in the context of interdisciplinary co-operation. In addition to archaeobotanists, researchers from other disciplines occasionally take part in the work of processing. Among others, the specialists listed below have all helped generously and have made several important contributions to that work: János Csapó, with his analytical chemical investigations of the nutrient content of ancient plants; Géza Facsar with the processing of seed remains, in particular grape remains; Miklós Kállay by addressing viticultural questions, Dezső Surányi, in addition to his theoretical research results, with his experience in the recultivation of ancient drupes, István Szabó with the historico-phytosociological evaluation of finds, Attila Szabó T. with his ethnobotanical results, András Terpó with his experience in the historical research of fruit species, and Endre Tóth with his assistance with the interpretation of finds.

Archaeobotany is more than an activity limited only to a museum setting. It is also connected with the work of many other institutions and organisations. Archaeobotany is important for the Nature Protection Office of the Ministry of Environmental Protection with respect to the discovery of the history of agrobiodiversity. At the gene bank of arable and garden crops in Hungary, the projects launched at the Institute for Agrobotany at Tápiószele are intended to discover the predecessors of old Hungarian regional varieties found in various archaeological periods. The Institute for Environmental Management of the Szent István University at Gödöllő, where work is in progress to discover and preserve the values of traditional (ecological) farming, has integrated among its tasks the study of historical agrobiodiversity, as well as research into the history of our indigenous Hungarian cultivated plants. By including archaeobotany among the subjects taught, the Institute of Archaeology of the Eötvös Loránd University (ELTE) and the Institute of Environmental Management of the Szent István University at Gödöllő have made active contributions to the training of specialists with modern, interdisciplinary, historico-ecological perspectives, and thereby indirectly to the development of archaeobotanical research. The Faculty of Genetics and Plant Improvement at the Szent István University and the Centre for Biotechnology at Gödöllő use the most modern biotechnological methods to research suitable seeds and hereditary material found in fruits.

The author has been conducting archaeobotanical research since 1983. At present he and his students: Andrea Torma at the Hungarian Agricultural Museum, Brigitta K. Berzsényi at the Matrica Museum and Orsolya Dálnoki at the Aquincum Museum in Budapest, are processing seed and fruit remains from excavations in Hungary.

Fig. 19. Imre Deininger (1844–1918), founder of Hungarian archaeobotany.

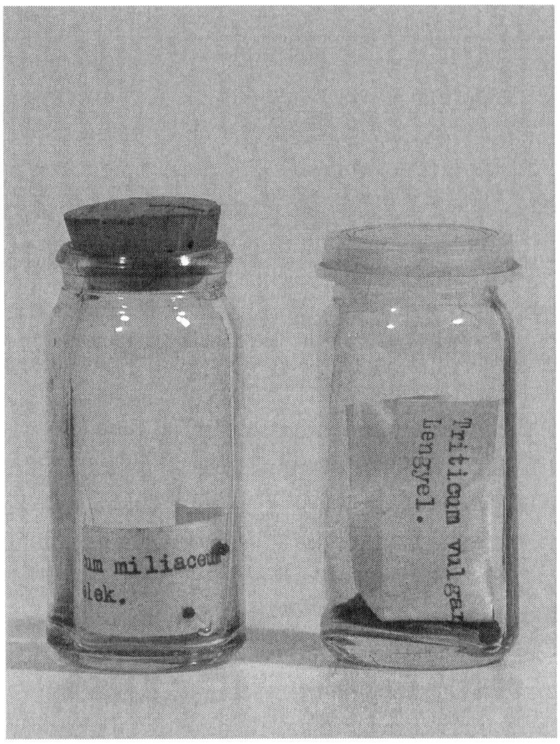

Fig. 20. Collected seeds from Aggtelek cave by Imre Deininger in the year 1872, Inventory of the Hungarian Agricultural Museum, Budapest.

Fig. 21. Miklós Füzes (Frech') (1931–1997), an outstanding representative of Hungarian archaeobotanical research in the near past.

Fig. 22. Samples from the proximity of fireplaces are rich in botanical finds (excavation site at Fonyód-Bélatelep lake dwelling settlement, 1964). Picture taken at the excavation.

3. The tasks of archaeobotany and its position within the system of sciences

Archaeobotany (also known as palaeo-ethnobotany, or carpology) is the science of the identification or tentative identification of plant remains and plant products. Its main area of investigation is the history of plant cultivation and vegetation. It studies the relationship between human beings and flora especially as they relate to human economic activities. In addition to the identification of cultivated plant remains, it monitors the transformation of wild species into cultivated plants and the spread of plant cultivation and agriculture in general. Furthermore, it evaluates the images of plants originating from various eras, the decultivation of plant species and the data of the various social sciences associated with plants.

Archaeobotany is a branch of botany. All the elements of botany, i.e. morphology, taxonomy, anatomy and geobotany are used in the course of identification of the materials from archaeological excavations. In addition, it plays a "bridging" role within the system of sciences as it is intimately connected with the science of archaeology as well (*Fig. 23*).

Archaeobotany, as the study of *plant macrofossils* (seeds and fruits) obtained from archaeological excavations, becomes particularly important when there is very little or no archaeological artifacts related to agriculture, written or iconographical material available about the cultivation of plants found. This is particularly the case in relation to prehistoric cultures of the Carpathian Basin.

Unlike pollen analysis, archaeobotany primarily studies plant remains that have been placed in the soil by human activity. Naturally, not all constituent species of ancient vegetation are reliably present in archaeological material, even characteristic species of individual plant associations may be missing. Due to numerous factors, plant diaspores that are preserved in the soil often exhibit differences relative to each other as well. Active factors include seed yield of individual species, resilience of seeds (e.g. hardness of husks), modes of dispersal of fruit/seed, climatic requirements of the plant, etc. The objects obtained from archaeological excavations (waste layers, pits, cesspits, etc.) primarily include remains of an era's field plant associations: a part (remains) of cultivated grains along with associated weed flora. In some instances, primarily in the case of wells and channels, the seeds of species of natural vegetation are also found. In recent past decades, results of paleobotanical study of natural vegetation were based on pollen analysis, but today study of seeds and fruits is more intensive, and extremely important with respect to the determination of fossil plant associations.

Archaeobotany is an interdisciplinary scientific field, an important tool for understanding prehistoric ways of life, indeed the only source of agricultural history until the Roman Age. Plant remains allow us to infer botanical knowledge, methods of farming, dietary habits and environments of people of those periods.

Plant remains are rarely found in the excavation of archaeological sites. Such organic materials are unstable and in natural conditions they are decomposed very quickly by microorganisms. However, they may be preserved in extraordinary conditions (e.g., charring by fire, immersion in water, extremely dry microclimates). Furthermore, the recognition of such remains requires extensive experience, systematic sampling and flotation (*Figs. 24–27*).

The finds of seeds and fruits from excavations are divided into "real" or "direct" and "pseudo" or "indirect" finds (WILLERDING 1970). "Direct" plant remains are individual organs or parts of plants that have more or less preserved their morphological characteristics and, in less frequent cases, their histological structures. All such cases are suitable for comparative study (*Figs. 28–33*). On occasion, identification requires cytological-microchemical (e.g. fibrous materials) or cytological-physical (e.g. study of fibrous materials in polarised light) processes. Directs plant remains are primarily collected using flotation.

The majority of archaeobotanical finds are direct plant remains: seeds and fruits. Some of them were deliberately placed in the soil (e.g. warehousing, settlement waste, cleaning after the burning of settlements, rituals, funerals), while some ended up in the soil accidentally (*Fig. 34*). Natural forces (wind, water, animal burrows, etc.) may also result in plant parts being deposited in deeper sections of the soil.

Seeds and fruits are the relatively more resilient parts of plants deposited in the soil (*Fig. 35*). They may be preserved due to edaphic, climatic and biotic factors: turfed into peat or, rarely, conserved thanks to heavy metal ions. They may also be encased in mortars or in extremely dry (e.g. burial vaults) or cold conditions (e.g. glacial ice) or coated by a chemical buffering agent against decay factors. Their most frequent occurrence, however, is in charred form (SÁGI – FÜZES 1966). The degree of preservation may differ due to variations in material structure, soil and climate, and associated human activities as well as by the age of the find. Accordingly, various methods may be required for preparation and processing.

In our climatic conditions, seeds and fruits are mostly conserved by fire charring, and to a lesser extent by the accumulation of humic materials (humification) or by turfing.

Indirect plant finds are created when seeds and fruits embedded in various materials are dissolved, burnt or perhaps digested by microorganisms. As a result, only their negative imprint remains. Their shared characteristic is that preservation of morphological information is typically the result of association with a matrix resulting from human activity. We distinguish three types of representations: imprints, negatives and casts. Imprints are made by almost planar formations. Negatives are imprints of three-dimensional objects (e.g. seeds, fruits). Casts ("stone seeds" or "stone fruits") are produced when the interior of the imprint or negative void is later filled by some other material (e.g. salt) and then solidified (FÜZES 1990).

Direct and indirect remains usually occur together. Careful examination of fragments of ceramics and fired clay usually reveals numerous plant imprints. Breaking up fired clay fragments typically further increases recovery of remains.

In the Carpathian Basin, from the earliest era of the Neolithic to the middle of the 20th century, clay mixed with vegetable matter was a common construction material. This was used to build wattle-and-daub walls and oven walls as well as floors. Neolithic houses were built of wattle-and-daub walls and clay-reinforced pile frameworks. Due to a high incidence of house fires, vegetable fragments mixed into the mud were often charred or burnt. The same result occurs when plant material incorporated in clay for ceramic production is fired. During the initial phase of pottery production, vegetable waste products (chaff, husks, spears, node fragments, etc.) from threshing, "throwing," i.e., winnowing, of cereal grains (e.g. grain seeds, chaff fragments, weed seeds) were added to lighten the clay (BALASSA 1963–64). Finer-grained, sand-free material was used for ceramics with a smaller proportion of plant matter than for wattle-and-daub walls or floors. It is expected, therefore, that more frequent imprints in the latter preserve more details. The significance of this evidence rests precisely in the fact that it allows us to make inferences regarding characteristics of grain cultivation and, in addition, weed flora as well.

In the early phases of plant cultivation, the presence of locally grown domestic plants is more strongly supported by imprints of discarded waste products at a site than by the presence of usable food remains. It is highly unlikely that a nomadic group would carry waste products of threshing any significant distance from the locale where the crop was harvested and processed.

The conventional process for making a "positive" of an imprint uses silicone rubber. This neutral material polymerises easily, fills cavities well and produces a surface that is rich in detail, making identification easier. The completed positives, which are easy to colour, are used for identification in accordance with standard archaeobotanical methods. Just as with seed and fruit finds, identification takes place under a stereo binocular microscope. The study of imprints requires great concentration as hardly any morphological characteristics suitable for identification are preserved. The identification of individual plant organs often requires creation of positive models.

The examination of imprints is essentially the same whether in Hungary or elsewhere. Only the technological processes used may differ. This work began in this country in the 1970s, using finds dating to the Neolithic (P. HARTYÁNYI – NOVÁKI 1971; FÜZES 1977). Unfortunately, those research projects, though they produced significant results, were discontinued after a few years due to the lack of researchers to continue the research.

In 1995, when we examined plant imprints on the ceramic remains found at the Neolithic site at Bicske-Galagonyás, we tried an Apple Macintosh-based image analysis system (GYULAI 1996a). The results indicate that computer image analysis can be used as a faster, more accurate and well-documentable method for studying plant imprints than the traditional rubber cast method.

AGE	TIME INTERVAL	FLANDRIA INTERGLACIAL PHASE	CLIMATE
Early Neolithic	6000–5500/5400 BC	Boreal/Atlantic	warm, humid
Middle Neolithic	5500/5400–5000/4900 BC	Atlantic	warm, humid
Late Neolithic	5000/4900–4500/4400 BC	Atlantic	warm, dry
Early Copper Age	4500/4400–4000 BC	Atlantic	cold, humid
Middle Copper Age	4000–3600/3500 BC	Atlantic	cold, humid
Late Copper Age	3600/3500–2700 BC	Atlantic/Subboreal	cold, humid
Early and Middle Bronze Age	2700–1400/1300 BC	Subboreal	cold, humid
Late Bronze Age	1400/1300–900/800 BC	Subboreal	cold, humid
Early and Middle Iron Age	900/800–450 BC	Subboreal	cold, humid
Late Iron Age	450 BC–1st century AD	Subatlantic	cold, dry
Roman Age	1st century-450 AD	Subatlantic	warm, dry
Early and Late Migration Period	450–895 AD	Subatlantic	warm, dry
Hungarian Conquest and Árpád Age	895–1301 AD	Subatlantic	warm, humid
Late Medieval	1301–1526 AD	Subatlantic	cold, humid

Fig. 23. Chronological classification of recent (Holocene) climatic changes in the Carpathian Basin.

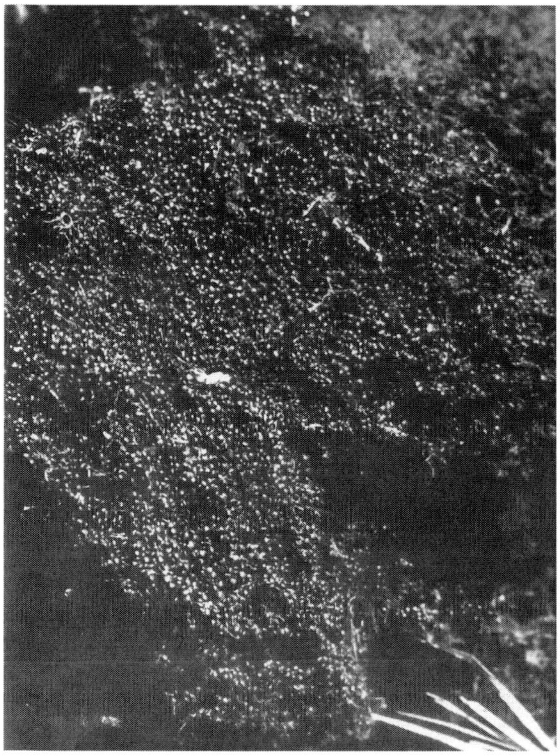

Fig. 24. Accumulation of seeds and bearings in turf (exploration of Fonyód-Bélatelep lake dwelling settlements, 1964). Picture taken at the excavation.

Fig. 25. Flotation of soil samples. Photograph by the author.

Fig. 26. Collecting soil samples at an excavation site. Photograph by the author.

Fig. 27. Drying of macroremains. Photograph by the author.

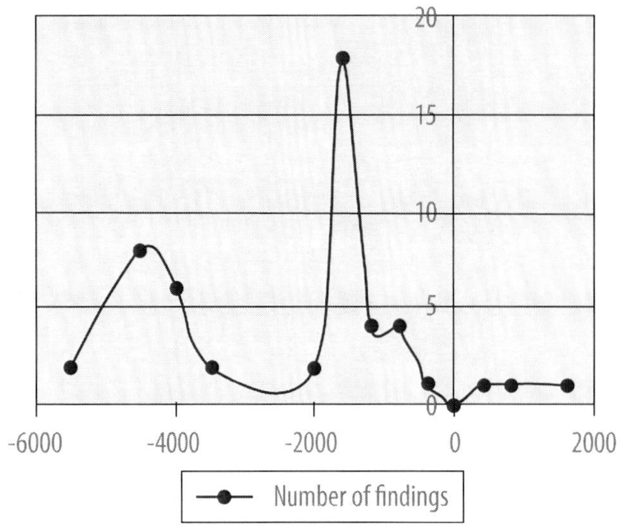

Fig. 28. Gene erosion of einkorn in the Carpathian Basin during the historical period, based on PÉNTEK – SZABÓ (1981) with supplements.

Fig. 29. Flowering einkorn field in Northern Hungary (Szarvasgede). Photograph by the author.

Fig. 30. Vitreous einkorn in Northern Hungary (Szarvasgede). Photograph by the author.

Fig. 31. Einkorn harvest in Northern Hungary (Szarvasgede). Photograph by the author.

Fig. 32. Einkorn harvest with a combine in Northern Hungary (Szarvasgede). Photograph by the author.

Fig. 33. Pithouse from the Early Árpádian Period. Reconstruction. The roof was covered with soil mixed with einkorn straw.

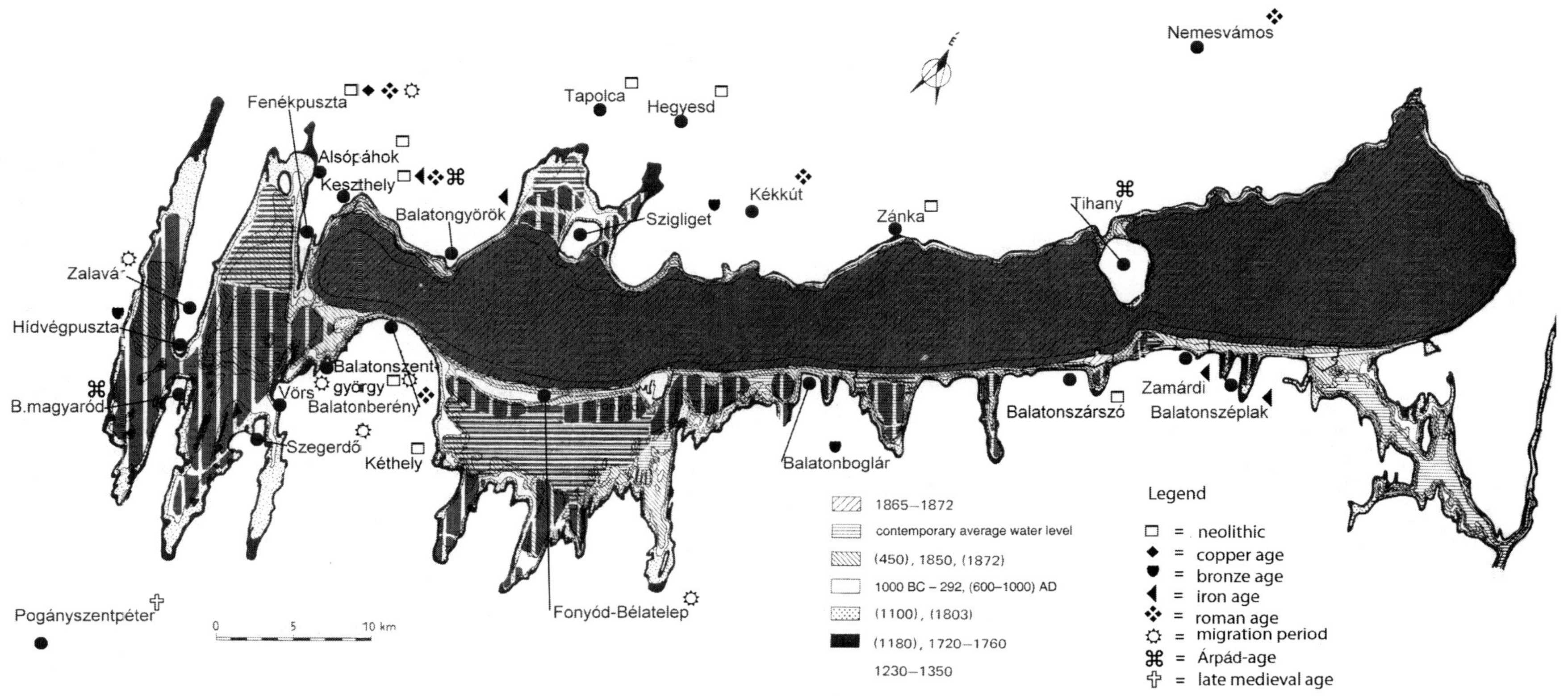

Fig. 34. Archaeobotanical sites at lake Balaton.

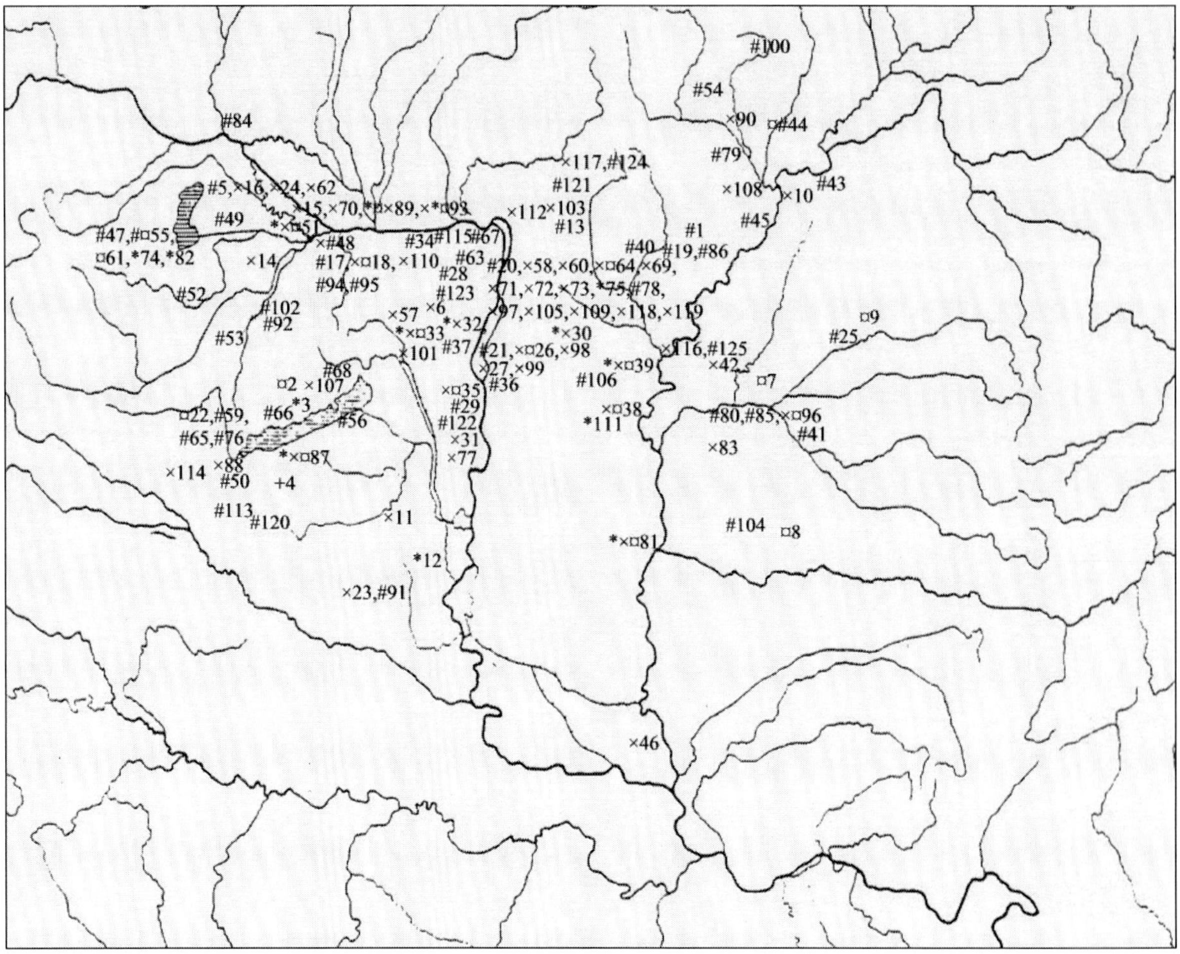

Fig. 35. Archaeobotanical sites of barley in Hungary.

* Two-rowed barley: 3; 12; 30; 32; 33; 39; 51; 74; 75; 81; 82; 87; 89; 93.
× Six-rowed barley: 6; 10; 11; 14; 15; 16; 18; 23; 24; 26; 27; 30; 31; 32; 33; 35; 37; 38; 39; 42; 46; 49; 51; 57; 58; 60; 62; 64; 69; 70; 71; 72; 73; 77; 81; 83; 87; 88; 89; 90; 93; 96; 97; 98; 99; 101; 103; 105; 107; 108; 109; 110; 111; 112; 114; 116; 117; 118; 119.
\+ Four-rowed barley: 4.
¤ Naked barley: 2; 7; 8; 9; 18; 22; 26; 33; 35; 38; 39; 44; 48; 51; 55; 61; 64; 81; 87; 89; 93; 96.
Barley (unspecified): 1; 5; 13. 17; 19; 20; 21; 25; 28; 29; 34; 36; 40; 41; 43; 44; 45; 47; 48; 50; 52; 53; 54; 56; 59; 63; 65; 66; 67; 68; 76; 78; 79; 80; 84; 85; 86; 91; 92; 94; 95; 100; 102; 104; 106; 113, 115; 120; 121; 122; 123; 124; 125.

Early Neolithic (6000–5300 BC) – 1. Füzesabony-Gubakút: barley

Middle Neolithic (5300–4700 BC) – 2. Hegyesd-Ágói dűlő: naked barley; 3. Zánka-vasútállomás: two-rowed barley; 4. Marcali-Lókpuszta: four-rowed barley; 5. Mosonszentmiklós-Pálmajor: barley; 6. Bicske-Galagonyás: six-rowed barley; 7. Dévaványa-Réthelyi dűlő: naked barley; 8. Battonya-Parázstanya: naked barley

Late Neolithic (4700–4500 BC) – 9. Berettyóújfalu-Szilhalom: (two-rowed) naked barley; 10. Tiszapolgár-Csőszhalom: six-rowed barley; 11. Lengyel: six-rowed barley; 12. Zengővárkony: two-rowed barley; 13. Aszód-Papi földek: barley; 14. Börcs-Paphomlok: six-rowed barley; 15. Lébény-Billedomb: six-rowed barley

Middle Copper age (4500–3500 BC) – 16. Mosonszentmiklós-Pálmajor: six-rowed barley; 17. Győr-Szabadrétdomb: barley

Late Copper age (3500–3000 BC) – 18. Győr-Szabadrétdomb: six-rowed barley, naked barley; 19. Kompolt-Kistéri tanya: barley; 20. Budapest, Bécsi út.: barley; 21. Csepel-Vízmű: barley; 22. Keszthely-Fenékpuszta: naked barley

Early Bronze age (3000–1600 BC) – 23. Pécs-Nagyárpád: six-rowed barley; 24. Mosonszentmiklós-Pálmajor: six-rowed barley; 25. Bakonszeg: barley; 26. Csepel, Hollandi út: six-rowed barley, naked barley; 27. Szigetszentmiklós-Vízmű: six-rowed barley

Middle Bronze age (1600–1300 BC) – 28. Solymár-Várhegy: barley; 29. Baracs-Bottyán sánc: barley; 30. Mende-Leányvár: two- and six-rowed barley; 31. Bölcske-Vörösgyír: six-rowed barley; 32. Százhalombatta-Földvár: two- and many-rowed barley; 33. Pákozd-Vár: two- and many-rowed barley, naked barley; 34. Süttő-Hosszúvölgy: barley; 35. Dunaújváros-Koszider: six-rowed barley, naked barley; 36. Dömsöd-Apaj: barley; 37. Százhalombatta-Téglagyár: barley; 38. Tiszaalpár-Várdomb: six-rowed barley, naked barley; 39. Tószeg-Laposhalom: two- and six-rowed barley, naked barley; 40. Jászdózsa-Kápolnahalom: barley; 41. Békés-Várdomb: barley; 42. Túrkeve-Terehalom: six-rowed barley; 43. Tiszaeszlár-Bashalom: barley; 44. Felsődobsza-Várdomb: barley, naked barley

Late Bronze age (1300–900 BC) – 45. Poroszló-Aponhát: barley; 46. Feudvar-Vojvodina: six-rowed barley; 47. Sopron-Krautacker: barley; 48. Győr-Szabadrétdomb: barley, naked barley; 49. Mosonmagyaróvár-Németdőlő six-rowed barley; 50. Balatonmagyaród-Hídvégpuszta: barley; 51. Börcs-Paphomlok: two- and six-rowed barley, naked barley; 52. Gór-Kápolnadomb: barley

Early Iron age (900–500 BC) – 53. Celldömölk-Sághegy: barley; 54. Aggtelek-Baradla barlang: barley; 55. Sopron-Krautacker: (two-rowed) naked barley, barley; 56. Siófok-Balatonszéplak: barley; 57. Fehérvárcsurgó-Eresztvényi erdő: six-rowed barley; 58. Rákoskeresztúr-Újmajor: six-rowed barley

Late Iron (Celtic) age (500 BC – 1st century AD) – 59. Keszthely-Fenékpuszta: barley; 60. Budapest-Corvin tér: six-rowed barley, naked barley; 61. Sopron-Krautacker: naked barley, barley; 62. Mosonszentmiklós-Pálmajor: six-rowed barley;

Roman age (1st century – middle 5th century AD) – 63. Budakalász-Luppa csárda: barley; 64. Budapest-Corvin tér: six-rowed barley, naked barley; 65. Keszthely-Fenékpuszta: barley; 66. Kékkút: barley; 67. Leányfalu-Móricz Zsigmond u.: barley; 68. Nemesvámos-Balácapuszta: barley; 69. Óbuda-Lajos u.: six-rowed barley; 70. Lébény-Billedomb: six-rowed barley; 71. Nagyszombat u.: six-rowed barley; 72. Budapest-Bécsi u.: six-rowed barley; 73. Budapest-Corvin tér: six-rowed barley; 74. Sopron-Beloiannisz tér: two-rowed barley; 75. Óbuda Corvin tér: two-rowed barley; 76. Keszthely-Dobogó: barley; 77. Dunakömlőd: six-rowed barley; 78. Budapest, Bécsi út 38–42.: barley

Barbaricum (1st century – middle 5th century AD) – 79. Szirmabesenyő-Sajóparti homokbánya: barley; 80. Gyomaendrőd: six-rowed barley; 81. Kiskundorozsma-Nagyszék: two- and six-rowed barley, naked barley

Early Migration period (middle 5th–6th century AD) – 82. Sopron-Városháza u.: two-rowed barley; 83. Eperjes-Csikóstábla: six-rowed barley; 84. Dévény: barley; 85. Gyomaendrőd: barley; 86. Kompolt-Kistéri tanya: barley

Late Migration period (6th–9th century AD) – 87. Fonyód-Bélatelep: two- and six-rowed barley, naked barley; 88. Zalavár: six-rowed barley

Hungarian Conquest and Árpád period (9th century – 1301 AD) – 89. Lébény-Billedomb (9th–10th century): two- and six-rowed barley, naked barley; 90. Edelény-Borsodi földvár (9th–10th century): six-rowed barley; 91. Pécs-Cella trichora (11th century): barley; 92. Pápa-Hantai u. (10th–11th century): barley; 93. Lébény-Billedomb (13th–14th century): two- and six-rowed barley, naked barley; 94. Győr-Gabonatér (12th–13th century): barley; 95. Győr-Szeszgyár (12th–13th century): barley; 96. Endrőd (12th–13th century): (two-rowed) naked barley, six-rowed barley; 97. Rákospalota-Újmajor (12th–13th century): six-rowed barley; 98. Csepel, Rákóczi Ferenc út (12th–13th century): six-rowed barley; 99. Szigetszentmiklós-Vízmű (12th–13th century): six-rowed barley; 100. Torna-Szádelővölgy (12th–13th century): barley

Late Medieval period (1301–1526 AD) – 101. Székesfehérvár-Palotai u. (15th century): six-rowed barley; 102. Külsővat (15th century): barley; 103. Szarvasgede (15th century): six-rowed barley; 104. Mezőkovácsháza (15-16th century): barley; 105. Budaújlak-Cserfa u./Lajos u. (15th–16th century): six-rowed barley; 106. Nyársapát (15th–16th century): barley; 107. Nagyvázsony-Csepely (15th–16th century): six-rowed barley; 108. Muhi (15th–16th century): six-rowed barley; 109. Budavár-Teleki palota (15th century): six-rowed barley; 110. Baj-Öregkovács-hegy (15th–16th century): six-rowed barley; 111. Lászlófalva-Szentkirály (15th–16th century): six-rowed barley; 112. Vác (15th–16th century): six-rowed barley; 113. Pogányszentpéter-kolostor (16th century): barley; 114. Bánokszentgyörgy-Szentkirály (16th century): six-rowed barley; 115. Esztergom-Vár (16th–17th century): barley; 116. Szolnok-Vár (16th–17th century): six-rowed barley; 117. Szécsény-Plébániatemplom (16th–17th century): six-rowed barley; 118. Budavár-Színház u. (török kor): six-rowed barley; 119. Óbuda-Medve u. (török kor): six-rowed barley; 120. Segesd (török kor): barley

New age (1526–17/18th century) – 121. Hollókő-Vár (17th century): barley; 122. Dunaföldvár-Öregtorony (17th century): barley; 123. Zsámbék-Zichy kastély kertje (17th century): barley; 124. Szécsény-Vörös tér (17th century): barley; 125. Szolnok. Gutenberg tér (17th–18th century): barley

4. Lineages of the most important cultivated plants in the Carpathian Basin

Among the cultivated plants of the Old World, cereals played a definitive role. It is probable that prehistoric people made acute observations of the development of plants and noted that they had different requirements. They selected better varieties that became primary cultivated plants produced in optimum locations to meet the requirements for each cultivated variety. However, weeds growing among protected plants did not escape their attention. Indeed some of them (e.g. rye) were specifically selected for development as cultigens. Many plants, e.g. fat hen (*Chenopodium album*) started along the path to domestication (secondary cultivated plants), but never became cultivated plants. Their development was arrested at a certain level and with the passage of the archaeological culture that attempted to domesticate them they reverted to the status of weeds.

Basically all plants cultivated in Hungary arrived from elsewhere. The following plants originate from the Fertile Crescent area (ZOHARY – HOPF 1988):

- all cultivated varieties of barley (*Hordeum*): two-rowed barley (*H. vulgare* subsp. *distichum*) and six-rowed barley (*H. vulgare* subsp. *hexastichum*). Six-rowed barley was always more significant than two-rowed barley. The variety of barley called "four-rowed" (*H. tetrastichon*) is in fact a type of six-rowed barley with a looser ear.
- the majority of the species in the wheat genus (*Triticum*): among others, the hulled einkorn (*T. monococcum* subsp. *monococcum*), emmer (*T. turgidum* subsp. *dicoccum*) and spelt (*T. aestivum* subsp. *spelta*).

Common bread wheat (*T. aestivum* subsp. *vulgare*), club wheat (*T. aestivum* subsp. *compactum*) and durum wheat (*T. turgidum* subsp. *durum*) have naked seeds. They do not originate from the Fertile Crescent. Other bread wheat species played much less significant roles in cultivation.

Common millet (*Panicum miliaceum*) is a cereal from Central Asia. Rye (*Secale cereale*) and oat (*Avena sativa*) were originally weeds of barley and bread wheat and became cultivated plants through a process of selection in Europe. Rice (*Oryza sativa*) and foxtail millet (*Staria italica*) were domesticated in the Far East. Sorghum (*Sorhum genus*) originates from tropical Africa.

Rice is a species of grass that was domesticated in South Asia. It is probably our oldest cereal. Phytolith examinations prove that it existed in the region of South China in the 8th millennium BC. The earliest preserved grains are from the 6th millennium BC (PRICE – FEINMANN 1993). It did not arrive in Europe until the beginning of the Middle Ages. The earliest seeds of foxtail millet were found in North China, dating to the 4th millennium BC. It is relatively common in the Bronze Age layers of South and Central Europe. The oldest seed of Grain sorghum (*Sorhum bicolor*) was found at its probable place of origin, the savannah of Africa, dating to the middle of the 3rd millennium BC. It has not been found in Europe.

Among leguminous plants, the common lentil (*Lens culinaris*), the pea (*Pisum sativum*) and the horsebean (*Vicia faba*) are often found in Central Europe. Chick-pea (*Cicer arietinum*), bitter vetch (*Vicia ervilia*), common vetch (*Vicia sativa*) and grass pea wine (*Lathyrus sativus*) are less common.

Horsebeans have been found as early as the middle of the 7th millennium BC in the pre-pottery Neolithic of the Middle East (PPN-B) (Nazareth, Jericho, Tell Abu Hureyra), but they occur only in the Eastern Mediterranean (primarily the area of Greece) from the 5th millennium BC. Chick-peas are also in evidence from an early age (PPN-A). This species was domesticated in the Middle East and quickly adapted to Mediterranean conditions. Bitter vetch, narrow-leaved vetch, summer vetch and vetchling are cultivated plants of the Middle East and the Eastern Mediterranean. Finds of those plants are known from the pre-pottery Neolithic: the oldest bitter vetch find occurred at Çayönü, while the oldest occurrences of narrow-leaved vetch and summer vetch are from Tell Abu Hureyra. The oldest find of vetchling is from Jarmo.

Among oil and fibre crops, flax (*Linum usitatissimum*), opium poppy (*Papaver somniferum*), hemp (*Cannabis sativa*) and gold-of-pleasure (*Camelina sativa*) played important roles in Central Europe at various times. Oriental sesame (*Sesamum indicum*), rape (*Brassica napus* var. *napus*) and common olive of Southern Europe (*Olea europaea*) occur less frequently.

Flax was domesticated in the Middle East. Its earliest occurrence (Tell Abu Hureya) is dated to the 8th–7th millennia BC. Opium poppy is an oil and drug plant of Eastern Mediterranean origin. Its seeds were found in large quantities in Central European Neolithic pile buildings, primarily in the south of Germany and in Switzerland. In contrast, hemp is a fibre plant from Central Asia. Its earliest remains are from Northeast China and from Central Europe, dated to the middle of the 3rd millennium BC. Its spread is strongly linked with the nomadic peoples of the steppes (e.g. Scythians, Sarmatians). Gold-of-pleasure, also originating from the steppes, was domesticated in Southwest Asia. It occurs frequently in the 2nd millennium BC sites of Central Europe (Switzerland, Poland). Oriental sesame cultivated in Southwest Asia and in the Mediterranean is probably from the Far East. Its earliest occurrences are in Armenia, from the second half of the 1st millennium BC. Other seed remains of uncertain age are known from ancient Mesopotamia. Although rape is a plant of very late domestication, wild cabbage (*Brassica rapa* subsp. *campestris*, syn. *B. campestris*) occurs quite commonly in several Central European settlements of the Late Neolithic. It is primarily a weed associated with cereals, but larger quantities of its seeds may have been suitable for obtaining oil. Common olive is an oil plant of Mediterranean descent. It already occurred in the area of Israel in the Natufi culture, but it became more common only from the 5th millennium BC. Its pits have also been found in Cyprus from that era.

The following are not cultivated plants of the Old World, but plants from America: maize (*Zea mays*), beans (*Phaseolus genus*), common sunflower (*Helianthus annuus*), potato (*Solanum tuberosum*), bush red pepper (*Capsicum annuum*), common tomato (*Lycopersicum aesculentum*), common tobacco (*Nicotiana tabacum*), gourds and squashes (*Cucurbita genus*), Jerusalem artichoke (*Helianthus tuberosus*). More recently, the introduction and cultivation of certain amaranth species (*Amaranthus caudatus*, *A. hypochondiacus*, *A. cruentus*) and of sweet potato (*Ipomoea batatas*) is attempted in Hungary as well.

Maize developed in the west highlands of Central Mexico, the genetic centre of its wild ancestor, Balsas teosinte (*Zea mays parviglumis* var. *parviglumis*). The earliest cultivated maize is an evolved but genetically incomplete transitional form recovered from archaeological sites in Oaxaca and Tehuacan in central Mexico. Direct AMS radiocarbon analysis of preserved remains from both regions date to the end of the 4th millennium BC, much younger than earlier radiocarbon dating studies indicated. Its first appearance in Europe is dated after the discovery of America. It reached Hungary by the last third of the 16th century, probably from Dalmatia. This is indicated by its old Hungarian name, "Turkish wheat".

Among gourd and squash varieties, winter squash (*Cucurbita maxima*), pumpkin (*C. pepo*) and its cultivated subspecies: feed-squash (*C. pepo* convar. *macrocarpa*), oil-squash (*C. pepo* convar. *styriaca*), spaghetti squash (*C. pepo* var. *oblonga*), ornamental squash (*C. pepo* convar. *microcarpa*), zuchetti (*C. pepo* var. *giromontii*) and patisson (*C. pepo* convar. *patissoniana*) originate from the tropical regions of South and Central America. They were brought to Europe mostly during the 15th and 16th centuries (CSELŐTEI – CSIDER – CSÁKY 1978). Gourds and squash occurring in ancient Egyptian and Roman texts are different from those plants of American origin. It is probable that they were domesticated varieties of cushaw, China squash or musky gourd (*C. moschata*) similar to zuchetti. Unlike the other species mentioned above, bottle gourd or calabash (*Lagenaria siceraria*) originate from old-world palaeotropical areas and can be regarded as indigenous in the Carpathian Basin. We have no data about the appearance of the above new-world squashes and gourds in Hungary.

The potato originates from the Andean region of South America, It arrived in Hungary in the second half of the 17th century. Bush red pepper and the bean plant are also ancient Precolumbian cultivated plants (GOWLETT 1993). They have been cultivated for at least eight thousand years. They first appeared in Hungary in the 16th–17th centuries. The grain amaranth species originates in South and Central America and Mexico among Precolumbian cultures. Today, introduction into cultivation is being attempted in Hungary as well (BAJI 2000). Sweet potato or yam has also been cultivated for some time.

Among vegetable plants, muskmelon (*Cucumis melo*) was probably domesticated in India around the 3rd millennium BC. It was known in all the empires of antiquity, but it was only cultivated on a larger scale from the late Roman imperial era. According to Egyptian wall paintings from the 2nd millennium BC, the variety cultivated there exhibits a green flesh (var. chate) (KEIMER 1924). The oldest find of its seeds is from the Bronze Age site of Tiryns in Greece (KROLL 1982) and from the Shahr Sokhta sites in Iran from the 3rd millennium BC (COSTANINI 1977). The garden eggplant (*Solanum melongena*) also originates from India. The watermelon (*Citrullus lanatus*) is an ancient cultivated plant. Its primary gene centre is believed to be in tropical Africa, with secondary gene centres in India and China. It was cultivated in ancient Egypt; its oldest seeds found in Nubia from the time of the 18th dynasty (middle of the 3rd millennium BC) (VAN ZEIST 1983). Similarly to the watermelon, the cultivation of the cucumber (*Cucumis sativus*) also began in India (SURÁNYI 1985). It has been cultivated for some three thousand years. Its name ("kisaim") occurs in the Bible. The ancient Greeks and the Romans also favored it.

Among the cultivated onion varieties, garden onion (*Allium cepa*) and garlic (*A. sativum*) are of Western Asian origin while leek (*Allium porrum*) originates from the Mediterranean. The earliest evidence of cultivation is provided by complete cloves of garlic and bulbs of onion found in an Egyptian burial vault of the 18th dynasty (TÄCKHOLM – DRAR 1954). Onions have also been found in Mesopotamia.

Headed cabbage (*Brassica oleracea* var. *capitata*) can be traced back only to the Middle Ages. Kohlrabi (*Brassica oleracea* var. *gongyloides*), on the other hand, has been a cultivated vegetable since antiquity. The famous wild cabbage of Pompeii may have been the ancestor of the kohlrabi.

The ancestor of garden carrot (*Daucus carota* var. *sativus*) is the wild carrot (*Daucus carota*), a species that is indigenous to Europe. Conflicting theories have been raised about the location and time of its domestication. The short-rooted varieties of the garden carrot appeared in the 17th century, believed to be the result of the plant improvement work of Dutch gardeners. The common garden parsley (*Petroselinum crispum*) is of Mediterranean origin and has been known since antiquity. The primary gene centre of the garden celery (*Apium graveolens* convar. *rapaceum*) ranges from the Caucasus to the Mediterranean. In Antiquity, the wild variety was consumed. We know little about the history of its cultivation. It has probably been cultivated in Hungary since the 15th–16th centuries. The garden parsnip (*Pastinaca sativa* subsp. *sativa*) is the longest-cultivated old-world vegetable. Its half-achenes occur regularly among the plant finds at Swiss and South German Neolithic lake dwelling settlements, and fragments of its vascular bundles have also been found during analysis of food remains on the interior walls of crockery from Swiss pile buildings. It was cultivated in Antiquity. Its significance was greater in the period before the garden carrot and the common garden parsley became prevalent (14th–18th centuries).

Lettuces include not only head garden lettuce (*Lactuca sativa* var. *capitata*), which appeared very late and is believed to be derived from cabbage-lettuce (*Lactuca serriola*), along with all plants whose leaves or stalks can be eaten as a salad: e.g. fruited cornsalad, purslane, ramsons, bear's-garlic, common dandelion, garden orache, common chervil, garden sage, chicory, etc.

The wild variety of garden sorrel (Rumex rugosus) and related species are indigenous to Europe. It is highly likely that it has been cultivated for a long time. On the other hand, prickly-seeded spinach

(*Spinacia oleracea*) was brought to Europe from Asia Minor around the 14th century. It has been cultivated in Hungary only since the 19th century.

Garden radish (*Raphanus sativus*) probably originated in the Middle East. The wild form of summer and winter radish was probably the wild radish (*Raphanus raphanistrum*). It has been cultivated for a long time in China. It was recorded among foods as a vegetable in ancient Egypt in 2700 BC. The ancient Greeks and Romans were also fond of it and attributed medicinal powers to it. During the Middle Ages, the Slavic Germanic peoples were the foremost cultivators of radishes.

Horseradish (*Armoracia lapathifolia*) originates from Western Asia or Southeast Europe. We have little information about its domestication and history in Hungary.

Among fruits, apple (*Malus domestica*) is the most cultivated fruit over the largest area and with the greatest diversity: almost ten thousand varieties are known around the world. With the exception of Northern Scandinavia, crab apple (*Malus silvestris* (L.) Mill.) can be regarded as indigenous to Europe. However our cultivated apples were not descended from the crab apple, but from the *Malus pumila* Mill. species, which originates from the Caucasus, the area of Afghanistan and Turkestan. The fruit of the crab apple, which is common in European forests, is in inferior in flavour to wild apple species of Western Asia (WILSIE 1969). Nevertheless, crab apples have always been eaten. During the excavation of the Robenhausen lake dwelling in Switzerland, a turfed apple and several apple-seeds were found (MESSIKOMMER 1913). Apples were found at Bronze Age sites in several locations around Europe as well. Written sources mentioning apples are known from ancient Egypt, Greece and Rome (in: SURÁNYI 1985). Varro and Pliny listed several varieties of apples. It seems that the highly sophisticated Roman culture of fruits included crab apples as well. It is certain that it was even cultivated in the Roman province of Germania (WILLERDING 1980). Apple finds are also known from the age of the Vikings (BEHRE 1983). Remains indicating consumption of crab apples exist from the Late Middle Ages (13th–14th centuries) as well. According to documentary evidence, increase in cultivated varieties occurred alongside the crab apple. The 9th-century work, entitled Capitulare de villis, lists sweet and sour, winter and summer apple varieties. WILLERDING (1983a) inferred from an analysis of 15th-century documents that during the Middle Ages, several morphological classes of apples existed in the area of Germany. In the 16th century, seventy apple varieties were recorded (KETTER 1985). Apples were a popular fruit in the Middle Ages. Examination of seeds from Central European cities supports the view that it had several morphological classes.

According to György MÁNDY (1972), cultivated pear (*Pyrus communis* L., syn. *P. domestica* Medik.) was obtained by crossing several wild species (*P. pyraster*, *P. amygdaliformis*, *P. salicifolia*, *P. elaeagrifolia*, *P. nivalis*). The gene centre of Asian species is Central Asia while that of European ones ranges from Asia Minor to Central Europe. Pear-seeds are much less frequent than apple-seeds in archaeobotanical finds. According to records, the wild pear grew only in the south, southeast and centre of Neolithic Europe. The first written record of pear cultivation is from the ancient Greeks. Pears were also mentioned by Theophrastus. Pliny listed some 30 pear varieties. In his Capitulare de villis, Charlemagne recommended the cultivation of pears (in: SURÁNYI ibid.). Pears were also cultivated in the Roman province of Germania ("Germania romana"). This is indicated by the pear find at the legionary garrison of Novaesium-Neuss. In the free part of Germania ("Germania libera"), a primitive form of pear was cultivated (WILLERDING 1980). In addition, remains of pears from the Roman Period were found at Nidderau-Heldenbergen (BAAS 1982). According to iconographic, written and archaeobotanical evidence, pears were cultivated in most of Europe in the Late Middle Ages.

There is no generally accepted view as regards the origin of cherry (*Prunus avium*) and sour cherry (*Prunus cerasus*). While VAVILOV (1934) claimed that both cherry and sour cherry originated from the Middle East, ZHUKOVSKY (1964) placed the primary gene centre of cherry and mahaleb cherry at the border of Europe and Siberia (cf. P. ERMÉNYI 1975–77; KÁRPÁTI – GÖRGÉNYI – TERPÓ 1968). According to MÁNDY (1972), the wild form of cherry is mazzard cherry (*Prunus avium* var. *avium*). In his opinion, sour cherry is either a hybrid of ground cherry, dwarf cherry and cherry or the autotetraploid form of cherry. According to La Baume (in: P. ERMÉNYI 1975–77), cherry came to the Mediterranean from Asia Minor and was then taken by the Romans to Central Europe. We assume that the cultivated black and soft sweet cherry and the cultivated crunchier sweet cherry are the results of mutation of wild cherry, whose fruit has a diameter of barely 10 mm. The primary change associated with the domestication of wild cherry is that the flesh grows heavier and more voluminous.

Neolithic finds of mazzard cherries are primarily known from Southeast and Southern Europe, Southern Germany and Switzerland. It was also consumed by the Celts. Cultivated cherries have been found in several locations within the boundaries of the Roman Empire. Sour cherry spread much later than cherry. It is not present among Central and Eastern European finds until the Early Middle Ages. According to evidence from urban layers of European cities dating to the Middle Ages that are rich in drupe kernels, cherries and sour cherries played an important part in people's diet.

According to Alphonse de Candolle and Nikolai Ivanovich Vavilov (in: MÁNDY ibid.) peach (*Prunus persica*) originates from China. MÁNDY (1972) placed the primary gene centre of the wild variety in Central and Northern China and a secondary one in Persia, Asia Minor and the Mediterranean. BRÓZIK (1962) assumed that peach was known in China around 5000 BC. It is well confirmed that it was very common from around 2000 BC. According to latest judgements, peach was created in North or Central China through crossing of *Persica davidiana* and *P. kansuensis* (SURÁNYI ibid.). From there, it was taken to Japan before the common era, then by trade to Asia and Persia. It found a new home on the southern slopes of the Himalayas and the Caucasus. From Persia, it was taken to Greece, then Egypt and Italy. In all likelihood, Theophrastus' mention of "Persian fruit" referred to peaches. In his work, entitled Georgica, Virgil (70–19 BC) speaks of honey-sweet peaches while Pliny called the fruit "malum persicum" (in: BRÓZIK ibid.). KEREKES et al. (1937) assumed that the ancient yellow clingstone variety of the Romans played an important role among European peach varieties and was the ancestor of the many later clingstone varieties. We note that the fruit is called free-stone if the mesocarpium separates easily from the putamen. If it does not, the fruit is called clingstone ("duránci" in Hungarian). Pliny called the peach with firmer, therefore more valuable flesh "duranicus". The term became common in Hungary after its use by LIPPAY (1664). The Latin "durantium" was transformed into "durazzo" (RAPAICS 1940). But, it is also possible that the term "duránci" was derived from the name of the Albanian town Durazzo or the ancient Greek town of Durrakhion (SURÁNYI 1985).

Pliny and Columella refer to a Gallic peach (in: P. ERMÉNYI 1975–77). Comparing the data of ancient authors, Magdolna P. ERMÉNYI (ibid.) came to the conclusion that peach arrived in Italy around 40–50 AD. It is only a conjecture that peach was already cultivated in Gaul before it arrived in Italy. The Celts may have played an important role in the distribution of the peach by moving it from the Pontus region via a Balkans-Danube route (P. ERMÉNYI ibid.). SURÁNYI (1985) raised the possibility that our "hairy peaches" and the "bald" or "rubber peaches" may not be instances of escape but a Celtic-Roman heritage. It has been established that peach was already widespread in France in the 8th–9th centuries with several varieties under cultivation. It spread around Europe in the 13th–17th centuries. Archaeobotanical remains confirm the above scenario of the spread of the peach.

The fruit of the peach plant is fleshy, sweet, and rarely bald. According to SOÓ – KÁRPÁTI (1968), varieties are classified as follows:

a) *Persica vulgaris* Mill. convar. *Laevis* (DC.) Janchen, the fruit is stone-free, yellow or red.
b) *Persica vulgaris* Mill. convar. *scleropersica* Dierb, the fruit is clingstone. Many varieties are grown, from those with white to those with deep red flesh. It sometimes grows wild.

Opinion is divided as regards the place of origin of the apricot (*Prunus armeniaca*): Armenia, Central Asia and China are all possible candidates (PATAY – SZ. PÓCZY 1964; SURÁNYI 1985). However, the source of the oldest apricot find is not one of those areas but the Bug-Dnieper area, from the 4th millennium BC (JANUSHEVICH 1975). Written evidence indiucates it was already cultivated in China in the 3rd millennium BC. Its spread may have been assisted by the Silk Road (SURÁNYI ibid.). The Greeks probably came in contact with the fruit as a result of the Asian campaign of Alexander the Great, though it was only cultivated in the region of Epiros. Theophrastus (327–287 BC) mentioned it. Romans may have received the apricot from the Greeks or during the war against the Armenians (69–63 BC). In his Historia naturalis (XV. 10–13), Pliny the Elder (23–79 AD) speaks of the profitability of cultivating the apricot, or, as he called it, "pomum armenicum" (Armenian apple) and the "armeniaca arbor" (Armenian tree). He states that in latifundia near Rome, such trees are leased for 100 thalers (in: NYÚJTÓ – SURÁNYI 1981). Apricots reached Central Europe with the Romans.

The cultivated plum (*Prunus domestica*) has a highly varied range of forms. Unfortunately, it has no uniform international classification system, which makes it difficult to classify plums found at archaeological excavations. Gams divided the *Prunus domestica* L. species into three subspecies (in: P. ERMÉNYI 1975–1977):

a) subsp. *insititia* (L.) Poiret – bullace plum
b) subsp. *italica* (Borkhausen) Gams. – greengage
c) subsp. *oeconomica* (Borkhausen) C. K. Schneider – garden plum

As regards the origins of plum, DE CANDOLLE (1894) questioned whether the plant was indigenous to Europe. In his opinion, plums originate from the east because the wild variety is found in all of Asia Minor, the southern slopes of the Caucasus and in the north of Persia. However, he did claim that the bullace plum was indigenous to Europe, as he found it in a wild form in Southern Europe. BERTSCH – BERTSCH (1949) believed that the "wild plum" came to Europe from Asia Minor. WERNECK (1955) concluded that plums were created as a cross between sloe, blackthorn and myrobalan plum in Central Europe. He assumed that various main forms of plums were common in the mixed oak forests of Neolithic Europe and that the spontaneous hybrids were selected and planted around settlements by people. Agreeing with Vavilov (in: P. ERMÉNYI ibid.) MÁNDY (ibid.) located the primary gene centre of plum in the Middle East and the secondary gene centre in the Mediterranean and in Europe.

The greatest problem for taxonomists is constituted by the primitive subspecies of cultivated plums. *P. domestica* is no longer a constituent of natural Central European forests. It only occurs in conjunction with fruit production. On the other hand, bullace plum is capable of surviving without human assistance. *P. domestica* L. subsp. *insititia* (Jusl.) Schneid. var. Juliana is particularly common around Budapest, Debrecen, Békés County, Győr, Kőszeg and Lake Balaton (SOÓ 1966).

The oldest known find of cultivated plum is from the south of the area that used to constitute the USSR, from New-Ruseshti, a settlement of the Tripolje culture (4th millennium BC) (JANUSHEVICH 1975). A plum stone dating to the 2nd–1st millennia BC from the area of the Caucasus (LISITSINA 1978) also seems to indicate the primary gene centre of plum. From there, nomadic peoples of the Neolithic may have carried the fruit towards Europe. The stone of Prunus domestica has also been found at a Neolithic settlement in Vedrovice in the Czech Republic (KÜHN 1979). The Romans, perhaps even the Celts, took part in spreading the plum around Europe. Pliny and Columella mention several varieties of plums: *cereolum* (myrobalan plum), *damasci* (Damascus plum), *onychium* (mirabella). Cato reports the propagation of plums by layering and grafting (in: SURÁNYI ibid.).

In the opinion of researchers, cultivated or garden plum was brought to the region of Pannonia by the Romans (RAPAICS 1940; SÁGI – FÜZES 1967). It is probable that as a result of selection that began in the Neolithic, the plum became a popular and common fruit by the Middle Ages. By that time, garden plum (*Prunus domestica* subsp. *oeconomica* (Borkhausen) C.K. Schneider), bullace plum (*P. domestica* subsp. *insititia* [L.] Poiret), myrobalan plum (*P. cerasifera* Ehrh.) and greengage (*P. domestika* subsp. *italica* [Borkhausen] Gams.) all developed numerous regional varieties, as exhibited by plum finds from the Middle Ages from Germany, the Czech Republic, Poland and Hungary.

Walnut (*Juglans regia*) has been known for a very long time. Its name is featured in Solomon's Song of Songs in the Bible. Theophrastus mentioned that it grew in the land of the Greeks, though it was cultivated. It was known in ancient India and China and is mentioned in German mythology. Walnut varieties are classified according to the shape of the fruit (PORPÁCZY – SZENTIVÁNYI – BRÓZIK 1955):

a) the fruit of the elongated walnut is of medium size, elongated, egg-shaped, conical at the top and often also at the bottom, with a rather smooth surface, a small rim and a rather thin shell which is easily broken. It is a highly valuable and popular variety;
b) the hard shell walnut is characterised by the vigorous formation of chambers;
c) the shell of the stone walnut is even harder;
d) the common round walnut is the most common form of the king walnut, in effect it is the basic type of the species. It has a regular spherical shape, a smooth surface, and the rim is also flat. It has a rather hard shell of medium thickness. The kernel fills the chambers quite well.

MÁNDY (1972) placed the primary gene centre of walnut in the Caucasian, Asia Minor and the Balkans. Hungary lies at the periphery of the habitat of the walnut. The earliest finds of walnuts are from Neolithic Europe, from various sites in the Laibach basin (SERCELJ – CULIBERG 1978). It is possible that the Celts already cultivated walnut. Evidence of walnut cultivation in the area of the Roman Empire comes from several sources. The writings of Cato (234–149 BC) on walnuts imply significant knowledge of cultivation. Varro had already noted the allelopathic effect of walnut on other plants. After the fall of the Roman Empire, finds of walnuts are known from most areas of Europe. According to palaeobotanical data, the walnut is indigenous to the Carpathian Basin (GREGUSS 1969). However, the pollen-analytic work performed at Lake Balaton (ZÓLYOMI 1952, 1980) and in the region of Pilisszentkereszt provided a different result: the Juglans genus is not indigenous to the Carpathian Basin (ZÓLYOMI – PRÉCSÉNYI 1985). Its pollen first occurs in minor quantities in the Bronze Age, but it only occurs frequently and in large quantities from the Roman Period on. It is highly likely that this plant of Asia Minor and the Balkans was brought into the Carpathian Basin by the Balkan peoples that lived along the Danube. Our cultivation of the walnut today is based on Roman practices (SÁGI – FÜZES 1967). Numerous remains from the Middle Ages indicate that walnut was a highly popular "dry" fruit during that time period.

4.1. Phylogenetics of the most important food plants

4.1.1. Wheat

Wheat is our most important cereal. Twenty percent of the calorie requirement of humans is derived from wheat. Cultivated species of wheat belong to the family of grasses (*Poaceae*) and the genera *Triticum* and *Aegilops*. Their original habitat was the area of the Fertile Crescent in the Middle East (KUCKUCK 1962).

All members of the wheat genus (*Triticum*) are believed to have descended from wild einkorn (*Triticum boeoticum* Boiss.). The position of wild diploid wheat (*T. urartu* Thum.) is less clear. It is probable that it is not a descendant of wild einkorn but, by a process of parallel evolution, of a shared ancestor species that has entirely disappeared. Einkorn (*T. monococcum*) developed directly from wild einkorn. On the other hand, wild emmer (*T. dicoccoides* Körn. auct. syn. Kcke.) and the range of the tetraploids were produced as a spontaneous cross between wild einkorn and goatgrass (*Aegilops speltoides* Tausch.).The entire hexaploid range, including club wheat and the core form of the species were formed from the spontaneous crossing of wild emmer or perhaps emmer (earlier called *T. dicoccum* Schrank) and the goatgrass known as *Aegilops squarrosa* L. (syn. *Ae. tauschii* Cosson) (Belea 1986; Belea – Fehér 1977; Belea – Kőszegi – Mrs. Kramarik 1991).

More recently, taxonomists have reclassified the Aegilops species to the *Triticum* genus due to their genetic similarity (Kimber – Feldman 1987).

The genomes of wheat species originate from the wild wheat of the *Triticum* genus (AA, BB) and goatgrass, *Aegilops squarrosa* L. (DD). According to the latest conjecture, *Aegilops searsii* Feld. et Kisl. (n=7) may be the donor of the B genome (Belea 1986).

J. Lelley – T. Rajháthy (1955), H. Kuckuck (1964), Gy. Mándy (1972), G. Láng (1976) and B. Burián (1983) expressed divergent views respectively about the origins of wheat species, but some recent finds were not available to them at the time they published their opinions.

Previously, hulled and naked wheats were described as separate species (Percival 1921). It was only later that it was discovered that some of them are genetically very close to each other. According to the latest phylogenetic studies, wheats are divided into three cytogenetic classes and represented in polyploid sequences (MacKey 1966; Zohary – Hopf 1988):

Traditional classification:	Modern classification:
1. Wild einkorn *T. boeoticum* Boiss. emend. Sciem. (fragile spindles, hulled grains) one grain per spiculum: *T. aegilopoides* (Link) Bal. two grains per spiculum: *T. thaoudar* Reuter *T. urartu* Tuman. 2. Cultivated einkorn *T. monococcum* L. (strong spindle, hulled grain)	Diploid (2n=14) einkorn AA genom name of both wild and cultivated variety: *T. monococcum* L. and subspecies
1. Wild emmer *T. dicoccoides* (Körn.) Aarons. (fragile spindle, hulled grains) 2. *T. ispahanicum* Heslot (wild, less fragile spindle, naked) 3. Chorasan wheat *T. turanicum* Jakubc. (wild, less fragile spindle, naked) 4. Cultivated emmer *T. dicoccum* Schübl. (strong spindle, hulled grains) 5. Durum wheat *T. durum* Desf. (cultivated, naked grains) 6. Emmer *T. turgidum* L. (cultivated, naked grains) and its "branching wheat" varieties: var. *mirabile,* var. *buccale* 7. Polish wheat *T. polonicum* L. (cultivated, naked grain)	Tetraploid (2n=28) emmer, durum, etc. AABB genom collective name: *T. turgidum* L. and subspecies

8. Four-rowed wheat *T. carthlicum* Nevski
   (= *T. persicum* Vav.)
   (cultivated, naked grain)

1. Wild Timofeyev wheat *T. araraticum* Jakubz. (fragile spindle, hulled grains) 2. Cultivated Timofeiev wheat *T. timopheevi* Zhuk. (strong earspindle, hulled grains)	Tetraploid (2n=28) Timofeyev bread wheat AAGG genom collective name: *T. timopheevi* Zhuk. and subspecies

1. Spelt *T. spelta* L. (hulled grains) 2. Macha wheat *T. macha* Dekr. et Men. (hulled grains) 3. Multiple-ear wheat *T. vavilovi* Tuman. (hulled grains) 4. Common bread wheat *T. aestivum* L. = *T. vulgare* Host. = *T. sativum* Lam. (naked grains) 5. Club wheat *T. compactum* Host. = *T. aestivo-compactum* Schiem. (naked grains) 6. Round-grain (naked) wheat *T. sphaerococcum* Perc.	Hexaploid (2n=42) common bread wheat AABBDD genom only exists in the cultivated form collective name: *T. aestivum* L. and subspecies

The Khamut wheat, which is found around the world and which many people believe to have originated from the time of the pharaohs, is in a peculiar position. It shows characteristics of both Polish and durum wheats. The latest studies, however, have shown it to be a version of durum wheat (*Triticum turgidum* subsp. *durum* var. *egyptiaca*).

As regards threshing, wheats fall in the hulled and naked categories (Zohary – Hopf ibid.). The hulled wheats are the more primitive forms: they include diploid einkorn, tetraploid emmer and hexaploid spelt (Kuckuck 1959). All have hulls tightly enclosing the grains. As a result, the end-product of threshing is not grains but spicula. The more developed forms are the naked wheats: tetraploid durum wheat and hexaploid common wheat. Their grains are only loosely enclosed by hulls and threshing produces naked grains.

The majority of the forms of cultivated wheat developed very rapidly in the Early Neolithic in the Middle East over a period of one or two millennia. The cultivated forms discussed below represent only the end results of that process, but not its beginnings (Hillman – Davies 1990).

Common bread wheat could not have developed in the Fertile Crescent, as the goatgrass (*Aegilops squarrosa*) that donated the DD genome was not present there. It probably developed in the area of the Caspian Sea around 6000–5000 BC.

Hexaploid common bread wheat (*T. aestivum* subsp. *vulgare*) is the most developed and most important wheat species. It is cultivated in 90% of all land producing wheat in the world (cca. 246 million ha.). Its wild form is not known. Today, it has some 450 varieties around the world. The greatest collection of bread wheat varieties is maintained at the Vavilov Institute in St. Petersburg where some seventy thousand samples of species and varieties are kept.

The addition of the D genome substantially improved the adaptability of common bread wheat. This allowed common bread wheat to survive the dry, cold winter and the rainy summer of the continental climate. The cultivated tetraploid bread wheat species have adapted more to the Mediterranean climate with its mild winter and dry and hot summer (ZOHARY – HOPF ibid.).

The shared characteristics of all cultivated forms are the larger grains and the rigid earspindle. Following ZOHARY – HOPF (1988), we divide hexaploid wheats into two categories:

a) hulled form:
- *T. aestivum* subsp. *spelta* (L.) Thell. (previously *T. spelta* L.) spelt,
- *T. aestivum* subsp. *macha* (Dekr. et Men) MacKey (previously: *T. macha* Dekr. et Men) Georgian or macha wheat. It is native to western Georgia.

When threshed, hulled wheats separate into spicula. In the case of einkorn and emmer, the bottom of the spiculum retains a fragment of the next earspindle (rachis) (coccoid fracture), but this is not the case with spelt (speltoid fracture).

b) naked form:
- *T. aestivum* subsp. *vulgare* (Vill.) MacKey (previously: *T. aestivum* Lyn. *T. vulgare* Host.) common bread wheat,
- *T. aestivum* subsp. *compactum* (Host.) MacKey (previously: *T. aestivum* subsp. *aestivo-compactum* Schiem.) club wheat. (Actually, the plant is not smaller but the ears are very compact.) The characteristically squat, "compactoid" grains have rounded corners. Their backs are protuberant. The oval scutellum is strongly curved.
- *T. aestivum* subsp. *sphaerococcum* (Perc.) MacKey: round-grain wheat, its grains are compact. It is cultivated in India and Pakistan.

Finds of hulled forms (*T. aestivum* subsp. *spelta*) are found where *Aegilops squarrosa* is also found: along the Kura river / Trans-Caucasia (5th or 6th millennium BC) (LISITSINA 1978), Arukhlo / Transcaucasia (4760 BC) (LISITSINA ibid.), Sakharova/Moldavia (4700 BC) (JANUSHEVICH 1984), Ovčarovo/North-Bulgaria (Gumelniţa culture: 3750 BC) (JANUSHEVICH 1978). This has ruled out the earlier theory that spelt developed in Central Europe, specifically in Southern Germany, during the Bronze Age (BERTSCH – BERTSCH 1949).

The naked form probably developed rapidly from the hulled form. It can be assumed present in the 5th millennium BC. It should certainly be sought in the temperate zone of Western Asia and Europe. One *T. sphaerococcum* type find is known from Belujistan in Pakistan (4000 BC) (JARRIGE – MEADOW 1980). (Its grain is difficult to distinguish from tetraploid naked wheat.)

Nevertheless, despite its greater yields, naked common bread wheat did not supplant hulled wheats (einkorn, emmer, spleta wheat) in cultivation for several centuries. According to archaeobotanical data and written records, einkorn and probably emmer, but certainly spelt was cultivated in the north of Hungary from prehistoric times until the end of the 19th century. Their cultivation continued until the middle of the 20th century in Transylvania.

4.1.2. Barley

Barley (*Hordeum vulgare* L.) is one of the most widespread cereals around the world due to its great toleration of various temperatures, its relatively short growing season and its extraordinary adaptability. It is one of our oldest cultivated plants, grown extensively in dry conditions, in soils with low nutrient contents and even in slightly salty soils. It is one of our most important food and fodder plants. In Europe it has been cultivated and used for making mush, bread and beer and as animal fodder since the Neolithic (SZKAZKIN 1979).

The primary gene centre of barley is the Fertile Crescent: Israel, Jordan, Southern Turkey, Iraq, Kurdistan and Southwest Iran.

Morphologically, barleys are divided into two-rowed and six-rowed (many-rowed) forms.

a) diploid form: *Hordeum vulgare* subsp. *distichum* Zoh. (previously *H. distichon* L.). Of the three spicula, only the middle one is fertile, the two side ones are degenerated. Since the fertile spicula, just like the seeds, are arranged in two rows, the ear is flattened on the sides (with two corners in cross-section). The diploid form is more ancient and the wild form is also diploid. The axis of the seeds is straight and symmetrical (SCHERMANN 1966). Diploid barley is usually grown as fodder or as malting barley.

b) hexaploid form: *Hordeum vulgare* subsp. *hexastichum* Zoh. (previously *H. hexastichon* L.). All spicula are equally developed. Each spiculum develops three seeds. The spicula are arranged in six rows so that the ear has a hexagonal cross-section. Hexaploid barleys are further classified into a looser ear type (with narrow and long earspindle sections – previously called four-rowed barley) and a more compact type (with short, squat earspindle sections – previously called six-rowed barley) (ZOHARY – HOPF 1988).

The hexaploid ear form is more developed than the diploid one. During domestication, a recessive characteristic became dominant, resulting in the two side spicula becoming fertile as well. The axis of the central seeds of hulled six-rowed barley is straight like that of two-rowed barley, but the axis of the side seeds is characteristically curved, best indicated by the curvature of sutura ventralis, which distinguishes it (SCHERMANN 1966).

Hordeum vulgare L. subsp. *spontaneum* (C. Koch) Körn. (previously: *H. spontaneum* (C. Koch) Körn.) is considered the ancestor of domesticated barley. This two-rowed, diploid (2n=14) spontaneous plant, which has a fragile ear spindle, is self-pollinating and is able to interbreed with all cultivated forms, is still common as a field weed alongside roads and in ploughed fields from the Mediterranean to Western Asia. However, its primary gene centre is the Fertile Crescent. It is sensitive to cold, rarely occurring at elevations over 1500 metres. It is almost entirely absent in the highlands of Anatolia and Iran. On the other hand, it is less sensitive to dry conditions, intruding far into the peduncular oak zone to the east, to the north and the west of the Syrian desert and further down in the valleys of the Euphrates and Jordan rivers (HARLAN – ZOHARY 1966).

The earliest finds of barley are the wild form, recovered from the pre-pottery layers of Northern Syria: 9000 BC: Tell Abu Hureyra (HILLMAN 1975); 8500–7500 BC: Ganj Dareh Tepe (VAN ZEIST – BAKKER-HEERES 1986); Tell Mureybit (VAN ZEIST 1970a); 7800–7300 BC: Tell Aswad/Damascus (VAN ZEIST – BAKKER-HEERES 1979); 7500–6750 BC: Ali Kosh/Iran (Bush-Mordeh phase) (HELBAEK 1969); 7000 BC: Çayönü/Turkey, oldest layers (VAN ZEIST 1972); 6700 BC: Beidha/Jordania (HELBAEK 1966; EHRICH 1992).

At the last three sites, wild barley is found along with cultivated forms. We note that at the above sites the wild form appears together with emmer. Based on the shape and anatomy of earspindle remains, KISLEV (1989) reports that the wild and domesticated finds of barley and emmer obtained from the above PPN-A sites are all wild forms.

Finds of cultivated forms: 7500 BC: Tell Abu Hureyra (Pre-Pottery Neolithic) (HILLMAN 1975); 7300–6000 BC: Tell Aswad (2nd phase) (VAN ZEIST – BAKKER-HEERES 1979); 7000–6500 BC: Jarmo/Iraq (HELBAEK 1959a). At these sites, the lowest layers contain "wild barley" while the top ones contain a cultivated form, indicating rapid domestication. This means that barley and wheat were domesticated at the same time and both became important cultivated plants of the Neolithic Middle East rather rapidly.

Six-rowed barley also appeared very early: the earliest find is from the pre-pottery layer of Tell Abu Hureyra in Syria dated to 7500 BC (HILLMAN 1975). In the older layer (7000 BC) of Ali Kosh in Iran, only two-rowed barley occurred initially, but from 6000 BC, sporadic occurrences of six-rowed and naked barley were also found (HELBAEK 1969). At the Çatal Hüyük and Haçilar sites in Anatolia, layers dated to 6000 BC contained both hulled and naked barley (HELBAEK 1970). Six-rowed barley type seeds were found at Tell es Sawwan, also in Anatolia, in a layer dated to 5800–5600 BC (HELBAEK 1964a). By the 4th millennium, the grain cultivation of Mesopotamia was primarily based on hulled barley. In fact, hulled barley is highly characteristic of the end of the Neolithic and the Bronze Age in the Middle East. It was found in larger quantities than wheats. In the meantime, barley became less sensitive to soil composition and climate, greatly enhancing its propagation and reaching areas less suitable for growing bread wheat.

Along with the other cultivated plants of the Fertile Crescent, barley quickly spread in all directions of the compass: towards the Indus and the Nile, to the areas beyond the Caucasus and the shores of the Mediterranean Sea (HARLAN – ZOHARY 1966). In the 6th–5th millennia it appeared in the Nile valley (DARBY – GHALIOUNGUI – GRIVETTI 1977), the area of the Caucasus (LISITSINA 1984) and in India (COSTANTINI 1981a).

In Europe, barley first reached the shores of the Aegean Sea. Its path of propagation spread through mainland Greece, Thessalia to the Balkans. In addition to einkorn and emmer, barley is regularly found in remains from the 6th–5th millennia in Greece. While these finds are characterised by two-rowed, six-rowed and naked barley, the two-rowed variety does not occur among the Neolithic barleys found in the Balkans. The majority of finds are of the easily threshed naked barley variety (the seed falls out of the hull during threshing), six-rowed barley occurs somewhat less frequently. The barley remains found at the sites of the early Neolithic Karanovo I culture in Bulgaria and the Starčevo culture in Serbia are also of this type.

Barley took less then 2000 years to reach Central Europe from Anatolia. Hungaray was one of the most important stations along the way. The peoples of the earliest Neolithic culture of Central Europe, namely the Körös or Criş culture of the Great Hungarian Plain and Transylvania were related to the early Neolithic agricultural cultures of the Balkans.

4.1.3. Rye

Rye (*Secale cereale* L.) is a cereal of the temperate zone of the Old World. It is primarily grown in the cooler regions of Northern and Western Europe. Compared to wheat, it is less sensitive to cold and to drought. Moreover, it can survive even in acidic and sandy soils, conditions that would be unsuitable for wheat. It is an allogamous, diploid (2n=14) grass pollinated by wind (EVANS 1976). Its seeds have high protein content and are suitable for making dough.

The primary gene centre of the *Secale* genus is Southwest Asia. *S. silvestre* Host. and *S. montanum* Guss. do not cross well with *S. cereale*, so they are unlikely to have been genome donors.

In fact, *Secale cereale* contains various wild and cultivated subspecies that may produce fertile hybrids with each other.

Four main races are distinguished (HOPF – ZOHARY 1988):

- cultivated plant: its earspindle is not fragile, its seeds are large (*S. cereale* subsp. *cereale*).
- weed-type wild plant: its earspindle is not fragile, it is formed from a weed, *S. segetale* (Zhuk.) Roshev. It also invades bread wheat fields, but in bad years it is the only harvest in Turkey, Syria, Iraq, Iran and the countries of the Balkans.
- weed-type wild plant with semi-fragile earspindle (only the top part of the ear is fragile). This race includes various populations. *S. afghanicum* (Vav.) Roshev. occurs in NE-Iran, Afghanistan and Transcaucasia.
- weed-type wild plant with fragile earspindle: *S. ancestrale* Zhuk., and *S. vavilovii* Zoh. occurs in various areas of the Middle East.

Neolithic and Bronze Age finds of rye are very rare:

- Tell Abu Hureyra/Northern Syria, epipalaeolithic layer yielded remains with fragile earspindles, probably *S. montanum* (HILLMAN 1975),
- Can Hasan III/Turkey, pre-pottery Neolithic layer yielded primitive cultivated form with non-fragile earspindle (HILLMAN 1978),
- Alaca Hüyük/Northern Middle Anatolia/Turkey: Bronze Age cultivated rye remains were found (HILLMAN ibid.).

It seems that rye is a twice domesticated cereal. It first appeared in the early Neolithic in Anatolia but disappeared during the Bronze Age (BEHRE 1992a). Then, in the course of the Early Iron Age in Central or Eastern Europe, it is again transformed from weed to crop plant.

4.1.4. Oat

Oat (*Avena sativa* L.) is an important cereal of the temperate zone. Its high protein (15–16%) and fat (8%) content makes it highly nutritious both for human and for animal consumption.

The genus is found in the Mediterranean. In addition to diploid wild forms, tetraploid and hexaploid forms also exist. *Avena sterilis* L. is regarded to be the ancestor of oat (SOÓ 1973; L. SZABÓ 1982). This spontaneous species was probably brought to Europe by migrating peoples during the Neolithic as a weed of spiciferous plants.

Common oat (*Avena sativa* L.) is hexaploid (6n=42), including various cultivated varieties. Its ear is not fragile. Although most of its cultivated varieties are hulled, some naked varieties also exist (var. *mutica*).

Based on the shape of the ears, the following types are distinguished:

- Pyramidal- or cylindrical-shaped, rounded clustered oat, with generally higher yields (subsp. *sativa*),
- bannered oat, which curves in one direction and has stronger straw (subsp. *orientalis*).

Thíe Neolithic and Bronze Age sites of the Middle East and the Mediterranean yielded only finds of wild oat (*A. sterilis*, *A. fatua*), but no cultivated oat. The oldest cultivated oat was found in the area of the Czech Republic (1st millennium BC) (TEMPÍR 1966).

This supports the contention that oat is a secondary cultivated plant; that is to say, a weed of bread wheat and barley that was later domesticated in the temperate zone through a process of selection (KÖRBER-GROHNE 1987).

4.1.5. Common millet

The origin and place of domestication of common millet (*Panicum miliaceum* L.) are yet to be established. Its wild form has not been determined with certainty. It may have descended from the Panicum spontaneum Lyssev ex Zhuk. species that occurs in Kazakhstan, Mongolia, Northern China and Afghanistan (DE CANDOLLE 1894; SOÓ 1973).

Following MANSFIELD (1986), the common millet genus is divided into three groups (in: BÁNYAI 1971):

1. wide-clustered common millets (*Panicum miliaceum* L. convar. *effusum* (Alef.) Mansf.),
2. side-curving, banner-clustered common millets (*Panicum miliaceum* L. convar. *contractum* (Alef.) Mansf.),
3. compact-clustered common millets (*Panicum miliaceum* L. convar. *contractum* (Alef.) Mansf.).

They can also be classified on the basis of the colour of their hulls: white, yellow, red, brown, grey.

Common millet is a tetraploid (2n=36), self-fertilizing cultivated plant. Its growing season is very short (60–90 days). It is sown in the spring, but a second, summer sowing also ripens. It tolerates extreme conditions (heat, poor soils, drought) well. During threshing and cleaning, the buds often break off. It is characteristic of the seed that the socket of the scutellum is shorter than half the length of the seed (SCHERMANN 1966).

Today, common millet has lost much of its significance. It has disappeared from Europe. It is primarily cultivated in Eastern and Central Asia, India and parts of the Middle East.

The oldest remains of common millet are from Eastern and Central Europe: Soroki/Ukrajna (Tripolje culture) (JANUSHEVICH 1976), Blahutovice/Czech Republic (TEMPÍR 1979) and Eizenberg/Thuringia (5th millennium BC) (ROTHMALER – NATHO 1957), Gomolava/Jugoslavia (4th millennium BC) (VAN ZEIST 1975), Northern Italy (3rd millennium BC) (VILLARET-VON ROCHOW 1958). Common millet has been shown to have existed in Central Asia since the 3rd millennium Bronze Age (LISITSINA – PRISEPENKO 1977). More recently it has been found at the Shortungha site in Afghanistan (end of 3rd, beginning of 2nd millennium BC) (WILLCOX 1991). Other uncertain finds may be common millet: Tepe Yahya/Iran (5th millennium BC) (COSTANTINI – COSTANTINI-BIASINI 1985), Georgia (Neolithic: 5th–4th millennium BC) (LISITSINA 1984), Northern China (Neolithic, Yang-Shao culture: 4th millennium BC) (HO 1977).

4.1.6. Common lentil

Common lentil (*Lens culinaris* L., syn L. *esculenta* Moench) is one of our most ancient cultivated plants. Many regional varieties of this diploid, self-fertillizing, soil-improving plant are cultivated. Its seed contains 25% protein but has a relatively low yield, 50–150 kg/ha. Particularly large quantities are consumed in India, Pakistan, Ethiopia and the countries of the Middle East and the Mediterranean.

There is still no uniform system of classification for the common lentil species. Taxonomy associates five species with the common lentil genus: *Lens culinaris*, *L. orientalis*, *L. nigricans*, *L. ervoides*, *L. odemensis*.

LADIZINSKY *et al.* (1984) recommended that the genus should be divided into two "biological species":

- *L. culinaris* subsp. *culinaris*, *L. culinaris* subsp. *orientalis*, *L. culinaris* subsp. *odemensis*,
- *L. nigricans* subsp. *nigricans*, *L. nigricans* subsp. *ervoides*.

We note that today an increasing number of biological sciences use the so-called "biological species" as opposed to the traditional concept characterised by external morphological (taxonomic) features (ANDREWARTHA – BIRCH 1954). The traditional, morphological approach, for instance, regards subspecies as morphological variations. As a result, this approach based on phenotypes is not entirely accurate, inasmuch as individuals classified in the same species do not always produce fertile descendants and problems of meiotic cell division may also arise. The concept of biological species regards a species to be a biological organism that is being shaped and is changing all the time, it can be conceived as a "phase" in the process of evolution. It is a higher level reproductive association whose members may exchange genes. It is an evolutionary unit within which morphological changes, microevolutionary processes may take place. Under the genetic approach, two populations may only belong to the same species if they are capable of producing fertile progeny.

From the practical perspective, the cultivated lentil (*Lens culinaris* L.) has two groups of forms (MÁNDY – KISS 1971):

- small seed lentil (convar. *microsperma* Baumg.): the pods are small, the diameter of seeds is 3–6 mm,
- big seed lentil (convar. *macrosperma* Baumg.): the pods are larger, the diameter of the seeds is 6–9 mm. The latter is the more developed form.

The small and large seed lentils are usually also distinguished taxonomically as subspecies: *Lens culinaris* L. subsp. *microsperma* Baumg. and *Lens culinaris* L. subsp. *macrosperma* Baumg.

For a long time, there was no established view as to the origin of the cultivated common lentil. A. DE CANDOLLE (1894) believed it to be indigenous to Greece and Italy. BARULINA (1930) found the greatest variability of the cultivated common lentil in India, Afghanistan and Turkmenistan. However, genetic investigations have ruled out the possibility of *L. nigricans* being the ancestor of the cultivated common lentil, as was previously supposed (RENFREW 1973). *L. orientalis* (Boiss.) Hand.-Mazz. is considered to be the wild ancestor of the cultivated lentil (LADIZINSKY 1979). That species shows the greatest similarity to the common lentil varieties grown in Central and Eastern Asia and produces fertile progeny with them (LADIZINSKY *et al.* 1984). Moreover, the oldest finds of cultivated lentil were found in the habitat of *L. orientalis*, the area of the Fertile Crescent. It has also been suggestged that some characteristics of the species, e.g. the yellow seminal leaf and the serrated stipules are derived from L. odemensis.

Lentils have been cultivated in Asia Minor since the early Neolithic (VAN ZEIST 1971). The oldest finds of common lentil are of the same age as barley and were found in the same locations: 9200–7500 BC: Mureybit (VAN ZEIST 1970a), Tell Abu Hureyra/Northern Syria (HILLMAN 1975) (pre-pottery cultures), 7000 BC: Jarmo/Northern Iraq (HELBAEK 1959a), Ali Kosh/Iran (HELBAEK 1969), Haçilar/ Turkey (HELBAEK 1970), Tell Abu Hureyra (HILLMAN ibid.), Jericho/Jordania (HOPF 1983) (Pre Pottery Neolithic B). Though the diameter of seeds is always small, only 2.5–3 mm, it is probable that they represent a cultivated species. Although the Palaeo-mesolithic common lentil seeds found in Franchthi cave, Greece, were attributed to L. nigricans by HANSEN (1978), it is conceivable that they are also from cultivated plants. The seeds of *L. nigricans* are very difficult to gather as the plant lives in cracks of rock.

Earliest finds of larger-sized lentil seeds, which are certainly domesticated, occurred at Tell Ramad/ Syria (6250–5650 BC) (VAN ZEIST – BOTTEMA 1966) and Tepe Sabz/Iran (5500–5000 BC) (HELBAEK 1969). The average diameter of these seeds was 4.2 mm.

4.1.7. Pea

The diploid (2n=14), self-fertilizing pea (*Pisum sativum* L.) has adapted perfectly to both Mediterranean and cooler temperate climates. The pea is a food plant while the field pea is a fodder plant (SCHERMANN 1966). The protein content of its seed is 22%. Many regional varieties are cultivated. The cultivated pea developed in the Mediterranean and in the Middle East from the ancient diploid, self-fertilizing *Pisum* L. genus.

According to the results of morphological, ecological and cytological studies, two species developed from the genus: *P. sativum* L. (the predecessor of the cultivated form) and *P. fulvum* Sibth. et Sm., found in the Eastern Mediterranean (DAVIS 1970).

A "steppe-type" short, but robust weed from the Middle East, *P. humile* Boiss. et Noä (syn. *P. syriacum* [Berger] Lehm) played an important role in the development of the wild *P. sativum* L. The Mediterranean *P. elatius* Bieb. may also have played a part as a gene donor (BEN ZE'EV – ZOHARY 1973).

The earliest pea finds from the Fertile Crescent (7500–7000 BC) are so small that it is difficult to decide with any certainty whether they are from wild or cultivated plants.

- 7500–7000 BC: Çayönü/SE-Turkey (probably wild), but cultivated from 6500 BC (VAN ZEIST 1972),
- 7500–6000 BC: Jarmo/N-Iraq (pre-pottery culture), Tell Aswad/S-Syria (VAN ZEIST – BAKKER-HEERES 1979), Jericho/Jordania PPN-B (Pre-Pottery Neolithic B) (HOPF 1983).

Finds become more frequent only after 6000 BC. Large quantities of peas have been found from that era at the sites at Çatal Hüyük in Turkey (5850–5600 BC) (HELBAEK 1964b) and at Haçilar (5400–5000 BC) (HELBAEK 1970).

5. The processing of botanical finds

5.1. Methodological foundations of processing

To achieve maximum results from archaeobotanical analysis, the work of the archaeobotanist should begin before the excavation of an archaeological site. The archaeobotanist and the archaeologist should consult with each other in developing a plan for field work. An effective sampling strategy for plant remains requires awareness of the age(s) of the site, cultural periods present, type and function of structures and features expected, types of soil, which areas will be excavated and what methods of excavation will be used. This allows the archaeobotanist to create a sampling plan appropriate to each specific site and to prepare to use the best techniques for differences of contexts, soil, probability of preserved remains, etc. Obviously, the sampling plan applied to the interior of an excavated structure will differ from that applied to areas outside a structure. The archaeobotanist must then discuss the plan with the archaeologist for further suggestions and refinement. The intended plan should then be communicated to participating experts from other disciplines to allow maximum multidisciplinary cooperation. Of course, the archaeobotanist explains the basic sampling strategy and recovery methods to the excavators, especially those who may directly assist in recovery of samples from the excavations.

Intensive work for the archaeobotanist begins the moment a sample is taken. The next step is flotation of the sample, preferably conducted as soon as possible to determine if the presence of plant remains meets expectations. The design and use of the equipment used for flotation must always take into account the number and size of samples, the type of site, dry or wet soil, and, not least, the financial resources available for the excavation. We usually use a flotation frame with a sequence of copper screens from 0.25 to 4 mm. In order to avoid contamination, we use water from the water-system for flotation wherever possible. Several Hungarian museums possess such frame flotation devices. The Budapest Historical Museum is capable of highly efficient work with its flotation device comprised of a tank and large screens effective for larger samples. It is similar to the "Cambridge machine" (cf. PEARSALL 1989).

Cleaned seeds and fruits arriving at the laboratory are identified by their morphological characteristics observed through a stereo binocular microscope. Identification handbooks and articles are also used for identification, but results are always compared with recent material.

The results of botanical processing may furnish answers to questions concerning the creation of the layer, the level of field cultivation, the ratio of cultivated to gathered plants and the environment of the period.

Traditionally, publications list the species present, frequency data and often include dimensions of recovered specimens. Authors may indicate whether the plants are spontaneous or cultivated, their archaeological contexts and, in some cases, also their origins. Anthropogenic categories corresponding to the utility of plants developed by ÁRENDÁS (1982) are highly informative. His categories, which were established from the perspective of human utilisation, effectively reflect the relationship between flora and humans.

Recently, attempts at reconstructing entire environments have become prominent. Due to specific habitat requirements of cultivated plants, their spread has lead to the transformation of nature. Today, such environmental changes can only be tracked through the complex use of various scientific investigations comprising environmental archaeology and landscape archaeology. The essence of the method of "Site Catchment Analysis" (settlement resource distribution survey) developed in England is the reconstruction of an ancient settlement, its environmental context including the location of its cultivated lands and its wider landscape suitable for hunting and gathering based on analysis of plant remains.

Although environments of individual archaeological periods have been inferred with some success, the reconstruction of plant associations seems to be an almost impossible task. It is certain that plant associations change with time. Further difficulties arise when the archaeological record does not contain all members of the flora of any given era. Even characteristic plants of an association virtually certain to exist in the environment may be missing. In addition, diaspores that appear in the soil are not preserved in the same way everywhere. The more contexts and features that are examined at a particular site (e.g. floors of dwellings, waste layers, storage locations, waste pits and cesspits, ditches, wells, cisterns, etc.), the greater the chance that sooner or later we will find seeds and fruits of the species that constituted the natural environment of the site at the time.

The archaeobotanical team of the Institute of Botany of the University of Basel has developed a computer program for the registration, electronic data processing and ecological evaluation of macroscopic archaeobotanical finds. This program allows for the registration of all excavation parameters associated with botanical finds and maintains a record of the results of identifications for each sample. The computer prepares species lists indicating the number of items, calculates relative frequency and continuity. With a sufficient number of species (at least 40) it is also possible to prepare a so-called thanatophytosociological analysis as well. The basis for this is an ecological databank that can be extended on a continuous basis.

The habitat-based ecological classification of Stefanie Jacomet allows subfossilized plant remains to be identified taxonomically according to various ecological features (Jacomet – Brombacher – Dick 1989). This analytical method, which involves computer data processing, also takes into account the fact that the composition of plant associations may change over time.

One of the more recent methods of archaeobotanical processing is based on computer assisted image processing and morphometric analysis (Rovner – Gyulai 2007). We have had promising results in the development of a computer-based seed identification method. In addition to performing measurements of seeds and fruits (area, perimeter, elements of details, curvatures), the computer image processing system has also proved suitable for "smart" classification (shape recognition, sorting). The images recorded are stored in a digital image database. During examination, the data represented as image information can be evaluated in an exact (numeric) manner. As a result, it is possible to identify species and types using techniques based on measurements. After several years of development, we shall soon complete a computerised seed identification system connected to a database containing the data of several thousand populations.

The development of chemical analytic methods over the last few decades has allowed the extension of certain analytic techniques to archaeobotanical material. Gas chromatography and atomic absorption spectrophotometry may assist us with establishing the nutrient contents of ancient cereals. Macro- and micronutrient, amino acid and fatty acid tests provide important information about the nutrient content of cereals that could not be obtained in any other way. It appears that their carbonised state is no obstacle to such tests. We applied the macro- and micronutrient and amino acid tests to the study of food and fodder grain seeds of various ages obtained from excavations around Lake Balaton (Gyulai 1995c) (*Figs. 36–40*). Accordingly, the micronutrient content of ancient cereals was significantly higher than that of the cereals currently in cultivation. In the case of some essential amino acids (e.g. lysine), the content of those old grains was several times higher than in today's cereals. Inasmuch as the amount of some amino acids decreases with time, measurement of them can be used for establishing the age of grain seeds from various archaeological eras that were otherwise preserved in identical conditions (*Fig. 41*).

The continuously developing toolkit of plant biotechnology and genetics allows the complete plant to be grown from even a single living cell (single-cell regeneration), as well as using the genetic material (DNA) in the seeds and fruits for species and type identification. We have tried both processes on seeds from a medieval well and obtained promising results.

5.2. Classification of botanical finds

Morphological examination plays a crucial role in processing seed and fruit remains. It takes precedence over all of the methods discussed above, regardless of the manner of conservation of the finds. Based on our own experience we must emphasise that while we used identification handbooks (BEIJERINC K 1947; BRECHER 1960; SCHERMANN 1966; SCHOCH – PAWLIK – SCHWEINGRUBER 1988) and articles for identification, our results were always checked against recent material, and in some cases model experiments were also required. This is essential because morphological characteristics may change with time and in the course of conservation. In our opinion, individual features may not only change as a function of geographical location, but even in the same area they may vary as a function of time.

It presents a further difficulty that colour and surface texture of these direct plant remains may change with time and suffer from intrusive reactions to the conservation process. On the other hand, the material may swell, distort or receive surface damage or corrosion as a result of chemical and biochemical variations.

Archaeobotanists must possess a highly detailed collection of recent seeds and fruits covering the cultivated and wild flora of Central Europe. (The compilation of such a reference collection is no small task, as today there are some 2400 flowering plants in Hungary.)

In describing species/types we used the naming convention of MANSFIELD (1986), the generally accepted archaeobotanical naming convention of ZOHARY – HOPF (1988) and the guidelines of the latest International Botanical Naming Codex (Tokyo Code 1993), while the names from SOÓ – KÁRPÁTI (1968) and PRISZTER (1986) were used for Hungarian equivalents.

We believe it is important to state that archaeobotany does not deal with dating plant remains. However, the grains of annual cereals are often used for radiocarbon dating (^{14}C testing). Yet with a certain amount of experience it is possible to tell (for instance on the basis of the coloration of the hull or the seminal leaf) whether there are any recent or near recent seeds and fruits among the archaeological material.

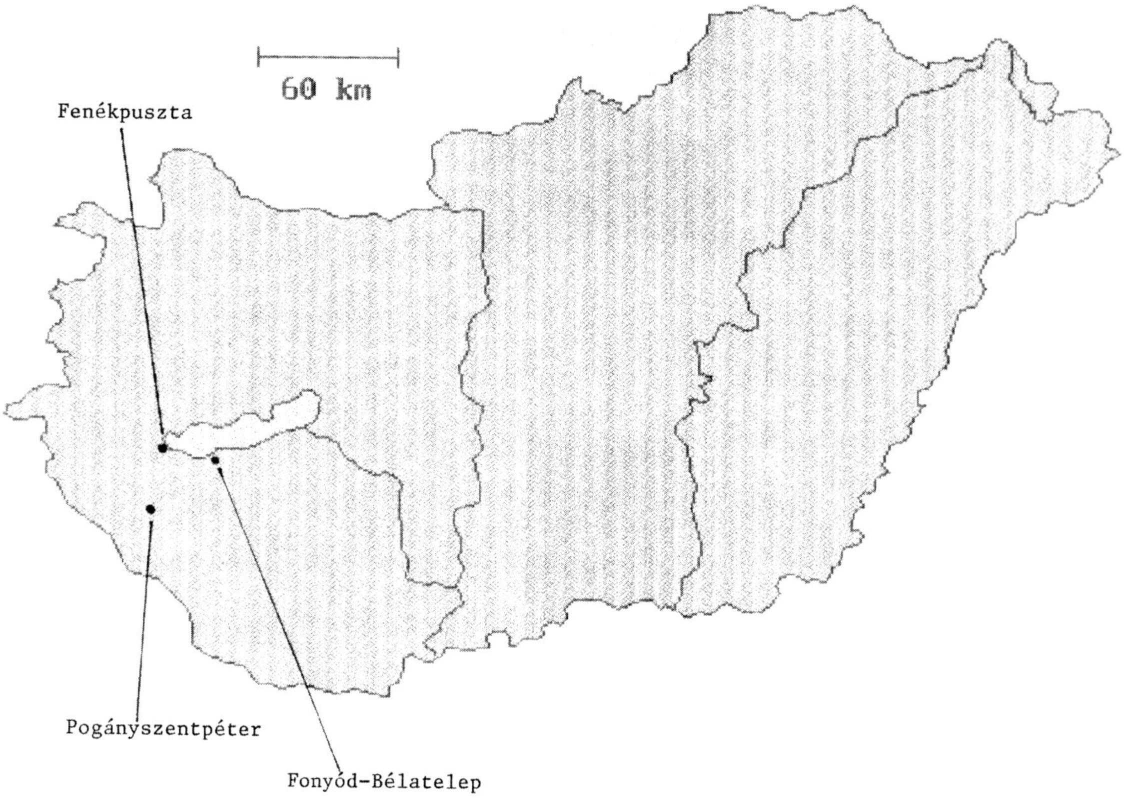

Fig. 36. Sites of carbonised cereals originating from different ages, subjected to analytical examinations. Map of the excavation sites.

ELEMENT CONTENTS MEASURED IN CARBONIZED GRAIN DIASPORA FINDS FROM VARIOUS ARCHAEOLOGICAL PERIODS

		Fonyód-Bélatelep 8th century Chenopodium album	Fenékpuszta 5th century Hordeum vulgare subsp. distichum	Fonyód-Bélatelep 8th century Hordeum vulgare subsp. distichum	Fonyód-Bélatelep 8th century Hordeum vulgare subsp. hexastichum	Fenékpuszta 5th century Secale cereale	Fonyód-Bélatelep 8th century Secale cereale	Pogányszentpéter 16th century Secale cereale	Fonyód-Bélatelep 8th century Panicum miliaceum
N	(%)	2,55	2,09	2,82	2,79	3,22	2,78	2,42	4,01
P	(%)	0,41	0,82	0,39	0,31	0,86	0,38	0,22	0,25
K	(%)	0,10	0,08	0,43	0,66	0,07	0,63	0,22	0,24
Na	(%)	1,46	0,18	0,27	0,26	0,04	0,25	0,31	0,12
Ca	(%)	1,134	3,044	0,879	1,839	3,648	1,722	4,510	0,934
Mg	(%)	0,331	0,816	0,103	0,323	1,385	0,389	0,792	1,197
Fe	(mg/kg)	1481	447	1350	1445	514	1735	511	1138
Mn	(mg/kg)	8,9	88,4	35,6	59,0	53,6	36,8	166,3	8,9
Zn	(mg/kg)	60,8	52,6	29,7	39,1	73,5	36,2	55,7	116,7
Cu	(mg/kg)	13,1	5,1	3,6	3,8	4,9	4,4	5,2	5,6

		Fenékpuszta 5th century Triticum monoccum	Fenékpuszta 5th century Triticum turgidum subsp. dicoccum	Fenékpuszta 5th century Triticum aestivum subsp. vulgare	Fonyód-Bélatelep 8th century Triticum aestivum subsp. vulgare	Pogányszentpéter 16th century Triticum aestivum subsp. vulgare	Fonyód-Bélatelep 8th century Triticum aestivum subsp. compactum	Pogányszentpéter 16th century Triticum aestivum subsp. compactum
N	(%)	3,84	4,26	2,76	3,74	2,74	2,74	2,63
P	(%)	0,15	0,13	0,70	0,27	0,28	0,23	0,25
K	(%)	0,09	0,01	0,10	0,68	0,06	0,07	0,3
Na	(%)	0,02	0,04	0,03	0,05	0,12	0,10	0,20
Ca	(%)	2,483	2,011	3,084	2,134	3,380	2,140	4,513
Mg	(%)	0,657	0,631	1,137	0,443	0,875	0,457	0,779
Fe	(mg/kg)	302	271	924	2370	347	1519	410
Mn	(mg/kg)	20,8	4,5	300,2	45,3	99,6	87,9	153,8
Zn	(mg/kg)	198,4	87,2	276,1	197,9	50,4	36,8	58,0
Cu	(mg/kg)	2,1	1,0	3,6	10,0	3,9	4,1	4,6

Fig. 37. Element contents measured in carbonized grain diaspora finds from various archaeological periods (after János Csapó).

AMINO ACID CONTENTS MEASURED IN CARBONIZED GRAIN DIASPORA FINDS FROM VARIOUS ARCHAEOLOGICAL PERIODS

Fenékpuszta 5th century Avena sativa			Fonyód-Bélatelep 8th century Avena sativa			Fonyód-Bélatelep 8th century Chenopodium album		
asp	0.033	10.03	asp	0.085	9.5505	asp	0.354	10.427
tre	0.015	4.5592	tre	0.033	3.7078	tre	0.174	5.1251
ser	0.012	3.6474	ser	0.032	3.5955	ser	0.195	5.7437
glu	0.03	9.1185	glu	0.107	12.022	glu	0.309	9.1016
pro	0.012	3.6474	pro	0.064	7.191	pro	0.212	6.2444
gli	0.027	8.2066	gli	0.063	7.0786	gli	0.441	12.989
ala	0.015	4.5592	ala	0.061	6.8539	ala	0.231	6.8041
cis	0.003	0.9118	cis	0.014	1.573	cis	0.008	0.2356
val	0.009	2.7355	val	0.053	5.955	val	0.238	7.0103
met	0.009	2.7355	met	0.017	1.9101	met	0.044	1.296
iso	0.015	4.5592	iso	0.024	2.6966	iso	0.12	3.5346
leu	0.009	2.7355	leu	0.058	6.5168	leu	0.223	6.5684
tir	0.038	11.55	tir	0.134	15.056	tir	0.365	10.751
pha	0.024	7.2948	pha	0.062	6.9662	pha	0.178	5.243
lys	0.033	10.03	lys	0.051	5.7303	lys	0.106	3.1222
hys	0.018	5.4711	hys	0.01	1.1236	hys	0.073	2.1502
arg	0.027	8.2066	arg	0.022	2.4719	arg	0.124	3.6524
tri	—	—	tri	—	—	tri	—	—
	0.329			0.89			3.395	

Fenékpuszta 5th century Hordeum vulgare subsp. distichum cv. nudum			Fonyód-Bélatelep 8th century Hordeum vulgare subsp. distichum cv. nudum			Fonyód-Bélatelep 8th century Triticum aestivum subsp. vulgare		
asp	0.052	11.158	asp	0.083	10.323	asp	0.064	11.169
tre	0.021	4.5064	tre	0.034	4.2288	tre	0.027	4.712
ser	0.022	4.721	ser	0.028	3.4825	ser	0.023	4.0139
glu	0.054	11.588	glu	0.098	12.189	glu	0.076	13.263
pro	0.039	8.3691	pro	0.062	7.7114	pro	0.037	6.4572
gli	0.036	7.7253	gli	0.062	7.7114	gli	0.048	8.3769
ala	0.032	6.8669	ala	0.053	6.592	ala	0.035	6.1082
cis	0.003	0.6437	cis	0.013	1.6169	cis	0.005	0.8726
val	0.029	6.2231	val	0.052	6.4676	val	0.034	5.9336
met	0.005	1.0729	met	0.01	1.2437	met	0.011	1.9197
iso	0.018	3.8626	iso	0.029	3.6069	iso	0.024	4.1884
leu	0.033	7.0815	leu	0.068	8.4577	leu	0.043	7.5043
tir	0.033	7.0815	tir	0.049	6.0945	tir	0.036	6.2827
pha	0.041	8.7982	pha	0.088	10.945	pha	0.037	6.4572
lys	0.028	6.0085	lys	0.044	5.4726	lys	0.035	6.1082
hys	0.006	1.2875	hys	0.01	1.2437	hys	0.007	1.2216
arg	0.014	3.0042	arg	0.021	2.6119	arg	0.031	5.4101
tri	—	—	tri	—	—	tri	—	—
	0.466			0.804			0.573	

Fig. 38. Amino acid contents measured in carbonized grain diaspora finds from various archaeological periods (after János Csapó).

Amino acid contents measured in carbonized grain diaspora finds from various archaeological periods

Pogányszentpéter 16th century Triticum aestivum subsp. vulgare			Fenékpuszta 5th century Triticum aestivum subsp. vulgare			Fonyód-Bélatelep 8th century Triticum aestivum subsp. compactum		
asp	0.179	11.215	asp	0.045	11.842	asp	0.074	10.898
tre	0.062	3.8847	tre	0.024	6.3157	tre	0.029	4.2709
ser	0.07	4.3859	ser	0.024	6.3157	ser	0.026	3.8291
glu	0.214	13.408	glu	0.047	12.368	glu	0.067	9.8674
pro	0.109	6.8295	pro	0.018	4.7368	pro	0.032	4.7128
gli	0.014	8.7719	gli	0.031	8.1578	gli	0.061	8.9838
ala	0.078	4.8872	ala	0.022	5.7894	ala	0.049	7.2164
cis	0.016	1.0025	cis	0.004	1.0526	cis	0.006	0.8836
val	0.082	5.1378	val	0.018	4.7368	val	0.039	5.7437
met	0.029	1.817	met	0.007	1.8421	met	0.015	2.2091
iso	0.086	5.3884	iso	0.014	3.6842	iso	0.034	5.0073
leu	0.109	6.8295	leu	0.025	6.5789	leu	0.055	8.1001
tir	0.094	5.8897	tir	0.023	6.0526	tir	0.051	7.511
pha	0.101	6.3283	pha	0.024	6.3157	pha	0.053	7.8056
lys	0.103	6.4536	lys	0.024	6.3157	lys	0.04	5.891
hys	0.023	1.4411	hys	0.008	2.1052	hys	0.016	2.3564
arg	0.101	6.3283	arg	0.022	5.7894	arg	0.032	4.7128
tri	—	—	tri	—	—	tri	—	—
	1.596			0.38			0.679	

Pogányszentpéter 16th century Triticum aestivum subsp. compactum			Fonyód-Bélatelep 8th century Hordeum vulgare subsp. distichum			Fonyód-Bélatelep 8th century Panicum miliaceum		
asp	0.181	12.388	asp	0.078	11.127	asp	0.095	12.45
tre	0.063	4.3121	tre	0.037	5.2781	tre	0.044	5.7667
ser	0.071	4.8596	ser	0.031	4.4222	ser	0.038	4.9803
glu	0.126	8.6242	glu	0.081	11.554	glu	0.074	9.6985
pro	0.063	4.3121	pro	0.039	5.5634	pro	0.03	3.9318
gli	0.142	9.7193	gli	0.052	7.4179	gli	0.057	7.4705
ala	0.095	6.5024	ala	0.03	4.2796	ala	0.065	8.519
cis	0.016	1.0951	cis	0.011	1.5691	cis	0.005	0.6553
val	0.071	4.8596	val	0.044	6.2767	val	0.043	5.6356
met	0.032	2.1902	met	0.017	2.4251	met	0.013	1.7038
iso	0.08	5.4757	iso	0.026	3.7089	iso	0.029	3.8007
leu	0.102	6.9815	leu	0.044	6.2767	leu	0.054	7.0773
tir	0.102	6.9815	tir	0.058	8.2738	tir	0.069	9.0432
pha	0.11	7.529	pha	0.079	11.269	pha	0.075	9.8296
lys	0.089	6.0917	lys	0.036	5.1355	lys	0.032	4.1939
hys	0.032	2.1902	hys	0.01	1.4265	hys	0.012	1.5727
arg	0.086	5.8863	arg	0.028	3.9942	arg	0.028	3.6697
tri	—	—	tri	—	—	tri	—	—
	1.461			0.701			0.763	

Fig. 39. Amino acid contents measured in carbonized grain diaspora finds from various archaeological periods (after János Csapó).

AMINO ACID CONTENTS MEASURED IN CARBONIZED GRAIN DIASPORA FINDS FROM VARIOUS ARCHAEOLOGICAL PERIODS

Fenékpuszta 5th century Secale cereale			Fonyód-Bélatelep 8th century Secale cereale			Pogányszentpéter 16th century Secale cereale		
asp	0.04	11.204	asp	0.077	11.492	asp	0.169	11.951
tre	0.017	4.7619	tre	0.034	5.0746	tre	0.083	5.8698
ser	0.019	5.3221	ser	0.03	4.4776	ser	0.07	4.0950
glu	0.035	9.8039	glu	0.08	11.94	glu	0.14	9.9009
pro	0.019	5.3221	pro	0.042	6.2686	pro	0.068	4.809
gli	0.032	8.9635	gli	0.049	7.3134	gli	0.116	8.2036
ala	0.021	5.8823	ala	0.043	6.4179	ala	0.081	5.7284
cis	0.004	1.1204	cis	0.006	0.8955	cis	0.017	1.2022
val	0.021	5.8823	val	0.042	6.2686	val	0.079	5.5869
met	0.007	1.9607	met	0.011	1.6417	met	0.017	1.2022
iso	0.013	3.6414	iso	0.022	3.2835	iso	0.048	3.3946
leu	0.021	5.8823	leu	0.053	7.9104	leu	0.082	5.7991
tir	0.024	6.7226	tir	0.045	6.7164	tir	0.105	7.4257
pha	0.035	9.8039	pha	0.06	8.9552	pha	0.157	11.103
lys	0.024	6.7226	lys	0.034	5.0746	lys	0.095	6.7185
hys	0.005	1.4005	hys	0.009	1.3432	hys	0.017	1.2022
arg	0.02	5.6022	arg	0.033	4.9253	arg	0.07	4.9505
tri	—	—	tri	—	—	tri	—	—
	0.357			0.67			1.414	

Fenékpuszta 5th century Triticum monococcum			Fenékpuszta 5th century Triticum turgidum subsp. dicoccum		
asp	0.045	12.064	asp	0.051	12.718
tre	0.018	4.8257	tre	0.018	4.4887
ser	0.018	4.8257	ser	0.024	5.985
glu	0.048	12.868	glu	0.051	12.718
pro	0.024	6.4343	pro	0.021	5.2369
gli	0.036	9.6514	gli	0.033	8.2294
ala	0.021	5.63	ala	0.027	6.7331
cis	0.006	1.6085	cis	0.006	1.4962
val	0.018	4.8257	val	0.017	4.2394
met	0.007	1.8766	met	0.007	1.7456
iso	0.013	3.4852	iso	0.011	2.7431
leu	0.033	8.8471	leu	0.023	5.7356
tir	0.027	7.2386	tir	0.027	6.7331
pha	0.019	5.0938	pha	0.031	7.7306
lys	0.016	4.2895	lys	0.027	6.7331
hys	0.009	2.4128	hys	0.009	2.2443
arg	0.015	4.0214	arg	0.018	4.4887
tri	—	—	tri	—	—
	0.373			0.401	

Fig. 40. Amino acid contents measured in carbonized grain diaspora finds from various archaeological periods (after János Csapó).

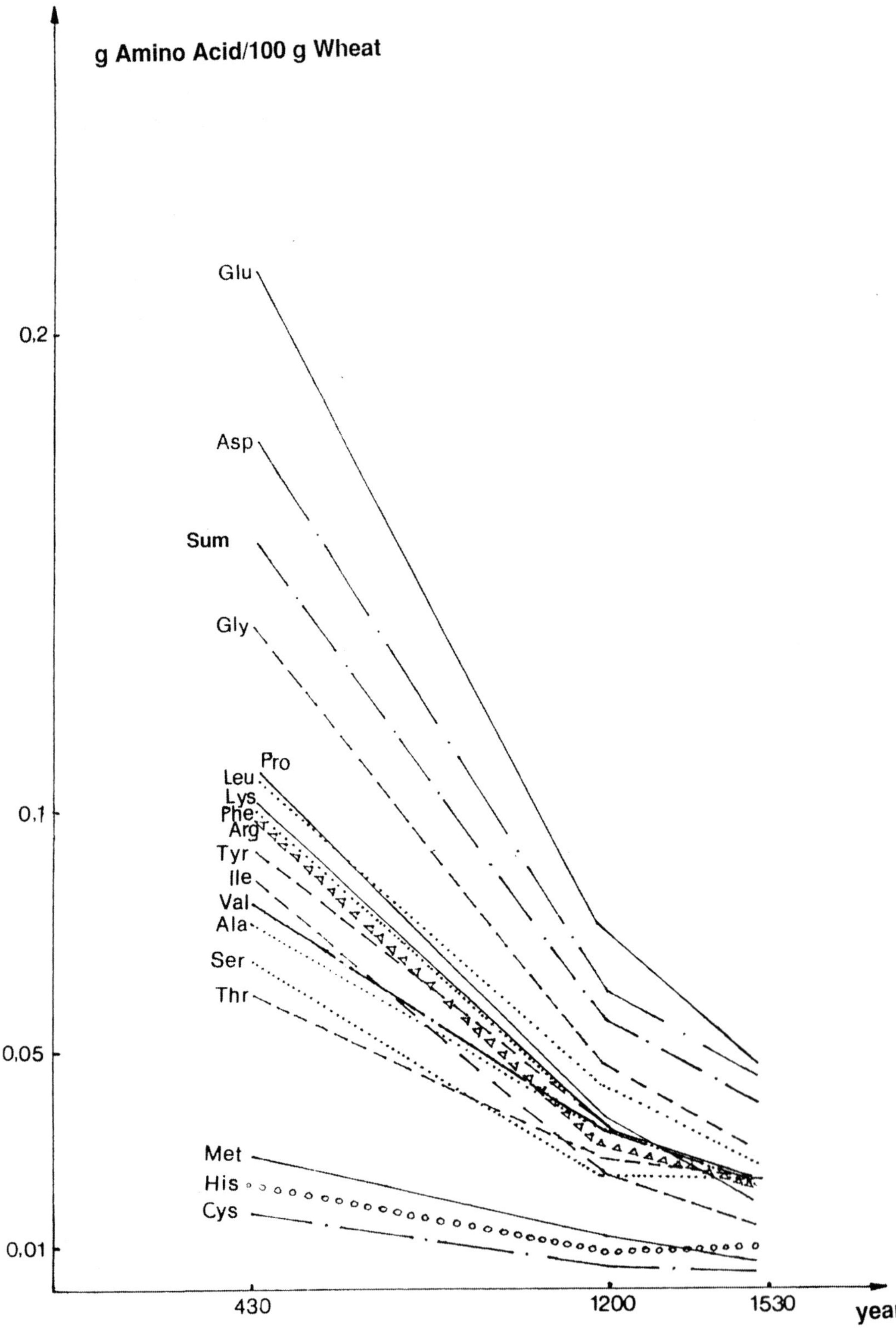

Fig. 41. Changes in the amino acid composition of common wheat (*Triticum aestivum* subsp. *vulgare*) remains by the lake Balaton region during the years (after János Csapó).

6. Collection and processing of botanical finds

6.1. The causes and main factors of seed/fruit preservation

Plant remains may be deposited in the soil naturally (coverage by soil, sedimentation of the remains of local vegetation) or as a result of human activity (burial, deposition).

If remains have been continuously immersed in water since their deposition (wet soil sites), seeds/fruits will remain intact, for in anaerobic environments microbial decomposition is inhibited. Such cultural layers are usually extremely rich in plant remains and the density of finds is very high. The concentration of seeds may exceed one thousand items per litre; or, in the case of wells and lake-side pile structures, it may even reach ten thousand.

Organic material deposited in mineral soils (dry soil sites) is partly or wholly absorbed and digested in aerobic conditions. Only carbonised material survives, so the density of finds is low.

Defining characteristics are preserved particularly well on non-carbonised items, as in most cases carbonised seeds/fruits expand and are deformed due to the heat.

In our geographical and climatic conditions, plant remains from archaeological eras are primarily preserved in carbonised form. Items may be carbonised naturally or by fire. The Hungarian Dictionary (1962) does not differentiate between the two concepts, but, following SÁGI – FÜZES (1966), a distinction is made between natural carbonisation (combined effect of pressure, temperature and time) and carbonisation by fire (e.g. roasting). In the case of plant remains from archaeological eras, it is the latter type of carbonisation that occurs (NOVÁKI 1981–83).

Carbonised seeds from archaeological periods were not carbonised naturally, but by fire. This happened slowly and gently. Natural carbonisation occurs as an effect of pressure and temperature over a very long period, e.g. the formation of mineral coal. Carbonisation by fire happens as a result of burning fires, the burning of settlements, cooking and roasting.

Microscopic items (pollen, spores) often survive attached to the surface of macroscopic finds (GREGUSS 1937). Their identification is the subject of palynology, a discipline that requires special training (BERTSCH 1942; ZÓLYOMI 1952; STRAKA 1975). In many countries, there has been a significant increase in the number of environmental reconstruction projects based on the study of the silica crystals present in the epidermis of some plant species, that is, silicophytoliths (BALL *et al.* 1996; ROVNER – RUSS 1992).

6.2. Identification and evaluation of finds

6.2.1. Quantitative evaluation

After flotation and drying, suitably labelled seed remains from floated samples are sent to an archaeobotanical laboratory where they receive temporary lab numbers. In the laboratory, seeds and fruits are initially separated from frequently occurring other organic remains (e.g. root fragments, insects and other arthropodous animals, cocoons, smaller snail shells, etc.) under a stereo binocular microscope with suitable lighting. All selection is performed using fine, flexible tweezers so as not to damage any very fragile organic remains. We keep a record of our findings. These materials, which are only “ballast” from the perspective of botanical processing, must also be packed in the hope that one day they might be identified by a specialist with suitable knowledge and may furnish important information concerning the fauna, environment and climate of the period.

The next step is the "fine selection" of the samples, which involves classification of seeds/fruits into groups according to their morphological properties. Finally, seed and fruit remains – depending on the quality and persistence of morphological characteristic usually as observed through a stereomicroscope – are identified down to botanical taxons (species, genus, family). Naturally, the diaspores thus categorised must be counted. If there are only a few pieces, this is easily done; but if the sample contains several thousand items, we use digital or mechanical hand counters to avoid mistakes. There also exist highly accurate automatic laboratory seed counting instruments.

An alphabetic list of species is prepared using the results of identification along with the numbers of seeds/fruits in individual samples. Computer data processing may furnish relative frequency (percentage ratio of each plant species within the sample), the percentage ratio of species relative to the total of all species, seed concentration (concentration of seeds and fruits in an individual layer or relative to the volume/mass of the sample). Those figures can answer questions; such as, which plant species were used frequently and which were not. What was the relationship between gathering and plant cultivation?

Under our climatic conditions, the majority of seeds and fruits found at excavations are carbonised. We assume that they are the same age as the cultural layer. The majority are seeds of cereals and associated weeds, with a smaller proportion consisting of the seeds/fruits of other plant species from the natural environment of the era. Their surface has a dull lustre, but due to corrosion they are usually spongy in structure. This implies that a larger quantity of grain was destroyed and the remains were thrown in the pits. During burning, the top layer of grain is completely oxidised (burnt to cinders), but the seeds in deeper layers are only carbonised due to indirect reduction (less air and more gentle heat).

For a limited period, humic acids are able to conserve seeds and fruits deposited in the soil and to protect them against microorganisms. Such seeds and fruits are characteristically brown and hard (subfossilised).

On very rare occasions, we find so-called calcinated seeds in the ashes of medieval fireplaces, which are a characteristic ivy colour, hard and fragile. They are primarily grape and fruit seeds and drupe kernels.

6.2.2. Qualitative (ecosociological) evaluation

During archaeobotanical processing, seeds and fruits are identified by their morphological properties. Again, publications provide an inventory of taxa present and variously include quantitative frequency and/or morphometric (i.e. measurement) data. The spontaneous or cultivated nature of plants is indicated along with their archaeological location and in some instances their origin. Recently, there has been significantly increased interest in attempts to reconstruct environments. In addition to plant sociology, ethnographic information and the results of experimental archaeology can also provide assistance with the evaluation of botanical finds. Obviously, knowledge of archaeological finds of a period and archaeobotanical parallels is required along with knowledge of any epigraphic (documents, other written sources) and iconographic data associated with plants.

The work of processing is made easier if plant species found at a site are arranged in groups. Naturally, plants whose consumed parts are not their seeds or fruits, such as some vegetables, mushrooms and pasture species, occur only very rarely.

From the perspective of human utilisation, plant species are classified as follows:

a) plants living in the wild: seeds and fruits of grasses, shrubs, trees;
b) gathered plants (plants that are wild but are gathered): edible plants, medicinal plants and herbs, dyes and tanning plants, symbolic/cultic and other "useful wild plants";

c) cultivated plants: plants that are used and cultivated for some reason, without any sign of domestication;
d) domesticated plants: plants deliberately cultivated by people , with visible signs of domestication (e.g. solid earspindle in cereals, synchronised ripening, large seeds) – in addition to cereals, leguminous plants, cultivated spices, fibre crops, fruits and grapes contribute to this category;
e) weeds: plants that have genetically adapted to association with domesticated plants, the size of their seeds/fruit is similar to domesticated plants and they ripen at the same time.

One of the most important questions following archaeobotanical identification is the extent to which the environment of a former settlement can be reconstructed. The basis for a qualitative evaluation is the knowledge of extant plant taxa, the spread of various species and plant associations. Investigation of flora always starts with plant associations that result from the cohabitation of various plants. For want of a better basis, we draw inferences concerning plant associations of ancient eras from the characteristics of present plant associations. Recent plant associations are well distinguished by habitat. Therefore, the best means of characterising a habitat ecologically is furnished by these associations, as the basic unit of plant sociology. Today's associations are the result of long development. For precisely that reason, however, they are not applicable as such to ancient eras. Our work is made more difficult by the fact that as a result of human activity at any given time, archaeological objects contain mixtures of the remains of various plant associations.

Environmental reconstruction is only feasible if a sufficient number (at least 40) species from the natural environment have been identified. Due to the numerous criteria involved, ecological evaluation is performed using a computer. The more species contained in the material from a specific excavation, the more accurate the result of the evaluation will be. Control is provided in the form of palaeo-phytocoenology.

Krasilov (1975) termed the botanical material obtained from excavations *"taphocoenosis"*. But, terminology, itself, does not resolve the problem that excavated materials do not contain all species of the vegetation of an era. Due to numerous factors described above, the diaspores that appear in the soil are not preserved in the same way everywhere.

Following Willerding (1983a), the entirety of plant remains deposited from the environment of a period under investigation is called "thanatocoenosis." This allows for numerous combinations, as it may include remains from more than one place (waste layer, waste pit, cesspit).

We gain a great deal more information if we are fortunate to find a localized feature or well-delineated context containing the remains of species of an ancient plant association: e.g. cereal grain remains together with associated weed flora. Following Merklin (1969), such finds are called "palaeo-biocoenosis." We distinguish autochthonous palaeo-biocoenoses when the combination of plants is furnished by species that all lived locally, and allochthon palaeo-biocoenoses, when the species in the plant association are from areas outside the site.

In addition to the seeds of cereals, remains of associated weed flora may also be found during botanical investigation along with sporadic occurrences of other elements of the ancient environment (cf. palaeo-biocoenosis). Therefore, the study of palaeo-biocoenosis created by humans provides an opportunity for gaining information about the lifeways and agricultural knowledge of ancient cultures. In the course of conducting "Site Catchment Analysis," the environment of a population is reconstructed on the basis of plant remains. We can establish the extent of arable lands surrounding settlements, their distances, places of secondary agricultural activities (e.g. pasture), as well as more remote areas where hunting and gathering took place (Renfrew – Bahn 1999).

Thanatocoenology is the science that analyses the ecological interrelations of plant remains found and identified in excavations. It reconstructs the flora and plant associations previously extant in an area, describing the botanical conditions of the environment of the site (WILLERDING 1979, 1983a, 1986b). The combination of classical Greek words (thanatos = death, coenosis = association, logos = science) refers to the archaic nature of the plant associations studied. It is an auxiliary science of archaeology, a marginal area; but, as a specialisation within botany, it is also an applied discipline. It is based on the habitat requirements of plants, tracking changes in species composition of plant associations as a function of both location and time. Thanatocoenology makes use of the ecological indicator role of plants as a fundamental principle to an even greater extent and even more stringently than the more recent science of synecology (ELLENBERG 1979).

The form, structure and dynamic changes of plant associations are shaped by ecological factors. Species (or their populations) usually change relative frequencies whenever a particular ecological factor varies (PRÉCSÉNYI 1980; FEKETE 1981). Although changing ecological factors may remain hidden, the resulting alterations to the floral assemblage may be observed. In effect, the resulting effect is a direct symptom of the cause. Indeed, according to some views, the evidence from plants is a "better" indicator than that obtained from scientific techniques used to measure ecological conditions directly. The appearance of an indicator plant is either directly associated with the factor it indicates or it is associated with a combination of factors that provides similar results. Based on their similar responses to several factors, species can be interpreted as belonging to a definable ecological assemblage of species. This was the path to the development of the indicator theory that considers indication a viable method of ecological research even though, in itself, it is not a method of direct investigation but a methodological theory (PRÉCSÉNYI ibid.).

According to the indicator system, a species is characterised by describing its biological spectrum, its adaptive lifeways, its habitat range and its ecological indicator values (i.e., L, T, K, F, R, N indication). We use Raunkiaer's lifestyle system for the analysis of the biological spectrum (in: ELLENBERG 1979). The main criteria of that system is constituted by the position relative to the soil surface of the wintering organs (seeds, sprouts, buds):

MM, M, N	=	phanaerophyton = trees and shrubs
Ch	=	chamaephyton = undershrubs
H	=	hemikryptophyton = plants that winter near the ground, under cover of leaf-litter, "semi-hidden"
G, HH	=	kryptophyton (geophyton + hydrato-helophyton) = plants that winter in the soil or in mud
TH	=	hemitherophyton = biennial
Th	=	therophyton = annual.

The members of the flora of a given region are categorised into flora component (area) types according to their current range, their path and time of migration and their place of origin. The following area types occur in Hungary: cosmopolitan, adventive, European circumpolar, Eurasian, European, Central European, continental, Mediterranean, Atlantic, Boreal, Alpine, Balkan, Carpathian, Pannonian, endemic and relict species. Their combinations also occur.

Many people have worked on ecological classification of plant species on the basis of the indication theory. Views and research methods differ and the content of ecological classifications is also varied. For the purposes of evaluation we use the ecological classification (classification of species according to their habitat requirements) advanced by ELLENBERG (1979): the T value indicates the geographical (climatic zone) area of the species, the W value indicates the water supply of the habitat, the R value indicates soil reaction, i.e. the acidity of the soil, the N value indicates the soil's nitrogen content. Additional scaled parameters indicating light requirement (L), physical structure of the soil (S) and the toleration of species to salt and disturbance (Bt) and being tread on (Tt) are also in use.

We have also used the anthropogenic categories developed by ÁRENDÁS (1982) to quantify the usability of plants for our evaluation. The essence of the method is that the plant remains are classified into artificial categories of origin. One plant may appear in more than one category. These artificial categories closely reflect the relationship between flora and humans.

During our processing work we have examined the following categories:

- cereals: the species in this category were cultivated as cereals, cereal substitutes or garden plants, and their grains, seeds were used for human nutrition;
- spontaneous plants: carbonised or turfed plant remains that occur on an accidental basis and for whose presence there is no particular explanation; remains of these plants, from the natural environment of the settlement occur in small numbers. Such macroscopic remains are usually found among groups of cereal items or in their immediate vicinity. In addition to their accidental occurrence, their practical significance is also taken into account (e.g. medicinal use, stimulants, implements, etc.). They are extremely significant because they furnish a great deal of information about the natural flora and climate;
- weeds: according to our present knowledge, they are segetal, fallow land, garden and ruderal species;

Weeds fall into two ecological categories (ELLENBERG 1979):

- *Secalietea* = the class of weeds of winter cereals,
- *Chenopodietea* = class of root and ruderal plants,
- *Polygeno-Chenopodietalia* = the order of hackfruits and spring cereal weeds,

The presence of cereal weeds allows us to infer the time of sowing and the method of cultivation of cereals as well as the method of harvesting (high or low, with a sickle or with a scythe).

Plant sociology and ecology allows us to arrange subfossil plant remains in a system according to various ecological criteria and finally to draw certain thanato-phytocoenological conclusions (based on JACOMET – BROMBACHER – DICK 1989):

1.1 = submerged aquaceous plants
1.2 = floating seaweeds, pondweeds
1.3 = diverse aquaceous plants
2.1 = reeds
2.2 = high sedge
2.3 = watershore pioneers
3 = diverse waterside plants
3.1 = marshland plants
3.2 = wet perennials
4.1 = wet fragmented forest
4.2 = fringing forests
5 = fresh and light mixed forest
6 = shady forest
7.1 = cleared forest
7.2 = moderate forestside
7.3 = arid forestside
8.1 = humid meadow
8.2 = moderate meadow
8.3 = arid meadow
9.1 = cultivated plants
9.2 = root or summercrop weeds

9.3 = cereal or wintercrop weeds
10.1 = humid ruderal plants
10.2 = moderate ruderal plants
10.3 = arid ruderal plants
Diverse = non-classifiable

6.2.3. The development of plant associations

KÜSTER (1985a) devoted an entire study to the origin and European spread of Secalietea species. Due to intensive industrial agriculture that has been adopted around Europe, many archaeophyton Secalietea species are doomed to extinction. Their influx is associated with certain well-defined eras. (Today, species of Neolithic cereal weed associations are relatively sparse even in the Middle East.)

Some segetal weeds (e.g. korn-rade, brome grass) arrived with Neolithic plant cultivators from the south-east. Later they migrated slowly towards the west accompanied by indigenous species of Central and Eastern Europe. For instance, field camomile (*Anthemis arvensis*) appeared in the Bronze Age and reached its maximum range in the Middle Ages; ball mustard (*Neslea paniculata*) appeared in the Migration Period and was most widely distributed in modern times; summer adonis (*Adonis aestivalis*) and shepherd's-needle (*Scandix pecten-veneris*) appeared and also reached their maximum area coverage in the modern period.

The species of cereal weed associations of Southern European origin were presumably brought to Central and Western Europe with the grain trade. White laceflower (*Orlaya grandiflora*), a plant of Mediterranean origin, first appeared as a cereal weed in Central and Western Europe during the Roman Period, but only reached its largest area coverage in the modern period. The Romans brought corn-buttercup (*Ranunclulus arvensis*), which, again, became wide-spread only in modern times. Bladder campion (*Silene vulgaris*), parsley-piert (*Aphanes arvensis*), wild radish (*Raphanus raphanistrum*), field pansy (*Viola arvensis*) also came to Central Europe from the Mediterranean during the Neolithic. They all achieved their greatest distributions in modern times. Wild radish is an exception, for that species was already widespread in the Middle Ages. Unfortunately, due to the small number of finds in Southern Europe, the exact places of origin of weed species is not possible to determine with greater accuracy.

Hairy tare (*Vicia hirsuta*), small toadflax (*Chaenorrhinum minus*), greater Venus's looking-glass (*Legousia speculum-veneris*) and wind bent grass (*Apera spica-venti*) are Secalietea species of Central European origin. Of those, the hairy tare appeared first in the era of the Neolithic linear band culture, but it only became widespread in modern times (WILLERDING 1986c). Small toadflax, greater Venus's looking-glass and wind bent grass appeared during the Roman Period or the Iron Age but also reached their greatest penetration in modern times. Mayweed camomile (*Anthemis cotula*) and field madder (*Sherardia arvensis*), of Western European origin, also became widespread in modern times. While field madder was already present as a cereal weed in the Iron Age, mayweed camomile appeared iinitially in the Middle Ages.

Rye brome (*Bromus secalinus*), black-bindweed (*Fallopia convolvulus*), cornflower (*Centaurea cyanus*), nipplewort (*Lapsana communis*), catchweed or cleavers (*Galium aparine*), false cleavers (*G. spurium*) and probably spring wild-oat (*Avena fatua*) were widespread cereal weeds of Neolithic Europe. Those species constituted Europe's earliest weed association, known as *"Bromo-Lapsanetum praehistoricum"*.

Following RADEMACHER (1968), WILLERDING (1986c) distinguished five phases within the development of weed associations:

- the 1st phase covers the initial period of grain cultivation in Central Europe. In addition to weeds introduced with sowing-seeds, elements of the previous flora were also present for some time. Domesticated species cultivated on arable land presented competition for the components of natural vegetation. For example, one of the indigenous species of the widespread mixed oak forest of the Neolithic, wood dock (*Rumex sanguineus*), remained in its earlier area and became a *Secalietea* species. Other plant species were less able to adapt to the changed conditions resulting from cultivation and subsequently disappeared. During the Middle and Late Neolithic, a whole range of foreign weed species migrated to the central regions of Europe.
- in the 2nd phase, low level cultivation of soil was spreading but weed flora was still very rich. The association of weeds of spring-sown cereals and root vegetables and the association of the weeds of autumn-sown cereals were not differentiated significantly. It was only after the spread of ploughing that weed associations of cereals and of root vegetables grew apart. However, intensive field agriculture mostly eradicated these weeds. KORNAS (1972) found surviving remains of the old weed association in the Polish Carpathians.
- in the 3rd phase of development of weed associations, maturing agricultural production and associated fertilisation and other soil maintenance procedures and watering schemes resulted in the appearance of "intensive weeds". Nitrophilous, creeping, shade-resistant species became dominant. They could only be eradicated with great difficulty.
- in the 4th phase, the prevalence of the combine harvester brought great changes in the species composition of segetal weed associations. Mechanisation and earlier, faster harvesting reduced the number of weed species, but on the other hand those that did survive adapted extremely well to the lifecycle of cereals.
- in the last, 5th phase, herbicides were deployed against the weeds that were believed to be invincible. After initial successes, some seed species began to develop *resistance*, so it is increasingly difficult to curtail their spread.
- Specimens of the subfossil species of anthropogenic plant associations are hidden in well-ventilated levels of the soil, partly burnt or in water sediment and silt (river- and lakesides, springs, wells, castle moats). Exact archaeological dating of the samples is feasible, the time of immigration of allochthonous species can also be determined. The species composition of an ancient plant association can only be studied directly in exceptional cases (e.g. threshing waste in a pit). In many cases, samples contain vegetable matter from several different sources. In such cases, the assemblages of finds can be classified according to their present phytocoenological behaviour, therefore only limited inferences can be drawn concerning the earlier composition of an association.

The human (anthropogenic) effect on the landscape (vegetation) is called synanthropisation that may apply to a taxon, an association, a cultivated landscape, a landscape, or anything that is affected by humans (TERPÓ 1999). Using the research results of ASCHERSON (1883), RIKLI (1903), NAEGELI (1905) and THELLUNG (1918–19), KORNAS (1968) developed a synanthropic system. In essence, plant species associated with human cultures are classified into geographical-historical categories: apophytes, archaeophytes and neophytes. Although the classification of plant species into those three geographical-historical (human) categories is based on ecological properties and associated characteristics, it also takes into account the ranges of individual plant species, the development of human society and the effects of social requirements on the flora. The class of apophytes contains indigenous, culture-adapted species. Their common property is that they more or less tolerate human activities, or may even be assisted by them. The archaeophytes are old-world plant species that appeared/were brought in during prehistoric times, Antiquity or the Middle Ages but whose cultivation has ceased. The neophytes (syn.

kenophytes) are the new-world (adventive) species brought in from America for cultivation, or those that escaped from America (neohemerophytes = escaped species). Archaeophytes and neophytes are species of foreign origin (anatropophytes). The synanthropic flora of an area – depending on where it is located geographically and which period we are studying – may vary. The number of species, i.e. diversity, may increase, coverage by hemerophyte species may increase, diversity may decrease and hemerobic species may gradually grow extinct. Migration of species between continents may intensify and original plant associations may lose characteristic species.

The synanthropisation of the Carpathian Basin has been continuous since the Neolithic. At the beginning of the Neolithic (Körös culture), new, foreign species associated with plant cultivation appeared in the landscape. At the beginning of cereal cultivation, einkorn, emmer and barley were the characteristic cereals with shorter growth-cycle common millet added in the Bronze Age. Along with cultivated plants, numerous wild ones also arrived, primarily from Asia Minor and the Mediterranean and to a lesser extent from Asia. These had lived in association with domesticated plants, often as wild relatives of domesticates in their places of origin. But, in a cultivated context, they were simply weeds. According to archaeological plant material, fan-hen or white goosefoot (*Chenopodium album*) was present in very large quantities, while maple-leaved goosefoot (*Ch. hybridum*), common orache (*Atriplex patula*), henbit deadnettle (*Lamium amplexicaule*) and common chickweed (*Stellaria media*) were also common. Pollen analysis has confirmed the expansion of grasses (*Poaceae*) during the Subboreal (Bronze Age) and the Subatlantic (Iron Age, historical eras).

TERPÓ (1999) raised the possibility that only some aspects of the origin and role of weed plants and their relationship with the cultivated (host) plants and human settlements have been discovered. He divided the synanthropic species occurring in the Carpathian Basin into four classes according to their levels of domination in the landscape:

1. Before 1500, the landscape was dominated by associations of archaeophytes and apophytes: archaeophytes: e.g. *Agrostemma githago*, *Centaurea cyanus*, *Echinocloa crus-galli*, *Papaver rhoeas*, *Setaria pumila*, *Sinapis arvensis*, *Stachys annua*,
   apophytes: *Artemisia vulgaris*, *Agropyron repens*, *Centaurea cyanus*, *Consolida regalis*, *Digitaria sanguinea*, *Portulaca oleracea*, *Raphanus raphanistrum*.
   K. BERZSÉNYI (2000) has refined the classification based on archaeobotanical finds. He has shown that, of the apophytes, fan-hen or white goosefoot (*Chenopodium album*) has been present continuously and in large quantities since the Neolithic. Field bindweed (*Convolvulus arvensis,*) that appeared in the Copper Age was present continuously with the exception of the Bronze Age. Knotgrass (*Polygonum aviculare*) appeared in the Bronze Age and has been present continuously ever since. Among the archaeophytes, korn-rade (*Agrostemma githago*) and black-bindweed (*Fallopia convolvulus*) are the earliest ones. With the exception of the Copper Age they have been present in all periods, sometimes in large quantities. Annual woundwort (*Stachys annua*) and shepherd's-purse (*Capsella bursa-pastoris*) appeared in the Bronze Age, though in much smaller quantities, just as they did in later eras.
2. Until 1800, archaeophytes, apophytes and neophytes (e.g. black locust, false acacia – *Robinia pseudo-acacia*) occurred with equal frequencies.
3. 1800–1945: the introduction and spread of neophytes accelerated, e.g. *Ambrosia elatior*, *Iva xanthiifolia*, *Sorghum halapense*.
4. From 1945, many neophytes began to act as invasive species: *Ambrosia artemisifolia*, *Amaranthus retroflexus*, *Coryza canadensis*, *Helianthus decapetalus*, *Galinsoga parviflora*, *Sorghum halapense*, *Asclepias syriaca*, *Ailanthus altissima*, *Aster* species. Cosmopolitan escaped domesticated plants: *Reynoutria japonica*, *Solidago canadensis*, *S. gigantea*, *S.*

inaequidens, *Heracleum sosnowskyi*, *H. mantegazzianum*, perennial *Helianthus* ornamental plant species, *Amaranthus deflexus*, *Eleusine indica*. Due to so-called hyper-anthropogenic factors, xenospontaneous associations are spreading in which foreign plants play dominant roles as characteristic species of many associations: e.g. urban forests: spontaneous black locust, false acacias, *Acer negundo* populations. Due to economic and social similarities, synanthropisation is also affected by globalisation. The only constraint is the climatic condition of the temperate zone. This process is by no means finished. New plants keep arriving, including many prospective food plants.

From the perspective of their macrohabitats, plant species accompanying human cultures are divided into ruderal and segetal species. Some segetal species appear with equal frequency in ruderal settings, especially if they have no competitors, e.g. quackgrass (*Agropyron repens*) and fan-hen or white goosefoot (*Chenopodium album*). The main cause of the decline of apophyte (indigenous) species is human activity (K. BERZSÉNYI 2000).

The history of weed associations in Hungary can be summarised as follows (for details, see the description of each period, respectively): the first segetal associations of the Neolithic and the Bronze Age had relatively small numbers of species (*Fig. 42*). At the beginning of the Iron Age, several new segetal species appeared: usually weeds with short stalks, corresponding to the method of iron sickle harvesting, i.e., cutting close to the ground. The differentiation of segetal vegetation began in the Roman Period. Weed associations reached their present species composition – as it was prior to industrial agricultural technologies and strong ruderalisation – during the Middle Ages.

The reconstruction of the composition of ruderal plant associations and the investigation of their secular development is even more difficult. Ruderal vegetation of the prehistoric period and the Middle Ages is well represented. Some Neolithic/Bronze Age remains of pastures and fields have also been found. Today's common field associations are much later developments, formed at the end of the Middle Ages. The number of prehistoric waterside and fringe forest species is relatively high and pioneer plants that respond to forest clearing are also present. Such changes are persistent and continuous and are therefore more likely the result of forest clearing for expanded farmland use rather than from climate change. We must note here that of approximately 156 archaeophyte weed species that have occurred at one time or another in Hungarian flora, many are on the brink of extinction today. Therefore, their study falls within the range of historical agrobiodiversity and their preservation is a task of national importance (UDVARDY 2000). Some of the ancient adventive species have already been placed on the list of protected species in Hungary where their elimination can result in legal penalties such as monetary fines. Such species include korn-rade (*Agrostemma githago*), wild pea (*Pisum elatius*), dyer's woad (*Isatis tinctoria*) and tassel-hyacinth (*Muscari comosum*). Some other species are of such significant aesthetic value that their seeds are already marketed as ornamental plants, e.g. field poppy, corn poppy (*Papaver rhoeas*), cornflower (*Centaurea cyanus*), cowherb (*Vaccaria hispanica*), while yet others are potential ornamental plants, e.g. field nigella (*Nigella arvensis*), forking larkspur (*Consolida regalis*), summer adonis (*Adonis aestivalis*) etc.

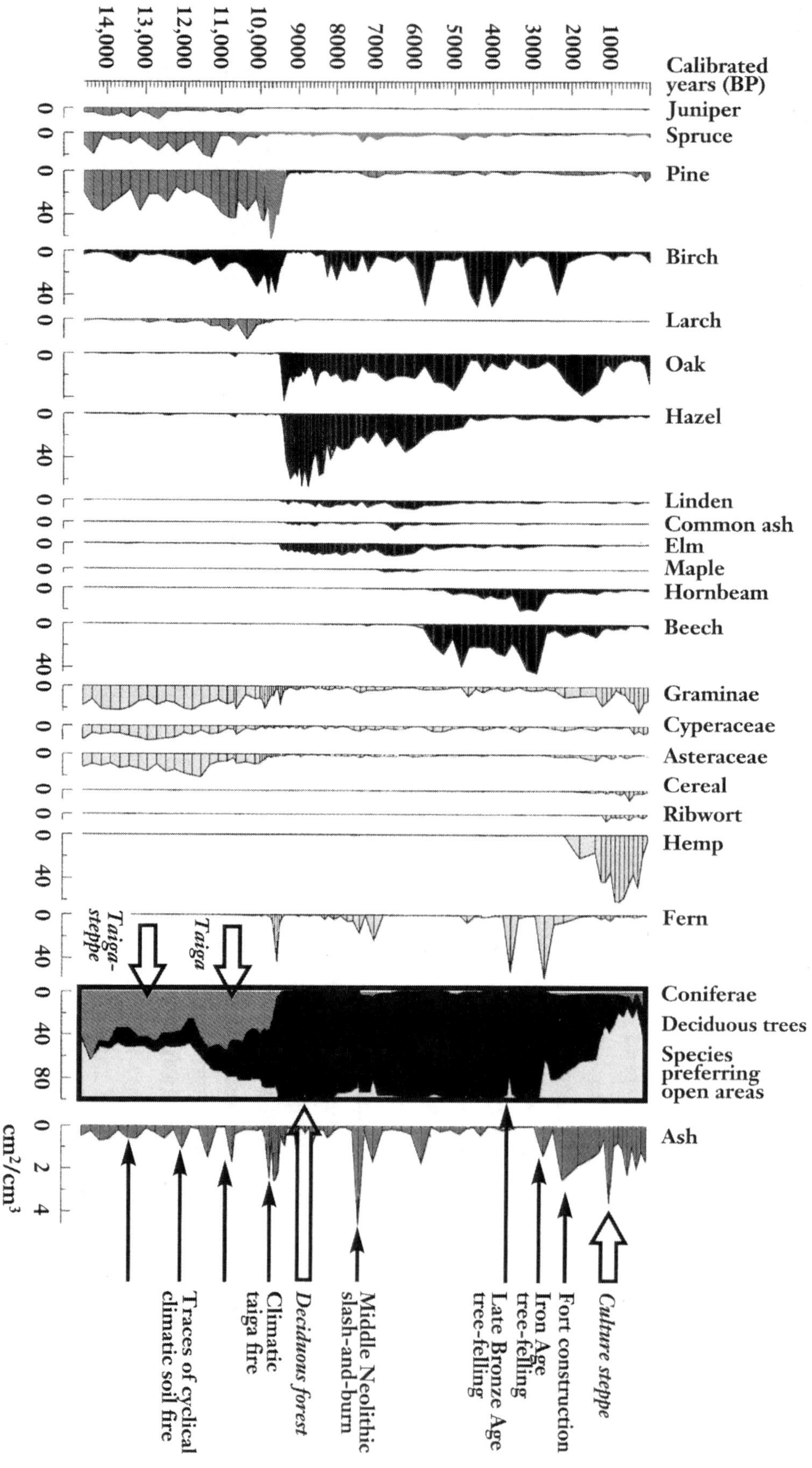

Fig. 42. Vegetation development by the end of Quarternary with human influence (Kelemér, Kis-mohos). After SÜMEGI, KERTÉSZ – RUDNER 2003.

7. The history of domesticated plants from the beginnings until the Early Modern Period

Many people have commented on the history of agriculture but few have studied it in sufficient detail and even fewer based their research on archaeological finds. Archaeobotanical finds were taken into account by only a handful of people. Yet they indicate that agriculture has been continuous sinc92 e the beginning of the Neolithic in the Carpathian Basin (*Tabs. 1–2*). Therefore we believe it is necessary to attempt a comprehensive description of the history of plant cultivation in Hungary based on the latest results of archaeobotany, drawing a parallel with finds of similar age from neighbouring countries to the extent required.

The plant finds described below are the results of over a hundred years of archaeobotanical processing work (*Tabs. 3–19*). Seed and fruit remains were identified in the period 1876–1891 by Imre Deininger, in the 1910s by Georg Lindau, in the period 1942–1964 by Ádám Boros, Emil Gubányi, Zoltán Zsák, János Maácz and Zdenek Tempír, in the period 1960–90 by Borbála P. Hartyányi, Miklós Füzes, István Skoflek, Veronika Árendás and by Géza Facsar. Since the middle of the 80s, the author and his students, Brigitta K. Berzsényi, Orsolya Dálnoki, Valéria Kovács, Katalin Fekete, Ferenc Gerócs and Andrea Torma have also joined the work of processing (*Tabs. 20–21*).

The manuscript was finished at in the end of the year 2003, but the database of plant remains have been updated through the end of the year 2006.

7.1. The Neolithic

7.1.1. The beginnings of interrelations between people and the environment

A little more than ten thousand years ago the Earth was ruled by hunter-gatherer populations. The beginnings of the relationship between humans and their environment (the hunter-gatherer lifestyle) go back to the Palaeolithic and the Mesolithic. We only have scarce information about the hunter-gatherer lifestyle and dietary habits of Central European Mesolithic populations. Based on botanical finds, the only plant said with certainty to have been consumed is the hazelnut (KÜSTER 1986a). No seed and fruit remains from those periods have been found in Hungary.

In the 10th millennium BC all of that began to change. In the Middle East, in Anatolia, domesticated animals and plants appeared, the utilisation of natural raw materials (stone, clay, earth colours, etc.) accelerated and the first permanent settlements were established. The plant species selected in various locations integrated quickly. Continuous production of plants, which furnished more secure livelihoods, lead to changes in the natural environment. The formation of the cultivated environment began. Those changes also brought changes of lifestyle: settlements became permanent, stock management and the production of pottery began. These events left well-discernible traces in the soil. Cultural layers found during excavations grew thicker, plant production is in evidence and other changes of the soil; such as, the appearance of ploughed fields and pits, etc., can also be registered.

We have very scarce information concerning the environment of human beings of those eras. The investigation of complex interactions between humans and the environment requires us to enrich our knowledge of the history of the environment. Such changes of the natural environment are the primary subject of historical ecology. In addition to its significance from the perspective of cultural history, such knowledge also helps us to better understand our present and expected future environment. Written sources (chronicles, legends, historical works) and iconographical sources (paintings, engravings etc.), religious beliefs, myths, ancient traditions, natural philosophies, legal and ethical categories contain many comments and references to the ancient environment, all of which need to be studied (R. VÁRKONYI 1992).

The transformation of the environment as a result of human intervention began with the appearance of domesticated plants and animals. That change – especially at the beginning – was extremely slow. The eradication of species that were deemed useless and were not cultivated, the exclusive cultivation of improved species resulted in the gradual extinction of the other species. Today, a significant proportion of natural plant associations have been transformed into artificial, cultivated landscapes. The majority of forests have been grubbed out, their place taken over by arable land, orchards, pastures and settlements. Increased use of chemicals has an effect on the biocoenosis of the soil and the ecosystems of our waters. As a result of all those factors, the number of remaining species has fallen below a critical level (GHIMESSY 1984).

It is a proven fact that plant cultivation increased at the expense of the natural vegetation. The relationship between cultivated plants and natural vegetation has been determined by a combination of the lifestyles of the populations that lived in the Carpathian Basin and by climatic conditions.

The Carpathian Basin has provided a home for human cultures as old as any other area in Europe. Plant cultivation has a history of eight thousand years in this region; vegetable cultivation has been going on for five thousand years, while fruits have been grown for two thousand years. The majority of cereals arrived in the Carpathian Basin with Neolithic populations, the majority of leguminous plants were brought in by the peoples of the Middle Bronze Age tell cultures, while fruit species and grapes appeared at the time of the Roman conquest. Eastern plant knowledge was brought by conquering Hungarians and cultivated plant species and cultivation know-how arriving from the West were added after the assumption of Christianity. All this manifold knowledge was forged into a comprehensive whole. Many plant species found a new home and established secondary gene centres.

7.1.2. The domestication of plants

Agriculture is a relatively recent branch of production – domestication began approx. 10–12 thousand years ago in Southwest Asia and 6–8 thousand years ago in Europe (ZOHARY – HOPF 1988). By domestication we mean social control over the plants and animals that are important to humans. It is becoming increasingly clear that the earlier notion that domestication was forced upon human populations by the decreasing quantities of game available for hunting is erroneous.

The first great and fundamental change in the history of human civilization was the "Neolithic revolution" (CLARK 1949). The creator of the concept, Gordon CHILDE (1936) believed that the chronological sequence of development was always as follows: pottery – agriculture – urbanisation. Discovery of the Pre-Pottery Neolithic period since than has modified the sequence: agriculture – pottery – urbanisation.

Humans used multiple selections from wild plant and animal populations of the time to establish the resources they needed. The nomadic, hunter-gatherer lifestyle was replaced by settled settlements supported by plant cultivation and animal husbandry.

Even in the earliest periods, people were served by the widest variety of raw materials. While in the Mesolithic – 13 thousand years ago in SE-Asia and 8–10 thousand years ago in Europe – humans used only the primary plant and animal biomass available in nature, In the Neolithic, following the transition to the new form of economic activity, the users of secondary biomass appeared for the first time in history.

The original inhabitants of the Middle East were so-called "harvesting populations", also known as "food collectors" (LIPS 1962; LA BAUME 1961). The first earliest plant-cultivating cultures appeared around the beginning of the 10th millennium BC in the area of the Middle East known as the Fertile Crescent, south of the Zagros Mountains, in the borders of the southern parts of Afghanistan, Iran and Anatolia (ZOHARY – HOPF 1988). More precise dating of the change is not possible, as the transition from gathering to settlement was a continuous, lengthy process, which played out differently in various geographical locations (WENKE 1999). The Fertile Crescent is the region of relevant gene centres where the great majority of the wild forms of the old-world cultivated plants were present in nature. At that time the region had much higher precipitation than it does today. The change is partly explicable by natural conditions: rich and easily obtainable food was available all through the year, which made stable, permanent settlement possible. People became accustomed to certain foods. As a result, the secondary process of population growth took off. The conversion to farming and animal husbandry was the beginning of the establishment of a manifold, highly complex system of interrelations between nature and humans.

The development of agricultural technology involved the appearance of tools that were used to cultivate and to thresh cultivated plants. Such tools already appeared in the late hunter-gatherer cultures of the Middle East. Certain agricultural tools: flint-edged sickles, mortars and pestles, as well as hand mills already appeared in the Late Mesolithic culture of Palestine, the so-called Natufi culture (according to ^{14}C dating, these finds are from 10,300–8,500 BC). The mortar even occurred in the earlier Kebaran culture (12,500–10,300 BC, non-calibrated data). Furrowing tools made from wood, antlers and bones were used to break up the soil.

The first domesticated forms of bread wheat and barley species appeared in the PPN-A, i.e. the Pre-pottery Neolithic period, also known as the Proto-Neolithic, around 8000–7000 BC. The earliest leguminous plants were pea and common lentil. It is difficult to judge whether they were cultivated forms, for the seeds of the small number of early finds were very small in contrast to domesticated seeds that are typically larger. People had no pottery at that time, but the domesticated forms of cereals were already present. Settlements consisted of mud dwellings with stone foundations. The domestication of plants was not a segregated process, it occurred in co-evolution with certain animal species (RINDOS 1984).

The earliest finds of the wild forms of einkorn, barley, rye and leguminous plants are from the late Natufi culture layers (9000 BC) at the Tell Abu Hureyra site in Northern Syria (HILLMAN 1975). This implies that in addition to hunting and fishing, the gathering of wild forms of the plants that were to be cultivated later was also important in the Natufi culture.

The invention of the sickle as a harvesting tool was one of the important elements of the "agricultural revolution". Compared to harvesting by hand, the efficiency of harvesting doubled with the introduction of bone or antler sickles with silex blades. The earspindles (i.e., rachis) of cereal species at the beginning of domestication were still very fragile, so fewer grains were lost if they were harvested with a sickle.

Why was it exactly those cereal and leguminous species that were domesticated? Was it easy or difficult? Those plants have good yields and are tasty. It is an advantage that bread wheat, barley, pea, common lentil and bitter vetch are diploidic, self-pollinating plants. People chose those plants from the natural environment by repeated selection (e.g. taking into account the fragility of earspindles), then sowed them in different locations (thereby avoiding back-crossing). The unavoidable crossings only extended the range available for selection.

Domestication was a very rapid process. Experiments suggest that a rigid earspindle can be developed in a few generations. While one layer of the Tell-Mureybit site in Northern Syria yielded wild einkorn and wild barley (8400–7500 BC) (VAN ZEIST 1970a), the cultivated form of emmer has been found in the I. A. phase of the Tell-Aswad site near Damascus (7800–7600 BC) (VAN ZEIST – BAKKER-HEERES 1979). Additional finds of cereals are known from Jericho, Tell-Es-Sawwan, Beidha and the Anatolian sites Çatal Hüyük, Haçilar and Çayönü (HELBAEK 1966, 1970; HOPF 1983; YAKAR 1991).

The main activities associated with the cultivation of plants are clearing land, breaking soil, sowing, harvesting and threshing. These work phases lead to certain changes in domesticated plants, which developed the so-called "features of domestication:" the grains are larger (larger seeds), the earspindle is no longer fragile, sections of the earspindle remain connected after ripening, which, incidentally, means that these species are no longer able to spread without human intervention. Subsequently, selection also resulted in uniform ripening time, increased size and number of grains and a reduction of the thickness of hulls.

The plants that migrated from the Middle East to the central regions of Europe – and towards Central Asia, the valleys of the Indus and the Nile – cover the entire range of the domesticated species originally developed there (mostly bread wheat, barley, etc.) (HARLAN – ZOHARY 1966). Along with the domesticated plants, many weeds also arrived in the course of the migration (WILLERDING 1988).

The knowledge of plant cultivation spread quickly towards Central Europe and arrived within two thousand years. Domesticated plants and animals reached the Balkans already by the end of the 7th millennium BC.

Domesticated plants first appeared in the Impresso-Cardium culture in the coastal regions of the Aegean, Mediterranean and Adriatic Sees. The oldest Neolithic settlement of Greece (Franchthi cave) is dated to the end of the 7th millennium. The oldest site in Italy is Passo di Corvo, while in Southern France it is Fontbrégoua.

7.1.3. Neolithic agriculture in Southeast and Central Europe

The concept of the "Neolithic" covers shared technologies of societies with various cultural historical characteristics. However, today the various Neolithic cultures are still distinguished on the basis of the formal and stylistic properties of pottery. Changes are imminent in this field precisely due to archaeobiological research. Neolithic lifestyle is understood increasingly well and the focus of research is shifting in that direction.

The beginning of the Neolithic in Hungary partly coincides with the beginning of the Atlantic climatic phase (according to earlier datings, 5500–2800 BC) (FERENCZY 1958; KALICZ 1993). This was more advantageous to plant cultivation compared to the warm, dry climate of the previous era, the Boreal phase, with its frequent continental extremes. This era in Hungary, also called the Atlantic oak phase, had a climate similar to the warm, humid and temperate submediterranean, which allowed the Carpathian Basin to play a bridging role in the spread of plant cultivation know-how from the region of the Middle East through the Balkans to Central Europe.

The rapid spread of domesticated plants and associated plant cultivation know-how is explained by the similarity of environments. Regarding arboreal coverage, for instance, the Great Hungarian Plain of the Atlantic was probably similar to the present-day Balkans and in particular to the Crimean peninsula. The Great Plain in the Atlantic Period was characterised by the common oak – hornbeam association (*Querco robori-Carpinetum*), which is regarded as an extension of the steppes of the Ukraine.

Based on pollen analyses from the Great Plain (KOMLÓI 1966) it became possible to reconstruct the vegetation of the Atlantic phase (LACZA 1991). Due to migrations of the River Tisza in the Holocene, the surface of the Great Plain is highly variable. The area, which is largely covered by loess, has always been floristically open to the north (Northern Hungarian Middle Mountains), the west (Transdanubia) and the south (Western and Eastern Illyria, Southern Russian Plain). With time, many climato-ecologically related species migrated or were brought into the area.

It seems that during the Holocene, the Atlantic phase was the richest in tree species. The pollen of birch (*Betula* spec.) was present in samples from almost all cores in sand-dunes. The juniper-poplar association (*Junipereto-Populetum*) of sandy areas may be regarded as characteristic and ancient. On the other hand, Scotch pine (*Pinus silvestris*) substantially receded. Hungarian narrow-leaved ash (*Fraxinus angustifolia* subsp. *pannonica*), a species originating from the Balkans, was probably very common. Hedge maple (*Acer campestre*) and in some locations plane-tree maple, sycamore maple (*A. pseudoplatanus*) were also common. The trees were home to lianas, e.g. old-man's-beard (*Clematis vitalba*). Along with hedge maple and common oak (*Quercus robur*), hornbeam (*Carpinus betulus*) also played a role in forming the canopy of forests.

The shrub layer was characterised by Tatarian maple (*Acer tataricum*) and blood-twig dogwood or red dogwood (*Cornus sanguinea*). New species also appeared: European black hawthorn (*Crataegus nigra*) and wayfaring tree (*Viburnum lantana*) migrated here from the Balkans. The grass layer was rich in species: e.g. ramsons, bear's-garlic (*Allium ursinum*), European wood anemone (*Anemona nemorosa*), European ginger (*Asarum europaeum*), lily of the valley (*Convallaria majalis*), English ivy (*Hedera helix*), summer snowflake (*Leucojum aestivum*), broad-leaved Solomon's-seal (*Polygonatum latifolium*), common lungwort (*Pulmonaria officinalis*), squill (*Scilla vindobonensis*) and figwort (*Scrophularia umbrosa*). Slender false brome (*Brachypodium silvaticum*), yellow dead-nettle (*Lamium galeobdolon*) and herb-Paris (*Paris quadrifolia*), which probably migrated here after the onset of the Flandrian interglacial, spread to a modest extent. The extent to which the ash-alder association (*Fraxino pannonicae-Alnetum*) was distinct from the strip of black alder marshy wood (*Thelypteridi-Alnetum*) association is still debated.

During his floristical research near Deliblat (today in Serbia), TUZSON (1915) already noted that the sandy plain had highly advantageous conditions for the migration of species towards the southeast and the north.

Therefore, under similar climato-ecological conditions, the knowledge of plant cultivation spread rapidly from the Middle East towards Central Europe (and, along another route, towards the Caucasus) and reached this region within 2000 years. One of the main routes of the spread of knowledge led along mainland Greece, through Thessalia towards the Balkans. Archaeological finds form the sites of the Karanovo I. culture in Bulgaria and those of the Starčevo culture in Serbia, which reached the Transdanubian region as well, support the above assumption. At the same time, the valley of the Strimon (Strumm) and the valley of the Marica-Tunja, which are parallel to the valley of the Vardar (Axios)-Morava are also possible candidates. The first farmers that reached the Great Hungarian Plain and Transylvania were the peoples of the Körös-Starčevo and the Criş cultures, relations of the early Neolithic farmer cultures in the Balkans.

The earliest farmer culture of Central and South-eastern Europe was the population of the Körös-Starčevo culture, which has been dated by the latest radiocarbon dating to the period 6000–4500 BC (*Fig. 43*). This population, which had connections with the Balkans and which conducted farming, animal husbandry and significant fishing along the great rivers, settled in the fertile open lands of the Great Plain near the Danube, avoiding the sandy areas that occur in the region. Their settlements were short-lived, but they kept returning, which can be attributed to the exhaustion of the soil and migratory farming.

The Great Plain, which constitutes almost half the country, is the western margin of the Eurasian forested steppe (TUZSON 1913). Archaeologically, it is one of the best-researched regions of the country.

The settlement and agricultural activities of the Körös culture are indicated by land snails (SÜMEGI *et al.* 1998) and plant remains in lake sediments of this period from the Great Plains suggesting forest felling and burning (WILLIS 1997). Frequent bone remains of cattle indicate a real "bovine domestication fever" (BÖKÖNYI 1959, 1974; BARTOSIEWICZ 1999). Rabbit bones found in open, grassy areas are also increasingly frequent, which is again connected to the increasing proportion of open areas due to felling of forests (VÖRÖS 1980).

Plant remains found in settlement excavations were mostly preserved as imprints of seeds, spicules, forks and glumes in burnt pieces of wattle-and-daub walls (in mud-flakes) and in pieces of pottery (P. HARTYÁNYI – NOVÁKI – PATAY 1967–68; P. HARTYÁNYI – NOVÁKI 1973–74; FÜZES 1990). No seeds of cereals have been found to date. Reasons can only be guessed at. The lack of research is obviously a contributing factor, but it is also highly likely that at the time of the beginning of plant cultivation in Hungary, people were very short of valuable cereal grains. Nevertheless, we can assert with certainty that they did grow cereals. The evidence consists of the imprints on the surface of pottery fragments, as ceramics were made lighter almost exclusively by the addition of chaff.

Imprints found on pottery fragments of Szeged-Gyálarét (Ottó Trogmayer's excavation 1964) and Röszke-Lúdvár (Ottó Trogmayer's excavation 1964–65), when cast in silicone rubber, proved to be largely derived from einkorn, and to a lesser extent, from emmer. Pottery fragments found at Battonya-Basarága (János Szabó's excavation 1975) and Gyomaendrőd (János Makkay's excavation 1990) did not bear imprints of any cereals other than einkorn (*Figs. 44–45*). This is attributable to difficulties of identifying imprints, to poor preservation of imprints and the failure to conduct such investigations extensively. Last but not least is the lack of luck in the case of Hódmezővásárhely-Gorzsa (Gyula Gazdapusztai's excavation 1956) and Szarvas-Szappanos meadows (Endre Krecsmerik's excavation 1912) In these cases, it was possible to establish the presence of wheat, but nothing more exact. Interestingly, no imprint of anything like barley was found. Nevertheless, we assume that barley was among the plants cultivated as well.

At Deszk-Olajkút (Ottó Trogmayer's excavation 1960) and Ószentiván (Ida Kutzián's excavation 1960), no plant imprints, only charcoal remains were investigated. According to non-calibrated radiocarbon data, Deszk-Olajkút dates to the late phase of the Körös culture (6605–6260 ± 100 BP) (QUITTA – KOHL 1969). Oak (*Quercus* spec.) was found at both sites, with elm (*Ulmus* spec.) in the latter one. Those are also important pieces of information concerning the ancient landscape. One textile imprint found at Gyomaendrőd was identified by Füzes as flaxen cloth. Even if this is not regarded as evidence that flax was being cultivated, it certainly proves that it was processed and woven. Pottery fragments of the Körös culture exhibit numerous textile imprints (MAKKAY 2001).

In addition to imprints, we now have recovered plant evidence that the people of the Körös- Starčevo culture were also gatherers. They were fond of tasty forest fruits: for example a finger-thick layer of hazelnut shells was found in one of the pits at the Méhtelek-Nádas site (Nándor Kalicz and János Makkay's excavation 1973). According to the research of Elvira BODOR Nagyné (1988), hazelnut was very common in the region of the country west of the Danube. That is by no means an accident, for hazelnut was the dominant shrub of the Boreal phase that preceded the Atlantic phase.

Cornelian cherry drupe kernels found at the Battonya-Basarága site are no less interesting. Szeged-Gyálarét is the only site of the Körös-Starčevo culture that has imprints other than those of cereals on pottery fragments. As they were associated with bread wheat seeds, it is highly likely that they may have been its weeds. At the same time, they constitute the earliest find of the Hungarian weed association: vetchling (*Lathyrus* spec.) and brome grass (*Bromus* spec.). Unfortunately, identification of the exact species was not possible on the basis of the imprints.

We may well call the natural make-up of the Carpathian Basin prior to human intervention a "raw landscape" (FRÁTER – BIRÓ – KÓSA 2000). The nature-transforming activity of humans became significant when they made the transition from the hunter-gatherer to the farmer-animal-keeper lifestyle. This involved settling down and conducting food production activity. The beginning of this lifestyle change coincided with the "most heavily forested" phase of the Great Plain, resulting from climatic effects.

The Carpathian Basin has been characterised by a sort of mosaic structure as regards its climate, soils and plant life since the Pleistocene. At the macro level, the area is the meeting-point of three different climates: the influence of the continental region decreases from east to west; from west to east the oceanic influence grows weaker; and, south to north the submediterranean effect diminishes. According to the divisions of Köppen (in: SÜMEGI 2000), climatic regions extend as follows: the region west of the Danube is predominantly under the influence of an oceanic climate, the Great Plain and the area between the Danube and the Tisza are largely in the forest-steppe zone while the environment of the Carpathians are largely under the influence of the sub-Carpathian climate; the region of the Bakony mountain is in a so-called transitory zone, while the south-western regions south of Lake Balaton are under submediterranean influence. Due to the fragmented configuration of climate zones, the plant regions are also fragmented. The northern and southern slopes of mountainous regions and the river valleys are less prone to those macro-effects, resulting in additional, characteristic fragmentation.

The sites of the Körös-Starčevo culture were found in the Great Plain and in the south of the region west of the Danube (*Fig. 46*). It seems that in addition to the Southern Balkans, the only place where the required ecological conditions for the production of domesticated plants and animals were present was that specific region of the Carpathian Basin. The advantageous conditions available due to a fortunate mixing of the western, eastern and southern climatic components were not available elsewhere for the plant species at the beginning of their adaptation. Climatic differences curtailed the further spread of the Körös-Starčevo culture (KERTÉSZ – SÜMEGI 2001).

The strongest climatic effects have always been cyclical in the Carpathian Basin: macrocycles every ten to one hundred thousand years, mezocycles every five to ten thousand years and microcycles with periods of five hundred to five thousand years. As a result of all that, the changes and development of plant associations were protracted both in space and in time. The plant species that were able to adapt to environmental changes made oscillating movements between the mountain ranges at the margins of the Carpathian Basin (Carpathicum, Illyricum, Noricum) and the central areas (Pannonicum). In the course of those movements, geobotanical fluctuation zones developed between the continental, the oceanic and the Balkanic areas, Precarpathicum, Preillyricum, and the Transylvanian corridor (SÜMEGI 2000).

The fragmented character of the Carpathian Basin, developed in the Quaternary period also affected the cultures that settled here. Populations migrating to the area from various climatic zones attempted to choose locations that were the most similar to their original environments. Obviously, these were areas were they could put their economic knowledge to the best use. This factor explains their configurations in space. As a result, natural zone boundaries also became boundary areas of archaeological cultures.

Transformation of the natural landscape of the Carpathian Basin began in the Neolithic. The primary change was the reduction of forested areas. This process began in the Great Plain and in the area west of the Danube around 6000 BC, while in the central mountain ranges it started around 5500 BC. According to the analysis of sampling cores at the Kis-Mohos marsh in Kelemér, clearly perceptible evidence of changes in geochemistry and vegetation were found from the time of the settlement of the first Neolithic human population (SÜMEGI 2000). Nevertheless, the area was repeatedly depopulated and in those periods the original vegetation quickly regenerated.

7.1.4. Agriculture in the Middle Neolithic

The earliest plant cultivator culture of Western Europe was the Linearband Ceramic culture (BAKELS 1979; WILLERDING 1983a). This culture, dated to the second half of the 6th millennium BC, largely occupied fertile loess land, the best soils available for plant cultivation. Among cereals, they grew barley, einkorn, emmer; as well as leguminous plants, including pea, common lentil. Among oil and fibre plants they used opium poppy and flax (HENNIG 1966; HOPF – BLANKENHORN 1984; KÜSTER 1989). In order to have a more varied diet, the population of the period gathered numerous wild plants in the vicinity of their settlements (BERTSCH 1941; KÜSTER 1986a). Seeds were well preserved.

Thanks to several decades of archaeological and scientific research, we know a great deal about these people. They lived in farmstead type small settlements ("villages") comprised typically of 4 to 6 houses and a population of perhaps 25 to 30 individuals (BOELICKE *et al.* 1988; LÜNING 2000; LÜNING – STEHLI 1989). Poughlands adjacent to such villages probably did not exceed six hectares. Houses, built close to each other, were encircled by walls providing security for the community against aggressive wild animals still living in extant forests and/or if relationships between villages became aggressive. Destruction of houses by fire was relatively common, either through accident or conflict. In any case, this probably contributed to site abandonment along with reduction in soil fertility through time. Early Neolithic villages were stable, that is, year-round settlements, but not necessarily of long duration. Abandonment of villages was common in favor of shifting to new land.

Early Neolithic people also utilized considerable materials available in their natural environment (branches, wattle, piles, clay, barley straw mixed with clay, common reeds, moss). While most domestic activities took place within the encircling village walls, felling and transportation of trees, building of new houses, working of fields, sowing and harvesting, supplemental hunting and gathering, etc., imposed intensive labour requirements on all functional members of such small communities. Extending cooperation between communities could act to counterbalance conflict with nearby communities.

Agricultural tools occurred only sporadically. Various furrowing tools were made from bone or stone, along with sickles, threshing sticks and stone hand-mills. People grew only hulled grains. Their cereals, in decreasing order of importance, were barley, club wheat, einkorn, emmer. They also used the following leguminous plants: pea and common lentil; and oil/fibre crops: opium poppy, flax, small quantities of gold-of-pleasure. Spelt, oat, rye, common millet, common lentil and horsebean were not yet present. In many cases threshing and cleaning refuse was found in houses or adjacent to them.

We note that the time of harvesting was by no means accidental. During the ripening process of cereals, dry material content increases and accordingly water content decreases (LÁNG 1976). The ripening of cereals is divided into four phases: green or milky ripeness, wax or yellow ripeness, full ripeness and overripening. At green ripeness, the water content of the grain is around 50%, its inside is milky white, nutrients flow in constantly. Grains at the wax ripening stage are still relatively soft and easily incised while water content is around 20–25%. The influx of nutrients usually comes to a halt at this point and only a small amount of starch may be deposited. From prehistoric times until the 1960s, the harvesting of cereals traditionally began in this phase of a typical two-phase harvesting by sickle or by scythe. At full ripeness, the water content of the grain drops to 13–15% and deposition of nutrients comes to a halt. Single phase harvesting (combined harvesting and threshing using a combine harvester) takes place in this phase. At this time the grains of common bread wheat fall out from among the husks more easily. In the overripe phase, the grains fall out, so that successful harvesting can never to be delayed to this phase that is still avoided today.

Based on the average yields of the previous century, FÜZES (1990) inferred the average yields of Neolithic cereals. According to his model experiments, the average Neolithic yield is obtained by subtracting a weight of 15–20% from the average yields of the 19th century ("acceleration percentage"), representing the measured growth of grain seeds over time since the Neolithic. Accordingly, the consumable (hull-free) yield of einkorn would have been 235.6 kg/ha. In order to verify these results, we successfully grew einkorn at the Szarvasgede biohistorical settlement for five years. Its processing (cleaning of hulls, milling and baking as bread) was the subject of continuous experiments (FEKETE 1999). We also tried to verify or refute the numerous, mutually contradictory archaeological, ethnographical and archaeobotanical theories associated with it. In our experiment, we achieved a five-year average yield that is 2.5 to 3 times higher than the above estimated figure. FÜZES (ibid.) estimated the Neolithic consumable yield of emmer at 345 kg/ha and that of spelt at 526 kg/ha.

Threshed grains were stored outside the house in grain pits or inside the house, (portioned for consumption) in pottery vessels, possibly in wicker or plank boxes. People used pottery for cooking and for the storage of foods and liquids. Their eating utensils were probably made of wood.

Successful farming and animal husbandry furnished a steady, secure source of food. Storage of large quantities of food, both as harvested grains and as "excess" farm animals, solved the problem of periods of scarce food, e.g., long winters or dry seasons when hunting and gathering were more difficult. A more secure food supply quickly led to rapid population increase estimated to be as much as 10% annually. Unlike hunters and gatherer who have little or no control over natural food productivity, farmers met this problem by expanding the size of fields and village and/or converting more forest land to farming and building new villages. Farming and animal husbandry became more intensive, communities had to be large enough for all members to provide sufficient labour to meet increased requirements.. However, there were strong fluctuations in population size; villages were created, abandoned or destroyed continually.

In the 5th millennium BC, i.e. the latest period of the Neolithic, a peculiar kind of pile architecture began to be practiced in Southwest Germany and Eastern Switzerland. While only very few settlements are known from 5000 BC, we know several dozen from the Neolithic and the Bronze Age from that period. There were numerous villages alongside each other. At least 500–2500 people lived around the Zurich lake at the time. It was precisely archaeobotanical research that has proven that the famous South German and Swiss pile structures were not built in water, as was previously thought (MESSIKOMMER 1913), but stood along the lakeside (JACOMET – BROMBACHER – DICK 1989). These houses were mostly wooden structures, characteristically oblong in shape, with clay ovens. The line of piles surrounding houses is also a characteristic feature. The "village," consisting of a few houses, was surrounded by a fence. When the water level, lower than it is today, increased, it flooded the settlements (BOLLINGER – JACOMET 1981).

Pottery may have functional (storage, cooking and eating vessels) and fashionable uses (stylistic indication of belonging to a social stratum). This makes a contribution to the social division of labour, therefore also reflects the history of society to some extent.

We have extremely scarce knowledge of Neolithic dress (clothing), essentially limited to textile processing and leather work. It is certain that all Neolithic cultures knew how to process flax and wool. Linseed grain and hull finds, flack-combs, spindles, weaving weights, thread and flaxen cloth remains and textile imprints constitute the evidence. However, twills were not yet used, so weaving frames probably did not have multiple heddles.

Unlike in Western Europe, there exist no early remains of arable fields in Hungary. Hardly any agricultural tools are known as they were mostly made of wood and decayed in the ground. Only large numbers of millstone fragments and flints, which may be associated with tool-making, indicate that any such tools existed. The only exceptions are bone furrowing tools and antler hoes (FÜZES 1990).

Excavation of the Great Plain Linearband Ceramic culture at Füzesabony-Gubakút in 1995 by László Domboróczky has unearthed archaeobotanical material from the earliest archaeological culture of the

Northern Great Plain (middle of 6th millennium BC) (DOMBORÓCZKY 1997). It is rendered even more significant for it is the first time direct or real cereal remains have been found from the Great Plain linear band pottery culture. Despite the large quantity of samples, we found a very small number of carbonised plant remains. This implies that their plant cultivation was on a modest scale. The people living here cultivated hulled cereals: barley and emmer. In that period, barley was a human foodstuff and the most important cereal here as it outnumber wheat grains by a ratio of four to one. Emmer seeds were bald, cleared of husks. Evidence of threshing was found in the form of a few forks (*furca bicornis*). The only weed species found was spring wild-oat (*Avena fatua*). Its hulled grains are important evidence of the origin of that archaeophyte, suggesting that it arrived in the region as a weed together with cereal species from the Middle East.

The Linearband Ceramic culture *(LBK)* of the region to the west of the Danube, dated to the Middle Neolithic, used the fertile loess lands. The material from the so-called *Transdanubian group* (Zánka train station, Tapolca parsonage garden, Hegyesd-Ágói meadow, Pápa-Vaszar, Alsópáhok-Kátyánalja meadow, Bazsi, Keszthely-Fenékpuszta Customs House, Keszthely-Zsidi Road, Magyaratád, Szenyér-Mesztegnyő, etc.) is closely related to the cultural region covering Western and Central Europe (FÜZES 1990, 1991). The majority of plant material was again preserved as imprints in pottery fragments and in mud-flakes. The number of sites that furnished plant imprints is almost four times the number of sites that yielded seeds as well.

Thanks to the indirect plant remains from the identification of imprints and the direct seed and fruit remains found in burnt-out houses and various (waste and storage) pits, our knowledge of Neolithic farming has grown significantly. We now see a culture that was familiar with plant cultivation technology. The number of cereals increased, new cultivated plants appeared including genetically more developed common and club wheat, perhaps also spelt and common millet as well as leguminous plants. The growing number of segetal weeds (korn-rade, brome grass, bedstraw, vetch, black-bindweed, fan-hen or white goosefoot) is also associated with the increasing significance of plant cultivation. The imprints of rush mats and common reed stalks indicate that the people living here strove to utilise plant species from their natural environment as well.

Let us enumerate the plant imprint sites. Between 1987 and 1989, at the Tapolca parsonage garden excavation lead by Károly Sági and Zoltán Törőcsik, a layer belonging to the oldest phase of the Transdanubian linear band culture was found under the foundation of a Roman wall. FÜZES (1990) identified plant imprints on the pottery fragments found there. After casting with silicone rubber, einkorn and emmer seeds along with earspindle section fragments and husks were identified.

In 1963 at Fenékpuszta, in part of the Roman castrum lying towards the Customs House, pits belonging to the late phase of the Transdanubian linear band culture were found (BAKAY – KALICZ – SÁGI 1966, site 1/20). The pottery fragments found here furnished many imprints of einkorn and emmer (FÜZES 1990).

In the outskirts of Bazsi, in the Csehi side field, during a land survey, pieces of pottery belonging to the more recent phase of the Transdanubian linear band population were found (BAKAY – KALICZ – SÁGI 1970, site 7/6). One vessel bottom fragment also showed a faint imprint of a woven mesh probably made using common cattail (*Typha* cf. *latifolia*) (FÜZES 1990).

In 1966, during the construction of a new road connecting Alsópáhok and Felsőpáhok, in the part called the Alsópáhok-Kátyánalja meadow, mud-flakes were collected as the construction cut into a pit from the more recent phase of the Transdanubian linear band pottery population (BAKAY – KALICZ – SÁGI 1966, site 1/20). The surface of the mud-flakes from the bottom of the pit showed imprints of cereal grains: einkorn, emmer, and common millet.

In 1965, along the Keszthely-Zsid road, pits belonging to the late, so called "Zselíz" phase of the Transdanubian linear band population were found. Einkorn seed imprints were found in a few places on the pottery fragments (FÜZES 1990).

Additional pottery fragments were found in the Tóret area near Pápa-Vaszar, at the excavation lead by Sándor Mithay (DAX *et al.* 1972). Miklós FÜZES (1991) found husk imprints of unidentified grasses (*Poaceae*) on the exterior surface of pottery fragments and the imprint of an einkorn fork (*furca bicornis*) on the internal surface of a larger fragment.

FÜZES (1990) identified the imprints found by Zsuzsa Miklós during a field survey on pottery fragments from the site at the Szob-Kilenec ditch in 1973 as einkorn forks.

In 1990, pottery fragments were collected at the Hegyesd-Ágó meadow for the Tapolca Town Museum. The fracture surface of one of the fragments showed the imprint of a grain of naked barley (*Hordeum vulgare caryopsis nuda*) (FÜZES 1991).

FÜZES (1990) observed the imprints of rush matting and perhaps the stem of the woodland European grape on a pot fragment from Magyaratád.

In comparison with the number of sites furnishing plant imprints that were listed above, the number of sites with direct or real seed remains is much smaller.

In 1964, during the reconstruction of the train station at Zánka, a number of Neolithic pits were unearthed. Unfortunately, no archaeological excavation took place, the site was only documented for a project on archaeological topography that was in progress at the time (BAKAY – KALICZ – SÁGI 1966, site 60/10). Mud-flake fragments taken from one of the pits were studied by Miklós Füzes (in: P. HARTYÁNYI – NOVÁKI – PATAY 1967–68) (*Figs. 56–60*). Unexpectedly large quantities of carbonised remains of many cereals were found: intact einkorn (*Triticum monococcum* subsp. *monococcum*) and emmer (*Triticum turgidum* subsp. *dicoccum*) grains (*caryopsis nuda*), carbonised fork (*furca bicornis*) fragments, two-rowed (?) barley naked and hulled grains (cf. *Hordeum vulgare* subsp. *distichun*). The site is the earliest known occurrence of seeds of common millet (*Panicum miliaceum*), common bread wheat (*Triticum aestivum* subsp. *compactum*) and club wheat (*T. ae.* subsp. *compactum*) in the Carpathian Basin. There were large numbers of imprints in the mud-flakes. According to mud-flake imprint evidence, the most important cereals here were hulled emmer and einkorn. In addition, Füzes also thought he had identified spicules of spelt and spicules of wild einkorn (*Triticum monococcum* subsp. *boeoticum*).

Many imprints on scattered pottery and mud-flake fragments found in the area between Szenyér and Mesztengő were successfully identified by FÜZES (1990) (*Fig. 54*). They include naked grains and forks of einkorn, naked seeds of emmer and the imprints of the seeds/grains of some segetal weeds: korn-rade (*Agrostemma githago*), foxtail millet (*Setaria* cf. *italica*) and downy brome (*Bromus tectorum*). The larger mud-flakes with common reed and branch imprints are probably from the walls of houses.

Even in this early period, weeds were hindering the work of plant cultivators. Carbonised remains of weeds belonging to the above cereal species were also found. The site is the first occurrence of korn-rade (*Agrostemma githago*) in Hungary. Sterile (?) brome grass (*Bromus* cf. *sterilis*), whose carbonised seeds were found, was also a weed of the above cereals. The mud-flakes from sidewalls, perhaps the roofs of houses, contained imprints of common reed (*Phragmites australis*) stalks and maces and imprints of leaves of narrow-leaved cattail (*Typha angustifolia*) identified by Miklós FÜZES (1990).

In 1968, at the Pári-Altäcker site excavated by István Torma, a layer of charred cereals was found in the bottom of a pit dated to the same period. According to the identification performed by Borbála P. Hartyányi, a decisive majority of the approximately 5 litres of carpological finds (*Figs. 47–50*) were emmer (P. HARTYÁNYI – NOVÁKI 1973–74). In addition, einkorn and small quantities of some weed species: brome grass (*Bromus* spec.), bedstraw (*Galium* spec.) and black-bindweed (*Fallopia convolvulus*), were identified .

In 1976, at the Becsehely-Újmajor field site, Nándor Kalich found a ditch belonging to the more recent phase of the Transdanubian linear band pottery population. There were carbonised seeds in the bottom of the ditch. The diaspores, processed by Miklós FÜZES (ibid.), included wild einkorn, einkorn and spelt, but the majority of the grains were emmer. Along with a few grains of common bread wheat, club wheat is also highly likely. Weeds are represented by only one vetch seed (*Vicia* sp.), swollen by heat.

In 1988, at Marcali-Lókpuszta, Károly Sági excavated a pit from the same period finding charcoals and carbonised seeds. Füzes identified the charcoal as being *oak* and the cereal grains as emmer and four-rowed (?) barley.

People of the Linearband Ceramic culture also consumed leguminous plants: after the flotation of the sample obtained at the site at Mosonszentmiklós-Pálmajor by András Figler, a small quantity of barley was found along with peas (*Pisum sativum*).

With some interruptions, excavations of a linear band pottery settlement continued in Bicske for several decades from 1930 to 1977, in the later years under the leadership of János Makkay. We have no information of plant remains. Reviewing the already restored so-called "note-head ornamented" pots from the 1976 excavation and identifying the imprints found on them, we have gained some information on plant cultivation of the inhabitants of the settlement. Plant imprints on the surface of pottery fragments were processed using a computer image processing system (GYULAI 1996a). Imprints of cereals included emmer and six-rowed barley along with their weeds, rye brome, fan-hen or white goosefoot (*Fig. 55*).

We also have a relatively larger number of plant remains of the Middle Neolithic from the eastern region of the country. At the site of the Szakálhát-Szilmeg group, barley and hulled wheats (emmer, einkorn) were primary food sources along with a new kitchen garden plant, small seed lentil.

István Ecsedi conducted an excavation in the Dévaványa-Réhelyi gát in 1970. Among the fragments of large storage vessels in two houses, large quantities of stockpiled two-rowed naked barley, einkorn and emmer seeds were found (P. HARTYÁNYI – NOVÁKI – PATAY 1967–68) (*Figs. 51–53*). Presumably, the relative quantities of seeds also indicate their relative significance in nutrition. A few seeds of small seed lentil were also found. Ten litres of cereal grains contained hardly any weed seeds. Only a few seeds of black-bindweed and fan-hen or white goosefoot were found. This indicates thorough cleaning or perhaps implies that in this early phase of cereal cultivation, there were few weeds and the cohabitation of host plants and weeds was just beginning.

No seeds, only charcoal identified as oak were found at the pit opened in 1963 by Nándor Kalicz at the Tarnazsadány-Sándorrésze site (FÜZES 1990).

7.1.5. Agriculture in the Late Neolithic

A pollen analytical study of sediments of Lake Balaton clearly indicates anthropogenic effects (ZÓLYOMI 1971, 1980). After sporadic unidentifiable cereal pollen appeared in the early Atlantic phase (6th pollen zone, Middle Neolithic), evidence of widespread cereal cultivation around Lake Balaton appeared in the Late Neolithic. The first identified wheat pollen is from sediment dated around 4500 BC. Based on the quantities of bread wheat pollen from the Late Neolithic, ZÓLYOMI (ibid.) inferred that cultivation of cereals increased. As the pollen are smaller than bread wheat pollen from more recent sediment layers, he believes them to be einkorn. Indirectly, the spread of cultivation of arable crops is also indicated indirectly by the increasing quantities of non-tree pollens (*Graminea*, *Cyperaceae*, *Artemisia*, *Chenopodiaceae*, etc.), which implies that forests were receding.

Late Neolithic sites in Hungary are relatively rich in plant remains. This is due in large part to increased attention in recent years by archaeologaists to take regular samples and to perform proper flotation. These archaeobotanical finds are generally carbonised recovered from burnt-out houses and various waste or grain storage pits. Thanks to such material, our knowledge of farming in the Late Neolithic has grown significantly.

Plant cultivation continued to progress in the Late Neolithic period. Increasing size and number of settlements has resulted in plant assemblages from this era that are relatively rich in species with large numbers of seeds. Botanical material from sites of the Tisza culture suggests a settled population cultivating cereals (einkorn, emmer, two- and six-rowed barley) and kitchen vegetables (common lentil, square vetchling, pea).

The older sites of the Tisza culture yielded only imprints of plants. The mud-flakes found at the Szeghalom-Kovácshalom site by Gyula Szeghalmy in 1940 had imprints of ears, spindles and hulls of cereals (in: FÜZES 1990). The 1957 excavation at Szegvár-Tűzköves by József Csalog and József Korek yielded fragmented imprints of emmer ears (P. HARTYÁNYI – NOVÁKI – PATAY 1967–68). In 1942, János Banner excavated a settlement of the Tisza culture at Hódmezővásárhely-Kökénydomb and found bread wheat grains stuck to the bottom of a pot, surrounded by other, unidentified carbonised seeds (BANNER 1942). No seeds, only charcoal remains are known from Hódmezővásárhely-Kopáncs (János Banner's excavation 1931) or Hódmezővásárhely-Kotacpart (János Banner's excavation 1933), both sites belonging to the Tisza culture (FÜZES 1990). Ferenc Hollendonner identified charcoal from both locations as common oak (*Quercus* cf. *robur*) and common elm (*Ulmus minor*).

In 1975, during the excavation of the prehistoric tell at the Battonya-Parázstanya site, Júlia G. Szénászky collected a significant quantity of botanical material from the pits and houses. Not only seeds, but also flour and meal for kitchen use were found. Botanical material of the Tisza culture was identified by Borbála P. HARTYÁNYI (1988–89). The presence of large quantities of seeds of seven cultivated plant species indicates manifold economic activities of the residents. Plant cultivation was of high importance to them. The majority of the fifty samples, all from different locations, contained emmer. Einkorn and naked barley were somewhat less common. Among leguminous plants, grass pea vine was the most common, with some common lentil. Pea was least frequent. Flax was also cultivated. Cornelian cherry was probably gathered. Only a very small number of weed seeds occurred among the cereals and leguminous plants stored for consumption: korn-rade, black-bindweed, fan-hen or white goosefoot.

Botanical material from tell settlements at Tisza-Herpály-Berettyóvölgy (Berettyóújfalu-Szihalom, Berettyóújfalu-Herpály) contained all hulled bread wheat species (einkorn, emmer). In the case of the Berettyóújfalu-Szilhalom tell excavation (by Márta Sz. Máthé, 1976), einkorn and emmer seeds occurred in roughly equal quantities in samples from various locations, but roughly five times that quantity of two-rowed naked barley (*Hordeum vulgare* subsp. *distichum* var. *nudum*) seeds were found (*Figs. 61–64*). According to observations by Borbála P. Hartyányi, emmer grains fall into two categories. The first, an elongated type, is very similar to emmer seeds from the Dévaványa-Réhelyi gát. The second, a broader type shows similarities with emmer from the Pári-Altäcker site. The significance of field pea (*Pisum sativum* subsp. *arvense*) is indicated by the fact that these seeds constitute about half the total quantity of finds (P. HARTYÁNYI – SZ. MÁTHÉ 1980). A few seeds of common bread wheat were also found. Complete, carbonised crab apple (*Malus silvestris*) fruits were gathered.

In addition to hulled wheats (einkorn, emmer), common millet was also found at the Berettyóújfalu-Herpály site (Nándor Kalicz and Pál Raczky's excavation 1978–82). Here again, the most important leguminous plant was field pea. According to ^{14}C testing, the estimated date of the Berettyóújfalu-Herpály site is 6570–6270 BP (HERTELENDI *et al.* 1997).

In 1981, during the excavation of one of the houses at Berettyóújfalu-Herpály, Kalicz and Raczky found grains stored for consumption. One hundred and fifty carbonised seeds are predominantly field pea seeds (*Pisum sativum* cf. subsp. arvense), with some bitter vetch (*Vicia ervilia*). That is the earliest known occurrence of bitter vetch in Hungary. It only became widespread later in Middle Bronze Age tells (e.g. Bölcske-Vörösgyír, Tiszaalpár-Várdomb). This vegetable had already appeared in Hungary in the Neolithic. Was it through their connections with the south that people obtained this ancient Middle Eastern and Mediterranean vegetable that was still a novelty to them? After the passage of the prehistoric period, the mildly poisonous bitter vetch no longer occurs in finds from Hungary. An earlier find, though outside Hungary, was reported at the site of the Baden culture at Jelšovce where a stock pit yielded bitter vetch (CHEBEN – HAJNALOVÁ 1997).

All botanical samples from the tell settlements of Tisza-Herpály-Berettyóvölgy contain peas. At Berettyóújfalu-Elmhalom, field pea provided almost half of all diaspores. Peas were also found at the excavation of the Late Neolithic (Herpály culture) tell settlement at Tiszapolgár-Csőszhalom in 1995. Plant remains were already observed at the excavation of the Tiszapolgár-Csőszhalom Late Neolithic tell settlement in 1957, when the excavation was led by Ida Kutzián: charcoal identified as oak was found in one of the houses and unidentified cereal grains were found in another. At the 1995 excavation, led by Pál Raczky, we collected a large number of soil samples from the floor levels of houses and various pits found in the Late Neolithic tell and in the settlement around the tell. According to radiocarbon dating, the estimated date of the Polgár-Csőszhalom site is 6700–6370 BP (HERTELENDI *et al.* 1997).

The botanical material obtained by flotation on site indicates a peasant population at the peak of plant cultivation know-how. Their main product was emmer. Einkorn and six-rowed barley (*Hordeum vulgare* subsp. *hexastichum*) played a less dominant role. Sporadic occurrences of naked bread wheat, spelt and common millet grains were also found. Due to the similarity of the seeds, it was not possible to decide whether the naked bread wheat grains were common wheat or durum wheat. As expected, the cereals were found together with leguminous plants: small seed lentil, pea, common vetch (*Vicia* cf. *sativa*). Carbonated stones of cherry, myrabolan plum (*Prunus* cf. *cerasifera* subsp. *myrobalana*) and dogwood (*Cornus mas*) were also found. These were probably gathered for consumption, just like oak acorns (*Quercus* spec.), blackthorn (*Prunus spinosa*) and crab apple (*Malus silvestris*). Many weed species indicate intensive cereal production. In addition to black-bindweed and fan-hen or white goosefoot, some weed species appear here for the first time in the Hungarian Neolithic: summer vetch (*Vicia angustifolia*), copse-bindweed (*Fallopia dumetorum*), false cleavers (*Galium spurium*), barnyard grass (*Echinochloa crus-galli*) and annual woundwort (*Stachys annua*). The site was also rich in other elements of the environment: danewort (*Sambucus ebulus*), hairy sedge (*Carex hirta*), water dock (*Rumex hydrolapathum*) and great bulrush (*Schoenoplectus lacustris*). This implies that there were watercourses near the tell settlement that intruded into arable land as well.

The Battonya-Vertán manor site is of roughly identical age with the one at Tiszapolgár, though it does not have the same plenitude of plant species. In 1982, during the excavation of a larger house, Júlia G. Szénászky found almost 20 g of carbonised plant remains. The majority were emmer seeds with some einkorn, as well as carbonised stones of dogwood and seeds of fan-hen or white goosefoot, danewort, amaranth (*Amaranthus* spec.) and members of the buckwheat family (*Polygonum* spec.) (P. HARTYÁNYI 1988–89).

Plant remains from sites associated with the Lengyel culture in the Transdanubian region indicate that the population still cultivated plants, but less intensively than in the previous era. At the Lengyel site, which gave the culture its name, Mór Wosinsky conducted several excavations between 1885 and 1890. The dating of the findings raised several subsequent problems (P. HARTYÁNYI – NOVÁKI – PATAY 1967–68). In 1890, Imre Deininger himself collected seeds there. His identification of the seeds from the vicinity of fireplaces, pits and from pots was published in 1892. He found the following species: many-rowed barley, common bread wheat, club wheat, einkorn, common millet, foxtail millet (*Setaria italica*), horsebean (*Vicia faba*), bitter vetchling (*Lathyrus sativus*), common lentil, spring wild-oat (*Avena fatua*), rye brome (*Bromus secalinus*), field brome (*B. arvensis*), barnyard grass (*Echinochloa crus-galli*), tufted vetch (*Vicia cracca*), common milkvetch (*Astragalus glychyphyllos*), korn-rade, cowherb (*Vaccaria hispanica*), pink (*Dianthus* spec.), ribwort-plantain (*Plantago lanceolata*), dodder (*Cuscuta* spec.), European barberry (*Berberis vulgaris*), plum (*Prunus* spec.), dogwood (*Cornus mas*), Scotch pine (*Pinus silvestris*), flax, sedge bicarpellat (*Carex* spec.).

The Lengyel culture site at Zengővárkony was excavated by János Dombay in the 1940s. The imprints of the mud-flakes from there were studied by Emil Gubányi in 1947 and identified as two-rowed barley and spelt (in: P. HARTYÁNYI – NOVÁKI – PATAY 1967–68). Reviewing the pottery fragments form the site, Miklós FÜZES (1990) found additional imprints. Among them he claims to recognise mazzard cherry (*Prunus avium* subsp. *silvestris*) and the imprint of a mat made from narrow-leaved cat-tail.

The pits belonging to the older phase of the Lengyel culture opened by Nándor Kalicz (1960–82) at the Aszód-Papi fields site yielded pottery fragments on which Miklós FÜZES (1990) found plant imprints again. He counted a total of 2500 (!) imprints of plant organs on the 14 fragments he reviewed, but only a small percent of those imprints allowed species to be identified. There were also husked grains of barley and some grains of naked oat (*Avena* spec. cf. *nuda*).

Only two species of cultivated cereals were found during the 1994 excavation by András Figler at Börcs-Paphomlok in the layers belonging to the 3rd phase of the Lengyel culture: six-rowed barley and common millet. Based on the total lack of bread wheat species, the population of the settlement is believed to have been animal keepers rather than plant cultivators. It is also possible that it was only a temporary settlement where they had insufficient time for growing bread wheat. It is also possible that the climate of the Neolithic-Copper Age transition period, which was inhospitable to plant cultivation (see below) was already making itself felt.

Charcoal remains are also known from the period of the Lengyel culture. FÜZES (1990) identified the pieces collected at the Letenye-Szentkeresztdomb site as oak Nándor Kalicz 1964) to silver fir (*Abies alba*).

We are also aware of other scattered Neolithic plant remains. In 1956, Gyula Gazdapusztai found a burnt-out Late Neolithic hut at the Hódmezővásárhely-Cukortanya site containing pots and cereal grains (P. HARTYÁNYI – NOVÁKI – PATAY 1967–68). However, these finds are nowmissing and no analysis conducted.

Miklós Füzes collected mazzard cherry stones from a burnt layer of the Late Neolithic (or perhaps Copper Age) flint mine excavated by László Vértes at Sümeg-Mogyorósdomb.

The fruit of the European woodland grape was collected in Hungary in the Neolithic. Its seeds were found at the Sé site (near Szombathely) of the Late Neolithic Lengyel culture (HORVÁTH 1975). The oak acorn, which is not exactly a fruit, should still be regarded as a "fruit of the forest". The nutrient-rich (roasted) oak acorn was used for several millennia as emergency food in times of famine or sometimes as a speciality for people working or gathering in the forest. This is certainly indicated by the Late Neolithic residential site excavated by Alán Kralovánszky in 1961 at Moha-Homokbánya (probably Lengyel culture), where carbonised oak acorns, found in one of the underground dwellings, were identified as pubescent oak (*Quercus pubescens*) (P. HARTYÁNYI – NOVÁKI – PATAY 1967–68).

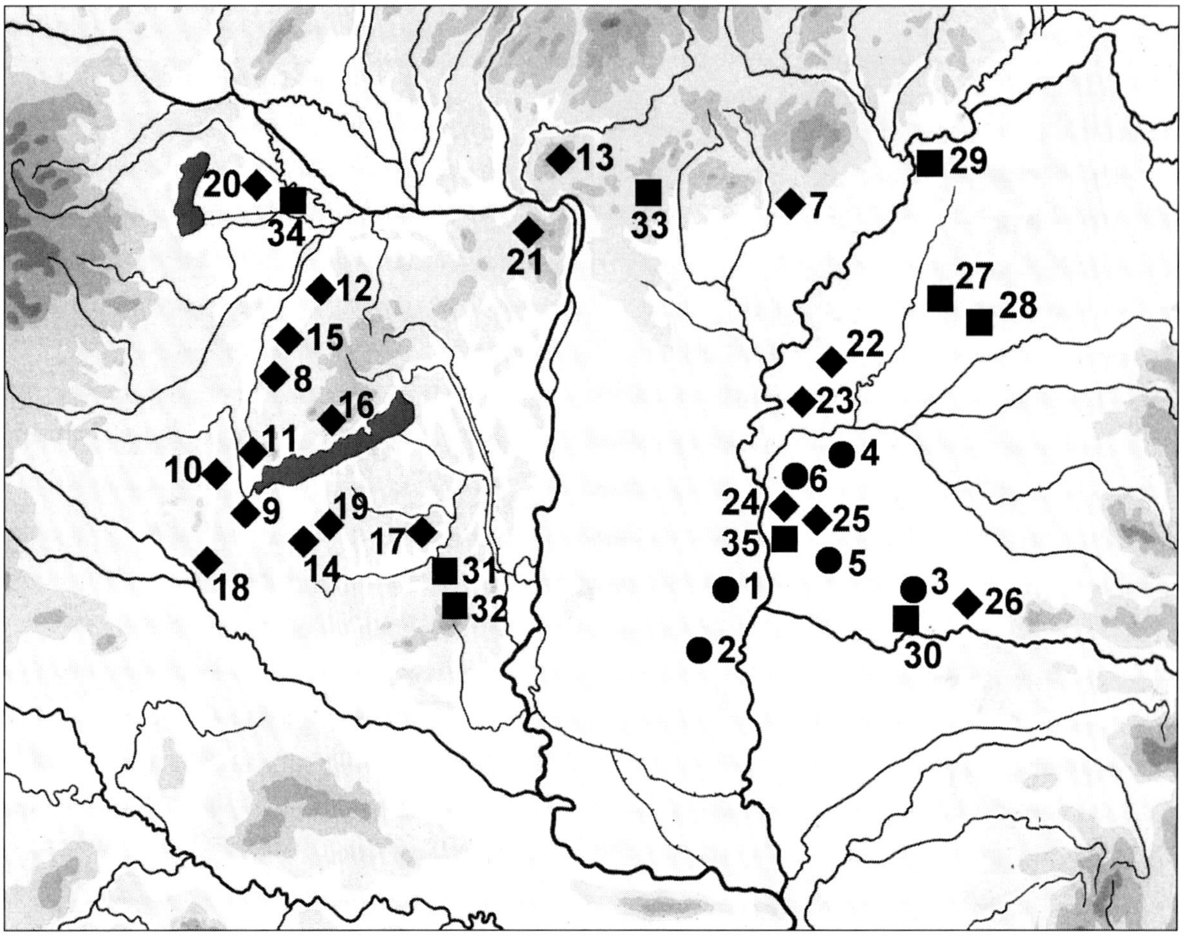

Fig. 43. Neolithic plant remains in Hungary.

● Early neolithic plant remains:
1 Szeged-Gyálarét: A, B
2 Röszke-Lúdvár: A, B
3 Battonya-Basarága: A
4 Gyomaendrőd: A
5 Hódmezővásárhely-Gorzsa: F
6 Szarvas-Szappanos: F

◆ Middle neolithic plant remains:
7 Füzesabony-Gubakút: K, B
8 Tapolca-Plébániakert: A, B
9 Fenékpuszta-Vámház: A, B
10 Alsópáhok-Kátyánalji dűlő: A, B, L
11 Keszthely-Zsidi út: A
12 Pápa-Vaszar: A
13 Szob-Kilenec: A
14 Szenyér-Mesztegnyő: A
15 Hegyesd-Ágói dűlő: J
16 Zánka-vasútállomás: A, B, G, L, D, E
17 Pári-Altäcker: B, A
18 Becsehely-Újmajor: B, A, C, D, E
19 Marcali-Lókpuszta: B, H
20 Mosonszentmiklós-Pálmajor: K, b
21 Bicske-Galagonyás: B, I
22 Dévaványa-Réhelyi dűlő: J, A, B, a
23 Szeghalom-Kovácsihalom: N
24 Szegvár-Tűzköves: B
25 Hódmezővásárhely-Kökénydomb: F
26 Battonya-Parázstanya: B, A, J, c, a, b

■ Late neolithic plant remains:
27 Berettyóújfalu-Szilhalom: J, A, B
28 Berettyóújfalu-Herpály: A, B, L, b
29 Tiszapolgár-Csőszhalom: B, A, I, D, C, L, a, b, d
30 Battonya-Vertán major: B, A
31 Lengyel: K, D, E, A, L, M, e, c, a
32 Zengővárkony: G, C
33 Aszód-Papi földek: A, B, K
34 Börcs-Paphomlok: I, L
35 Hódmezővásárhely-Cukortanya: N

Cultivated plant remains:
A = einkorn (*Triticum monococcum*)
B = emmer (*Triticum dicoccum*)
C = spelta wheat (*Triticum spelta*)
D = common wheat (*Triticum aestivum*)
E = club wheat (*Triticum compactum*)
F = wheat (*Triticum* spec.)
G = two-rowed barley (*Hordeum distichon*)
H = four-rowed barley (*Hordeum tetrastichum*)
I = six-rowed barley (*Hordeum hexastichon*)
J = nacked barley (*Hordeum nudum*)
K = barley (*Hordeum* spec.)
L = common millet (*Panicum miliaceum*)
M= foxtail millet (*Setaria italica*)
N = cereals (*Cerealia*)
a = small seed lentil (*Lens culinaris*)
b = field pea (*Pisum sativum*)
c = grass pea wine (*Lathyrus sativus*)
d = bush vetch (*Vicia sativa*)
e = horsebean (*Vicia faba*)
f = common flax (*Linum usitatissimum*)

Fig. 44. Cornelian cherry (*Cornus mas*) stone from the Körös culture settlement of Battonya-Basarága. After FÜZES 1990.

Fig. 45. Einkorn (*Triticum monococcum*) spicelet (furca bicornis) imprint on a sherd found in the Körös culture settlement of Gyomaendrőd. After FÜZES 1990.

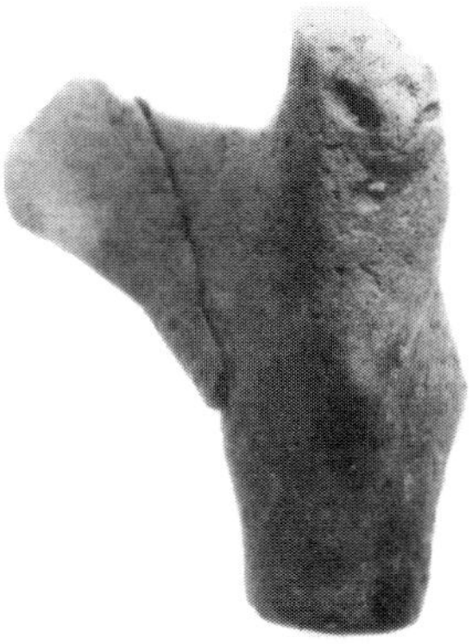

Fig. 46. Emmer (*Triticum turgidum* subsp. *dicoccum*) imprints like eyes. Clay altar. Balatonszentgyörgy, Starčevo culture. After FÜZES 1990.

Fig. 47. Emmer (T*riticum turgidum* subsp. *dicoccum*) naked grains from Pári-Altäcker dülő, Linearbandceramic culture, Inventory of the Hungarian Agricultural Museum, Budapest.

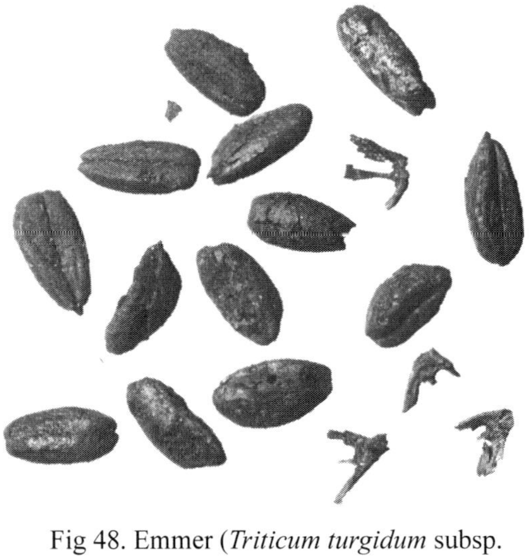

Fig 48. Emmer (*Triticum turgidum* subsp. *dicoccum*) naked grains from Pári-Altäcker dülő, Linearbandceramic culture, Inventory of the Hungarian Agricultural Museum, Budapest.

Fig. 49. Emmer (*Triticum turgidum* subsp. *dicoccum*) naked grains from Pári-Altäcker dülő, Linearbandceramic culture, Inventory of the Hungarian Agricultural Museum, Budapest.

Fig. 50. Wheat (*Triticum* spec.) spicelets (*furca bicornis*) from Pári-Altäcker dülő, Linearbandceramic culture, Inventory of the Hungarian Agricultural Museum, Budapest.

a.

b.

Fig. 51. a. Einkorn (*Triticum monococcum*), b. Emmer (*Triticum turgidum* subsp. *dicoccum*) naked grains from Dévaványa-Réhelyi gát, Tisza culture, Inventory of the Hungarian Agricultural Museum, Budapest.

Fig. 52. Barley (*Hordeum vulgare*) grains found in a storage vessel from Dévavanya-Réhelyi gát, Tisza culture, Inventory of the Hungarian Agricultural Museum, Budapest.

Fig. 53. Naked barley (*Hordeum vulgare* var. *nudum*) from Dévavanya-Réhelyi gát, Tisza culture, Inventory of the Hungarian Agricultural Museum, Budapest.

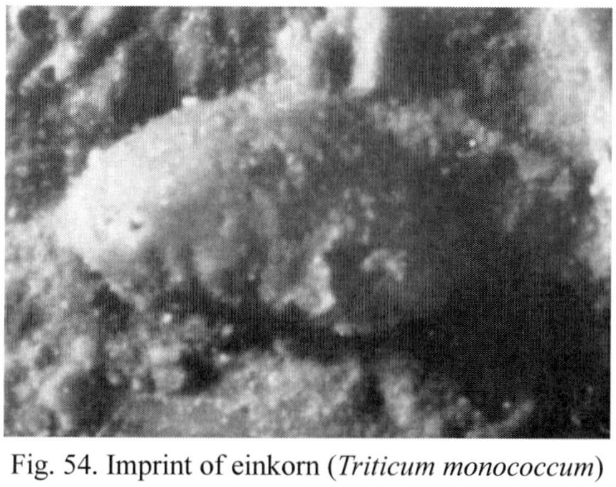

Fig. 54. Imprint of einkorn (*Triticum monococcum*) grain from Szenyér-Mesztegnyő. After FÜZES 1990.

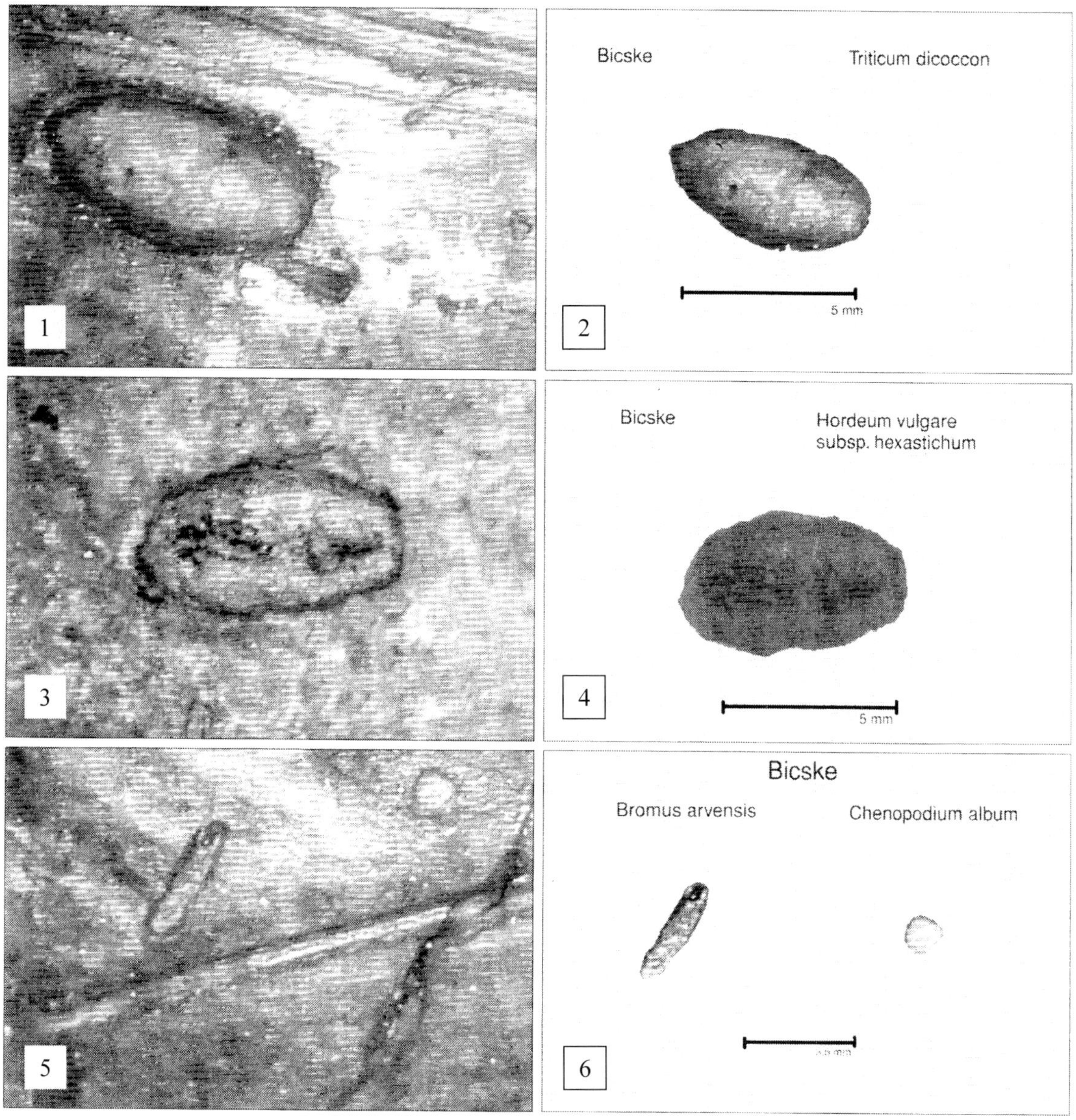

Fig. 55. Bicske-Galagonyás. Digitized images of seed imprints found on potsherds (left) and the positive image obtained by computer (right):

1. Emmer (*Triticum turgidum* subsp. *dicoccum*) naked grain imprint on the wall of a pot fragment found at the Bicske-Galagonyás Neolithic site. Photograph by the author.
2. Computerised inverse image made from the emmer (*Triticum turgidum* subsp. *dicoccum*) imprint found at the Bicske-Galagonyás Neolithic site. Photograph by the author.
3. Six-rowed barley (*Hordeum vulgare* subsp. *hexastichum*) imprint found at the Bicske-Galagonyás Neolithic site. Photograph by the author.
4. Computerised inverse image made from the six-rowed barley (*Hordeum vulgare* subsp. *hexastichum*) imprint found at the Bicske-Galagonyás Neolithic site. Photograph by the author.
5. Brome grass (*Bromus arvensis*) grains and fan-hen (*Chenopodium album*) seed imprints on the wall of a pot fragment found at the Bicske-Galagonyás Neolithic site. Photograph by the author.
6. Computerised inverse image made from the brome grass (*Bromus arvensis*) and fan-hen (*Chenopodium album*) imprints found at the Bicske-Galagonyás Neolithic site. Photograph by the author.

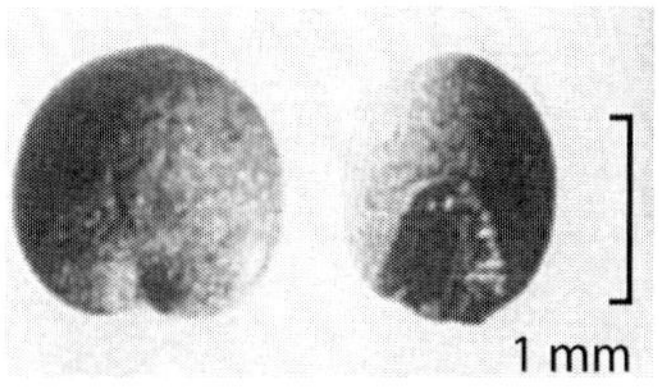

Fig. 56. Common millet (*Panicum miliaceum*) naked grains from the Linearbandceramic culture settlement in Zánka. After FÜZES 1990.

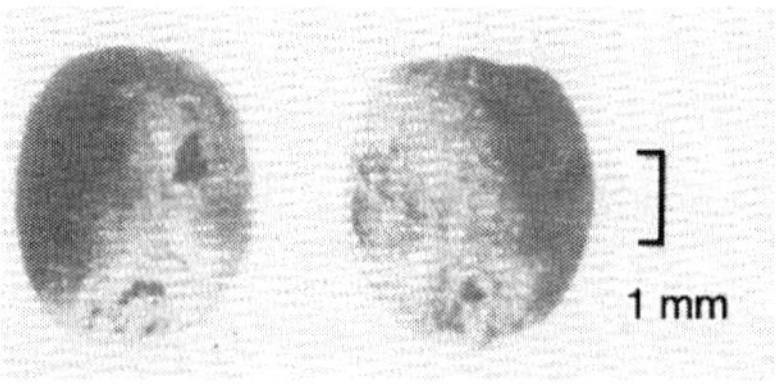

Fig. 57. Club wheat (*Triticum aestivum* subsp. *compactum*) grains from the Linearbandceramic culture settlement in Zánka. After FÜZES 1990.

Fig. 58. Korn-rade (*Agrostemma githago*) seed from the Linearbandceramic culture settlement in Zánka. After FÜZES 1990.

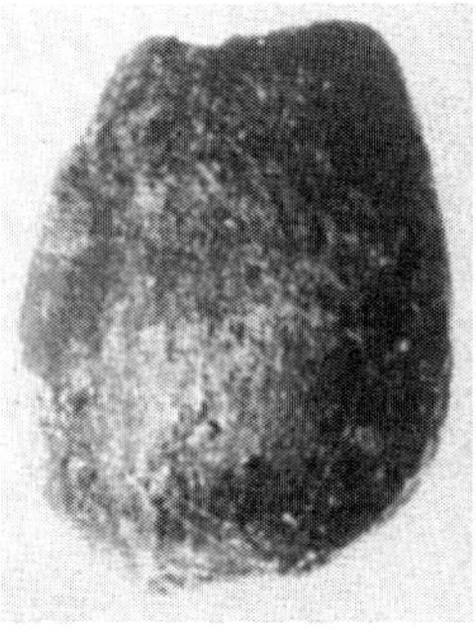

Fig. 59. Hazelnut (*Corylus avellana*) shell from the Linearbandceramic culture settlement in Zánka. After FÜZES 1990.

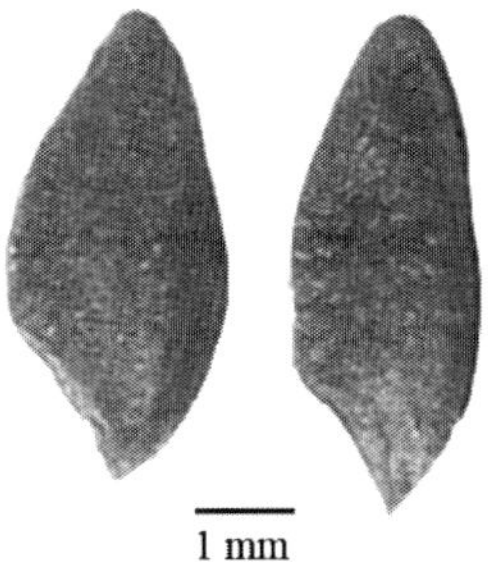

Fig. 60. Einkorn (*Triticum monococcum*) grains from the Linearbandceramic culture settlement in Zánka. After FÜZES 1990.

Fig. 61. Field pea (*Pisum sativum* var. *arvense*) seeds from the Late Neolithic settlement of Berettyóújfalu-Szilhalom. After P. HARTYÁNYI – SZ. MÁTHÉ 1980.

Fig. 62. Einkorn (*Triticum monococcum*) grains from the Late Neolithic settlement of Berettyóújfalu-Szilhalom. After P. HARTYÁNYI – SZ. MÁTHÉ 1980.

Fig. 63. Crab apple (*Malus silvestris*) charcoal fruits from the Late Neolithic settlement of Berettyóújfalu-Szilhalom. After P. HARTYÁNYI – SZ. MÁTHÉ 1980.

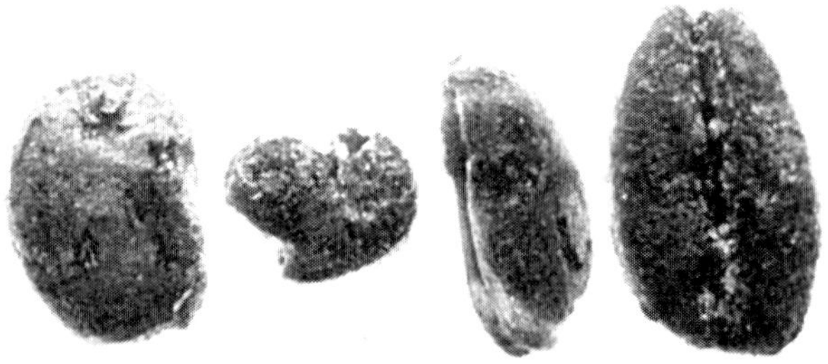

Fig. 64. Common bread wheat (*Triticum* cf. *aestivum* subsp. *vulgare*) grains from the Late Neolithic settlement of Berettyóújfalu-Szilhalom. After P. HARTYÁNYI – SZ. MÁTHÉ 1980.

7.2. The Copper Age

7.2.1. Botanical finds from Copper Age cultures

Toward the end of the Neolithic, the warm and humid, temperate climate gradually grew worse. Species composition of the extensive mixed oak forests of Central Europe (oak, elm, linden, ash, with hazelnut in the shrub layer) which had developed on the loess soils and also provided a good basis for agricultural cultivation, began to change. According to pollen analysis performed by ZÓLYOMI (1980), the period since 8200 BC, had the highest level of forestation in the Carpathian Basin. The Copper Age, which began around 4500 BC, shows a reduction of cereal pollen counts relative to the Neolithic.

According to pollen analyses, the slow but distinct change to a colder and wetter climate resulted in a significant spread of European beech in many areas of Europe (JACOMET – BROMBACHER – DICK 1989). In parallel, the oak, birch, elm and hornbeam species, along with silver fir and Norway spruce in higher-lying areas, began to recede. The pollen of walnut (Juglans regia) also appeared as a result of the spread of floristic relationships with the Balkans.

Since an oak canopy allows sunlight to penetrate, an oak forest has a rich shrub layer. In contrast, European beech has closed canopy, therefore its shrub layer consists of only a few species. Alder and hornbeam, which formed gallery forests and fragmented forests, continued to play a role, albeit a lesser one. In parallel with the cooling, the level of forestation remained at the maximum until the end of the Copper Age. Initially, the ratio of hornbeam and European beech increased with the cooling, reaching 30% of the total of non-arboreal pollens (NAP). The changes stabilised by the beginning of the Bronze Age.

The number of Copper Age sites investigated archaeobotanically is very small: Zalaszentbalázs, Rákoskeresztúr-Újmajor, the vicinity of Győr, Kompolt, Óbuda and Csepel Island. For that reason, we are as yet unable to form a comprehensive image of plant cultivation of the Copper Age cultures of Hungary. What is clear is that the farming lifestyle developed in the Neolithic changed in the first half of the Copper Age. Plant cultivation continued, but its significance relative to animal husbandry grew smaller.

Animal bone finds also indicate that the Copper Age had cool and rainy weather. There are many hunted animals among the bones, including the unusual presence of lion (Panthera leo cf. persica) at Tiszaluc-Sarkad (VÖRÖS 1982–83). Nevertheless, the majority of bones are from domesticated animals. The Bolerázi culture site at Győr-Szabadrétdomb also indicates a predominance of animal husbandry. The site is dated to 3388–3042 BC (FIGLER *et al.* 1997).

During the Copper Age, probably as a result of southeastern influences, the proportion of sheep and goat bones increased, which indicated nomadic shepherding (BARTOSIEWICZ 1995). Horse also appeared. A most extensive and varied range of prehistoric cattle as regards body size and breed was also found at sites dating to the Copper Age (VÖRÖS 1982–83). A number of ritual cattle burial sites are also known from that period, which could only have been possible with intensive farming of livestock (ENDRŐDI – VÖRÖS in print). The predominance of animal husbandry may have been a direct result of the changed, cooler and rainier climate. Among cereals, barley and hulled wheats (primarily emmer) and in some instances two-rowed naked barley are present. The situation changed at the end of the Copper Age, when the climate once again changed.

The quantity of seeds and the composition of cerealst in the Middle Copper Age clearly indicate a reduction in the production of arable crops. Excavation of an Early and Middle Copper Age settlement at Zalaszentbalázs-Szőlőhegy (late Lengyel culture – Balaton-Lasinja culture, Eszter Bánffy's excavation 1992) yielded no cereal grains at all. Longitudinal and transversal cells from the seed wall of hulled bread wheat (*Triticum* spec.) were found in the food remains at the site (GYULAI 1995a). On the other hand, the appearance of a cemented, cleaned mass of common bulrush (*Schoenoplectus lacustris*) is quite informative (*Fig. 65*). It is clear that it was used to replace cereals, which were difficult to grow due to adverse weather. As those nutlets are rich in nutrients, it is probable that this plant, which is still common in streams, marshes and higher-lying parts, was used as a sustenance food.

The extent to which the first half of the Copper Age was characterised by the decline of cereal cultivation and plants in general is unknown. In any case, it is food for thought that the Öcsöd-Kendereshalom site of the Middle Copper Age Bodrogkeresztúr culture also lacks the remains of cultivated plants. In 1956, József Csalog's, excavation of pits and graves of the Bodrogkeresztúr culture at a residential site of the Körös culture, 20 cl of carbonised seeds were found. They were, almost without exception, charlock (*Sinapis arvensis*) seeds (P. HARTYÁNYI – NOVÁKI – PATAY 1967–68) (*Fig. 66*). Is it possible that the seeds of this cruciferous plant, which to us is only a weed, were stocked in such quantities to obtain oil?

During an archaeological survey between 1991 and 1994 prior to the extension of the M1 motorway around Győr, András Figler found a site containing material from several periods of the Copper Age (Mosonszentmiklós-Pálmajor, Győr-Szabadrétdomb). This was our first opportunity to track the changes of this transformational agricultural period of the Middle-Late Copper Age. Although the Middle Copper Age Ludanice culture samples from Mosonszentmiklós-Pálmajor yielded only a few grains of six-rowed barley and emmer, barely sufficient proof of their cultivation. The samples from Győr-Szabadrétdomb of the Ludanice – Balaton-Lasinja culture contained plant remains indicating that the cultivation of cereals was restricted to two species during that era, barley, unspecified due to fragmented seeds, and common millet.

During the excavation of the Rákoskeresztúr-Újmajor site, the soil samples from the Ludanice culture pits only yielded a few dozen seeds of six-rowed barley in the company of a few seeds of water and marsh plants (excavation by Zoltán Bencze and Zsuzsa M. Virág 1995–96). We note that the partial absence of bread wheat indicates some serious problems with plant cultivation, perhaps due to a nomadic, pastoral lifestyle. Permanent settlements in the region during this era are known, and animal bone remains found indicate a predominance of animal husbandry. Thus, we may assume that the absence of wheat is attributable either to a lifestyle that includes keeping of large animals but not the cultivation of wheat; or, to extreme weather unfavourable for plant cultivation. Yet barley and fast-growing, spring-sown and rather undemanding common millet was still possible and important in those conditions.

Climatic reasons for the decline of plant cultivation (long-term cooling, rainy weather) are supported by archaeobotanical remains found in a Middle Copper Age pit at Albertfalva in 2001. A large pit of the stab and drag culture, excavated by Anna Endrődi, which also contained beautifully ornamented pottery fragments, obsidian, mud-flakes and animal bones, yielded a single, unidentifiable, carbonised seed fragment with some food remains. The total absence of cereal remains in that era of the Copper Age has been accepted as a fact. As a result of the extremely unfavourable climate, the cultivation of cereals receded or disappeared entirely in many locations.

The extreme, cool and rainy climate improved in the Late Copper Age. Material recovered from the Bolerazi layer of Győr-Szabadrétdomb contains many more cereal remains. Six species of cereals occur: six-rowed barley, naked barley, common millet, einkorn, emmer and club wheat. Over one hundred barley grains make it quite clear that its cultivation was of definitive importance for the Bolerazi culture. Unfortunately, the barley grains are highly fragmented, so only a few have been identified as six-rowed barley and naked barley. Among wheats, the more ancient hulled emmer and einkorn were cultivated. As the sample contained many emmer and einkorn earspindle fragments, it is conceivable that threshing waste was deposited in the pits. Occasionally, seeds of the genetically more developed naked club wheat were also found. The third hulled wheat species, spelt, was also detected. Common millet was an important meal plant. Among leguminous plants, only chickpea (*Cicer arietinum*) occurred. Weeds also indicate cereal cultivation. Their number is low (wild turnip, fan-hen or white goosefoot), which, again, indicates that the cereal grains had been cleaned, perhaps ventilated or selected by hand before they were burnt. The number of species from the natural environment of the era is also low and primarily remains of plants, i. e. fruits, gathered and eaten. Sloe, blackthorn and dogwood were collected from forest fringes and danewort from forest clearings. They are all so-called "wild fruits," known to be tasty and rich in vitamins.

In 1994, Andrea Vaday, working on the M3 motorway project, found carbonised seeds during the excavation of a pit oven at a Late Copper Age site at the Kompolt-Kistér farmstead. Eszter Bánffy (in: BÁNFFY – T. BÍRÓ – VADAY 1997) quoted Nándor KALICZ (1991) in attributing the site to the Proto-Boleraz culture. The material was dominated by gathered wild fruits: hazelnut (*Corylus avellana*), dogwood (*Cornus mas*) and woodland European grape (*Vitis vinifera* subsp. *silvestris*) (GYULAI 1997a) (*Figs. 67–69*). All of this reinforces our assumption that these remains are of plants stored for some purpose. The seeds/fruits of other species occurred sporadically in the sample. Nevertheless, it is certain that barley was an important cereal for that culture.

Among prehistoric sites in Hungary, Kompolt is the only location where both dogwood and woodland European grape were found in large quantities. While it is possible to make both jam and a drink from dogwood, woodland European grapes can only be used to make wine. The wine made from this grapes has a fine taste. The fruit of the dogwood can also be fermented into an excellent wine with the addition of honey and water. In our opinion, both species were used to make drinks. Those remains are also the earliest indirect evidence for the fermentation of alcoholic drinks.

Gathering and consumption of fruit of the woodland European grape are attested by seed remains from a number of Neolithic settlement sites both in and outside Europe. The area of the woodland European grape, which adapted well to the quaternary postglacial period, probably extended further in the Mesolithic. The oldest evidence of this grape was found at the Grotta dell'Uzzo Mesolithic site near Trapani in Italy (COSTANTINI 1981b). In a sediment from Lake Balaton dated to the Atlantic phase, Mária FARAGÓ Mihálchné (1983) and Elvira BODOR Nagyné (1988) found grape pollen that was not possible to identify more precisely. In addition to the pollen material, macroscopic remains are also known from that era. At the late Neolithic settlement of Sé near Szombathely, Ernő HORVÁTH (1975) found seeds of woodland European grapes. Miklós FÜZES (1990) identified an imprint on the side of a net-weight from the Neolithic settlement at Magyaratád as grape, more precisely woodland European grape. Those Neolithic finds of grapes indicate that the Atlantic phase was characterised by a very mild climate which facilitated the spread of the woodland grape. After the cooling of the first half of the Copper Age, rising annual average temperatures from the middle of the Copper Age again allowed it to gain ground.

Now that we have found the earliest evidence for the preparation of wine in Hungary at the Kompolt site, the opinions that reject finds, such as the large "fermenting" vessel of the Nagyrév culture, which allow us to infer cultivation of grapes and making of wine (NOVÁKI 1975) should be revised.

Plant remains have been found in the Late Copper Age Baden culture in Hungary on two occasions: from the landfill of one of the pits at Budapest, Bécsi Road No. 38–42 (ENDRŐDI 1998); and from improvised fireplaces found at the excavation prior to the construction of a gas pipeline within the area of the Csepel Waterworks and from the furnace chamber of a fireplace surrounded by the fragments of a special, gynaecomorph ("female-shaped") pot (Anna Endrődi's excavation 1999). According to those finds, the Baden culture primarily cultivated hulled cereals, barley, einkorn and emmer (*Fig. 71*). These village sites were populated by a settled population whose primary activity was the cultivation of bread wheat. Naked club wheat (*Triticum aestivum* subsp. *compactum*) occurs sporadically, while common bread wheat (T. ae. subsp. vulgare) does not occur at all. Einkorn and emmer fork fragments (*furca bicornis*) found along with weed remains indicate that cereals were cleaned after the harvest, probably by wind, and that the waste from this operation was burned (ENDRŐDI – GYULAI 1998–2000).

The Csepel Waterworks site yielded a large quantity of carbonised oak acorns (*Quercus* spec.) (*Figs. 70, 72–73*). This indicates that during the era of the Baden culture, there must have been significant mixed oak forests still in the area. Cracked acorns were collected and probably stored for consumption. We found no cups or husks, which is further evidence of the practice of cleaning. Based on phytogeological considerations, it is probable that the acorns were fruits of the common oak (*Quercus robur*). Oak acorns have been known as a sustenance food since ancient times. Ethnographical data suggest that people travelling in forests were still fond of eating them in the 19th century. According to ethnographic analogies, if oak acorns are buried in the soil for an extended period, they loose some of their bitter taste. Cracked acorn can be roasted over fire and eaten as a delicacy or even ground into flour. The same is also applicable to European beech acorns (*Fagus silvatica*) found at the same site (a delicacy when roasted) and the once tasty fruit of the single-seed hawthorn (*Crataegus monogyna*), which occurs at the f forest edge locationss (*Figs. 74–75*).

Unlike the Hungarian sites, those of the Baden culture in Slovakia are archaeobotanically well researched (CHEBEN – HAJNALOVÁ 1997). Comparison of the botanical material from nine published sites with the cultivated species found at the Csepel Waterworks site, a number of important observations arise:

- there is a similarity of culitavted cereals found at the Slovakian and the Hungarian sites. The predominant sources of carbohydrates were hulled cereals. Spelt and common millet occur in Slovakia, but are missing in Hungary. In Slovakia (e.g. Kamenín) emmer is most frequent with a variable complement of einkorn. This is reversed in the case of the Csepel Waterworks. Both locations have barley and club wheat, though they were not significant at either site. It is highly likely that plants, at least wheats, were sown mixed.
- It seems to be an important difference that in the area of the Csepel Waterworks, we found no seeds of any leguminous plants in the samples from any of the four fireplaces. This is certainly not the typical situation, it is more likely to be attributable to the differences ication of the soil samples, While the Slovakian samples were from waste pits or houses, Hungarian samples were collected from the vicinity of fireplaces, where only grains of cereals and oak acorns were dried and roasted, while leguminous plants were not. The cultivation of cereals always assumes the presence of row crops (leguminous plants, root vegetables). That was probably the case in the Baden culture as well. For example, a stock pit found at the Jelšovce site contained peas and bitter vetch in addition to cereal grains,. The same location also yielded seeds of the oil crop, flax.

Although the valuable archaeological collection of the Balaton Museum at Keszthely was destroyed almost completely by an air raid during World War II while it was being removed from the building, a small quantity of archaeological material did survive.. This includes cereal remains collected by Árpád Csák at Fenékpuszta in 1905, within the area of the fort and identified as Bronze Age remains. Due to such circumstances, we can only accept this dating provisionally, and we do not know which era of the Copper Age the remains originate from. Grain remains that survived the war were processed by Miklós FÜZES (1990). According to his findings, cultivated hulled wheats dominared, most of all einkorn and emmer. Grains of other cereals were also found sporadically: spelt, naked barley, rye. The last of those would be the oldest find of rye in the Carpathian Basin. There was an unusually high proportion of weed seeds among the cereal grains: korn-rade (*Agrostemma githago*), rye brome (*Bromus secalinus*), barren brome (*Bromus sterilis*) and some other, unidentifiable variety of brome grass (*Bromus* spec.), fan-hen or white goosefoot (*Chenopodium album*) and long-pod poppy (*Papaver dubium*).

Fig. 65. Common bulrush (*Scheonoplectus lacustris*) from a pit of the Lengyel-Balaton-Lasinja culture in Zalaszentbalázs-Szőlőhegyi mező. Photograph by the author.

Fig. 66. Charlock (*Sinapis arvensis*) seeds from Öcsöd-Kendereshalom, Bodrogkeresztur culture settlement, Inventory of the Hungarian Agricultural Museum, Budapest.

Fig. 67. Cornelian cherry or dogwood (*Cornus mas*) stones from the Kompolt-Kistéri hamlet Late Bronze Age pit-oven. Photograph by the author.

Fig. 68. Woodland European grape (*Vitis vinifera* subsp. *silvestris*) seeds from the Kompolt-Kistéri hamlet Late Bronze Age pit-oven. Photograph by the author.

Fig. 69. Hazelnut (*Corylus avellana*) shell fragments from the Kompolt-Kistéri hamlet Late Bronze Age pit-oven. Photograph by the author.

Fig. 70. Oak (*Quercus* spec.) whole acorns from the Csepel Waterworks Late Copper Age Baden culture pit fireplace. Photograph by Ferenc Hegyi.

Fig. 71. Einkorn (*Triticum monococcum*) grains from the Csepel Waterworks Late Copper Age Baden culture pit fireplace. Photograph by Ferenc Hegyi.

Fig. 72. Oak (*Quercus* spec.) half acorns from the Csepel Waterworks Late Copper Age Baden culture pit fireplace. Photograph by Ferenc Hegyi.

Fig. 73. Oak (*Quercus* spec.) acorns fragments from the Csepel Waterworks late Copper Age Baden culture pit fireplace. Photograph by Ferenc Hegyi.

Fig. 74. Beech (*Fagus* spec.) acorns from the Csepel Waterworks Late Copper Age Baden culture pit fireplace. Photograph by Ferenc Hegyi.

Fig. 75. Single-seed hawthorn (*Crataegus monogyna*) stones from the Csepel Waterworks Late Copper Age Baden culture pit fireplace. Photograph by Ferenc Hegyi.

7.3. The Bronze Age

7.3.1. Early Bronze Age plant cultivation

Around 3000 BC a new era of the Holocene, also known as the Flandrian interglacial (which had started around 8200 BC), commenced and lasted until 800 BC. The era is called the Subboreal phase, which, from the archaeological perspective, included the end of the Copper Age and the entire Bronze Age (WILLERDING 1983a). The initially cool and rainy weather of the Subboreal, or as it is called in the Carpathian Basin, the 1st European beech phase, became somewhat drier by the middle of the period. Among tree species, the proportion of European beech decreased while oak and hazelnut increased.

The Great Plain in the Copper-Age – Bronze Age transitional period was characterised by alternating deciduous forests with thick undergrowth and steppes with gallery forests. As a result of the longer period of cooler, wetter climate, steppes with gallery forests were slowly replaced by oak mixed with European beech. The mass intrusion of hornbeam and European beech began and extensive flood plains of rivers developed shrubby alder gallery forests (LACZA 1991).

The areas west of the Danube were covered by extensive European beech mixed forests, with some intrusions of oak, elm and linden in places. Their closed foliage allowed only little light to penetrate, so their undergrowth was sparse. Shrub and herb layers grew richer only where gaps in the foliage were made by fallen trees.

Infiltration of the Carpathian Basin by the great Pitgrave culture began at the time of the Late Copper Age Baden culture and ended with the arrival of the Early Bronze Age Makó culture. We can assume that the locals and the immigrants formed a symbiotic relationship. Groups of the Somogyvár-Vinkovci culture were penetrating from the south to the north. The Makó culture developed the early Nagyrév cultural complex in the center of Hungary. Concurrently, the Nyírségi culture persisted in the east, while west of the Danube the Somogyvár-Vinkovci culture survived. The western Bell-Baeker Csepel Group with their bell-shaped pots made their first appearance along the Danube near Budapest (KALICZ-SCHREIBER 1974; KALICZ-SCHREIBER – KALICZ 1998).

At present, the only known botanical material from the era of the Somogyvár-Vinkovci culture is from Pécs-Nagyárpád (P. HARTYÁNYI – NOVÁKI – PATAY 1967–68) (*Fig. 76*). All the cultivated plants found (einkorn, emmer, club wheat, six-rowed barley, common lentil, pea) were already known and cultivated in the Neolithic. Early Bronze Age plant remains from the northeastern part of the country were only found at the 1993–94 excavation at Mosonszentmiklós-Pálmajor; including six-rowed barley, emmer, common bread wheat and pea.

Of the two highly significant Early Bronze Age cultures of the Great Plain, the Nagyrévi culture had connections to the south, while the Hatvan culture had connections to the east. The tell settlements of the Hatvan culture placed emphasis on plant cultivation, though they also had significant animal husbandry. Cultures to the west of the Danube present a more uniform picture regarding their lifestyles, if not their origins. The Kisapostag culture combined animal husbandry and plant cultivation.

As a result of recent research, we are now able to report two botanical finds associated with the Bell-Baker Culture population (*Fig. 77*). Archaeobotanical samples obtained from the excavation of Rózsa Kalicz-Schreiber in Csepel, at Hollandi st. No. 33 in 1973 were finally processed after a quarter century in storage. Fortunately, storage in a cool, stable context apparently had no negative effects on preservation of remains. This material is dominated by cereal grains (GYULAI 2002). Barley and bread wheat grains occur in roughly equal quantities. The majority of the barley is six-rowed, but there are also some naked barley without husks. The only representative of wheat is emmer. The material contains seeds/fruits of a large number of water/marsh plants. There were from regularlyflooded areas near the settlement. Common bulrush grows among common reeds; water pepper is a waterside plant; and, bog-rush is a plant of marshy fields. It is possible that the large quantity of common bulrush seeds were used as a cereal supplement (c.f. the Zalaszentbalázs-Szőlőhegy field).

In 1999, Anna Endrődi found a new site containing bell-shaped pottery in the area of the Szigetszentmiklós Waterworks. We flotated the soil samples and recovered grains of three cereals, six-rowed barley, common millet and emmer (GYULAI 2002). Carbonised wheat-worm galls found at the site will provide an important piece of information for the history of plant protection in Hungary. These galls, filled with larvae, develop where the wheat-worm (*Tilenchus* or *Anguina tritici*) inject grains .

In 1999, Gábor Lőrinczi found an Early Bronze Age cemetery in the right-of-way of an extension of the M5 motorway at Kiskundorozsma. Archaeobotanical material from this burial ground is modest, but highly significant, for it is dated to the Early Bronze Age of Hungary. One cereal grain, found in fragments precluding species identification, must have been highly significant for the deceased. It was placed in the grave in a separate vessel implying a puzzling ritual importance conferred on a single roasted grain in burial vessel.

The 2001 excavation at Albertfalva, led by Anna Endrődi, has furnished new information about prehistoric lifestyles. The majority of houses, workshops and pits found originate from the Early Bronze Age, the more recent period of the Bell-Baeker Csepel Group (last third of the 3rd millennium BC) (ENDRŐDI 2002). The cool, rainy weather of the Subboreal grew somewhat milder during this period. Following the previous period of more mobile animal husbandry, stable settlements based on the production of cereals and other plants were restored.

Flotation of two dozen soil samples from various contexts resulted in identification of almost fifteen hundred remains from 55 plant species (GYULAI 2002). The majority of diaspores were cereals, the most important being einkorn followed by emmer and barley. Some naked barley, clearly two-rowed variety rather then the result of threshing, was found among remains of hulled barley grains. Sporadic occurrences of hexaploid common bread wheat were also found. It is highly likely that the people living here were not cultivating individual species of cereals separately, but in the ancient, mixed fashion. The assemblage also contained seeds of high protein-content leguminous plants: *pea* and *horsebean,* adding clear evidence of intensive agricultural activities (*Fig. 78*).

The majority of plant remains from the soil samples of two "workshops " were quite recognisable threshing waste: husks, earspindle fragments, weed seeds, etc. The high number of weed species (17 species!) indicates that the cereal was not cleaned when it was burned. The following are only some of the weed species found: korn-rade (*Agrostemma githago* L.), rye brome (*Bromus secalinus* L.), quackgrass (*Agropyron repens* (L.) P. B.), black-bindweed (*Fallopia convolvulus* (L.) A. Löve), barnyard grass (*Echinocloa crus-galli* (L.) P. B.), false cleavers (*Galium spurium* L.), red horned-poppy (*Glaucium corniculatum* (L.) Rudolph), field cow-wheat (*Melampyrum arvense* L.), ball mustard (*Neslea paniculata* (L.) Desv.) and field madder (*Sherardia arvensis* L.).

In addition to the large quantity of weed remains we also found intact, small "escapee" grains of some cereals (primarily einkorn and emmer) in the threshing waste. The tailings were probably brought from the threshing yard to use as fuel. The composition and arrangement of the remains suggests that threshing and the removal of husks took place in a single process, using a method that is at present unknown. This inference is based on the fact that, during threshing, ears of hulled cereals separate into spicules, and additional mechanical action is required to remove the husks – unlike in the case of naked wheats, e.g. common bread wheat, where the husks come off during threshing. We do not know the process, but it is easy to imagine that the cereal, harvested using a sickle, was placed in a threshing yard. The large number of fork (*furca bicornis*) remains indicates that the removal of husks was effected using some sort of sharp, strong mechanical tool, such as a threshing sledge with silex blades. The force used to remove husks is indicated by the presence of barley spicules which are more difficult to process in addition to the einkorn and emmer forks. Tailings and weed seeds were separated by winn owing. The remaining tailings can be used as fodder, for lightening clay used for pottery and, as evidenced here, as fuel.

The large number of weed species is surprising and unusual in this early era. Moreover, many of them occur for the first time in the history of cereal production in Hungary. This is a strong counterargument to the earlier notion that at the beginning of cereal cultivation through to the Early Bronze Age, the *Scleranthо-Secalitea* association (cereal weeds) only had a small number of species. The majority of archaeobotanical finds are from grain pits and from houses, where the cereals had already been cleaned. Here, we see quite the contrary – the refuse of threshing/winnowing. For this reason, the Albertfalva site is particularly significant from the perspective of the history of synanthropisation. Among the many weed species present is the earliest occurrence in Hungary of wild einkorn (*Triticum* cf. *boeoticum* Boiss.) This species can by no means be regarded as a component of the natural flora, but rather is evidence of status of agricultural progress of these prehistoric harvesting peoples. Despite the long history of crop selection and improvement, the wild ancestor still occurs among cultivated einkorn.

On the other hand, the number of species present from the natural environment is small. When looking at remains of threshing waste, this is understandable. The only wild species whose remains could occur here are those that actually grew spontaneously in the threshing yard or in the settlement itself. The habitat requirements of such species found here provide clues about the environment and the arable land(s) of the period (*Figs. 79–81*). Arable land(s) where plants were cultivated had an average rather than extreme water supply. However, lower-lying areas existed which were flooded, if only periodically. The common spike-rush (*Eleocharis palustris* agg.) is characteristic of higher-lying areas, while the tasteless water-pepper (*Polygonum mite* Schrank) is a waterside pioneer and marshland/flood plain weed. Harvested cereals were taken from the fields around the settlement to a threshing yard, where threshing, the removal of husks and winnowing were performed. The threshing yard could only have been in periodic used for many other field species, both those needing an average water supply and those preferring dry settings, were found in the samples including brown knapweed (*Centaurea jacea* L.), hare's-foot clover (*Trifolium arvense* L.) and basil thyme (*Calamintha acinos* (L.) Clairv.). The site also furnished the earliest occurrence of goatgrass (*Aegilops* spec. cf. *cylindrica* Host) as a constituent of Hungarian flora. The remains of a ruderal association may originate from a threshing yard or any other trodden areas (paths, sides of ditches, the vicinity of houses). Their number is small. Dwarf mallow (*Malva neglecta* Wallr.), knotgrass (*Polygonum aviculare* agg.), wild barley-grass (*Hordeum murinum* L.) stinking nightshade (*Hyoscyamus niger* L.), and cat-mint (*Nepeta cataria* L.), which all belong in this category, also attest to a turn to a drier climate.

In the summer of 2002, as the excavation of the Bell-Baeker Csepel-Group settlement at Albertfalva continued, about a dozen carbonised fruits of crab apple (*Malus silvestris*) were found in a pot. The dried fruit, which was once stored for consumption, was probably gathered in the surrounding forests.

7.3.2. Middle Bronze Age plant cultivation

In view of the relatively rich archaeobotanical material recently recovered from the Middle Bronze Age in Hungary, it is necessary to attempt a summary of the plant cultivation along with associated issues . As it is not possible to consider plant cultivation separately from the natural environment, the study of climate and vegetation are both indispensably important. At the same time, we also try to draw inferences concerning the lifestyle of Bronze Age people.

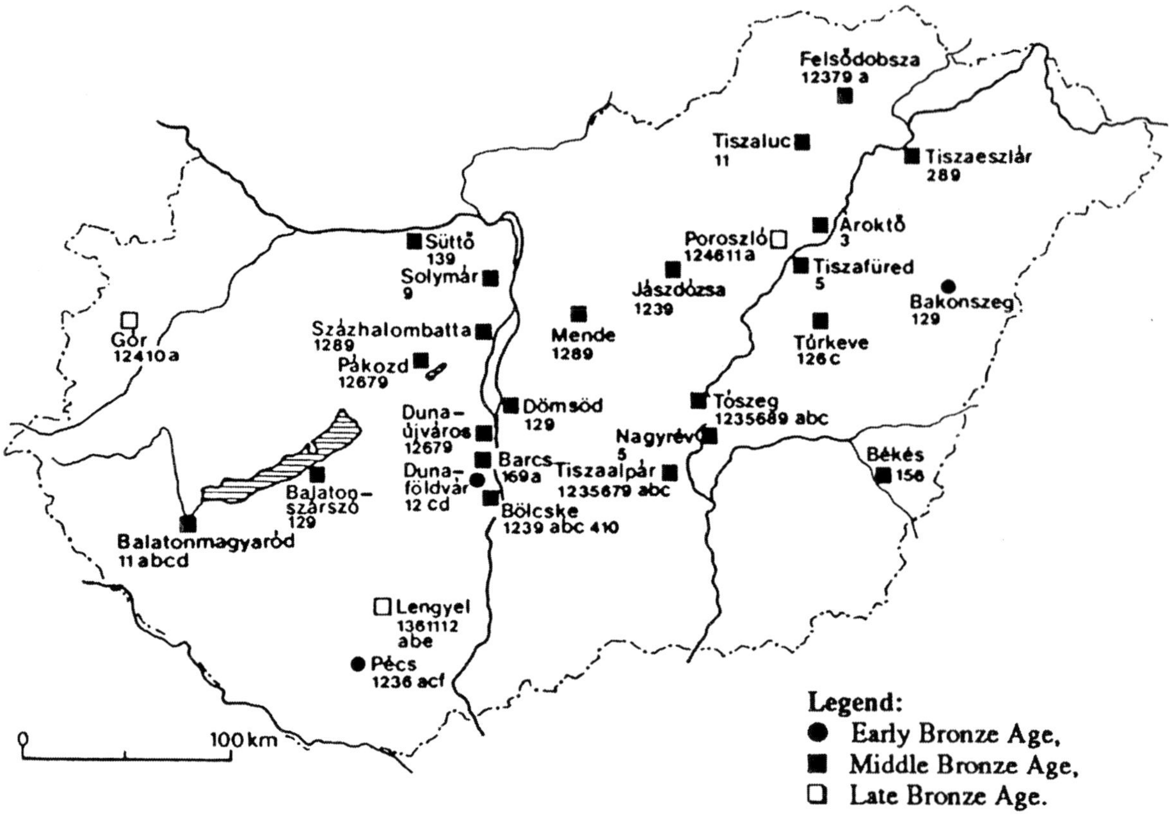

Species register:

1 = eincorn *(Triticum monococcum)*
2 = emmer *(Triticum turgidum subsp. dicoccum)*
3 = common and dwarf wheat
(Triticum aestivum subsp. aestivum and Triticum aestivum subsp. compactum)
4 = spelt-wheat *(Triticum aestivum subsp. spelta)*
5 = wheat species *(Triticum sp.)*
6 = common barley *(Hordeum vulgare subsp. hexastichum)*
7 = common barley *(Hordeum vulgare var. nudum)*
8 = long-eared barley *(Hordeum vulgare subsp. distichum)*
9 = barley *(Hordeum vulgare)*
10 = rye *(Secale cereale)*
11 = true millet *(Panicum miliaceum)*
12 = flax *(Linum usitatissimum)*
a = lentils *(Lens culinaris)*
b = chickling vetch *(Lathyrus sativus)*
c = peas *(Pisum sativum)*
d = ervil *(Vicia ervilia)*
e = horsebean *(Vicia faba)*
f = gold of pleasure *(Camelina sativa)*

Fig. 76. Plant finds from the Bronze Age.

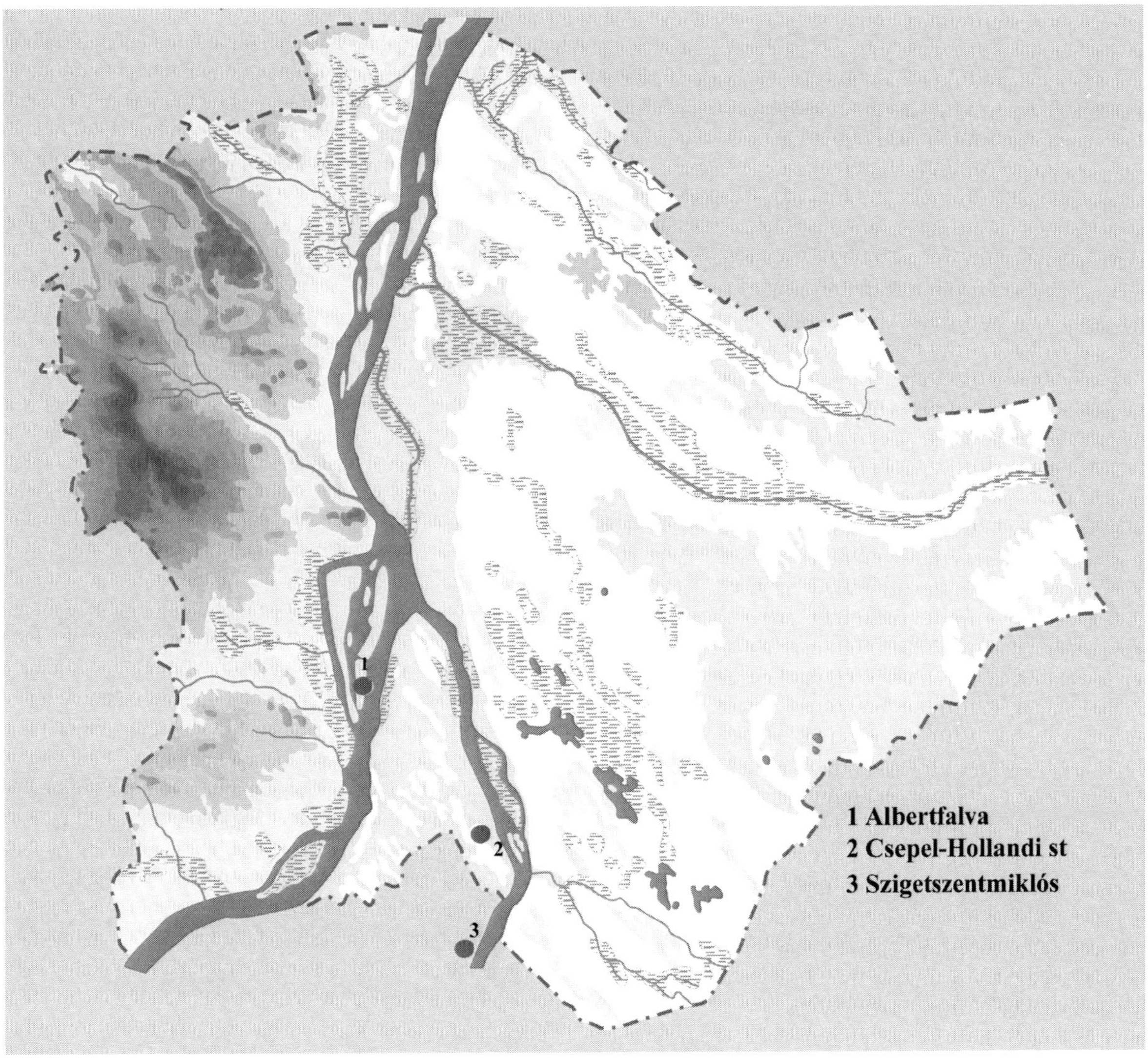

Fig. 77. Settlements of the Bell Beaker culture.

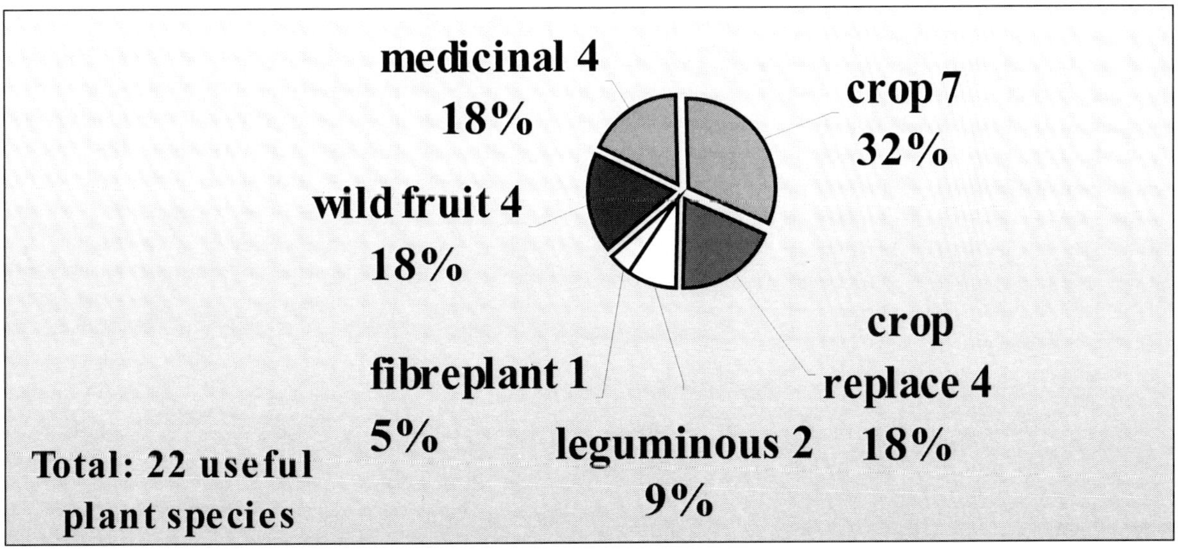

Fig. 78. Distribution of useful plant species in the Bell Beaker settlements.

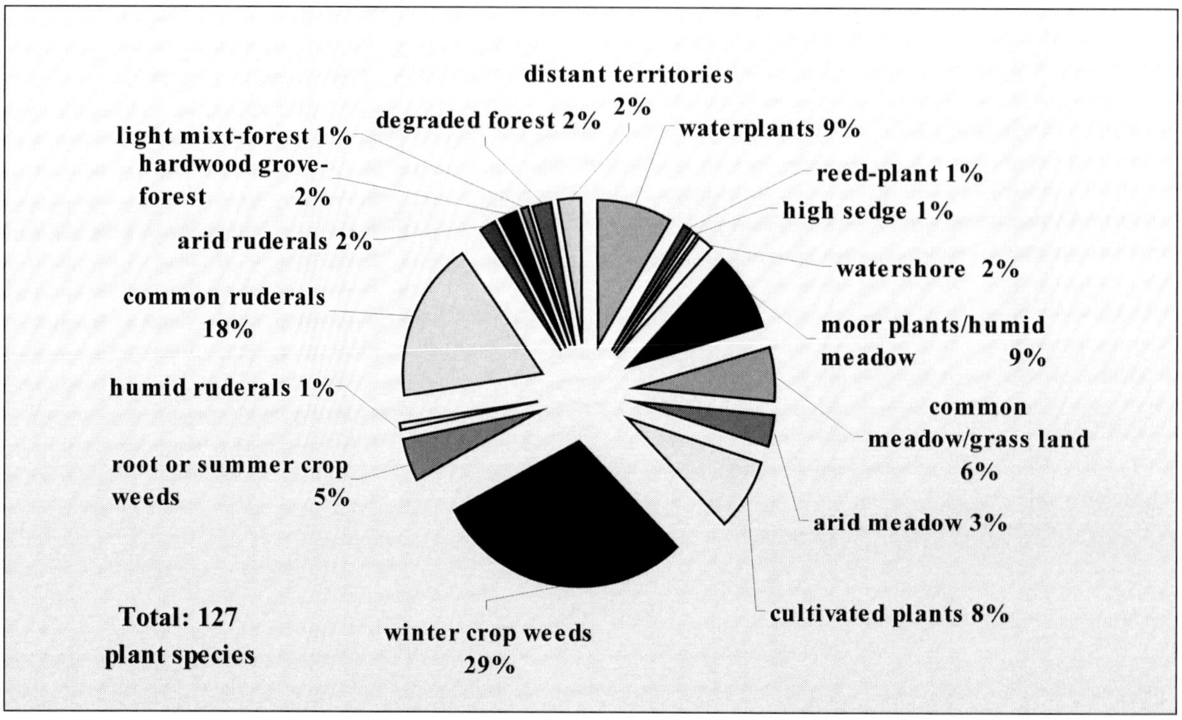

Fig. 79. Ecological distribution of species in the Bell Beaker settlements.

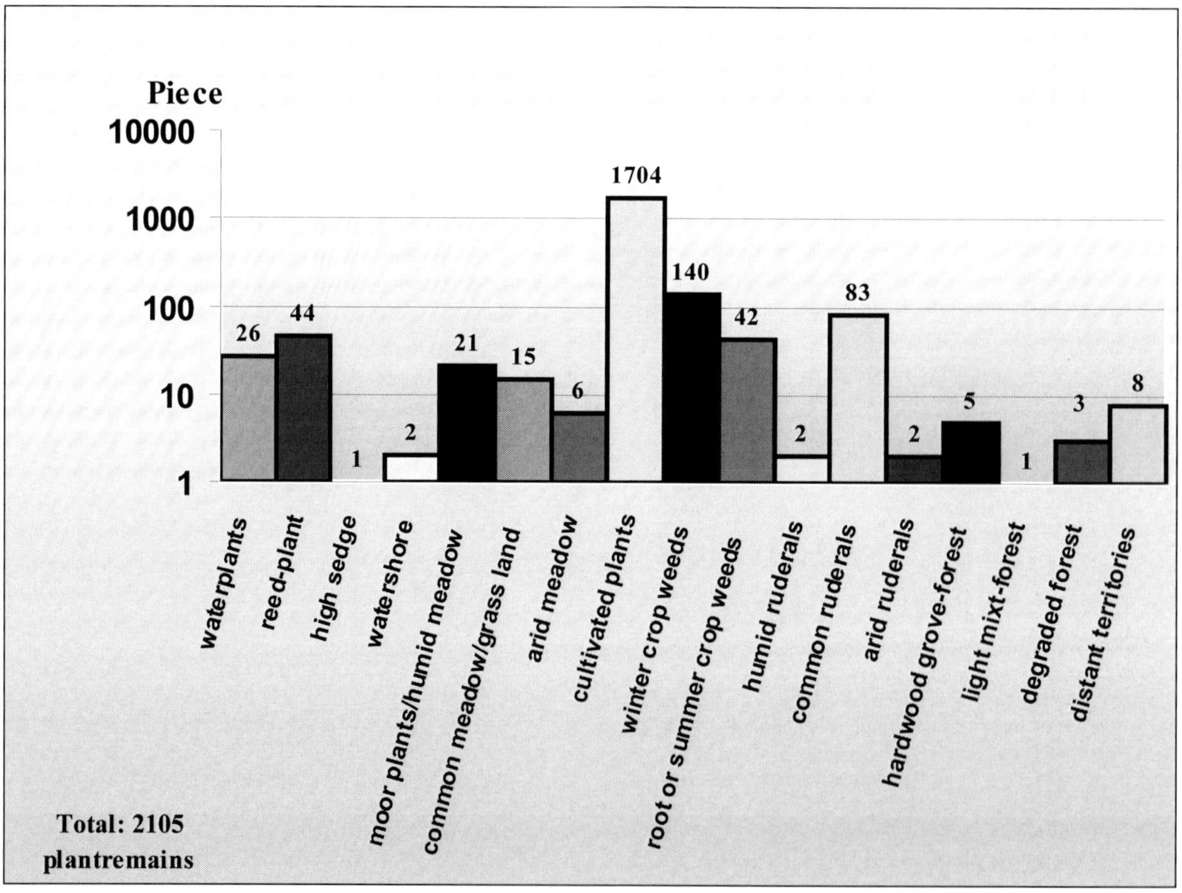

Fig. 80. Ecological distribution of the plant remains in the Bell Beaker settlements.

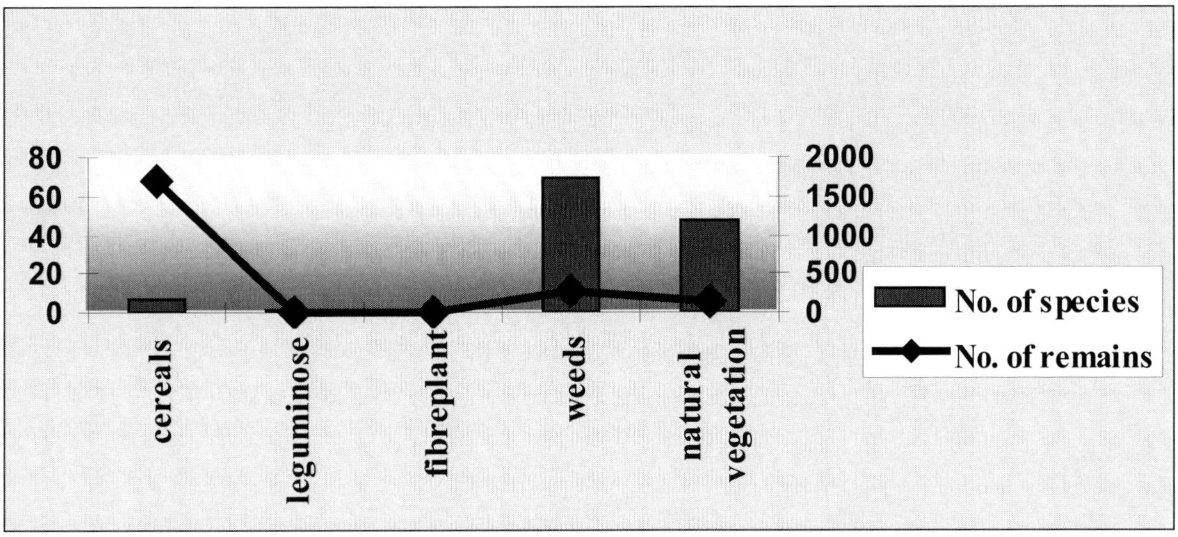

Fig. 81. Distribution of plant remains from the Bell Beaker settlements.

Fig. 82. Common bread wheat (*Triticum aestivum* subsp. *vulgare*) grains from Ároktő-Dongóhalom, Hatvan-Füzesabony culture, Inventory of the Hungarian Agricultural Museum, Budapest.

Fig. 83. Creeping thistle (*Cirsium arvense*) acheniums from Ároktő-Dongóhalom, Hatvan-Füzesabony culture, Inventory of the Hungarian Agricultural Museum, Budapest.

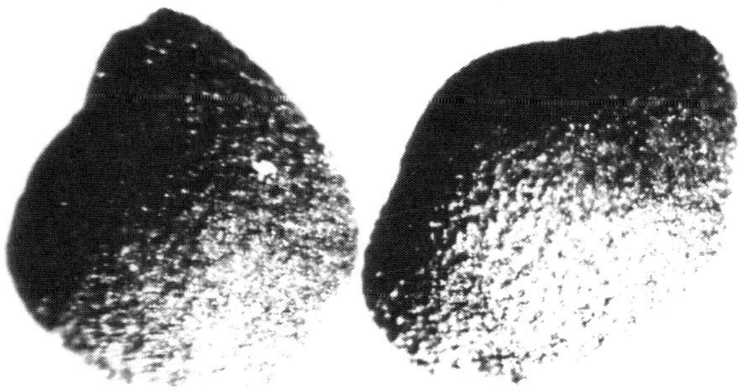

Fig. 84. Field bindweed (*Convolvulus arvensis*) seeds from Ároktő-Dongóhalom, Hatvan-Füzesabony culture, Inventory of the Hungarian Agricultural Museum, Budapest.

Fig. 85. Hedge buckwheat (*Fallopia dumetorum*) nutlets from Ároktő-Dongóhalom, Hatvan-Füzesabony culture, Inventory of the Hungarian Agricultural Museum, Budapest.

By the Middle Bronze Age, in addition to mobile animal husbandry that had characterised the Early Bronze Age, the settled, plant cultivator and animal keeper lifestyle came to predominate. The change began first in the Great Plain reaching the region west of the Danube somewhat later, but still limited to the eastern part of the region.

Conditions for plough agriculture developed in the Carpathian Basin by the middle of the 3rd millennium: segetal plants and domesticated beasts of burden were both present, while carts hitched teams of draught animals were used. Therefore it is a justified assumption that people were using ploughs by the end of the millennium (BALASSA 1973). In the Balkans as well as in East-Central Europe, einkorn was the main bread wheat species, while in Central and Western Europe it was emmer.

Individual cultures living agricultural/plant cultivating lifestyles are specifically associated with soil types and topographic areas. The Nagyrév culture, and its descendant, the Vatya culture, inhabited the central areas of the Great Plain and the eastern edge of the region west of the Danube and the Hatvan culture in the northern part of the Great Plain, with its descendant, the Füzesabony culture all used fertile chernozem soils that were highly suitable for agricultural cultivation (*Figs. 82–85, 88–90*).

In their choices of settlement location, the peoples of Bronze Age cultures showed a distinct preference for using naturally protected high-ground areas, loess highlands protected by the double system of terraces of the Danube bank. The earthworks of the Vatya culture do not cross the boundaries of the "Mezőség" area with its fertile soils and the northern part of the region between the Danube and the Tisza rivers, areas that are highly suitable for agriculture, a phenomenon that is again indirectly related to plant cultivation.

The majority of the botanical material from the Middle Bronze Age earthworks described below were archaeobotanically processed by Borbála P. Hartyányi and Gyula Nováki (P. HARTYÁNYI – NOVÁKI – PATAY 1967–68), with important additions by Brigitta K. Berzsényi and the present author in recent years (K. BERZSÉNYI – GYULAI 1998). As a result of all this research, our image of the plant cultivation cultures of the Middle Bronze Age has been considerably refined (GYULAI 1993).

The fortified Bronze Age settlements along the Danube have been known since the 19th century. These settlements, protected by horseshoe-shaped battlements and earthworks have an estimated area of 16–20 hectares. The majority of the botanical finds from the earthworks of the Vatya cultures were found in sporadic occurrences: Baracs-Bottyán battlement, Dunaújváros- (Dunapentele) Kosziderpadlás, Mende-Leányvár, Pákozd-Vár, Solymár-Várhegy, Százhalombatta-Téglagyár (*Fig. 86*). The majority of those finds are from such special contexts as fireplaces, from pits and from pots, so they can by no means be regarded as representative. Systematic sampling took place only in the case of three excavations: Bölcske-Vörösgyír, Százhalombatta-Földvár and Tiszaalpár-Várdomb, yet a comparison of the cultivated species found at all eight sites is instructive (ENDRŐDI – GYULAI 1999) (*Figs. 90–120*). The majority of the seeds and fruits found are from cereal species, followed by the remains of leguminous plants. They also include a number of remains from weed and wild fruit species as well as species of the natural vegetation of the period (*Figs. 121–131*). These earthworks-dwellers conducted lively agricultural activities. They did not specialise in the production of a single cereal species but based their crop production "on several pillars". The importance of cultivated plants within the lives of the settlements is attested best by the "stock", consisting of large numbers of cereal grains that were found at all the sites, in particular around fireplaces, in pots and in storage pits. The most important cereals were six-rowed barley, einkorn and emmer (*Figs. 184–188*). It is certain that barley was used for human consumption. Only barley was found at Solymár-Várhegy, while 98% of all the seeds were again barley at Baracs-Bottyánsánc. Most of the grains were six-rowed barley, but there were also some occurrences of two-rowed barley (e.g. Mende-Leányvár).

Among wheats, husked wheats, in particular einkorn and emmer, played a dominant role. Gyula NOVÁKI (1969) made an interesting observation after charting the bread wheat remains on a map (*Fig. 87*). He distinguished two large areas. The first includes the eastern and north-eastern corner of the region west of the Danube and the northern part of the region between the Danube and Tisza rivers. The dominant bread wheat species is usually einkorn, with emmer and common bread wheat playing only a secondary role. The reason for this configuration is as yet unknown. However, we are able to provide further evidence for his proposal, as in the case of Bölcske-Vörösgyír almost three-quarters of the cereal remains found are einkorn (K. BERZSÉNYI – GYULAI 1998). The only counterexample was furnished by Százhalombatta-Földvár, where the number of emmer grains was twice that of einkorn grains.

The second area covers the areas on the right bank of the Tisza and the eastern part of the region between the Danube and Tisza rivers. The dominant bread wheat species here is emmer. It is followed by common bread wheat, while einkorn usually only plays a minor role.

In the area where the above two large areas overlap (Tószeg, Mende, Tiszaalpár) einkorn and emmer occur in equal proportions (*Figs. 135–138, 140*).

All of this indicates that there were some geographical divisions in the bread wheat species cultivated that are not evidence of cultural differences, as the Bronze Age cultures in Hungary were not really isolated from each other. This is quite an achievement by the Bronze Age plant cultivators surprising astute in adapting to the microclimatic and edaphic properties of their environments. They were well aware of the most promising plants to cultivate in any given setting.

The prehistoric settlement of Felsődobsza-Várdomb was excavated six times in the period 1857–77. Seeds from those excavations were identified by Imre Deininger. H e found einkorn, common bread wheat, barley and the common lentil (*Figs. 132–134*). Those species that were previously assigned to the Neolithic or the Late Neolithic (in: BERTSCH – BERTSCH 1949) were reclassified after the verifying excavation of Ferenc Tompa (1934–35) to the Middle Bronze Age (cf. P. HARTYÁNYI – NOVÁKI – PATAY 1967–68). At that time, grains of emmer were also found. Imre Deininger, who had studied the excavations of Oswald Heer near Zurich, also found the grains of the species he simply called "ancestral wheat" (Triticum vulgare antiquorum Heer). Indeed, he found it not only at this site, but also at the Late Neolithic site at Lengyel. The ancestral wheat described by the pioneers of archaeobotany, who were later very simply regarded as inexperienced, were clearly labelled a mistake by the archaeobotanists of later years who pronounced the grains to have come from the hexaploid club wheat. It was KISLEV (1979–80) who pointed out that this bread wheat is significantly different from club wheat. It did exist, if only for a short period, and it was identical with the ancient naked wheat from the earliest archaeological eras (7th–9th millennia BC) that also only existed for a brief period. The oldest known naked bread wheat is called Triticum parvicoccum Kislev, and it was tetraploid, not hexaploid in form.

The sites at Bölcske-Vörösgyír (excavation led by Gyula Nováki) and at Százhalombatta-Földvár (excavations led by Ildikó Poroszlai) also yielded grains of the third hulled bread wheat, spelt, but only in low proportions, 1–3% (*Figs. 139, 141–147, 149*). The same applies to common bread wheat (Bölcske-Vörösgyír, Tiszaalpár-Várdomb, Százhalombatta-Földvár) and club wheat (Tiszaalpár-Várdomb, Gyula Nováki and István Bóna's excavation) (*Fig. 148*). These wheat species were not widespread, but were cultivated to a limited extent. With such a low number of grains it is quite possible that people cultivated some sort of "mixed cereal". We note here that we also found grains of spelt in one of the Bronze Age pits opened by Attila Horváth in Corvin Square in the 1st District of Budapest in 1997.

At the Middle Bronze Age site of Ároktő-Dongóhalom, in one of the pits of the tell settlement belonging to the Füzesabony culture excavated by Tibor Kemenczei, rye grains were found (in a ratio of 1:10) alongside common bread wheat (P. HARTYÁNYI – NOVÁKI – PATAY 1967–68). That is the second occurrence of rye in the Carpathian Basin (cf. the Copper-Age site at Keszthely-Fenékpuszta).

Archaeobotanical remains rarely provide evidence to determine if various cereal species were grown together or separately. We usually find them mixed together. But, it is highly likely that, in the past, mixed sowing of various bread wheat species and perhaps even other cereals was common (see our comments in section 7.9.1). Mixed sowing promoted the propagation of certain weed species, which allowed for their selection (see secondary cultivated plants in the next chapter).

There is plentiful evidence for the continued practice of the mixed cultivation of cereal species. The villagers of the Taurus mountain in Asia Minor were still growing mixed wheat and barley in the first half of the 20th century (TAGÁN 1941). It is interesting that the practice of mixed growing of various cereal types did not disappear entirely in the second half of the 20th century. On the Swedish island of Gottland, the fields, largely populated by common bread wheat, still had some einkorn, emmer and spelt (DANERT et al. 1981). According to László Holly (personal communication), only a decade ago, common bread wheat and durum wheat were grown mixed in areas of Syria near the Mediterranean shore. In humid areas, common bread wheat dominated, while more arid areas were dominated by durum wheat. In Algeria, self-sufficient family farms of the Atlas mountains still grow two-rowed and six-rowed barley mixed with common bread wheat and durum wheat.

Common millet only occurs in two sites: Bölcske-Vörösgyír and Százhalombatta-Földvár, and even there only sporadically (GYULAI 1996b) (*Fig. 150*). That was the period in which the cultivation of common millet began to spread in the Carpathian Basin.

Small seed lentil, pea, bitter vetch, horsebean and common vetch were leguminous plants of the earthworks (*Figs. 152–154*). Among the leguminous plants, small seed lentil are the most numerous, in some instances contributing half the total number of seeds (Bölcske-Vörösgyír, Tiszaalpár-Várdomb). They are followed in order of importance by (wild) *pea*. Its cultivation was widespread, for in addition to Bölcske-Vörösgyír, Százhalombatta-Földvár and Tiszaalpár-Várdomb, it was also found among the botanical remains from the Pákozd-Vár site.

It is interesting to note that wherever there are large numbers of barley grains, the number of field pea seeds is also large (e.g. Tiszaalpár-Várdomb). It is possible that this indicates the roots of the tradition, still alive on many peasant farms, of sowing peas over barley.

Bitter vetch (*Vicia ervilia*), this leguminous plant that is poisonous when raw, was a widespread species in the Middle Bronze Age, but was no longer cultivated after the end of the Bronze Age. We found very many *bitter vetch* seeds at the tell settlement of Bölcske-Vörösgyír, but they were also found at the Pákozd-Vár and the Százhalombatta sites. The same locations also yielded horsebean (*Vicia faba*). Seeds of common vetch (*Vicia sativa*), which is also poisonous when raw, were found only at the Baracs-Bottyánsánc site.

Although we have only documentary evidence, we must accept provisionally that in 1876, common lentil (*Lens culinaris*) and chick-pea (*Cicer arietinum*) were also found at the Tószeg-Laposhalom site (PULSZKY 1897). We have more reliable information about seeds found in the same location in 1906 by Georg LINDAU (1917): common lentil, pea and grass pea wine were found. Seeds excavated in 1928, belonging to the end of the Nagyrévi culture were identified by Zoltán Zsák in 1949. Reviewing the material, Zdenek TEMPÍR (1964) also found bitter vetch.

We can infer cultivation and multipurpose utilisation of flax (*Linum usitatissimum*) from thread remains found in a grave of the Bronze Age burial ground at Szőreg (GREGUSS 1937) and a seed find from Tiszaalpár (P. HARTYÁNYI – NOVÁKI – PATAY 1967–68). We note that seeds of the flax plant can provide not only oil, but also a pigment, and may even be prepared as food. And flax was not the only oil crop of the era. In other areas of Europe, Bronze Age people cultivated gold-of-pleasure and opium poppy to make oil. The seeds of gold-of-pleasure (*Camelina sativa*) were found at the Pécs-Nagyárpád site of the Somogyvár-Vinkovci culture (P. HARTYÁNYI – NOVÁKI – PATAY ibid.). Remains of opium poppy (*Papaver somniferum*) have not yet been found.

At the Bölcske-Vörösgyír site we had the advantage of being able to establish the temporal sequence of samples. This allowed us to observe differences of plant cultivation within the prehistoric culture (K. BERZSÉNYI – GYULAI 1998). Seven cereal species were obtained from the first phase of the Vatya culture: two-rowed and six-rowed barley, naked barley, common bread wheat, emmer, einkorn, spelt). Emmer was the dominant species, while einkorn is diminished.

The species composition of cereals from the second phase of the Vatya culture differs. Two-rowed barley is no longer present alongside six-rowed and naked barley. In addition to large numbers of hulled wheat grains, such as einkorn, emmer and spelt, common bread wheat also appears. We have found a very special botanical phenomenon, the "two-grained" einkorn in both phases. Although the basic form of the species is more or less characterised by the "two-grained" feature, this is the first archaeobotanical proof of its existence in Hungary.

From the second phase of the Vatya culture, the quantity of cereal species increased. The majority of the grains are einkorn, followed by emmer. Spelt is again only present in small proportions. Only a few grains of common bread wheat were found.

The largest number of cereal species was found in the third phase of the Vatya culture. All species present in earlier phases were still present, and the number of cultivated cereal species expanded. The most important addition is a very important meal plant, common millet. In addition to six-rowed barley, two-rowed barley was at least as important. This evidence projects an image of a population with sophisticated agricultural/plant cultivation know-how. Their versatile use of cereals meant that they had a secure livelihood even in years of bad harvests.

No seeds of leguminous plants were found in the first phase of the Vatya culture. Common lentil and pea, along with bitter vetch, which had not been previously cultivated, appeared in the second phase and became commonplace in the third. The surprisingly large numbers of common lentil seeds were classified into small seed and large seed categories.

As glumes and earspindle fragments of cereals and weed seeds occur in very large numbers at some of the sites (Bölcske-Vörösgyír, Százhalombatta-Földvár), it is conceivable that those samples are from threshing or processing waste. (In the case of hulled wheats, threshing only breaks the earspindles up into spicules, the husks usually remain on the grains.) After threshing, the cereals were probably cleaned by winnowing.

In contrast to the sites along the Danube, which are rich in plant remains, we found only one site in the west of the region to the west of the Danube that contained seeds. The samples from the 1990 excavation of András Figler at Győr-Szabadrétdomb also yielded small seed lentil and flax seeds.

7.3.2.1. Weeds of the Middle Bronze Age

The seed and fruit remains of segetal weed associations carry a great deal of information. Weeds adjust their lifecycles to cultivated plants. Their presence or absence may be used to establish the sowing time of cultivated plants, the height of harvesting and we can also infer the nutrient supply of the arable land (e.g. possible use of manure) and its climatic conditions, changes of rotation systems, etc.

The remains of weed seeds indicate that in prehistoric times people harvested at medium height, halfway between the ear and the ground, and they only rarely cut the cereals close to the ground (KNÖRZER 1985; WILLERDING 1988). Experimental archaeological tests with a bronze sickle have confirmed this (BERENOVÁ 1991). As emmer does not grow to a uniform height and the culms are usually short, it is not possible to harvest right under the ear. If the cereal laid down for some reason, they had to harvest low. However, they were able to harvest emmer twice as fast as common wheat, probably because its earspindle is less solid.

The remains of weeds found at Middle Bronze Age sites also indicate sophisticated cereal production. At sites where systematic sampling took place, identified weeds are all of tall growth, which indicates that cereals were harvested at about two-thirds of the height of the stalks, probably using a sickle.

Spring wild-oat (*Avena fatua*), downy brome (*Bromus tectorum*), green and rough bristle-grass (*Setaria viridis, S. verticillata*) and fan-hen or white goosefoot (*Chenopodium album*) are members of a weed association connected to hackfruits or spring cereal weeds (*Polygeno-Chenopodietalia*). The species listed may have been the weeds of spring-sown cereals (e.g. barley, common millet) and legumes (e.g. common lentil, pea, horsebean, bitter vetch). The weed seeds found indicate calciferous, highly nutritious, but not manured loess soils.

We note that the fan-hen or white goosefoot seeds and downy brome grains found in the samples could also have come from ruderal weed associations of decaying soils. Such ruderal weed associations are found in all areas affected by human activities (by the side of ditches and roads, embankments, fallow ground, around buildings) where the soil is rich in nitrogen or where it is perhaps manured.

In the case of the Bronze Age earthworks at Bölcske-Vörösgyír, the number of the seeds of two weed species is so high that it cannot be explained by any propagation of the seeds caused by some unusual event (*Fig. 151*). The proportion of seeds of field brome (*Bromus arvensis*) and rye brome (*B. secalinus*) compared to cereals (einkorn, emmer, spelt, two-rowed and six-rowed barley) is extremely high: 3.75%. As they are all naked, or more precisely, their husks have been removed, and as they occur together with cleaned cereals, it is very likely that they were grown together with the cereals. The may have been ground together with the cereals and used as food.

The se *Bromus* species were obviously in the process of domestication as so-called secondary domesticated plants, but before they could be truly domesticated, they were removed from cultivation with the passage of the Bronze Age.

In 1990, at the excavation of the tell settlement at Túrkeve-Terehalom from the time of the Ottomány culture (approx. 16th century BC) by Marietta Csányi and Judit Tárnoki, the remains of a plant association and associated weed flora (palaeo-ethnobiocenosis association) were recovered from samples collected from floors of houses that had burned (GYULAI in print) (*Figs. 158–162, 164–182*). The grain and fruit remains, which had been stored for human consumption, indicated a sophisticated culture of plant cultivation. This included some weeds that were probably also consumed or even deliberately cultivated. Similarly to Middle Bronze Age tells near the Danube, husked wheats, einkorn and emmer, were the most important bread cereals at Túrkeve, but naked common wheat was also present. The total absence of common millet is surprising. Cereal weeds indicate that wheat was sowed in autumn. Legumes (pea, common lentil, bitter vetch) may have played a very important role in everyday consumption After threshing, cereals were cleaned and stored for consumption in the two houses excavated. However, the process did not provide comprehensive protection against the seeds/fruits of weed plants: spring wild-oat, rye brome, black bindweed, etc.

A sample from the burnt floor level of one of the houses of the tell contained cleaned fan-hen or white goosefoot (*Chenopodium album*) seeds, indicating that the plant was consumed, perhaps even cultivated (*Fig. 174*). Fan-hen or white goosefoot, which is considered a ruderal plant, is found in the material of almost all excavations at least sporadically. It may also have played a role as a segetal weed (*Fig. 155*). Species of the goosefoot genus are weeds that prefer nitrogenous soils, grow around stables, houses and by the sides of ditches. According to written records, fan-hen or white goosefoot seeds were collected in large quantities during famines of the Middle Ages due to their high protein content, and/or young leaves of the plant were cooked to make a sauce. Moreover, it is possible that fan-hen or white goosefoot was cultivated as a cereal replacement. One species of the genus, good-King-Henry (*Chenopodium bonus-henricus*), was cultivated during the Middle Ages (DANERT *et al.* 1980).

As an archaeobotanical parallel, we may mention the cleaned seeds of many-seeded goosefoot (*Chenopodium polyspermum*) found in one of the houses of the tell settlement covering several eras found at Feudvár at the edge of the Titel highlands near Mošorin in Voivodina, as they may also have been prepared for consumption, or perhaps as fish-bait (?) (KROLL 1990a). The sisymbrium (*Sisymbrium* spec./*Descurainia* spec.) and marsh spurge (*Euphorbia palustris*) seeds also found that may have been associated with healing (Kroll 1990b).

The appearance of the safflower (*Carthamus tinctorius*) is highly interesting; other than Túrkeve, it only occurs at the Mošorin-Feudvár site (KROLL 1990b). Its occurrence was connected with cosmetic use (make-up). In our opinion, it is more likely that it was used as a food colouring. According to István Már, it is still cultivated in Eastern Hungary for that purpose. The range of cultivated plants found at Mošorin-Feudvár is very similar to the botanical material obtained from Túrkeve-Terehalom (*Fig. 163*).

7.3.2.2. Remains of Bronze Age flora

At sites where samples were also collected from the environs (Százhalombatta-Földvár, Bölcske-Vörösgyír), we found species that were original constituents of the natural flora. The species and the numbers of their seeds/fruits both indicate that botanical material of the Vatya culture represents a typical *palaeo-ethnobiocenosis* association, i.e. the majority of the material is composed of cultivated plants with their weeds, along with the remains of some gathered plants. If species found at Middle Bronze Age sites are compared with one of the richer Neolithic sites, it is clear that the number of forest-fringe species decreased. As a result of climatic changes we have already outlined previously, vegetation of forest fringes grew sparser, there were fewer shrubs with edible seeds or fruits growing around the earthworks.

Although the number of constituents of the original vegetation found in the archaeological material is certainly too small for a comprehensive reconstruction of the environment, we can still draw inferences concerning the environment of the Middle Bronze Age sites mentioned. These species were mixed into the samples spontaneously, although in the case of some species, such as dewberries and crab apples found at Százhalombatta, it is possible that these wild fruits were also gathered.

The environment of the Százhalombatta earthworks was probably varied. Dewberry and crab apple indicate a gallery forest. Wall germander (*Teucrium chamaedris*) indicates a light mixed forest. Several meadow species are also present; including medic (*Medicago* spec.), sheep's sorrel (*Rumex acetosella*), meadow clary (*Salvia pratensis*), meadow fescue (*Festuca pratensis*), common sorrel (*Rumex acetosa*), common birdsfoot-trefoil (*Lotus corniculatus*), bur medic (*Medicago minima*) and bulbous buttercup (*Ranunculus bulbosus*). These plant remains from Százhalombatta have been used as a guide for planning the environmental reconstruction of the archaeological park constructed there (MÜLLER 1997).

The environment of the Bölcske tell was probably also forested. The forest was the probable source of the oak acorn (whose exact species could not be identified), while blackthorn and single-seed hawthorn grow in moderately moist or dry forest fringe environments. The meadow that once spread out next to the settlement had a moderate water supply, with rather dry areas in some places. The meadow was the source of soft brome (*Bromus mollis*, syn. *B. hordeaceus*) seeds and the seeds of black medic, hop clover and yellow trefoil (*Medicago lupulina*).

The vitamin-rich wild fruits (blackthorn, dogwood, single-seed hawthorn) whose remains were found at the Bölcske-Vörösgyír site are still popular foods (*Fig. 153*). They may have been consumed raw, but also as jam or a cordial. These species also have medicinal effects. Naturally, this does not necessarily mean that they were used as medicinal plants, though it cannot be excluded. For example, the "acorn coffee" made from roasted oak acorns was once given to sickly children, and it was also used to cure enteritis in old people (RÁPÓTI – ROMVÁRY 1983). The medicinal effect of blackthorn has been known for a long time. Tea made from its flowers (*Pruni spinosae flos*) is a mild diuretic and laxative. The infusion of its fruit (Pruni spinosae fructus) was consumed to fight enteritis. After the first frost, its fruit was also used to make an obstruent jam (RÁCZ – RÁCZ-KOTILLA – LAZA 1984). The flowers of the single-seed hawthorn (*Crataegi flos*), along with its leaves (*Crataegi folium*), buds (*Crataegi summitas*) and its red, but still hard fruit can be dried to obtain a soothing tea that helps the coronary system of the heart and bolsters the heart muscles (RÁCZ – RÁCZ-KOTILLA – LAZA 1984).

The following ruderal species were found at the above two earthworks: fan-hen or white goosefoot (*Chenopodium album*), wild barley-grass (*Hordeum murinum*), green bristle-grass (*Setaria viridis*), hairy sedge (*Carex hirta*), white campion (*Silene alba*, syn. *Melandrium album*), knotgrass (*Polygonum aviculare*), creeping buttercup (*Ranunculus repens*), thyme-leaved sandwort (*Arenaria serpyllifolia*), wild carrot (*Daucus carota*), common mallow (*Malva silvestris*), hare's-foot clover (*Trifolium arvense*) – these species indicate areas under strong human influence and a large settlement.

7.3.2.3. Agriculture in the Middle Bronze Age

Thanks to recovery of plant remains, we know a great deal about the farming activities of Bronze Age people. The area selected for cultivation was first burned, then cleared of plant residues. Ploughing was introduced in this era, but no effort was made as yet to maintain the fertility of the soil. Land was used until it was exhausted. Then new areas, perhaps pastureland was taken into cultivation. Old cultivated areas were left fallow. After a while they returned to those fallow lands.

At the beginning of the Bronze Age, the southeastern and central parts of Europe had different agricultures. In the Balkans, in particular in the area that is northern Bulgaria today, flooding of arable land was used to restore the fertility of the soil (TODOROVA 1982). The softened soil was furrowed and cereals were sowed in the rows thus obtained.

In Central Europe, fields were turned using wooden ploughs drawn by oxen. The oldest specimens of ploughs with soles are from the shores of the Mediterranean Sea, dating to the end of the 3rd and the beginning of the 2nd millennia BC. We have data about a plough sole in Denmark from the middle of the 2nd millennium BC. From Northern and North-Western Europe, marks of Bronze Age ploughing are known along with the remains of ploughs (BALASSA 1973; ROWLEY-CONWY 1987). During excavation of a Neolithic burial mound at South-Street in the South of England, marks made by an ard were observed (FOWLER – EVANS 1967).

According to GOLDMANN (1982), people consumed around 200 kg of cereal per person per year in prehistoric times. However, we believe that his estimate of the minimum yield at 150 kg/ha of bread wheat or equivalent cereal is far too low. Yet he bases his calculations on that figure to arrive at the conclusion that 1000 km^2 of land could support some 80 thousand people, which is the equivalent of an average population sustenance capacity of 75 people/km^2. The calculation includes non-cultivated land as well. However, population density was well below that value: it was probably around 6.4 persons/km^2 in Denmark (POULSEN 1983). This implies that there was no obstacle whatsoever to a sudden demographical explosion.

Now we see how important it is to determine yields, as they are the basic values that drive the division of labour, demographic growth, etc. In recent years, experimental archaeological settlements have been established to study the history of plant cultivation. According to results of harvest experiments performed at the Butser Ancient Farm in Britain, emmer cultivated using Celtic tools yielded 2.5 tonnes (!) per hectare on average (REYNOLDS 1991). We ourselves performed a cultivation experiment with einkorn at the biohistorical settlement at Szarvasgede (GYULAI 1999c). We attempted to prove or disprove the innumerable archaeological, ethnographical and archaeobotanical theories associated with the species. Over a five-year period, our population of einkorn produced an average yield of 1.6 t/ha without replenishment of the soil and without any chemicals. We are conducting continuous experiments to process it (removal of husks, grinding, baking bread) (FEKETE 1999; MÁTHÉ 1999). It can be seen that in both cases, yields were much higher than the figures used in the literature to assess archaeological conditions and the history of plant cultivation.

Archaeologists excavating Bronze Age settlements found a very large number of pits. Based on auxiliary material found, the majority are identified as clay pits or waste pits, but others are interpreted as grain storage pits. Relatively deep pits – usually without any mudwork, but sometimes fired, and which have narrow necks and regular shapes – were used to store threshed and cleaned cereals for various lengths of time. WILLERDING (1970) doubted whether the pits of the late Copper Age Baden culture found near Ossarn were really storage pits. However, the late Bronze Age pits found at Common Galingaleu and Burgschleinitz were certainly used to store plant seeds, though in the case of the latter we can read about a strong mixture with legume seeds (Werneck 1949). At the Urnfield culture settlement of Ichtershausen in Germany, a "storage pit" with daubed walls was found, with cereal grains at the bottom (SCHULZE-MOTEL – GALL 1967).

It is probable that storing grains in a pit was not the only viable option. At least this is the conclusion KNÖZER (1972) came to when he found carbonised bread wheat and barley grains in the pile holes of the Bronze Age settlement at Langweiler in Germany and inferred that he had found the remains of a granary.

Just as in the case of "storage pits" of neighbouring countries, there are also pits of regular shape in Hungary whose soil infill has been shown by flotation to contain cereal grains. NOVÁKI (1981–83) made a comprehensive study of grain pits in Hungary. Of the large number of pits found at the early Bronze Age settlement of the Somogyvár-Vinkovci culture at Pécs-Nagyárpád, only a single pit free of daubwork or burning seems to be a storage pit, though even in that pit, the seeds of various plants were mixed together. One of the pits of the Middle Bronze Age settlement of the Füzesabony culture at Ároktő-Dongóhalom also contained carbonised cereal grains in vessels. The Vatya culture tell settlement at Bölcske-Vörösgyír also has some regularly shaped pits, which are also probably grain pits, with no sign of daubwork or burning.

7.3.2.4. Consumption of grapes and fruit in the Late Bronze Age

As a result of the Atlantic phase climatic optimum, the area of the woodland European grape extended far beyond the Carpathian Basin. We have numerous prominent finds indicating that this fruit was already being gathered in several places around Europe in the Neolithic and the Copper Age: in Northern Italy, in Belgium, in Switzerland (STUMMER 1911), in Greece and Spain (HOPF 1978), in Germany (SCHIEMANN 1953), in Moravia (OPRAVIL 1977), in Denmark (TROELS-SMITH 1944) and in Sweden (HJELMQVIST 1955). The seed find from Auvernier (Switzerland) already belongs to the phase of domestication: it is a transitional form toward the wine grape (VILLARET-VON ROCHOW 1971). Its seeds were also found dated to the Atlantic climate phase in Hungary, in the late Neolithic Lengyel culture settlement at Sé near Szombathely (FACSAR – JEREM 1985).

Other findings of woodland European grapes from Bronze Age Europe come from the flood plain of the Danube, the Po plain, the areas of Garda Lake and Parma, Bosnia and Germany (STUMMER 1911; AMMERMANN *et al.* 1978; BERTSCH – BERTSCH 1949). The Bronze Age material from the Balkans and the shores of the Mediterranean plays an important role in tracking the development of the wine grape. According to some opinions, cultivated grapes had already appeared in the Neolithic in Italy (Grotta dell'Uzzo, Trapani) (COSTANINI 1981b) Seeds of early grapes were found at Grotte de l'Abeurador, a Neolithic settlement in France (ERROUX 1981).

Although only the seeds of woodland European grapes were found at sites of the late Dimini period (Arapi, Sesklo, Pefkakia) and in Iolkos, they may have participated in the process of the development of the wine grape (RENFREW 1966). We believe that the Late Neolithic grape seed find from Dimini, near Volos, Greece is very important. According to the identification of KROLL (1979), those seeds constitute a transitional form between the woodland European grape and the cultivated grape, also known as the wine grape. Are they perhaps the first evidence of grape cultivation in Thessalia? The earliest known cultivated grape seed of Southern Europe was found at the Early Bronze Age Dikili-Tash site (2400 BC) (KROLL 1983).

György MÁNDY (1972) believed the *Vitis vinifera* subsp. *silvestris* and the *V. vinifera* subsp. *caucasica* taxons (the latter particularly for Asian varieties) to be the ancient forms of the cultivated wine grape, and declared its primary gene centre to be located in the Mediterranean: Armenia and Persia. That area in the Caucasus, the Trans-Caucasus and Dagestan is one of the primeval centres of cultivated plants and possesses the greatest wealth of forms (LISITSINA 1978). During evolutionary development, the area of the Caucasus was one of the places where grapes were domesticated, and from where the knowledge of their cultivation spread to the west and the east.

According to the opinion of Géza Facsar, the place of origin should be sought in Palestine, Egypt and Syria (in: FACSAR – JEREM 1985). He neither refutes, nor supports the cultivated nature of the grape seeds found in Kastanas, Northern Macedonia (KROLL 1983). The truth is that the identification of seeds, in particular grape seeds based on their morphological and metric indices requires tremendous experience.

The oldest seeds of wine grapes are from the Somu-tepe (Azerbaijan, Khazan area) and the Sulaveri (Georgia) settlements, from the turn of the 5th and 4th millennia BC. In later periods, the rich finds of grape seeds from the shores of the Caspian Lake indicate continuous cultivation (LISITSINA – PRISEPENKO 1977). From that era we also know grape seeds from New-Rusteshti in the Bug-Dnieper area (second half of the 4th millennium BC, end of the middle Tripolje culture) (JANUSHEVICH 1975), and from Sahr-i-Sokhta from the Seistan desert (3000 BC) (TOSI – GERSTER 1983).

The grape seed finds mentioned above indicate that the wine grape developed from the woodland European grape in a polygenetic manner in several locations but more or less at the same time. The woodland European grape constitutes a single species, but it does have three geographical subspecies. The dioecious woodland European wine grape, with its geographically distinct area, developed into the monoecious wine grape in a separate area as a result of selection.

The number of species in the *Euvitis* subgenus is as high as seventy. They are assigned to three area centres: 1st – North America (e.g. *V. labrusca*, *V. riparia* = *V. vulpina*, *V. cordifolia*, *V. aestivalis*, *V. rupestris*, *V. berlandieri*), 2nd – Eastern Asia (e.g. *V. amurensis*), 3rd – the Mediterranean (e.g. *V. silvestris*) (KÁRPÁTI – GÖRGÉNYI – TERPÓ 1968).

Regarding the development of the wine grape, we are interested only in one of these species: the original Mediterranean woodland European grape. The range of the dioecious (sometimes monoecious) woodland European grape, with its usually black, rarely white, grapes, is very large – from Western Europe through the Mediterranean to Western Asia. In Central and Western Europe, var. *silvestris (typica)* is also present, South-Eastern Europe and the Balkans have the var. *balcanica*, while Western Asia has the var. *aberrans*. The woodland European grape also occurs in Hungary, primarily along rivers, though its range has receded recently. (It is not to be confused with the ornamental *Parthenocissus* shrub, whose leaves also redden in autumn, or the American and Japanese forest grapes!)

The first scientific system of wine grape plants was developed by ANDRASOVSZKY (1925) who distinguished 5 species of wine grapes: *V. alemannica* (Burgundy types), *V. byzantina* (chasselas types), *V. mediterranea* (jackdaw grapes), *V. deliciosa* (Muscatel grapes), *V. antiquorum* (acorn grapes).

Today, the innumerable varieties of wine and dessert grapes are all placed in a single subspecies (Vitis vinifera subsp. vinifera) and its three cultivated varieties, which at the same time indicate their origin (KÁRPÁTI – GÖRGÉNYI – TERPÓ 1968):

a) *convar. alemannica* or *proles occidentalis* area group. It originally inhabited Western Europe. Wine grapes. The grapes are round, white or blue, full of juice, acidic: e.g. Ruländer, Rhine Riesling;
b) *convar. byzantina* or *proles orientalis* area group. It originally inhabited Central Asia and the Caucasus area. Primarily dessert grape varieties, e.g. Afuz Ali, Muscat Ottonel;
c) *convar. mediterranea* or *proles pontica* area group. It originally inhabited Pontus and the shores of the Black Sea: Asia Minor, Greece, Bulgaria and Bessarabia. Hungary also belongs here. Mostly wine grapes, some dessert grapes: e.g. Ezerjó, Furmint.

Naturally, there are very many hybrids of the individual groups, and they are now generally cultivated in locations other than their original ones.

While the wine grape is an anthropic, archaeoadventive, Mediterranean – Central Asian species in Hungary, the woodland European grape is an indigenous species (SOÓ 1966).

According to written records, grapes became entirely commonplace in Greece by the 8th–12th centuries BC. According to the descriptions of Homer and Theophrastus, several varieties were cultivated both freely and trained on props and trees. Around 2900 BC, grapes were being planted in the towns of Lagas, Sumer and Uruk. The depictions of grapes in the Egyptian pyramids, the Book of Kings and the Bible all speak of extensive cultivation of grapes and the cult of wine.

In contrast with the above Bronze Age areas, we have no seed finds to support the notion of an early viticulture, although a carbonised vine-shoot was found on the outskirts of Békés, in the Gyulavarsánd culture layer of the Békés-Várdomb site (VALKÓ – STIEBER 1969; STIEBER 1971; FÜZES 1971). According to FÜZES – SÁGI (1968) that and a few other archaeological finds (the large "fermenting (?)" vessel of the Nagyrév culture found at Diósd, a bronze sickle with a hose from Transylvania) suggest that grapes were already cultivated in Hungary in the Bronze Age, though their cultivation may even have begun in the Copper Age. In our opinion, those finds are from the woodland European grape, because the earliest wine grape seeds of the Carpathian Basin are from the transition period between the Late Bronze and Iron Ages (see below). That, however, does not rule out knowledge of the procedure for making wine (see the Copper Age site at Kompolt).

Reviewing the plant remains from Bronze Age sites in Hungary, we can find plenty of evidence that, in order to make their diet more varied, prehistoric people were glad to collect vitamin-rich forest fruits and berries. Let us list some finds indicating human consumption, storage, perhaps preparation for consumption (jam, cordial, etc.): pear (*Pyrus* spec.) – Tószeg-Laposhalom (LINDAU 1917); dewberry (*Rubus caesius*) – Alpár; common elder/red elder (*Sambucus nigra/racemosus*) – Ároktő-Dongóhalom, Jászdózsa-Kápolnadomb; dogwood (*Cornus mas*) – Jászdózsa-Kápolnadomb, Lengyel; crab apple (*Malus silvestris*) – Pákozd-Vár; blackthorn (*Prunus spinosa*) – Bölcske-Vörösgyír (P. HARTYÁNYI – NOVÁKI – PATAY 1967–68; P. HARTYÁNYI – NOVÁKI 1973–74), woodland European grape (Mosonmagyaróvár-Németdőlő).

Unfortunately we have no data about the gathering and consumption of mushrooms, but we certainly assume that it did happen.

7.3.3. Plant cultivation of the late Bronze Age

According to pollen analysis, the climate took a turn for the worse again in the Late Bronze Age (1450–1300 BC) and then stabilized until the beginning of the Iron Age (800–600 BC). This is the period of late Subboreal cooling. It is the last period of the Subboreal phase which had begun around 3000 BC.

In the Late Bronze Age, the climate grew cooler and wetter relative to the Middle Bronze Age, though still without any extremes. The water level of lakes and rivers and the level of groundwater increased. The deterioration of the climate led to changes in the continental character of the flora. Along with European beech (*Fagus silvatica*), hornbeam (*Carpius betulus*) was a common species.

When investigating the flora, effects of human activity must be taken into account together with climatic changes. In contrast to the mixed oak forests of the Neolithic, the European beech forests of the Late Bronze Age, with a smaller number of species, did not provide enough food for the people gathering there. As a result, they started large-scale deforestation to create open land for the production of cereals and legumes to afford them a more stable livelihood. A substantial quantity of timber may have been required for pyre funerals (personal communication from Zsófia Medzrihradszky). The second period of major deforestation of the Hortobágy occurred in this period.

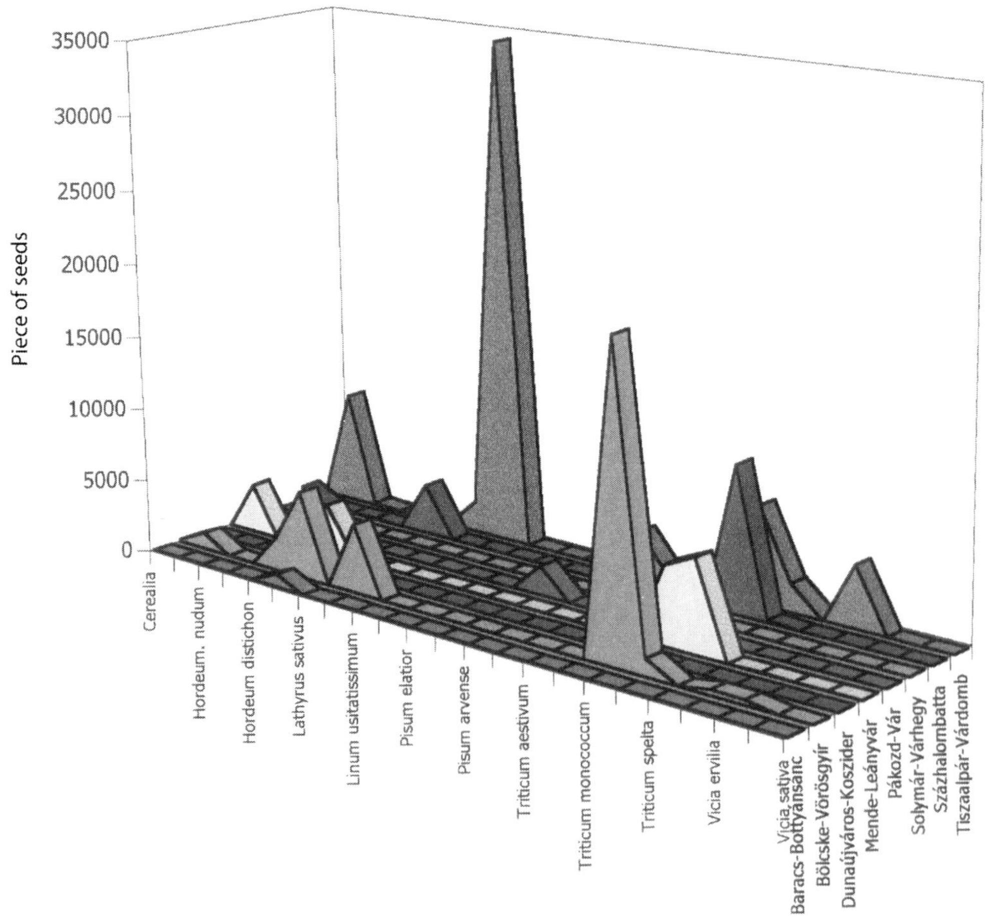

Fig. 86. Cultivated plants of the Bronze Age tell settlements.

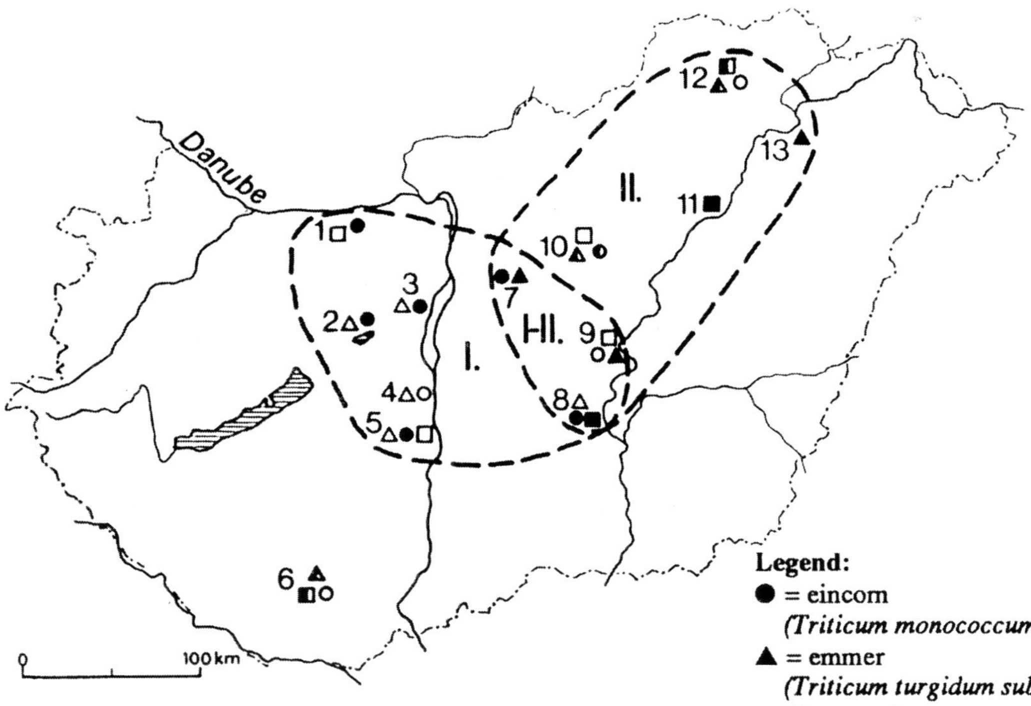

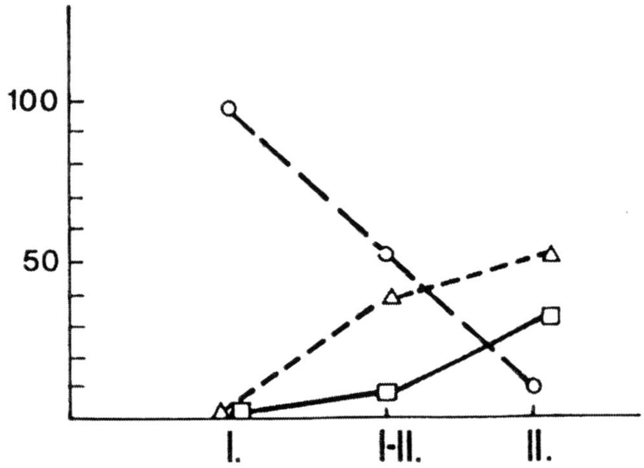

Legend:
● = eincorn *(Triticum monococcum)*,
▲ = emmer *(Triticum turgidum subsp. dicoccum)*,
■ = common/dwarf wheat *(Triticum aestivum subsp. aestivum* and *Triticum. aestivum subsp. compactum)*.
Full symbols stand for 98-100 % contribution,
half- full symbols stand for 30-60 % contribution,
empty symbols indicate occurrences equal or smaller than 1 %.

Roman numerals stand for the main distribution areas and their overlap respectively.

The diagram shows the geographical distribution of the summarized percentages of wheat species. See p. 25 for detailed exploration.

Site register:

1 = Süttő - Hosszúvölgy
2 = Pákozd - Vár
3 = Százhalombatta - Földvár
4 = Dunaújváros - Kosziderpadlás
5 = Bölcske - Vörösgyir
6 = Pécs - Nagyárpád
7 = Mende - Leányvár
8 = Alpár
9 = Tószeg - Laposhalom
10 = Jászdózsa - Kápolnahalom
11 = Ároktő - Dongóhalom
12 = Felsődobsza - Várdomb
13 = Tiszaeszlár - Bashalom
14 = Bakonszeg - Kádárdomb
15 = Dunaföldvár - Kálvária
16 = Poroszló - Aporhát

Fig. 87. Distribution of wheat species of Bronze Age sites (updated map after NOVÁKI 1969).

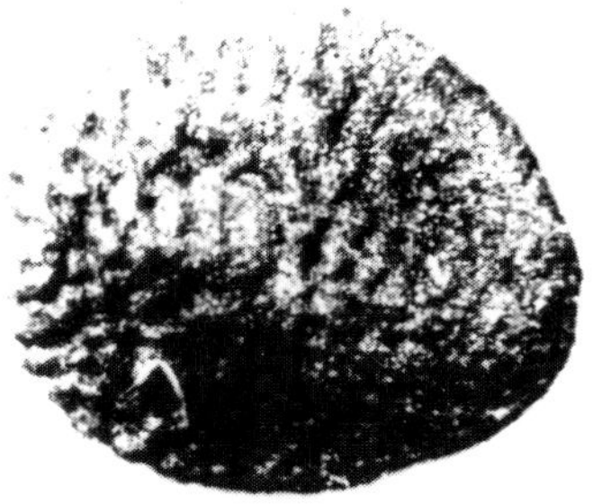

Fig. 88. Danewort (*Sambucus ebulus*) seed from Ároktő-Dongóhalom, Hatvan-Füzesabony culture, Inventory of the Hungarian Agricultural Museum, Budapest.

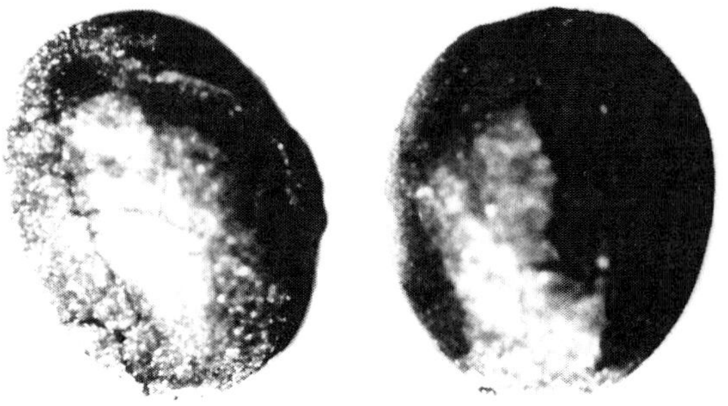

Fig. 89. Vetch (*Vicia* spec.) seed from Ároktő-Dongóhalom, Hatvan-Füzesabony culture, Inventory of the Hungarian Agricultural Museum, Budapest.

Fig. 90. Six rowed barley (*Hordeum vulgare* subsp. *hexastichum*) hulled grains from Alpár-Várdomb, Vatya culture, Inventory of the Hungarian Agricultural Museum, Budapest.

Fig. 91. Barley (*Hordeum vulgare*) ear spindles from Alpár-Várdomb, Vatya culture, Inventory of the Hungarian Agricultural Museum, Budapest.

Fig. 92. Barley (*Hordeum vulgare*) ear spindles fragments from Alpár-Várdomb, Vatya culture, Inventory of the Hungarian Agricultural Museum, Budapest.

Fig. 93. Four-rowed barley (*Hordeum tetrastichum*) hulled grains from Alpár-Várdomb, Vatya culture, Inventory of the Hungarian Agricultural Museum, Budapest.

Fig. 94. Many rowed barley (*Hordeum polystichum*) hulled grains from Alpár-Várdomb, Vatya culture, Inventory of the Hungarian Agricultural Museum, Budapest.

Fig. 95. Many rowed barley (*Hordeum polystichum*) hulled grains from Alpár-Várdomb, Vatya culture, Inventory of the Hungarian Agricultural Museum, Budapest.

Fig. 96. Naked barley (*Hordeum vulgare* var. *nudum*) grains from Alpár-Várdomb, Vatya culture, Inventory of the Hungarian Agricultural Museum, Budapest.

Fig. 97. Barley (*Hordeum vulgare*) spiculum fragments from Alpár-Várdomb, Vatya culture, Inventory of the Hungarian Agricultural Museum, Budapest.

Fig. 98. Einkorn (*Triticum monococcum*) naked grains from Alpár-Várdomb, Vatya culture, Inventory of the Hungarian Agricultural Museum, Budapest.

Fig. 99. Einkorn (*Triticum monococcum*) naked grains from Alpár-Várdomb, Vatya culture, Inventory of the Hungarian Agricultural Museum, Budapest.

Fig. 100. Emmer (*Triticum turgidum* subsp. *dicoccum*) naked grains from Alpár-Várdomb, Vatya culture, Inventory of the Hungarian Agricultural Museum, Budapest.

Fig. 101. Emmer (*Triticum turgidum* subsp. *dicoccum*) naked grains from Alpár-Várdomb, Vatya culture, Inventory of the Hungarian Agricultural Museum, Budapest.

Fig. 102. Emmer (*Triticum turgidum* subsp. *dicoccum*) naked grains from Alpár-Várdomb, Vatya culture, Inventory of the Hungarian Agricultural Museum, Budapest.

Fig. 103. Emmer (T*riticum turgidum* subsp. *dicoccum*) longshaped grains from Alpár-Várdomb, Vatya culture, Inventory of the Hungarian Agricultural Museum, Budapest.

Fig. 104. Emmer (*Triticum turgidum* subsp. *dicoccum*) short and thick-set grains from Alpár-Várdomb, Vatya culture, Inventory of the Hungarian Agricultural Museum, Budapest.

Fig. 105. Common bread wheat (*Triticum aestivum* subsp. *vulgare*) grains from Alpár-Várdomb, Vatya culture, Inventory of the Hungarian Agricultural Museum, Budapest.

Fig. 106. Club wheat (*Triticum aestivum* subsp. *compactum*) earspindles from Alpár-Várdomb, Vatya culture, Inventory of the Hungarian Agricultural Museum, Budapest.

Fig. 107. Club wheat (*Triticum aestivum* subsp. *compactum*) earspindles from Alpár-Várdomb, Vatya culture, Inventory of the Hungarian Agricultural Museum, Budapest.

Fig. 108. Grass pea wine (*Lathyrus sativus*) seeds from Alpár-Várdomb, Vatya culture, Inventory of the Hungarian Agricultural Museum, Budapest.

Fig. 109. Small seed lentil (*Lens culinaris* subsp. *microsperma*) from Alpár-Várdomb, Vatya culture, Inventory of the Hungarian Agricultural Museum, Budapest.

Fig. 110. Field pea (*Pisum sativum* var. *arvense*) seeds from Alpár-Várdomb, Vatya culture, Inventory of the Hungarian Agricultural Museum, Budapest.

Fig. 111. Flax (*Linum usitatissimum*) seeds from Alpár-Várdomb, Vatya culture, Inventory of the Hungarian Agricultural Museum, Budapest.

Fig. 112. Wild strawberry (*Fragaria vesca*) seeds from Alpár-Várdomb, Vatya culture, Inventory of the Hungarian Agricultural Museum, Budapest.

Fig. 113. Crab apple (*Malus silvestris*) seeds from Alpár-Várdomb, Vatya culture, Inventory of the Hungarian Agricultural Museum, Budapest.

Fig. 114. Brome grass (*Bromus arvensis*) naked grains from Alpár-Várdomb, Vatya culture, Inventory of the Hungarian Agricultural Museum, Budapest.

Fig. 115. White goosefoot or fan-hen (*Chenopodium album*) seeds from Alpár-Várdomb, Vatya culture, Inventory of the Hungarian Agricultural Museum, Budapest.

Fig. 116. Field gromwell (*Lithospermum arvense*) nutlets from Alpár-Várdomb, Vatya culture, Inventory of the Hungarian Agricultural Museum, Budapest.

Fig. 117. Blackberry or bramble (*Rubus fruticosus*) seeds from Alpár-Várdomb, Vatya culture, Inventory of the Hungarian Agricultural Museum, Budapest.

Fig. 118. Raspberry (*Rubus idaeus*) from Alpár-Várdomb, Vatya culture, Inventory of the Hungarian Agricultural Museum, Budapest.

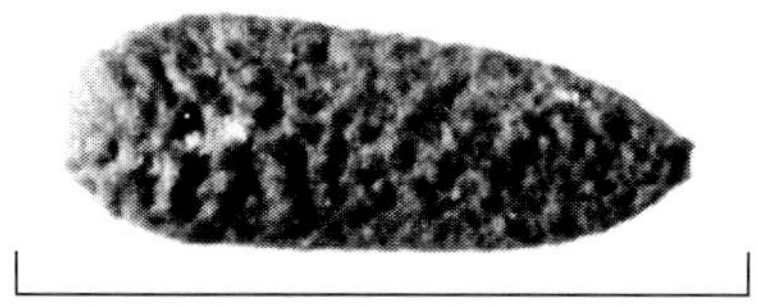

Fig. 119. Common elder (*Sambucus nigra*) seed from Alpár-Várdomb, Vatya culture, Inventory of the Hungarian Agricultural Museum, Budapest.

Fig. 120. Woody nightshade (*Solanum dulcamara*) from Alpár-Várdomb, Vatya culture, Inventory of the Hungarian Agricultural Museum, Budapest.

Fig. 121. Einkorn (*Triticum monococcum*) naked grains from Balatonboglár, Middle Bronze Age, Inventory of the Hungarian Agricultural Museum, Budapest.

Fig. 122. Emmer (*Triticum turgidum* subsp. *dicoccum*) naked grains from Balatonboglár, Middle Bronze Age, Inventory of the Hungarian Agricultural Museum, Budapest.

Fig. 123. Barley (*Hordeum vulgare*) grain from a house in Baracs, Nagyrév culture, Inventory of the Hungarian Agricultural Museum, Budapest.

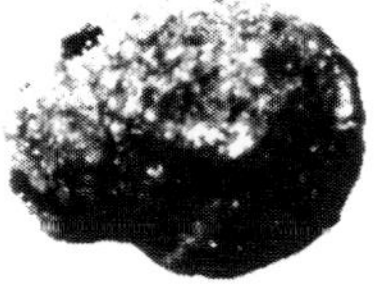

Fig. 124. Clover (*Trifolium* spec.) seed from a house in Baracs, Nagyrév culture, Inventory of the Hungarian Agricultural Museum, Budapest.

Fig. 125. Korn-rade (*Agrostemma githago*) seeds from a house in Baracs, Nagyrév culture, Inventory of the Hungarian Agricultural Museum, Budapest.

Fig. 126. Barren brome (*Bromus sterilis*) hulled grains from a house of Baracs, Nagyrév culture, Inventory of the Hungarian Agricultural Museum, Budapest.

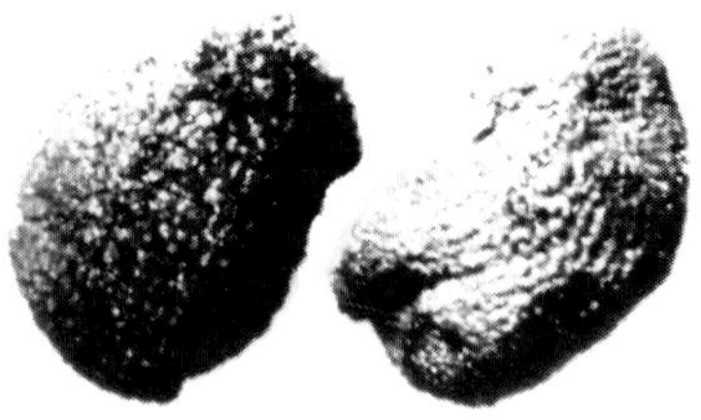

Fig. 127. Hare's-ear mustard (*Conringia orientalis*) seed from Baracs-Bottyán sánc, Nagyrév culture, Inventory of the Hungarian Agricultural Museum, Budapest.

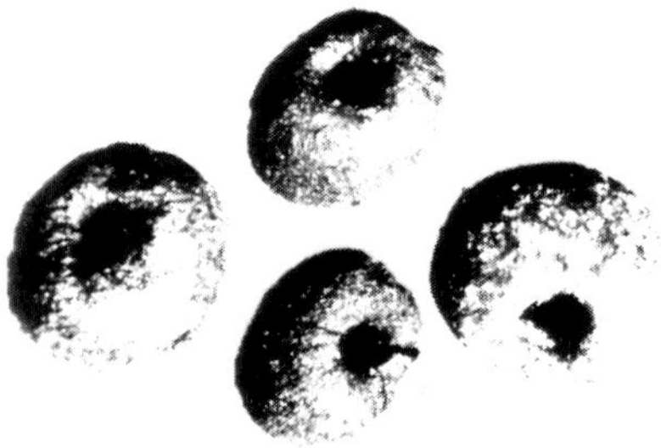

Fig. 128. False cleavers (*Galium spurium*) achenes from Baracs-Bottyán sánc, Nagyrév culture, Inventory of the Hungarian Agricultural Museum, Budapest.

Fig. 129. Redshank (*Polygonum persicaria*) nutlets from Baracs-Bottyán sánc, Nagyrév culture, Inventory of the Hungarian Agricultural Museum, Budapest.

Fig. 130. Einkorn (*Triticum monococcum*) naked grains from the Bronze Age settlement Dunaújváros (Dunapentele), Inventory of the Hungarian Agricultural Museum, Budapest.

Fig. 131. Einkorn (*Triticum monococcum*) naked grains from the Bronze Age settlement Dunaújváros (Dunapentele), Inventory of the Hungarian Agricultural Museum, Budapest.

Fig. 132. Einkorn (*Triticum monococcum*) naked grains from the Bronze Age settlement Felsődobsza, Inventory of the Hungarian Agricultural Museum, Budapest.

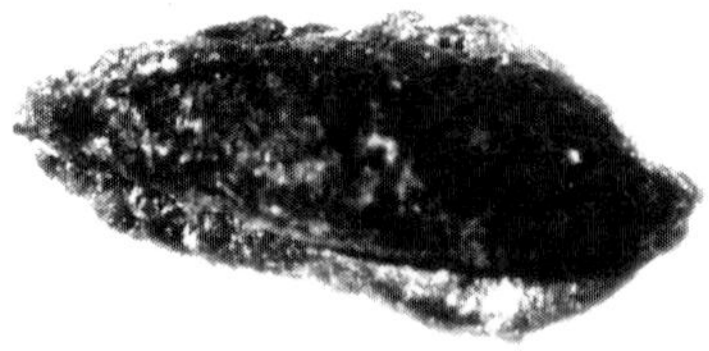

Fig. 133. Einkorn (*Triticum monococcum*) naked grain from the Bronze Age settlement Felsődobsza, Inventory of the Hungarian Agricultural Museum, Budapest.

Fig. 134. Einkorn (*Triticum monococcum*) naked grain from the Bronze Age settlement Felsődobsza, Inventory of the Hungarian Agricultural Museum, Budapest.

Fig. 135. Einkorn (*Triticum monococcum*) naked grains from the Middle Bronze Age settlement Jászdózsa-Kápolnahalom, Inventory of the Hungarian Agricultural Museum, Budapest.

Fig. 136. Einkorn (*Triticum monococcum*) naked grains from the Middle Bronze Age settlement Jászdózsa-Kápolnahalom, Inventory of the Hungarian Agricultural Museum, Budapest.

Fig. 137. Emmer (*Triticum turgidum* subsp. *dicoccum*) naked grains from the Middle Bronze Age settlement Jászdózsa-Kápolnahalom, Inventory of the Hungarian Agricultural Museum, Budapest.

Fig. 138. Einkorn (*Triticum monococcum*) naked grains from the Bronze Age settlement Mende-Leányvár, Inventory of the Hungarian Agricultural Museum, Budapest.

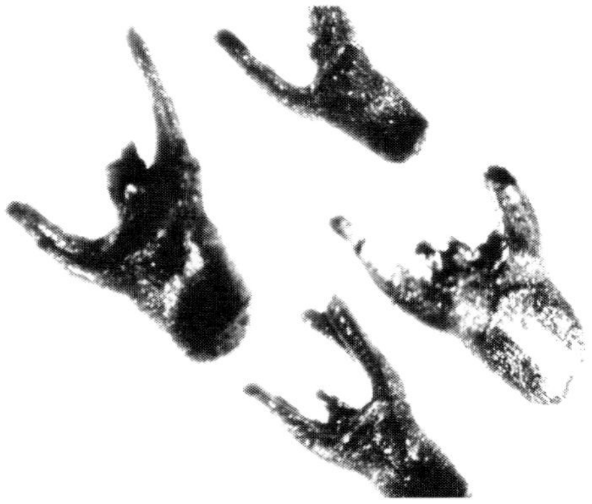

Fig. 139. Einkorn (*Triticum monococcum*) spikelets from the Middle Bronze Age tell settlement Százhalombatta-Téglagyár, Vatya culture, Inventory of the Hungarian Agricultural Museum, Budapest.

Fig. 140. Emmer (*Triticum turgidum* subsp. *dicoccum*) naked grains from the Bronze Age settlement Mende-Leányvár, Inventory of the Hungarian Agricultural Museum, Budapest.

Fig. 141. Einkorn (*Triticum monococcum*) naked grain from the Middle Bronze Age tell settlement Százhalombatta-Téglagyár, Vatya culture, Inventory of the Hungarian Agricultural Museum, Budapest.

Fig. 142. Emmer (*Triticum turgidum* subsp. *dicoccum*) naked grains from the Middle Bronze Age tell settlement Százhalombatta-Téglagyár, Vatya culture, Inventory of the Hungarian Agricultural Museum, Budapest.

Fig. 143. Brome grass (*Bromus arvensis*) naked grains from the Middle Bronze Age tell settlement Százhalombatta-Téglagyár, Vatya culture, Inventory of the Hungarian Agricultural Museum, Budapest.

Fig. 144. Einkorn (*Triticum monococcum*) naked grains from the Bölcske-Vörösgyír Middle Bronze Age tell settlement. Photograph by Tibor Kádas.

Fig. 145. Einkorn (*Triticum monococcum*) naked grains from the Bölcske-Vörösgyír Middle Bronze Age tell settlement. Photograph by Tibor Kádas.

Fig. 146. Emmer (*Triticum turgidum* subsp. *dicoccum*) naked grains from the Bölcske-Vörösgyír Middle Bronze Age tell settlement. Photograph by Tibor Kádas.

Fig. 147. Emmer (*Triticum turgidum* subsp. *dicoccum*) naked grains from the Bölcske-Vörösgyír Middle Bronze Age tell settlement. Photograph by Tibor Kádas.

Fig. 148. Club wheat (*Triticum aestivum* subsp. *compactum*) grain from the Bölcske-Vörösgyír Middle Bronze Age tell settlement. Photograph by Tibor Kádas.

Fig. 149. Barley (*Hordeum vulgare*) grains from the Bölcske-Vörösgyír Middle Bronze Age tell settlement (Photograph by Tibor Kádas)

Fig. 150. Common millet (*Panicum miliaceum*) naked grains from the Bölcske-Vörösgyír Middle Bronze Age tell settlement. Photograph by Tibor Kádas.

Fig. 151. Field bindweed (*Convulvulus arvensis*) seeds from the Bölcske-Vörösgyír Middle Bronze Age tell settlement. Photograph by Tibor Kádas.

Fig. 152. Bitter vetch (*Vicia ervilia*) seeds from the Bölcske-Vörösgyír middle Bronze Age tell settlement. Photograph by Tibor Kádas.

Fig. 153. Sloe or blackthorn (*Prunus spinosa*) stones from a house in the Bölcske-Vörösgyír middle Bronze Age tell. Photograph by the author.

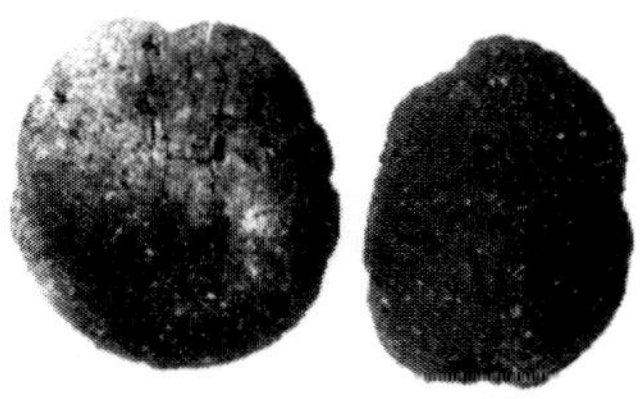

Fig. 154. Horsebean (*Vicia faba*) seeds from the Bölcske-Vörösgyír Middle Bronze Age tell settlement. Photograph by Tibor Kádas.

Fig. 155. Danewort (*Sambucus ebulus*) seeds from a house in the Bölcske-Vörösgyír Middle Bronze Age tell. Photograph by the author.

Fig. 156. Common or red elder (*Sambucus nigra* or *racemosa*) seeds from the Bölcske-Vörösgyír Middle Bronze Age tell settlement. Photograph by Tibor Kádas.

Fig. 157. Rye brome (*Bromus secalinus*) grains from a house in the Bölcske-Vörösgyír Middle Bronze Age tell. Photograph by the author.

Fig. 158. Emmer (*Triticum turgidum* subsp. *dicoccum*) naked grains from a house in the Túrkeve-Terehalom Middle Bronze Age tell. Photograph by Tibor Kádas.

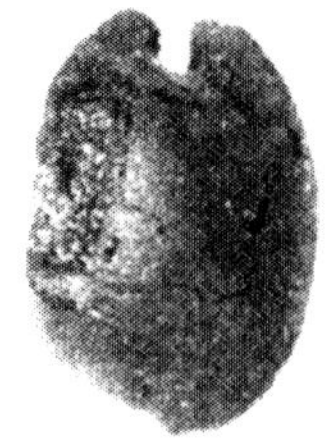

Fig. 159. Common bread wheat (*Triticum aestivum* subsp. *vulgare*) grains from a house in the Túrkeve-Terehalom Middle Bronze Age tell. Photograph by Tibor Kádas.

Fig. 160. Einkorn (*Triticum monococcum*) spikelets from a house in the Túrkeve-Terehalom Middle Bronze Age tell. Photograph by Tibor Kádas.

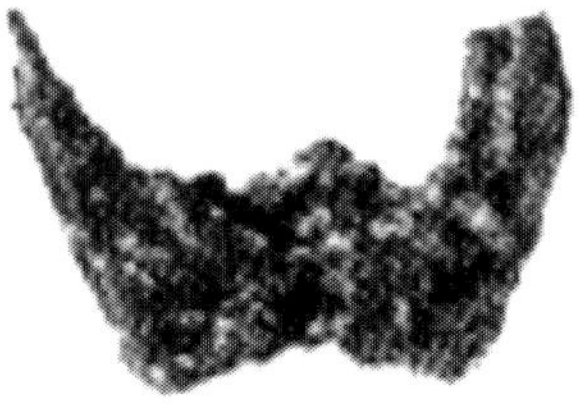

Fig. 161 Emmer (*Triticum turgidum* subsp. *dicoccum*) spicelet (furca bicornis) from a house in the Túrkeve-Terehalom Middle Bronze Age tell. Photograph by Tibor Kádas.

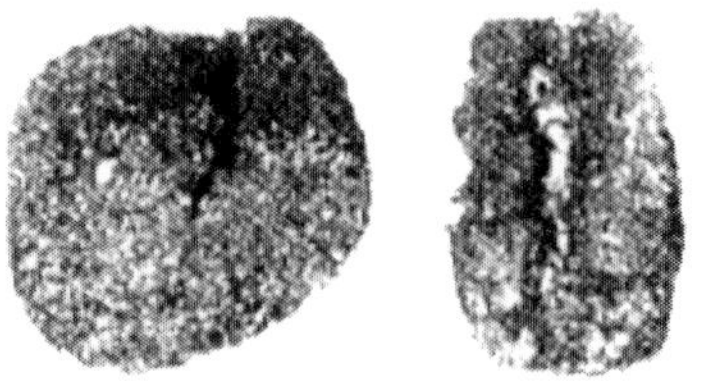

Fig. 162. Barley (*Hordeum vulgare*) grains from a house in the Túrkeve-Terehalom Middle Bronze Age tell. Photograph by Tibor Kádas.

Fig. 163. Safflower (*Carthamus tinctorius*) acheniums from a house in the Túrkeve-Terehalom Middle Bronze Age tell. Photograph by Tibor Kádas.

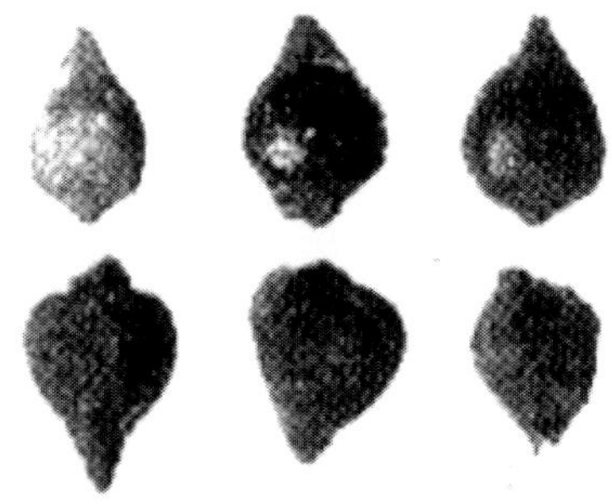

Fig. 164. Knotgrass (*Polygonum aviculare*) nutlets from a house in the Túrkeve-Terehalom Middle Bronze Age tell. Photograph by Tibor Kádas.

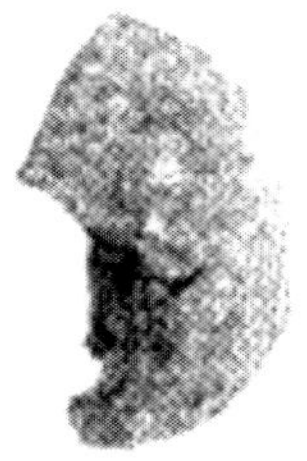

Fig. 165. Danewort (*Sambucus ebulus*) seed fragment from a house in the Túrkeve-Terehalom Middle Bronze Age tell. Photograph by Tibor Kádas.

Fig. 166. Small seed lentil (*Lens culinaris* subsp. *microsperma*) seeds from a house in the Túrkeve-Terehalom Middle Bronze Age tell. Photograph by Tibor Kádas.

Fig. 167. Pea (*Pisum sativum*) seeds from a house in the Túrkeve-Terehalom Middle Bronze Age tell. Photograph by Tibor Kádas.

Fig. 168. Bitter vetch (*Vicia ervilia*) seeds from a house in the Túrkeve-Terehalom Middle Bronze Age tell. Photograph by Tibor Kádas.

Fig. 169. Narrow-leaved vetch (*Vicia angustifolia*) seed from a house in the Túrkeve-Terehalom middle Bronze Age tell. Photograph by Tibor Kádas.

Fig. 170. Rye brome (*Bromus secalinus*) grain fragment from a house in the Túrkeve-Terehalom Middle Bronze Age tell. Photograph by Tibor Kádas.

Fig. 171. White mustard (*Sinapis alba*) seed from the Túrkeve-Terehalom Middle Bronze Age tell settlement. Photograph by Tibor Kádas.

Fig. 172. Spring wild-oat (*Avena fatua*) naked grains from the Túrkeve-Terehalom Middle Bronze Age tell settlement. Photograph by Tibor Kádas.

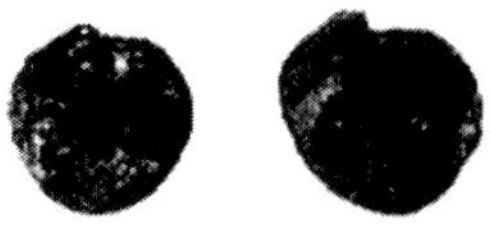

Fig. 173. Common orache (*Atriplex patula*) from a house in the Túrkeve-Terehalom Middle Bronze Age tell. Photograph by Tibor Kádas.

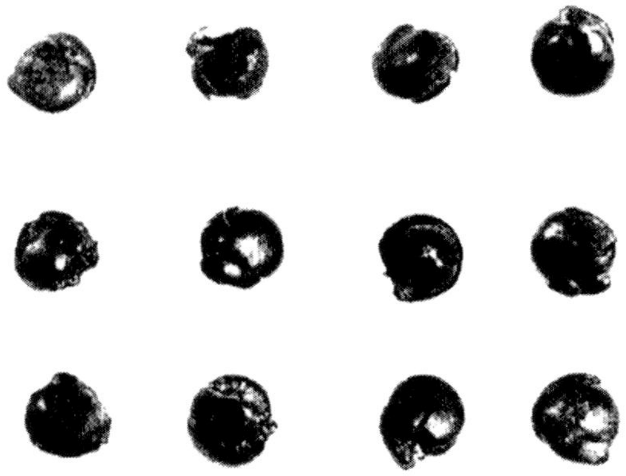

Fig. 174. White goosefoot or fan-hen (*Chenopodium album*) seeds from a house in the Túrkeve-Terehalom Middle Bronze Age tell. Photograph by Tibor Kádas.

Fig. 175. False cleavers (*Galium spurium*) achenes from a house in the Túrkeve-Terehalom Middle Bronze Age tell. Photograph by Tibor Kádas.

Fig. 176. Brome (*Bromus* spec.) grain fragment from a house in the Túrkeve-Terehalom Middle Bronze Age tell. Photograph by Tibor Kádas.

Fig. 177. Curled dock (*Rumex crispus*) glans from a house in the Túrkeve-Terehalom Middle Bronze Age tell. Photograph by Tibor Kádas.

Fig. 178. Common sorrel (*Rumex acetosa*) glans from a house in the Túrkeve-Terehalom Middle Bronze Age tell. Photograph by Tibor Kádas.

Fig. 179. Meadow-grass (*Poa* spec.) naked grain from the Túrkeve-Terehalom Middle Bronze Age tell settlement. Photograph by Tibor Kádas.

Fig. 180. Gipsywort (*Lycopus europaeus*) seeds from the Túrkeve-Terehalom Middle Bronze Age tell settlement. Photograph by Tibor Kádas.

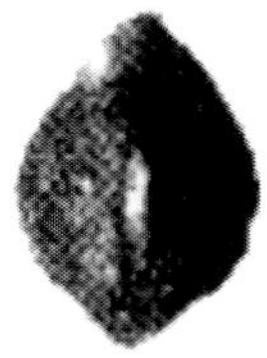

Fig. 183. Knotgrass (*Polygonum aviculare*) nutlet from the Túrkeve-Terehalom Middle Bronze Age tell settlement. Photograph by Tibor Kádas.

Fig. 185. Two-rowed barley (*Hordeum vulgare* subsp. *distichum*) hulled grains from a pit of Tiszaeszlár-Bashalom, Inventory of the Hungarian Agricultural Museum, Budapest.

Fig. 187. Spring wild-oat (*Avena fatua*) naked grains from a pit of Tiszaeszlár-Bashalom, Inventory of the Hungarian Agricultural Museum, Budapest.

Fig. 181. Field pansy (*Viola arvensis*) seed from the Túrkeve-Terehalom Middle Bronze Age tell settlement. Photograph by Tibor Kádas.

Fig. 182. Black-bindweed (*Fallopia convolvulus*) nutlets from the Túrkeve-Terehalom Middle Bronze Age tell settlement. Photograph by Tibor Kádas.

Fig. 184. Barley (*Hordeum vulgare*) grains hulled grains from a pit of Tiszaeszlár-Bashalom, Inventory of the Hungarian Agricultural Museum, Budapest.

Fig. 186. Black-bindweed (*Fallopia convolvulus*) nutlets from a pit of Tiszaeszlár-Bashalom, Inventory of the Hungarian Agricultural Museum, Budapest.

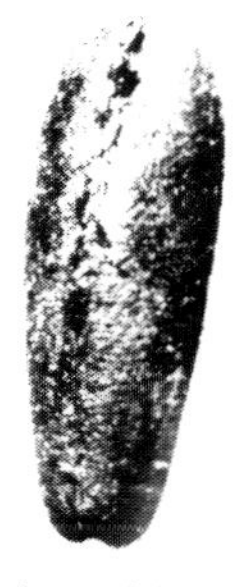

Fig. 188. Spring wild-oat (*Avena fatua*) naked grain from a pit of Tiszaeszlár-Bashalom, Inventory of the Hungarian Agricultural Museum, Budapest.

Fig. 189. Two-rowed barley (*Hordeum vulgare* subsp. *distichum*) hulled grains from Poroszló-Aponhát, Inventory of the Hungarian Agricultural Museum, Budapest.

Fig. 190. Emmer (*Triticum turgidum* subsp. *dicoccum*) naked grains from Poroszló-Aponhát, Inventory of the Hungarian Agricultural Museum, Budapest.

Fig. 191. Spelta wheat (*Triticum aestivum* subsp. *spelta*) grains from Poroszló-Aponhát, Inventory of the Hungarian Agricultural Museum, Budapest.

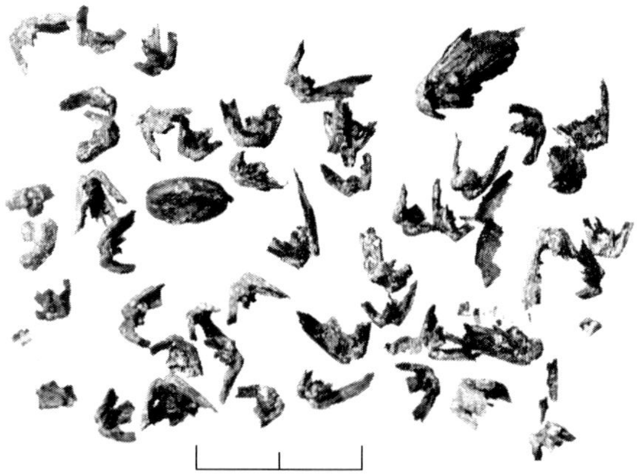

Fig. 192. Spelta wheat (T*riticum aestivum* subsp. *spelta*) spicelets (*furca bicornis*) from Poroszló-Aponhát, Inventory of the Hungarian Agricultural Museum, Budapest.

Fig. 193. Common bread wheat (*Triticum aestivum* subsp. *vulgare*) grains from Poroszló-Aponhát, Inventory of the Hungarian Agricultural Museum, Budapest.

Fig. 194. Common millet (*Panicum miliaceum*) naked grains from Poroszló-Aponhát, Inventory of the Hungarian Agricultural Museum, Budapest.

According to palynological research conducted around Lake Balaton, cereal pollen began to predominate in the second half of the Bronze Age indicating more extensive arable lands (ZÓLYOMI 1980). Pollen of segetal weeds such as knapweed (*Centaurea* spec.), plantain (*Plantago* spec.), buckwheat family (*Polygonum* spec.) and the dock family (*Rumex* spec.) also indicate agricultural activities.

7.3.3.1. Agriculture of the Urnfield culture

Around 1300 BC, tell settlements, built on top of each other for centuries, were *d*epopulated. From the west, the peoples or the cultural influences of the Tumulus culture, and its descendant, the Urnfield culture, arrived in the Carpathian Basin. They built large, fortified settlements on the highest topographical locations. Today, there exists no clear explanation for the sudden change of the form of settlements and lifestyles. The increasingly wet climate and the fear of attacks from other peoples or other settlements surely contributed. These fortified settlements, which survived for longer periods, were also centres of specialized functions, i.e. commercial and industrial centres and also the centres of a system of settlements that we do not know much about today. The settlement structure of the Late Bronze Age was partly constructed from such large-scale "fort-towns", and innumerable smaller peasant farms. According to the archaeobotanical finds, people's activities consisted of a mixture of animal husbandry and plant cultivation (GYULAI 1996d).

The Urnfield culture was a cultural trend and a material culture that extended to a large part of Europe (HARDING 1987). Agriculture was the foundation of an efficient production method. People conducted self-sufficient crop production. New plant species appeared, plants that were only rarely cultivated before becoming widespread. Among the cereals, common millet, among legumes, horsebean, were characteristic plants of the era (JÄGER 1965). This was also the period when *rye* appeared in Western Europe (CHAMBERS – JONES 1984). Oil crops that had been produced since the Neolithic were augmented with gold-of-pleasure.

An increase in the number of cultivated plant species is connected to a demographic explosion that occurred during the Urnfield culture. More cultivated species made production more secure. Exhaustion of soil fertility and the deteriorating climate may have contributed directed to diminished productivity of existing crops augmented by the introduction of new species due better adapted to the new regime .

At the beginning of the 1st millennium BC, people in Northern Europe began to build stonewalls to separate fields. Many of these remains can still be seen in Scandinavia and in Holland, but there are no signs of such a practice in Central Europe (MÜLLER-WILLE 1965). The construction of stonewalls, which is very demanding work, may indicate power relationships but also crop rotation. On the other hand, it is clearly an effect of a deteriorating climate that the higher-lying lands of the English Plain, which had been separated by heaps of stones, were abandoned and lower-lying lands were cultivated instead (BRADLEY 1978).

Late Bronze Age humans divided cultivated areas into arable fields, gardens, pastures, orchards, forests and lake areas (GOLDMANN 1987). Areas were constantly converted to new primary functions, e.g., shallow lakes could even be drained to serve as fertile land for cereals. Topographically flat areas were preferred for agriculture so as to protect against erosion that destroyed the humus layer. Hillsides were more likely to be used as pasture. A plant sociological evaluation of archaeobotanical finds from as Urnfield culture site alongside Lake Constance confirms well illustrates the above functional categories (RÖSCH 1991).

7.3.3.1.1. Plant cultivation of the Urnfield culture in areas near the Carpathian Basin

According to the botanical finds from the Urnfield culture site at Burkheim, Southern Germany, large quantities of common millet and foxtail were grown in addition to the bread cereals (emmer, einkorn, barley, spelt, common bread wheat) (KÜSTER 1988b). This was by no means a unique phenomenon during that age, for both plants were also found at late Bronze Age sites at Ouroux-Marnay in France (HOPF 1985), at Wasserburg-Buchau (BERTSCH – BERTSCH 1949), Untermainbach and Landshut (KÜSTER 1988a) in Germany, and at Zug-Sumpf (NEUWEILER 1931) and Greifensee-Böschen (KÜSTER 1988b) in Switzerland. As the flour of common millet and foxtail millet is unsuitable for making bread, those cereals were primarily used to make meal. However, the sites listed above yielded no rye or oats. Opium poppy and flax were popular, widespread crops. Both were oil crops, but the latter could also be used as a fibre crop. It is also possible that in the Late Bronze Age, opium poppy, which occurs in very large quantities, was also used to obtain opium in Southern Europe (KÜSTER 1988b). Pea is present everywhere. The cultivation of this protein-rich legume has a beneficial effect on the soil, for the bacteria inhabiting their root nodules (*Azotobacter*, *Clostridium*, *Rhisobium*) bind nitrogen from the air, increasing the easily depleted nitrogen content of the soil.

According to the archaeobotanical material described above, plant cultivation was uniform at the foothills of the Alps. This is all the more interesting because the ecological conditions of the sites were in some cases significantly different (KÜSTER 1991). The elevated settlements of Burkheim and Stillfried feature the same range of cultivated plants that characterises the lakeside sites of Champréveyres, Greifensee-Böschen and Hagnau-Burgen or the Danubian island of Straubing-Öberau.

According to a comprehensive study performed by KÜSTER (1991), during the Urnfield culture period the cereals cultivated in the vicinity of the Alps were spelt, barley and common millet, or, in western regions, foxtail millet. Legumes – horsebean, pea, common lentil and bitter vetch – also played an important role in people's diets. Rye does not occur at all, but common oat (*Avena sativa*) does. Opium poppy grew sparser relative to the Neolithic, though it was still a significant cultivated plant. Pollen, and at the Greifensee-Böschen site, the fruit of the European chestnut (*Castanea sativa*) was also found. Bitter vetch and the above-mentioned European chestnut indicate Mediterranean connections. There were striking changes in the range of weeds. Alongside tall weeds: black-bindweed and fan-hen or white goosefoot; shorter species, such as Scarlet pimpernel (*Anagallis arvensis*) and field pennycress (*Thlaspi arvense*) also appeared. This is connected to the low method of harvesting (e.g. reaping of common millet).

The archaeobotanical finds of the region around the Alps also confirm that horsebean was consumed (KARG 1989). In all probability, like at other late Bronze Age settlements, gold-of-pleasure was used in addition to flax to make oil (BUURMAN 1988).

The plant species found at the Stillfried site of the Urnfield culture (around 800 BC), on the March-Morava River, close to Hungary, are similar to those found at sites of the Urnfield culture in Hungary west of the Danube (SCHNEIDER 1991). According to the results of the carpological survey of samples from storage pits, the inhabitants primarily grew hulled cereals and six-rowed barley. The quantity of einkorn somewhat exceeds that of emmer. Other cereals present include spelt, bread and club wheat as well as common millet. The segetal seeds shown also indicate highly developed, autumn-sown hosts. It is possible that the strawberry ground cherry (*Physalis alkekengi*) was meant for consumption given the high vitamin C content of its stone, while hazelnut (*Corylus avellana*) was also a popular food. Plant species from the environment of the site indicate a warm, continental climate of the Subatlantic phases which is similar to today's climate.

Late Bronze Age sites are rich in weeds that are "precursors" of weed species that appear in the Iron Age. For instance, field pennycress (*Thlaspi arvense*) makes its first appearance here (KÜSTER 1988b). It is interesting that against expectations, thermophilous species do not occur. All these species require a moderately warm climate.

The ancient cultivators preferred dry, calciferous loess soils. Weeds indicate that the nutrient supply of ancient fields varied. For example, while goosefoot species and annual meadow grass (*Poa annua*) prefer nutrient-rich soils, lady's bedstraw (*Galium verum*) and sheep's sorrel (*Rumex acetosella*) prefer leaner soils. Those extremes could be connected to the keeping of cattle (KÜSTER 1985b). We assume that the Late Bronze Age already saw the appearance of harvesting near the ground (at an approximate height of 50 cm from the ground). This method of harvesting became predominant only in the Iron Age (WILLERDING 1986c).

7.3.3.1.2. Plant cultivation of Late Bronze Age cultures in the region west of the Danube

In contrast to the rich plant findings from Middle Bronze Age tells, the settlements of the Late Bronze Age are much poorer in plant remains, and are largely concentrated in the region west of the Danube (GYULAI 1993). Fewer settlements have been excavated and, of these, environmental archaeological excavations have been performed only for the last decade (Sopron-Krautacker, Gór-Kápolnadomb). Yet, based on botanical material from the Late Bronze Age sites discussed below, we are confident in asserting that despite all contrary opinions, the quality of plant cultivation in the Middle Bronze Age did not suffer sudden deterioration. Indeed, the range of species identified is clear evidence of further development.

According to seeds found at the Poroszló-Aponhát site of the *Gáva culture*, hulled wheats (einkorn, emmer) continued to play an important role in the Great Plain. Among the bread cereals, einkorn retained its leading place. The cultivation of barley also remained important, but common millet was also a popular food (P. HARTYÁNYI – NOVÁKI – PATAY 1967–68) (*Figs. 189–194*).

Seeds, found in the middle Urngrave layer (12th to 9th centuries BC) of the Mošorin-Feudvár tell settlement near the Tisza River, provide an excellent picture of the plant cultivation culture of the Late Bronze Age – Early Iron Age (KROLL 1990a). In addition to einkorn and emmer, many-rowed barley, rye, common wheat, common millet, common lentil, grass pea wine, flax and opium poppy were also cultivated. Sporadic occurrences of spelt and common bread wheat grains were also found. Weed species found indicate a calciferous, nutrient-rich loess soil that was not manured. Wild forest strawberry (*Fragaria vesca*), danewort (*Sambucus ebulus*), crab apple (*Malus silvestris*) and common elder/red elder (*Sambucus nigra/racemose*) remained popular gathered plants. Water-chestnuts (*Trapa natans*) were collected from river backwaters for food and ornamental purposes.

In 1987, during the excavation of the Tumulus culture settlement of Balatonmagyaród-Hídvégpuszta, organic remains, including seeds, were found at the bottom of a refuse pit. Based on rrecovered pottery fragments, the archaeologist László Horváth dated the remains to around 1200 BC. Organic materials floated from the Late Bronze Age refuse pit were identified as leftover foods: including peas, bitter vetch and grass pea wine seeds as well as common millet meal remains (GYULAI 1996c) (*Figs. 195–203*). Those food remains were placed in the pit at virtually the same time. Perhaps they were burnt during cooking or in some other way, maybe along with other objects. The archaeozoologist István Takács identified several bones of fish and small birds with marks of roasting in the samples. During processing of the finds we found several slightly porous carbonised fragments of variable size from a single larger food item. Archaeobotanical processing, the macroscopic analyses of Max Währen and the microscopic studies of Benno Richter as well as the instrumental analytical investigation performed by János Csapó (macro- and trace elements, amino acids and fatty acids) drew the conclusion that the fragments are remnants of a wild strawberry cake made using baking industry standard bread wheat and common millet flour with the addition of pig fat.

In the period 1973–83, an environmental archaeological excavation was performed at the Sopron-Krautacker site (JEREM *et al.* 1984, 85). Except for a brief interruption, the site was inhabited from the Urnfield culture through to the Late La Tène period. Plant remains were systematically collected. Flotation of the samples from pits of a variety of purposes, houses and graves yielded botanical material with a rich composition of species. The species composition from a waste pit from the Urnfield culture layer (9th–7th centuries BC) supports our inferences drawn from the Gór-Kápolnadomb material that we shall describe below (FACSAR – JEREM 1985). All this indicates that the level of plant cultivation and plant knowledge of the Urnfield culture was uniform. Here again, the main product was emmer, with barley playing a much less significant role. Naked wheat was not cultivated.

From 1989 to 1993, we floated a significant quantity of Late Bronze Age botanical material at an excavation by Gábor Ilon at Gór-Kápolnadomb. The samples from prehistoric pits proved very rich in plant remains (GYULAI – TORMA 1993). The inhabitants of the settlement lived in a varied environment and conducted lively agricultural activities. Ninety-four percent of the remains were from cultivated plants. They primarily cultivated hulled wheats with emmer as most important. It was followed by einkorn and hulled spelt. The difficult task of identification was made easier by the fortunate presence of spelt forks (furca bicornis), which can be identified with relatively certainty. A small number of barley seeds indicates that it was not a very significant cereal. The same is also applicable to common millet. It may have been sowed independently, but it is also possible that it was used as an after-seed. Naked, i.e. hulled common millet grains indicate that it was also processed. More developed naked common bread wheat and club wheat were also present but only sporadically. We believe it to be extremely significant that a few grains of rye were also found.

Sophisticated cereal production is usually accompanied by sophisticated kitchen garden plant cultivation. Peas and common lentil seeds prove that this was certainly the case in this settlement. Their oil crop was gold-of-pleasure.

Korn-rade (*Agrostemma githago*), field and rye brome (*Bromus arvensis* and *B. secalinus*), false cleavers (*Galium spurium*), black-bindweed (*Fallopia convolvulus*) and annual woundwort (*Stachys annua*) were the weeds of autumn-sown cereals (emmer, einkorn, autumn barley). The weed flora associated with row crops and spring-sown cereals is also rich: darnel (*Lolium temulentum*), yellow bristle-grass (*Setaria lutescens*), common chickweed (*Stellaria media*), field pennycress (*Thlaspi arvense*), white and maple-leaved goosefoot (*Chenopodium album* and *Ch. hybridum*), curltop lady's-thumb (*Polygonum lapathifolium*) and black nightshade (*Solanum nigrum*) were all found. Einkorn, emmer and spelt earspindle fragments found together with the remains of weeds indicate that after hulling, the grains were cleaned, probably by winnowing. This is also indicated by the fact that we found exceptionally high numbers of emmer and spelt forks (*furca bicornis*). This may be explained by the fact that they were unable to remove all contaminants and weed seeds from the cereals they stored. The negative imprints of burnt husks and earspindle fragments, threshing and perhaps hulling/winnowing waste found in the many mud-flakes located during the excavation only reinforce the assumption that hulled grains were produced. Species from the natural environment indicate a varied environment: there were open water areas, marshland, meadowland and forest in the vicinity of the settlement.

The plant cultivation of Gór in the Late Bronze Age – Early Iron Age is very similar to that of Krautacker of the same period. Among the cultivated plants found here, the proportions of cereals are similar, though there is much less of them. The grains are mixed with the seeds of legumes. The only relatively pure deposit of barley was found in one of the pits at Sopron-Krautacker.

Both excavations yielded a relatively low quantity of seeds, which is not only attributable to well-kept fields, but also to the time of harvesting. In the past, cereals were harvested before they dried and the seeds were easy to thresh while also eliminating some late-ripening weed species. This was worth the effort even if the grains had to be dried carefully afterwards to avoid generating excessive heat.

Since the majority of a meagre crop could be lost due to the rough movements of a scythe, harvesting by grabbing handfuls of grain and cutting them near the ear using a sickle was more effective.

These excavations yielded only a very small number of legume seeds (horsebeans, common lentils, peas). This raises the possibility that perhaps the row crops were not cultivated separately but that they served as "useful contaminants" of grains. As we know from Aristophanes, common lentil was the food of the poor in the classical world (DE CANDOLLE 1894). Its seeds were too small to deal with separately. It was not very long ago that ground lentils were still being mixed with bread cereals. Common lentil and pea can even occur in autumn cereals, for they are able to survive milder winters.

Between 1994 and 1996, during an archaeological survey preceding the extension of the M1 motorway near Győr, several Late Bronze Age sites were excavated under the management of András Figler: Börcs-Paphomlok, Mosonmagyaróvár-Németbánya, Mosonmagyaróvár-Németdőlő. All cereals cultivated in that period were found: six-rowed, two-rowed and naked barley, common millet, einkorn, emmer, common and durum wheat. Quantity of cereal grains occurred in the following ranking: emmer, six-rowed barley, common millet, common bread wheat. The remaining species occurred sporadically, in quantities that were too small for evaluation. All of this indicates a settled population with sophisticated agricultural/plant cultivation know-how. Their use of varied cereals provided a reliable subsistence even in bad harvest years. These people conducted highly varied and therefore secure cereal production, implying long-lived agricultural settlements.

A Late Bronze Age well was found unexpectedly at the Mosonmagyaróvár-Németdőlő site. The wet, oxygen-free environment of the well preserved organic remains in the landfill. Thanks to the careful collection work and flotation, the site can now be regarded as one of the sites of the Carpathian Basin with the richest range of species. It is also one of the oldest Hungarian well-sites to be processed. A large number of seeds from a segetal weed association of autumn-sown species indicates an autumn host (primarily emmer, common bread wheat, perhaps autumn barley): e.g. parsley-piert (*Aphanes arvensis*), black-bindweed (*Fallopia convolvulus*), narrow-fruited cornsalad (*Valerianella dentata*), annual knawel (*Scleranthus annuus*) and annual woundwort (*Stachys annua*). However, the number of species belonging to the spring-sown weed association (*Polygeno-Chenopodietalia*) is even higher. There were weeds of spring-sown cereals (primarily common millet, and perhaps spring barley; spring wheat although rare in this period did occur) and garden vegetables. They included common orache (*Atriplex patula*), common chickweed (*Stellaria media*), fan-hen or white goosefoot (*Chenopodium album*), maple-leaved goosefoot (*Ch. hybridum*), henbit dead-nettle (*Lamium ampexicaule*), red dead-nettle (*L. purpureum*), the small-flowered fumitory (*Fumaria schleicheri*), cotton thistle (*Onopordium acanthium*), curltop lady's-thumb (*Polygonum lapathifolium*) and black nightshade (*Solanum nigrum*).

The fruits of many plants found here have a pleasant taste and were probably consumed The vitamin-rich crab apple (*Malus silvestris*), dewberry (*Rubus caesius*) and woodland European grape (*Vitis vinifera* subsp. *silvestris*) were collected in the nearby gallery forest. They were surely also aware that leaves of the goosefoot species may be used as salads while the seeds are good grain substitutes. Some of the species may also have been used as natural dyes, e.g., danewort (*Sambucus ebulus*) was used for dying leather in the past.

We also found medicinal plants among the carpological remains of the well, including common yarrow (*Achillea millefolium*), woolly burdock (*Arctium tomentosum*), stinking nightshade (*Hyoscyamus niger*), knotgrass (*Polygonum aviculare*), tormentilla cinquefoil (*Potentilla erecta*), dewberry (*Rubus caesius*), broad-leaved dock (*Rumex obtusifolius*), soapwort (*Saponaria officinalis*), annual woundwort (*Stachys annua*), common chickweed (*Stellaria media*), danewort (*Sambucus ebulus*), woody nightshade (*Solanum dulcamara*), wall germander (*Teucrium chamaedris*) and common vervain (*Verbena officinalis*).

Naturally, this does not mean that any or all were used as medicinal plants, but it cannot be excluded, either.

Some strongly poisonous species also occur: hemlock (*Conium maculatum*), stinking nightshade (*Hyoscyamus niger*) and black nightshade (*Solanum nigrum*). It is highly likely that the inhabitants were familiar with their effect and it is also possible that they used some of those plants, which contain hallucinogens, as drugs/poisons. At any rate, they are highly dangerous if put in food.

Plant species found at the site also allow us to draw inferences about the ancient environment. There must have been a large body of water very near the settlement. A considerable quantity of common bulrush (*Schoenoplectus lacustris*) achenes had been taken to the settlement from a reed-bed near the water. It may have been used as a grain substitute, cf. similar to the bulrush stock of the Late Neolithic – Early Copper Age site at the Zalaszentbalázs-Szőlőhegy field. The following waterside plants were found: fox sedge/prickly sedge (*Carex vulpina/muricata*) and gipsywort (*Lycopus europaeus*). The open water had adjacent marshlands, with, among others, yellow sedge/slender tufted sedge (*Carex flava/gracilis*) and tormentilla cinquefoil (*Potentilla erecta*). Given their great similarity the achenes could belong to either of the two species. Higher ground in the topographic succession was the source of achenes of tussock sedge/yellow sedge (*Carex elata/gracilis*) and common spike-rush (*Eleocharis palustris*).

Meadowland, used for animal pasturage, was probably variable and, given the number of species, rather large. Its wetter regions may have furnished ragged-robin (*Lychnis flos-cuculi*), creeping cinquefoil (*Potentilla reptans*), common self-heal (*Prunella vulgaris*) and tall violet (*Viola* cf. *elatior*). Based on the number of species, most of the meadow had a moderate water supply. These areas were the habitat for musk thistle (*Carduus nutans*), smooth meadow-grass (*Poa* cf. *pratensis*), common yarrow (*Achillea millefolium*), garden daisy (*Bellis perennis*) and pale sedge (*Carex pallescens*).

It is probable that some of their fields were obtained by clearing woodland, as danewort (*Sambucus ebulus*) and sermountain (*Laserpitium latifolium*) usually prefer to grow in cleared woodland areas. Hemlock (*Conium maculatum*) came from a forest fringe with moderate water supply.

The number of the seeds/fruits of ruderal plant species is almost as high as that of seed plants. It is true that fan-hen or white goosefoot can be classified both as a ruderal plant and as a weed of row crops. The majority of those species are from typical habitats: woolly burdock (*Arctium tomentosum*), knotgrass (*Polygonum aviculare*), creeping buttercup (*Ranunculus repens*), broad-leaved dock (*Rumex obtusifolius*), soapwort (*Saponaria officinalis*), white campion (*Silene alba*) and common vervain (*Verbena officinalis*). A smaller number are from arid habitats: wild carrot (*Daucus carota*), stinking nightshade (*Hyoscyamus niger*), cat-mint (*Nepeta cataria*), silver cinquefoil (*Potentilla argentea*) and welted thistle (*Carduus acanthoides*). All this means that they chose a site for their settlement that was a somewhat drier area at the time.

In the summer of 2000, Anna Endrődi, Attila Horváth and László Horváth found five Late Bronze Age wells with wattle work and plank linings, dug using bucket technology at the Dunakeszi-Székesdűlő site, slated for construction of an Auchan supermarket. The mud of the wells contained an extremely high quantity of organic remains. Among them, well no. 315 was the richest in plant remains (GYULAI 2003).

In addition to cereals and seed species, the number of species from the natural environment of the era is very high (46 species!). Not counting a part of the diaspores of cereal grains and their weeds, the majority of the grains and seeds were preserved in a non-carbonised, so-called "sub-fossil" state due to their continuous immersion in water.

The species composition and proportions of the cereals found here are consistent with cultivated plants found at sites of Late Bronze Age cultures in Hungary (GYULAI 2001a). Common millet occurs in the largest quantity, but barley is also present. Only hulled wheat grains were found (emmer and einkorn). All grains were found in a cleaned state, stored for consumption. They are the produce of settled, crop producing inhabitants.

Weed associations of autumn-sown cereals (*Secalietea*) indicate an autumn-sown host (bread wheat, perhaps autumn barley). They include field brome (*Bromus arvensis*), black-bindweed (*Fallopia convolvulus*), common fumitory (*Fumaria officinalis*) and annual woundwort (*Stachys annua*). They are tall weed plants, which indicates that cereals were harvested using a sickle, at about two-thirds of the height of stalks. Of the weed association of hackfruits or spring cereal weeds (*Polygeno-Chenopodietalia*), the only species that occurs is the green/rough bristle-grass (*Setaria viridis/ verticillata*). This indicates the cultivation of spring-sown cereals, primarily common millet, perhaps spring barley, while spring wheat was rarely sown at that time.

With respect to gathered plants, the site is very similar to the Mosonmagyaróvár-Németdűlő site described above. Again, we find crab apple, dewberry and woodland European grape with the new additions of strawberry ground cherry (*Physalis alkakengi*), hazelnut (*Corylus avellana*), dogwood (*Cornus mas*) and common or red elder (*Sambucus nigra/racemosa*). The seeds of fan-hen or white goosefoot and danewort were also found here as well.

The herb, wild marjoran (*Origanum vulgare*), and medicinal plants, common fumitory (*Fumaria officinalis*), white horehound (*Marrubium vulgare*) and common nettle (*Urtica dioica*) were also found. Strongly poisonous species also occurred: stinking nightshade, black nightshade and woody nightshade.

There must have been a large body of water, swamp and regularly flooded area very near the settlement. The majority of plant species are derived from such a habitat. The seeds/fruits of pondweed (*Potamogeton* spec.) and various floating weeds such as yellow water-lily (*Nuphar lutea*) and amphibious bistort (*Polygonum amphybium*) came from open water. But, an even larger number of remains are of plants that grow in waterside reed-beds, including Scottish dock (*Rumex aquaticus*) and tabernaemontanus bulrush (*Schoenoplectus tabernamontani*). We found particularly large quantities of half-achenes from water fennel (*Oenanthe aquatica*) and achenes from common bulrush (*Schoenoplectus lacustris*). Higher-lying areas next in the topography were the sources of water germander (*Teucrium scordium*) and common spike-rush (*Eleocharis palustris*) achenes. We also found the remains of a number of waterside plants: brown galingale (*Cyperus fuscus*), carpet cinquefoil (*Potentilla supina*), blister buttercup (*Ranunculus sceleratus*). The achenes of bristle club-rush (*Schoenoplectus sataceus*) and bog-rush (*Schoenus nigricans*) indicate that there was marshland at the border of the body of water and dry land.

There may have been a meadow/pasture near the settlement. Its wetter section was the source of creeping cinquefoil (*Potentilla reptans*). Given the number and variety of species, the larger part of the meadow had a more moderate water supply. Other plants from this area include pale sedge (*Carex pallescens*), common bird's foot-trefoil (*Lotus corniculatus*) and ragged-robin (*Lychnis flos-cucoli*).

The material contains a fairly large proportion of species from a so-called ruderal association. The number of seeds/fruits of ruderal species exceeds that of weeds, therefore the settlement itself must have been of considerable size. Such ruderal weed associations are found in areas affected by human influences (sides of roads, ditches, embankments, fallow areas, around buildings) where the nitrogen content of the soil is high, or where perhaps the soil is manured. Individual species, such as common nettle (*Urtica dioica*) or stinking nightshade (*Hyoscyamus niger*), indicate wet as well as dry areas of "ruined soil". The majority of species are from a typical natural habitat, which was the most characteristic environment around the settlement. These include fan-hen or white goosefoot (*Chenopodium album*), knotgrass (*Polygonum aviculare*), creeping buttercup (*Ranunculus repens*), broad-leaved dock (*Rumex obtusifolius*) and white campion (*Silene alba*).

Almost fifty kilos of mud samples from the well were taken from five different depths. Thus, some stratigraphic inferences can also be drawn from the plant remains found in them. The sample from -240 cm has hardly any plant species, which indicates the period of use of the well. The remains that do occur are largely cereals and gathered plants. The next sample, from -200 cm, was similar, with the difference that the number and seed concentration of cultivated and wild plants was higher. This implies that the well remained in use during that period, but the economy of the settlement became more intensive, possibly due to expansion of the settlement?. The sample from -160 cm had a large number of cultivated plants, but some marshland plants also appeared, perhaps as a result of a wetter period. The sample from -120 cm hardly contains any cultivated plant remains, while the seeds/fruits of open water species appeared. It is probable that at that time the well was flooded by expansion of a small lake, still extant, in a nearby depression. The sample from -100 cm has the seeds/fruits of only a few species. As it silted up, the well did its job: it preserved a message from the past coded in seeds.

In contrast to the above sites in the west of the country, which are rich in plant remains, so far we have only one Late Bronze Age site from the eastern part of the country that yielded plant remains. Excavation of a prehistoric settlement covering several tens of hectares began in 2001 at the Ludas, Varjú-dűlő site. The majority of objects found during the excavation led by László Domboróczky, Csilla Ács, Károly Tankó and Simon Gall belong to the Late Bronze Age Kyjatice culture, though a small number of Copper Age, Celtic and Sarmatian objects were also found. Many soil samples were collected from the various pits, Urnfields and vessels, which were floated on site. Processing is still in progress, but we can already draw some inferences from the review of the almost 200 Late Bronze Age samples.

Soil samples from pits contain carbonised grains of various cereals: common millet, emmer, einkorn and many-rowed barley. Relative to the cereals, we found a large quantity of weed remains: black-bindweed, barnyard grass, catchweed, cleavers, false cleavers, fan-hen or white goosefoot, common fig-leaved goosefoot, maple-leaved goosefoot, spring wild-oat, common mallow, creeping buttercup, knotgrass, green/rough bristle-grass, annual ryegrass, tasteless water-pepper, spiked star of Bethlehem, meadow fescue, danewort, lady's bedstraw, hare's-foot clover, charlock.

The seeds of legumes also occur: field pea, small seed horsebean, small seed lentil. The single opium poppy seed we found was so fragmented that it was impossible to tell whether it was from the cultivated or the wild variety.

Fewer samples were taken from fireplaces. Obviously, common millet and emmer grains found there must have fallen down the side of the fireplaces during cooking and were slowly carbonised by the radiating heat.

The soil samples from Urnfields also contain some plant remains. They are primarily smaller and larger fragments of charcoals with a few grains of six-rowed barley, emmer, club wheat and a few seeds of weeds: white melilot, false cleavers, narrow-leaved vetch, summer vetch and even remains of common bunt and stinking smut. They are all carbonised. It is unusual, but even the soil washed from some vessels contained diaspores: the fragmented grains of common bread wheat and some other cereals.

Before the onset of winter, a granary, i.e. a grain storage pit sunk in the ground was also found at the excavation, with several kilograms of carbonised emmer wheat mixed with soil.

Fig. 195. Carbonised seeds and bearings among inorganic residues (for the most part leguminous) waiting for selection. Flotated sample from the Late Bronze Age pit in Balatonmagyaród-Hídvégpuszta. Photograph by the author.

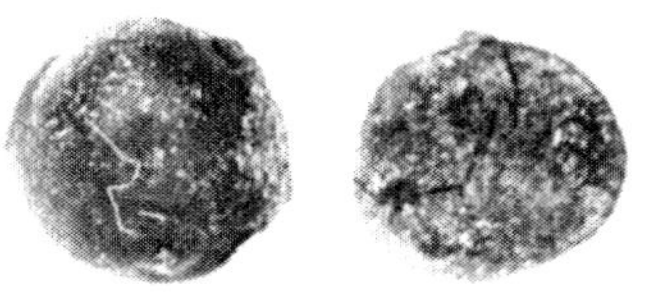

Fig. 196. Small seed lentil (*Lens culinaris* subsp. *microsperma*) seeds from the waste pit of the Late Bronze Age settlement of Balatonmagyaród-Hídvégpuszta. Photograph by Tibor Kádas.

Fig. 197. Barley (*Hordeum vulgare*) grains hulled grain from the Late Bronze Age settlement of Balatonmagyaród-Hídvégpuszta. Photograph by Tibor Kádas.

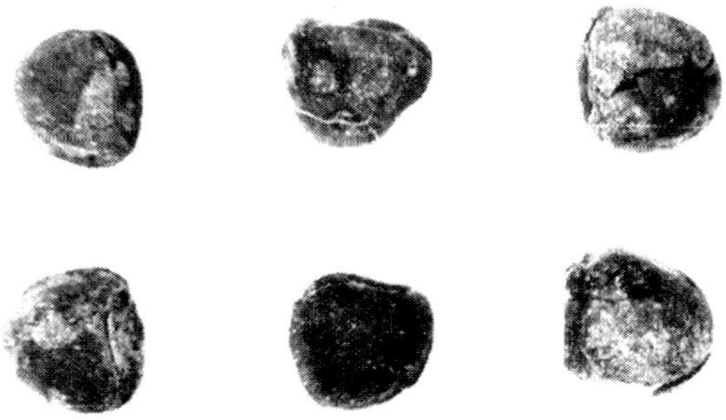

Fig. 198. Bitter vetch (*Vicia ervilia*) seeds from the waste pit of the Late Bronze Age settlement of Balatonmagyaród-Hídvégpuszta. Photograph by Tibor Kádas.

Fig. 199. Pea (*Pisum sativum*) from the waste pit of the late Bronze Age settlement of Balatonmagyaród-Hídvégpuszta. Photograph by Tibor Kádas.

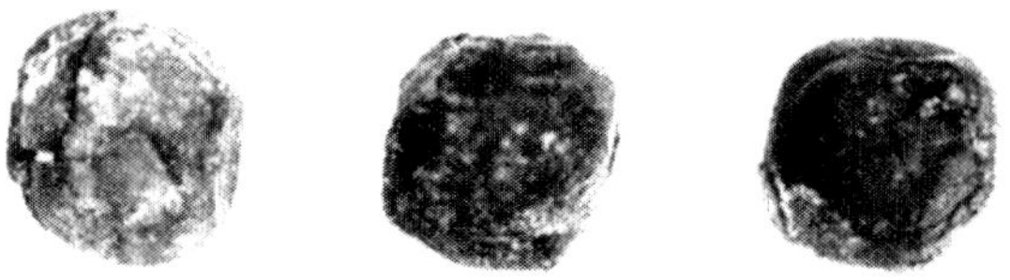

Fig. 200. Grass pea vine (*Lathyrus sativus*) seeds from the waste pit of the late Bronze Age settlement of Balatonmagyaród-Hídvégpuszta. Photograph by Tibor Kádas.

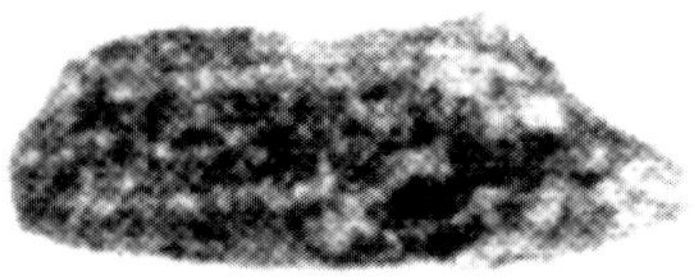

Fig. 201. Spring wild-oat (*Avena fatua*) naked grain from the waste pit of the Late Bronze Age settlement of Balatonmagyaród-Hídvégpuszta. Photograph by Tibor Kádas.

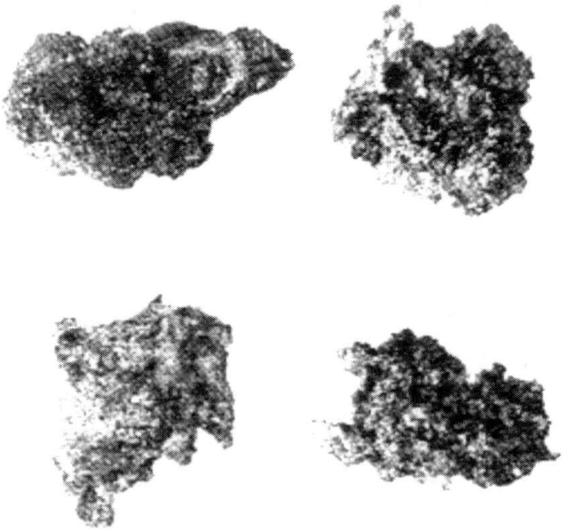

Fig. 202. Gruel fragment from the waste pit of the Late Bronze Age settlement of Balatonmagyaród-Hídvégpuszta. Photograph by Tibor Kádas.

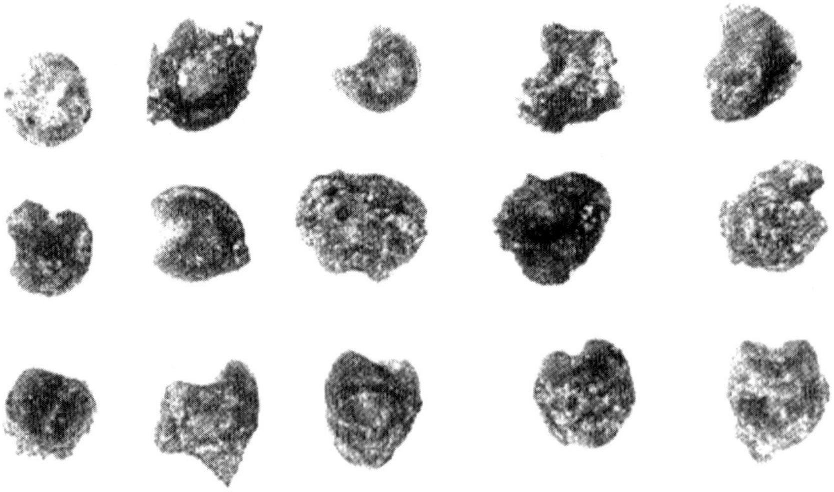

Fig. 203. Common millet (*Panicum miliaceum*) naked grains from the waste pit of the Late Bronze Age settlement of Balatonmagyaród-Hídvégpuszta. Photograph by Tibor Kádas.

7.4. The Iron Age

Around 800–600 BC, a new era of the Holocene, the Subatlantic phase, also known as the Flandrian interglacial, began around 8200 BC and continues to the present. During the era, known as the Iron Age from the archaeological perspective (9th pollen zone, 900–0 BC), the climate turned drier. Its continental character was somewhat enhanced and groundwater levels dropped. The cool and rainy, though temperate climate of the Subboreal became milder. Blytt – Sernander put the beginning of the Subatlantic phase at 800 BC while Nejstadt defined the era as beginning in 600 BC and designates it as late Holocene (in: SOMOGYI 1984).

As precipitation decreased, European beech began to recede. Due to the dryer climate and the retreat of forests, the formation of chernozem soils began in the Great Plain. European beech gained ground in mountainous regions but receded elsewhere. Hornbeam also became rarer. The climate favoured the spread of oak. The contrasting process of the reduction of the population of oak trees is attributable to anthropogenic effects. An increase of non-arboreal pollen (NAP) indicates regular plant cultivation. Bladderwort (*Utricularia* spec.) and water-crowfoot (*Batrachyum* spec.) appeared in waters, while yellow iris (*Iris pseudacorus*) and flowering rush (*Butomus umbellatus*) appeared in reedbeds. Pollen analysis of sediments of Lake Balaton performed by ZÓLYOMI (1971, 1980) indicated anthropogenic effects quite clearly. In the Subatlantic phase, the increase of birch pollen may be attributed to slow degradation resulting from the clearing of woodlands. Deforestation, the spread of agriculture, grazing and treading resulted in soil erosion.

7.4.1. Iron Age agriculture on the peripheries of the Carpathian Basin

As a result of the environmental archaeological excavations of recent decades, our knowledge of Iron Age plant cultivation is consistently increasing. The Hallstatt and La Tène eras (around eight hundred to a thousand years ago, depending on the location) are very important periods of cultural history (*Fig. 207*).

The sophisticated plant cultivation of the Late Bronze Age deteriorated at the beginning of the Hallstatt Age, and recovered slowly. Remains from the La Tène era allow us to infer a return to sophisticated agriculture. The level of plant cultivation is nearly uniform, though there are some differences between various areas and periods. Cultural trends had a stronger effect than the pre-existing climatic differences on the agriculture of populations living here.

There are no significant differences between the Hallstatt Period (700–420 BC) plant remains from the Czech Republic and from Slovakia (WASYLIKOWA *et al.* 1991). Among cultivated cereals, emmer is in first place, followed by einkorn, many-rowed barley, common millet, pea, common lentil and horsebean. Flax tends to occur more in eastern parts and spelt was found in the western part of Slovakia. Bohemia and Moravia are characterised by common bread wheat. Naked barley and gold-of-pleasure were found in Slovakia (HAJNALOVÁ 1989a). Common oat was cultivated in the area that is the Czech Republic today, while its cultivation began in Slovakia in the early Celtic period.

The following wild species were consumed: hazelnut, crab apple, sloe, blackthorn, and bramble. Oak acorns should also be included. Pear seeds found could have come from wild or cultivated plants. Some cultivated (domesticated) fruit species are also known; e.g., plums, peaches. Hallstatt Period sites are particularly rich in weed plants related to the restoration of diversity of cultivated species. In some cases the number of species exceeds one hundred (!) (KÜHN 1977).

The Hallstatt site at Pod, Bosnia, provided one of the earliest cases of the cultivation of oats (KROLL 1991). Although the majority of the finds are wild oat (*Avena sterilis* and *A. fatua*), common oat (*A.*

sativa) also occurs. First century BC plant remains from Krivina in Bulgaria (HAJNALOVÁ 1979b) and Svetjina in Serbia (Borojevic 1987) indicate the spread of common wheat and durum wheat and the beginning of the spread of rye.

The Hallstatt sites of Romania (1200–450/300 BC) have furnished all hulled and naked wheats with the exception of durum wheat, along with six-rowed barley and oak (WASYLIKOWA et al. 1991). In La Tène period (Geto-Dacian culture, end of 1st century BC), emmer, naked six-rowed barley, common millet, rye and foxtail millet were added to the earlier cereals. Among the legumes, seeds of common lentil, pea and horsebean have been found. Opium poppy and gold-of-pleasure were used as oil crops, while hemp was grown for fibre. The seeds of wine grape also occur in small quantities (CARCIUMARU 1983 a, b).

Cereal remains from Hallstatt sites in Poland indicate that plant cultivation culture was by no means uniform (KLICHOWSKA 1984). They cultivated six-rowed barley, emmer, spelt, common and club wheat and common millet. The dominant species, defined by yield, were different in different areas. Based on the number of seeds, it seems that barley played the leading role on one occasion (Bnin, 99% of all grains), while emmer was in first place twice (Biskupin, Smuszew, 45% of all grains), spelt once (Kamieniec, 94% of all grains) and common and club wheat once (Grzybiany, 51% of all grains) (TOMCZYNSKA – WASYLIKOWA 1988). Gold-of-pleasure and opium poppy were also in cultivation, probably along with wild turnip. Rye and oats were found in many Hallstatt pollen profiles. Modest quantities do not allow us to infer independent cultivation, therefore it was probably grown mixed with other bread wheat species. Hazelnut, bramble, wild raspberry and mazzard cherry were wild fruits, but they may have been planted for their fruit in some instances.

The foothills of the Alps (like other areas of Europe) have many fewer plant remains from the Hallstatt Period (starting in 800 BC) (KÜSTER 1991). The Dillingen-Steinheim site (Hallstatt C/D) still exhibited the cereals of the Urngrave culture, but to a much more modest extent. On the other hand, common millet occurred in large quantities. The Niederlbach site is particularly rich in leguminous remains.

The northern half of Germany, in particular the Rhine area, provides an exception. The number of Iron Age sites with plant remains is twice the number of such Bronze Age sites (KNÖRZER 1991). The area of the Lower Rhine is particularly rich in Iron Age plant remains (800–0 BC). Having reviewed the botanical material from 50 sites, KNÖRZER (ibid.) reached the conclusion that the main production crop was barley, in particular naked barley. It was followed by emmer. Einkorn was not a significant bread cereal.

The carbonised carpological remains from the Boomborg site along the lower section of the River Ems constitute a good representation of Iron Age plant cultivation of the area (BEHRE 1970). Barley was the most important cereal here as well. They also cultivated flax, gold-of-pleasure and horsebean. Naked barley was also cultivated on the island of Archsum/Sylt (KROLL 1987)

According to the plant remains found at La Tène sites of the Rhine area in Germany, plant cultivation intensified further. Common millet (Stelle/Hamburg) and oats (Rullstorf/Lüneburg) were added to the cereals of the Hallstatt Period (naked and hulled barley, emmer) (KROLL 1980).

All the protein-rich legumes known from that era were found at the Hallstatt Age site of Langweiler/Düren: horsebean, pea and common lentil (GÖBEL – KNÖRZER 1973).

Botanical finds from the Hallstatt site at Niederlbach in the Isar Basin are dominated by naked barley (KÜSTER 1988c). Other cereals found at the site include many-rowed barley, emmer, common bread wheat, einkorn, common millet and a variety of oat that could not be identified more precisely. Such large-scale occurrence of naked barley is a unique phenomenon in that period. Among legumes, horsebean, pea and common lentil were found. Shells of hazelnuts indicate gathering. Carbonised seeds and fruits were not found in storage pits but in samples taken from various locations in the culture layers of the settlement.

A set of plant remains covering a number of archaeological periods is known from the Stuttgart-Mühlhausen site (PIENING 1988a). Compared to La Tène, the Urngrave culture was poor in species. They cultivated wheat (spelt), barley and pea. In contrast, in La Tène Period the number of the species as well as the volume of cultivated species increased. Some species indicating the exploitation of grasslands were also found: ribwort-plantain (*Plantago lanceolata*) and red clover (*Trifolium pratense*) (perhaps members of an ancient *Molinio-Arrhenatheretea* association?).

In the south of Germany, spelt became the most important bread cereal by the Early Celtic Period (e.g. the Lauffen/Neckar site) (PIENING 1983). In the Middle and Late Celtic Periods in Southern Bavaria (e.g. Manching) barley and spelt were cultivated, though in some instances spelt was replaced by emmer (KÜSTER 1991). Rye and common oat also occurred, although in small quantities. Their cultivation could not have been significant. There are some legumes, but not in the quantities usual for Urnfield sites. In some cases, arable lands were within the settlement fortifications.

The excavation of the Christenberg and Goddelau sites near Münchausen in Northern Germany yielded considerable plant remains (KREUZ 1992–93). According to the research, the medium-dry climate of the era had a beneficial effect on agriculture. According to dendrochronology, the site is dated to 420 BC (La Tène A). Among cereals, barley, spelt, emmer, common/durum wheat and common millet were found. The remains also contained some oats. Peas, lentils, horsebeans and bitter vetch were cultivated. Gold-of-pleasure and flax were grown as oil/fibre crops.

Most Late Bronze Age sites of the lower Rhine area contain large quantities of spelt (KNÖRZER 1991). Common oat and rye occur very rarely. Cultivation of common millet was widespread, foxtail millet also occurred in some locations. It is conceivable that barnyard grass was also taken into cultivation. All the legumes known at that time (pea, common lentil, horsebean, bitter vetch) also occur here. Flax and opium poppy occur sporadically. Seeds of gold-of-pleasure are commonly found at Celtic sites. Seeds of root vegetables are also found (carrot, fruited cornsalad). It is possible that some of these wild species were planted in gardens near houses. The seeds/fruit of several wild fruit species were brought in from the surrounding forests. Increasing size of pastureland is indicated by a large number of plant species from such habitats. These areas mostly yield the seeds/fruits of low-growing grasses, important evidence indicating that grass was mowed, and mowed low, probably using scythes.

Analysis of the remains from Late Bronze Age arable lands between Ems and Weser furnished a surprising result. It seems that the Celtic farmers living there had improved the soil of their arable lands (Sylt, Flögeln-Haselhörn) using humus-rich soil from remote locations (ZIMMERMANN 1976). Similar indications have been found at Vaarsee in the Netherlands (BRONGERS 1976). Lands thus improved were primarily used to grow barley. These are some of the earliest indications of soil fertility maintenance (BEHRE 1980).

Plant remains from the Late La Tène site at Porz-Lind near Cologne (approx. 100 BC) reinforce the assumption that plant cultivation was uniform during that period (KNÖRZER 1987). Emmer, barley and rye were their cereals. They grew significant quantities of common millet. Common millet, though its cultivation had begun in the Rhine area in the Late Bronze Age, became significant only by the Late Iron Age. It lost its significance again by the Roman Period. The most important Iron Age oil crop of the area was gold-of-pleasure. But, there is also evidence of flax and opium poppy. It is highly questionable whether oil was pressed from those plants. It is much more likely that they obtained the oil by boiling. The weed species found here belong to the weed associations of autumn-sown cereals and hackfruits or spring cereal weeds associated with garden vegetables.

A different range of cereals was found in the south of Germany. The most important cereal at the Freiberg-Geisingen/Ludwigsburg site was naked barley. In addition, the grains of all the other bread wheat species (einkorn, emmer, spelt, common and club wheat) were found (PIENING 1988b).

In contrast to the relatively rich plant findings of the Hallstatt layers of Poland, La Tène material has little in the way of plant remains. All the more important cultivated plants do occur (barley, emmer, spring/club wheat, spelt, rye, common millet, oat (*Avena* spec.), horsebean, pea, hemp) but only in small quantities (KLICHOWSKA 1984).

In the area of the Czech Republic and Slovakia, the primary cereal of La Tène Period (420–0 BC) was emmer, but smaller or larger amounts of einkorn, spelt, many-rowed barley, common millet, pea, lentil and horsebean were also found (WASYLIKOWA *et al.* 1991). The proportions of common bread wheat and rye are so small that it is questionable whether they were independently cultivated. It is certain that common oat, two-rowed and many-rowed barley as well as hemp were cultivated (TEMPÍR 1968; HAJNALOVÁ 1989a). In La Tène Period, the number of weed species decreased. According to the cereal weeds that were found, people harvested at medium height. Cereal grains and the seeds of legumes were ritually burned (HAJNALOVÁ 1979a).

7.4.2. Crop production of the Early Iron Age in Hungary

We only have a comparatively modest number of findings from the Early and Late Bronze Ages in Hungary, as opposed to the aforementioned wide variety of plant residues in Western and Central Europe. The primary reason for this is a lack of archaeobotanical explorations of these sites.

Albeit in pits at various sites containing Late Bronze and Early Iron Age layers (Celldömölk-Sághely, Gerjen-Várad, Bakonyszentkirály-Zöröghegy) carbonised seeds occur every now and then, but it seems that these are not genuine grain pits (P. HARTYÁNYI – NOVÁKI – PATAY 1967–68). These grains appear to be secondary waste material. However, it can be concluded from these sporadic bread wheat, barley and legume (common lentil, horsebean) finds that during the Late Bronze Age – Early Iron Age crop production in these places – though on a limited scale – must have remained continuous. Early Iron Age bronze reaping-hooks excavated in Kőszeg and many other sites indicate crop production as well (NOVÁKI 1965–66).

The exact ages of archaeological layers explored at different places in the Aggtelek-Baradla cave system between 1876 and 1877 are almost impossible to determine (*Figs. 208, 210*). According to P. HARTYÁNYI – NOVÁKI – PATAY (1967–68), graves in the "Bone house" and "Corridor" and the seeds found therein are from the Early Iron Age. Their opinion is based on radiocarbon tests of common millet: 2560 BP ± 75 years (a report by Mebus A. Geyh from 1967 in: P. HARTYÁNYI – NOVÁKI – PATAY 1967–68). Imre Deininger classified seeds these contextss into ten cultivars; such as, common bread wheat, einkorn, common millet, naked barley, grass pea vine, horsebean, pea, common lentil, gold-of-pleasure. He also lists a naked grain, which he termed "tiny wheat" (in: NYÁRY 1881). Dimensions for the seeds are also provided. In addition to cultivars, 14 species of weeds were found. Due to the amusing Hungarian equivalents of contemporary botanical names, these are listed by names: mohar sertike (today: zöld muhar, green bristle-grass) (*Setaria viridis*), tisztest bibetyű (today: sármányvirág, iron woundwort or ironwort) (*Sideritis montana*), tompa lórom (today: réti lórom, broad – leaved dock) (*Rumex obtusifolius*), szulák cikkszár (today: szulákkeserűfű, black – bindweed) (*Fallopia convolvulus*), lóromlevelű cikkszár (today: lapulevelű keserűfű, curltop lady's–thumb) (*Polygonum lapathifolium*), pokolvar libatop (maple-leaved goosefoot) (*Chenopodium hybridum*), óra ziliz v. dinnyehibik (today: varjúmák, flower-of-an-hour) (*Hibiscus trionum*), ragadós galaj (catchweed or cleavers) (*Galium aparine*), tejoltó galaj (lady's bedstraw) (*G. verum*), tóparti galaj (today: mocsári galaj, common marsh-bedstraw)

(*G. palustre*), disznóparéj (today: szőrös disznóparéj, hairy amaranth) (*Amaranthus retroflexus*), földi bodza (ground elder) (*Sambucus ebulus*), mezei zsálya (field clary) (*Salvia* cf. *pratensis*), lándzsás útifű (ribwort-plantain) (*Plantago* cf. *lanceolata*). One can assume, however, that emmer, common millet, slender pea (*Pisum elatius*) and vetchling (*Lathyrus* spec.) seeds – coming from the cave section called The Sultan's Sofa – explored by Ottokár Kadic during the 1930s are older, coming from the Late Bronze Age Kyjatice culture.

The possibility cannot be excluded that some of the long lived housing settlements of the funeral-urn culture survived until the Celtic conquest. This could by no means be a general feature, however. Based on palynological investigations, the ratio of herbaceous NAP (non-arboreal pollen) and *Chenopodiaceae* (goosefoot family) diminished in human settlements around Balaton at the advent of the Hallstatt Period, while the number of *Fagus* (European beech) pollen increased. The conclusion can be drawn that the number of settlements must have been reduced during that period (ZÓLYOMI 1980).

Taking all this into account, we find that common conditions are best reflected by the Sopron-Krautacker site (JEREM *et al.* 1984 1985). Here, people abandoned the settlement at the end of Hallstatt B Period followed by a hiatus of approximately 300 years before it was again colonised . Its abandonment must have been influenced by climatic conditions as well. People living here might have moved away due to the unfavourable, cold and wet climate of a transient period during the Subboreal-Subatlantic phase (app. 800–500 BC).

A cool, humid climate adversely affecting human settlement in the Alpine regions of Central Europe is confirmed by the short-lived expansion of glaciers (Burga 1988). Colonisation was not restored until the short, but warm and dry period of the Late Hallstatt. Following this, at the beginning of La Tène Period (ca. 300 BC), the climate became humid again until the advent of the Roman Period. According to the carpological investigations of Géza Facsar, major expanses of open water and marshes might have been present in the proximity of the site during the Late Hallstatt and Early La Téne of Sopron-Krautacker. The deterioration of the climate just added to the number of plant species invading from the woodland and shrubbery (FACSAR – JEREM 1985).

Significant changes in crop production at Sopron-Krautacker during the Late Hallstatt Period (6th century BC), can be seen compared to that of the funeral-urn culture. The number of barley grains (including bare barley) is greater by two orders of magnitude than that of wheat. On the other hand, bare grain common bread wheat and common oat appear as well. Furthermore, the seeds of common lentil are also found in substantial numbers.

We have more botanical finds from the Early Iron Age, though they are sporadic: Miklós Füzes collected a few grains of common millet and barley from the Early Iron Age burnt layer at Siófok-Balatonszéplak, and identified barley imprints in an Early Iron Age daub fragment at Keszthely-Dobogó (in: P. HARTYÁNYI – NOVÁKI – PATAY 1967–68). P. Hartyányi found carbonised, unspecified bread wheat (*Triticum* spec.) and einkorn grains, earspindles fragments, as well as rye brome (*Bromus secalinus*) and bristle-grass (*Setaria* spec.) berries in daub fragments coming from the Sopron-Várhely fortified settlements of the Hallstatt C/D Period (P. HARTYÁNYI – NOVÁKI 1967–68). The aforementioned findings provide evidence only for the presence of specified crop species, not for the quality of crop production.

Éva F. Petres, Béla Jungbert and Tibor Kovács excavated nine Early Iron Age barrow graves in the territory of the Fehérvárcsurgó-Eresztvény forest between 1983–87. Earth samples collected there were processed in 1996 (K. BERZSÉNYI – GYULAI in print). The burnt graves contained mainly carbonised grains of cereals. Six-rowed barley and emmer bread wheat found likewise characterise Early Iron Age culture in Hungary. Another typical wheat type of the Bronze Age, einkorn, also occurred, but from the relatively small number of the grains found, it could not have been very significan. The same applies to common millet. Grains of spelt were also present, albeit sporadically.

7.4.3. Crop production east of the Danube River in the Early Iron Age

The Rákoskeresztúr-Újmajor site excavated in 1996 by Anna Endrődi, has thus far provided the only Scythian botanical findings from the Carpathian Basin (*Figs. 205–206*). The most important cereal identified was six-rowed barley followed by common millet. It must be noted that the dominance of barley and common millet appears to be a typical feature of migrating, quickly moving nomadic peoples dealing with animal husbandry. Other grain crops were grown but they were by no means significant. Emmer is the dominant wheat cultivar. Einkorn, naked barley and rye grains account for not more than 1% of the total finds. Sporadically, in a way typical for peoples of Oriental origin, the characteristically cornered grains of club wheat also appear. Particularly notable are the few grains of rye found here All this demonstrates that the Scythians, although they did use wheats and grew them in a kind of ancient mixed grain, were preoccupied with the production of barley and common millet that better suited their way of life and traditions.

Along with residues of weeds, furcae bicornis of emmer and barley spike fragments demonstrate that grain was cleaned and presumably sliced as well. Their legumes were small seed lentil and pea.

7.4.4. Crop production in the Late Bronze Age in Transdanubia

Archaeological finds from the Celtic Period in Hungary refer to two distinctly different settlements. Hilltops were fortified with the so-called oppidums. We do not know much of the life of these communities. One might assume that farming was absent here, but was conducted in the peasant settlements that had much fewer inhabitants. The area selected from the wilderness was grubbed up first. Land was cleared of trees and bushes by case axes and long knives. Ploughing was made by a burrower plough equipped with a long bladed hose plough-iron. Garden produce was probably cultivated using a hoe. Ripe grains were reaped by hooks while grass was cut using a short scythe. No implement to be associated with vineyards is known from the Hungarian Celtic material. At any rate, the array of implements connected with farming and crop production is very modest, which does not confirm widely held assumptions about a high level agriculture associated with the Celts (Müller 1982).

Similar conclusions were suggested by archaeobotanical investigations made on earlier excavations from the Celtic Period. As to grains, earlier Celtic botanical finds are modest, rather sporadic, which does not support any conclusions about significant farming activities (Nováki 1975). The apparent contradiction was resolved by settlement discoveries in recent years that were reinforced and complemented by archaeobotanical investigations.

Unfortunately, Celtic samples collected and floated by Miklós Füzes in 1983 at Fenékpuszta, on the site excavated by István Erdélyi, containing a significant amount of carbonised grains, still await processing.

The species-rich botanical finds coming from Sopron-Krautacker in the Late Hallstatt – Early La Tène kitchen waste-pit and houses are of a transitional character (Jerem *et al.* 1984, 1985). Crop production features of the Hallstatt Period described earlier are still prevalent, yet a few characteristic species are present showing signs of a more elaborate crop production system and vegetable cultivation method. The dominant position of barley is about to cease and it is found in equal amounts with emmer wheat (Facsar 1987). To a smaller extent, the more advanced type naked grained common bread wheat is present as well. Even the most ancient hulled bread wheat, einkorn persists. Oil seed crops include flax, gold-of-pleasure and opium poppy. As to legumes, small seed lentil, horsebean and pea were consumed. Sopron-Krautacker is the earliest discovery of cucumber (*Cucumis* cf. *sativus*) in Central Europe. Earlier, Mikulčice in Southern Moravia held that record with a site from the 7–8th centuries (OPRAVIL 1972).

Of the plant species found here, hornbeam (*Carpinus betulus*) suggests a cooler climate. The tasty fruit of sloe (blackthorn) (*Prunus spinosa*) might have come from nearby forest edge. The knowledge about and use of herbs is also assumed. Maybe this was the reason that common gromwell (*Lithospermum officinale*) was collected from the surrounding meadows and oak woods. It was possibly used as a medicinal tea brewed from its flowers and leaves that cleanse the blood, alleviating conditions of the liver, stomach, and intestines or hives (VARRÓ 1991).

Testing of plant residues from different installations (mainly houses and graves) of the continuously inhabited settlements of Sopron-Krautacker in the Celtic Period (5th century BC – 1st century AD) concluded that a change of grain crops occurred in the Early La Tène (LT/B). Common bread wheat became the key staple crop. Although emmer was retained, the number of grains found was only a third of the former. The same can be said of barley (naked barley). As to garden-produce, legumes (lentil and pea), and among vegetables, cucumber was grown.

It was also stated with regard to the botanical finds from Mosonszentmiklós-Pálmajor (András Figler's excavation 1993–94), that the composition of grain crops changed in the Celtic Period. Hulled wheat varieties were "neglected", and aside from six-rowed barley, common bread wheat and common millet are the staple crop. Small seed lentil was eaten as a legume.

Grain crops from the Middle La Tène (LT/C) at Sopron-Krautacker are represented mainly by emmer and barley (JEREM et al. 1984, 1985). No naked barley was encountered here or in the Late Celtic Period. The lack of common bread wheat cannot be regarded significant. Flax was also cultivated here, being equally suitable for both fibres and oil. Dwarf cherry (*Prunus fruticosus*) stones, probably from human consumption, were found in the debris of a burnt house (JEREM *et al.* 1985).

The samples taken from one of the farm buildings and a grave in Sopron-Krautacker from the Late La Tène (LT/C2-D) proved to be extremely rich in species. Most of the finds belong to common bread wheat as the grain crop. The number of emmer and barley grains is almost identical, yet it is only half of that of common wheat. Sporadically, common millet is also found. Garden produce legumes are dominated by lentil and pea. After a long period of absence, opium poppy appears again. Weeds identified: annual woundwort, false cleavers, black-bindweed; all refer to autumn sowing of grain crops and an increase in arable crops. The presence of ruderal, nitrophil species: fan-hen or white goosefoot, common elder seeds, suggest a denser cultural occupation.

It is a well known assumption that the Celts – before the Romans – already engaged in fructiculture and viticulture (P. ERMÉNYI 1975–77). In all probability they indeed may have had certain fruit-farming, yet no extensive fructiculture or viticulture can be demonstrated. At any rate, shells of sweet cherries and bullace plum unearthed in Schwäbisch-Hall suggest this possibility (BERTSCH 1941).

JACOMET (1983) has good reasons to assume that the Celts grew walnuts. Indeed, "big fruit" walnut shells were found in the Celtic settlement of Magdalenensberg/Klagenfurt (WERNECK 1969). Subsequently, walnut has been found only in Haithabu from the Viking era (BEHRE 1983). These are rather sporadic occurrences. Whether this is the case in Transdanubia remains to be seen in future research.

Similar difficulties emerge when the issue of potential vine growing in the Iron Age is addressed. In fact, two fragments of grape-seeds were found in a grave in Sopron-Krautacker during the Late Hallstatt – Early La Tène layer (JEREM *et al.* 1985) (*Figs. 204, 209, 211*). According to meticulous investigations carried out by Géza Facsar, one of the seeds has the traits of a cultivar, which he evaluated as black wine grape (FACSAR – JEREM 1985). He also demonstrated on the basis of comparative analysis that these grape seeds show a unique level of advancement as opposed to other grape seeds from the Carpathian Basin. He concluded that local viticulture must have been in its initial stage. Three further grape seeds were unearthed in a pot from a Hallstatt Period (7th century BC) tumulus in Zagersdorf, Burgenland, a mere 7 kilometres from Sopron. According to the determination made by Facsar, these bear the traits of cultural convergence, albeit very rudimentally (in: KAUS 1987).

The very first Hungarian occurrence of common plums is the recovery of almost fifty stones from a grave of Late Hallstatt – Early La Tène age in Sopron-Krautacker. Their value at the time is demonstrated by the fact that they were found beside a richly furnished corpse, put on a small tray. Most probably, the Celts participated in spreading common plums in Europe (RAPAICS 1943; SÁGI – FÜZES 1966). This plum find supports the assumption that along with the Balkans, the Carpathian Basin belongs to the secondary gene centre of plums (Jerem *et al.* 1985). It has to be noted here that plum finds (*Prunus* species), without any closer identification of the species, are known from Lengyel's prehistoric site as well (DEININGER 1892). However, these – probably wild plum – stones have not yet been processed appropriately.

During the excavation works at the Regöly motte from the Late Celtic Period (LT-D), stones of peaches were encountered (FACSAR – JEREM 1985). According to the examinations made by Géza Facsar, their shell (stone) expressed a great deal of similarity to the finds of apricot from Linz (WERNECK 1955). This thin, narrow half shell is the earliest peach residue found in the Carpathian Basin, well ahead of those from the Roman Period (cf. Győr-Homokgödör, Budapest IIIrd district) (P. HARTYÁNYI – NOVÁKI 1973–74), and Ménfőcsanak.

In past few years, additional Celtic settlements were unearthed in the territory of Budapest. In order to clarify the way of living and husbandry at the time as well as the environment, large amounts of soil samples were taken at these excavation sites for the purposes of interdisciplinary investigations.

Botanical residues taken at Corvin Square in Budapest in 1997–98, dated to the Late Celtic Period (La Tène C/D, ca. 1st century BC.) confirm the picture drawn about the advanced agriculture of the Celts. The botanical assessment of the samples taken from the Celtic structures at the excavation led by Tibor Hable was made by the author, then by Stefanie Jacomet and Orsolya Dálnoki.

The overwhelming majority of the seeds and crop yields found at the settlement inhabited by the *Eraviscus* are cereals (DÁLNOKI 2000). Most of them are spelt and emmer. Somewhat less, but still considerable is the amount of barley grains. The majority of weeds indicate a grain crop host in autumn sowing: korn-rade, black-bindweed, rye brome, field brome, soft brome, downy brome and upright brome, green/rough bristle-grass. A significant amount of nitrophil ruderalia occur as well, mostly goosefoots. Relatively rarely, yet the seeds of oil and fibre crops appear as well: flax, hemp, rape. Seeds and gatherings of the wild fruits of forest edges are also encountered in a relatively great number: wild pear, cornelian cherry, dogwood, sloe (blackthorn), hawthorn, hazelnut, bramble, elderberry and grape. Unfortunately, the stone of the last was found only in a minor fragment which cannot be subjected to any more detailed examination.

At the site of the "Campona" shopping centre built in Budapest-Nagytétény in 2001, yet another Celtic settlement was excavated. We had the opportunity to analyse more than a dozen soil samples taken from the various structures with different designations found at the excavation led by Gábor Elmas.

Preponderant in the diasporas was grain crop. Those living here must have produced a surprising variety of different cereals, yet only a few of them were of decisive significance. As for bread wheats, mainly the hulled ones (einkorn, emmer, spelt) were cultivated. It seems from the ratio of different grains that the most important kind of bread wheat must have been emmer, although einkorn was also known. Relatively many grains of barley were found, yet it was not always possible to decide whether they belonged to the six-rowed (multiple rowed) or the two-rowed cultivar. The large amounts of barley are justified by the fact that barley in this period was also used for human consumption. It was considered very valuable, beside common mille. The great number indicates how much this porridge-plant was loved. In the Celtic period, one can also presume the production of rye, as it happened here.

Surprisingly, only a little amount of *spelt*, typical for the Celts living in the areas of Western European countries, was demonstrated in the Hungarian Celtic finds. This apparently has climatic and ecological reasons. Of the archaeological periods in the Carpathian Basin, only in the Celtic period was spelt cultivated, even if not to a great extent.

No far-fetched conclusions can be drawn from the sporadic presence of naked *aestivum* bread wheat grains (common bread wheat, club wheat). Nevertheless, the sowing seeds of hulled wheats must have been an inhomogeneous population as beside common bread wheat the similarly naked and characteristically cornered club wheat grains are also encountered.

It was not possible to decide in the case of a few oat grains whether they were cultivars of common oat or merely larger than usual seeds of oat-grass, which appeared as a weed. It would be of great significance if it turned out to be the first option as the first appearance of common oat in Europe can be dated to this age. Legumes included small seed lentil and pea.

Wheats must have been sown in the autumn, just as rye was. Barley could be sown both in the autumn and spring. Common millet was sown in the spring. Unfortunately, the majority of the cereals were so fragmented that no closer identification was possible.

The large number of weeds suggests that the grain crops produced were not subjected to cleaning in all cases when ripened. There are several samples, however, with explicit contents of thrashing such as glumella and ear-spindles fragments. Autumn grain crop weeds are almost twice as many as those of cereals sown in the spring of in intertilled crops. This again supports the assumption discussed above that the population who lived here mainly preferred cereals sown in the autumn. Crop weeds detected were tallish. Their presence gave rise to the conclusion that reaping was made at about two thirds of the height of the stalk, in other words not much below the ears, probably using a hook.

A larger amount of the weeds can be associated with winter cereals (wheats, rye, maybe winter barley) (*Secalietea*): quackgrass, barnyard grass, black-bindweed, field cowwheat, rye brome, smooth finger-grass, annual woundwort, wild oat, purslane. Rough and green bristle-grass, red fingergrass, common mallow, ribbed melilot are mainly species from spring sowing cereals (common millet, maybe spring barley), in other words hackfruit weed associations (*Polygeno-Chenopodietalia*).

Species were also introduced from the natural environment of the time. These might have been accumulated in the samples through natural processes such as wind drift, runoff and animal impact. First of all, the local environment around the settlement was a waterlogged landscape with common reed. Namely, common bulrush is a characteristic species of the common reed zone (sedge marshes). Meadows and pastures were also present in the neighbourhood. Relatively many remains of species originating in the meadows and pastures were found. Although wet parts might have been present in this meadow or pasture, where for instance creeping cinquefoil might have come from, this was not the typical picture. Species preferring average water cover are in great abundance in the finds: lady's bedstraw, large hop clover, parsley-pier, sheep's sorrel, hare. At certain places, it might have been explicitly arid, as indicated by black medic, hop clover, yellow trefoil.

Although the number of species originating in ruderal weed associations is considerably low, only fan-hen or white goosefoot, maple-leaved goosefoot, knotgrass belonging here, if one looks at the actual number of seeds detected in the finds, the number of seeds belonging to this group make up almost half of all the finds. These ruderalia come from areas subjected to human impact, such as ditches, roads, yards, buildings, embankments, any place where soil is rich in nitrogen or was fertilised by manure (maybe accidentally). They belong to the ruderal weed association with average water supply.

It can be seen from all this that the Pannonian-Illyrian inhabitants of Transdanubia possessed high-level farming skills even before the Roman conquest, and based on the finds material, they can be assumed to have had fructiculture and viticulture as well.

7.4.5. Reconstruction of agriculture in the Iron Age

7.4.5.1. Crops produced, ploughland and farming implements

The largest part of Central Europe from the period of the 5th century BC up to the Roman Period belonged to the "Celtic Empire", ranging from the Mediterranean Sea, through the Alps up to North-Western Europe, where the Hessen region lies today. La Tène Period farming is characterised by species diversity. This diversity cannot be interpreted only in terms of climatic conditions or chronological sequence. Celtic lifeways were not homogeneous throughout its extensive area of distribution. Just as Celtic culture shows diversity of clothing style, ceramic decorative patterns and funerary ritual from region to region, significant aspects of crop production varied in regional configuration and emphasis. The shared characteristic of Celtic agriculture was a high level of complexity and sophistication in all regions.

Climatic deterioration in the La Tène Period brought about changes in the quality of land cultivation and technology as well. In a humid and cold climate farming had to facilitate drainage. At the same time, rye and oats, which adapted well to climatic deterioration increased in quantity and importance, but required deeper ploughing (BALASSA 1973).

The nature of crop production implies an advancement of agriculture. Let us begin a short review of the Iron Age in Europe with its associated cereal species. During the Iron Age, einkorn gradually lost its significance as a direct result of advancements in farming methods. It occurs only sporadically in the Late Bronze Age, more as a weed. Emmer, however, remained a key bread wheat type. Spelt became increasingly important following the Early Iron Age. Although both were cultivated, common wheat and club wheat were losing their significance. Coming from the East, rye appears in Western Europe during the Iron Age for the first time. In a site in Thunau/Kamp, Austria from the Hallstatt Period (8th century BC) large grain rye was already present (WERNECK 1954). At La Tène sites in Vlineves/Melnik, Bohemia (3rd–2nd century BC) (TEMPÍR 1968) and in Stanz/Schwaz, Tirol (WERNECK 1961) both large and small grain rye were found. These finds are associated with cultures of different origin. Growing of barley is significant everywhere and sometimes is the most important cereal grown. In the Netherlands, at Gees/Oosterhesselen in an Early Iron Age settlement, 98.8% of all residues were barley, while emmer made only 1.2% (VAN ZEIST 1970b). Common oat was produced only from the beginning of the Late Bronze Age. Common millet was significant mainly in Central-Eastern Europe. Buckwheat (*Fagopyrum esculentum*) is known only in this period from Eastern Europe (Jelabuga/Russia) (OPRAVIL 1974).

Primarily pea, common lentil and horsebean production account for legume crops. Planting of gold-of-pleasure used for obtaining oil increased in this period. Opium poppy appears only sporadically while flax remained a significant oil and fibre crop, probably the latter, in spite of the fact that hardly any textile fragments were recovered.

Plant species found at Celtic settlements point towards an agriculture built on solid foundations. Maybe the great number of cultivated plant species served to minimize the risk and uncertainty inherent in growing crops. Diversity of this kind (pea, common lentil, bitter vetch, horsebean, spelt, common millet) is otherwise more typical in Central and Southern Europe.

There are a number of ancient ploughland remains known in England (Gwithian, Cornwall), the Scandinavian countries (Ribe, Jütland), The Netherlands (Velsen Hoogovens, Monster, Vogelenzang) and Northern Germany (Oudemolen, Feddersen Wierde, Groningen) called "Celtic fields" (ZIMMERMANN – KÖRBER-GROHNE – KROLL 1984). These lands were separated from each other by approximately 90 cm high stone walls or ridges, some extant to the present day. Their shape is flared with a side length of 10–50 m, square or sometime triangular or hexagonal. Around the settlements were plots without ridges or stone piles, which might have been used for crop rotation. These fields are characteristic features on flat land and hillsides, because mountain ranges were only sparsely populated at the time. In certain areas, periodical grazing can also be assumed in addition to crop production on arable land (WILLERDING 1980).

Tests indicated that soil phosphate content is higher around these walls than average, which points to the accumulation of organic matter, maybe deliberate fertilising. However, the fertilising effect must have come from kitchen garbage – as indicated by an abundance of pot fragments, and manure mixed with straw that played only a secondary role. Zimmermann raised the possibility that in the northern German regions, Late Bronze Age inhabitants might have fertilised the land by marling, well before its use by Romans (ZIMMERMANN – KÖRBER-GROHNE – KROLL op.cit.). His statement is based on the presence of marl pits found here from the Late Bronze Age. It is strange that in Northern Germany, but mainly in the seaside regions (Goddelau and Christenberg), the number of plant species grown is fairly low (KREUZ 1992–93). This might be attributable to nutrient supply, maybe through use of sea kelp for manuring. On the other hand, local preferences of regional cultures might also account for it.

In addition to loess soils, which are easy to till and were therefore broken for new ground as early as the period of the Linearband Keramik culture, sandy and slightly clayey soils were also broken and tilled in the Iron Age (WILLERDING 1980). At least, this is indicated by the different acidity of weed species detected from the Iron Age. Beside neutral species such as black-bindweed (*Fallopia convolvulus*), nipplewort (*Lapsana communis*), the presence of annual knawel (*Scleranthus annuus*) and field spergula (*Spergula arvensis*) indicates acidic soils. The fact that sporadically occurring limy soils were also used for growing cereals can be concluded from the presence of the alkali false cleavers (*Galium spurium*). Certain areas were probably tilled for a longer period. Maybe it was discovered that a correlation exists between feeding and genetic improvement. It is possible that similarly to the Romans (see the books by Columella written on agriculture), they fed horses on barley and bitter vetch. Further investigations are needed in order to clarify the quantitative relations of natural grazing and feeding.

From the Hallstatt Period, farming implements were made of iron. These ploughshares could also be fixed in a diagonal position (BALASSA 1973). Such hooked and skid ploughs could only be used for superficial scratchy tillage. However, plough blade finds made of iron are rather sparse. Those made of wood must have remained widely in use and continued to be used until the Middle Ages, which is supported by several Northern European, mainly Danish finds (Hendriksmose/Jütland, dated: 350 ± 100 years BC) (STEENSBERG 1976).

Celtic ploughshares from Central Europe can be divided into two main types: long blade hose and shovel form. The former group is defined by blades found in England (Bigbury, Bloxham), Germany (Gotha, Steinsburg, Trifels), Bohemia (Brnicko, Lipany, Uhersky Brod), Austria (Königshof, Kaiserbrunn), Croatia (Idrija) and Romania (Racari-Dolj). Ploughshares coming from sites in the Carpathian Basin (Galis-Lovacska, Munkács, Szalacska, Kaposmérő, Ugra), Kornwestheim (Germany), Kaiserbrunn (Austria), Idrija (Croatia), Kolin, Novy Bydzov (Bohemia) are short and shovel-form. The latter shape could be widespread in the Carpathian Basin as well (Velem-Szentvid, Szalacska) (BALASSA op.cit.). Blades in Northern Europe were shaped like an arrow (Mors/Limfjord, dated: 160 BC) (LAEGDSMAND 1968).

Different types of blades occur at the same time in the same area, suggesting that they were employed for different uses. Increasingly more land was required for the needs of a growing population and intensive agricultural activities can be assumed in the oppidums (BALASSA 1973). Forest and forest steppe soils might have been broken using spiked ploughshares. These must have been very useful in breaking land cleared by fire in order to till between the leftover roots and stones. However, this plough type was suitable for breaking small pieces of land only (up to approximately 35 x 35 m). Shovel-shaped blades might have been used for tilling existing ploughland. They were supplemented by a coulter (skife) and a fore-carriage. Such ploughs muzzled the earth sideways in both directions. At any rate, they were suitable for tilling deeper, wider and in a larger area. The remains of such a symmetric blade plough were found in the Feddersee site (300 BC). In the German seaside region (Holssel), ploughing imprints of a burrower plough are known, while ploughing tracks of a plough equipped with a breast appear in the Feddersee region.

Further farming implements for the Celtic Period are the iron clad wooden spade, sickle (hook) and scythe. Reaped grain was usually stored in grain pits, but in the Northern German areas above ground barns were used (in: ZIMMERMANN – KÖRBER-GROHNE – KROLL 1984).

7.4.5.2. Developing of weed associations

Several new segetal species appeared in Europe at the beginning of the Iron Age. These were usually short stalked weeds, corresponding to the new type of harvesting (iron sickle, reaping close to the soil surface) (WILLERDING 1986c).

According to KNÖRZER (1989), in the Lower Rhein region the number of species encountered in a pasture or grazing land during the Neolithic (57 species) quadrupled by the Celtic Period (224 species), more than 60% of these being thermophytes. He connects this observation with the retreat of forests and the growing amount of grazing land around settlements, indicating an increase of stabled cattle farming. Investigations by BEHRE (1988) concur with this proposed configuration. Analysis of macroscopic botanical finds found in the Lower Ems region suggests that rivers here were still bordered by extensive forests in the 7th century BC. As a result of deforestation by the newly arrived population, the size of the forested area diminished. Spelt, barley, horsebean, flax and gold-of-pleasure were grown here in large areas. Stables were built for cattle as early as in the Early Iron Age. Livestock grazed on pastures adjacent to the settlements. Based on the analysis of manure residues, the dominant species populating the pastures of the time were mainly smooth/rough meadow-grass (*Poa pratensis/trivialis*), water foxtail (*Alopecurus geniculatus*), white clover (*Trifolium repens*), and rattleweed (*Rhinanthus* spec.). Environmental conditions changed around 300 BC (Dünkirchen I). Settlements were flooded and abandoned to be repopulated during the Roman conquest.

UNTIL 1976 WILLERDING (1986c) processed weed species finds in Central Europe producing some 850 publications accounting for approximately 300 weed species from the Palaeolithic up to the Late Middle Ages. According to his investigations, the following species appear during the Iron Age for the first time: bishop's goutweed (*Aegopodium podograria*), wind bent grass (*Apera spica venti*), slender parsley-pier (*Aphanes microcarpa*), wood burdock (*Arctium nemorosum*), woolly burdock (*A. tomentosum*), common wormwood, absinthium (*Artemisia absinthium*), winter-cress (*Barbarea vulgaris*), black mustard (*Brassica nigra*), welted thistle (*Carduus acanthoides*), musk thistle (*C. nutans*), chaffweed (*Centunculus minimus*), field mouse-ear (*Cerastium arvense*), red goosefoot (*Chenopodium rubrum*), hemlock (*Conium maculatum*), smooth hawk's-beard (*Crepis capillaris*). In addition, small toadflax (*Chaenorrhinum minus*), greater Venus's looking-glass (*Legousia speculum-veneris*), wind bent grass (*Apera spica-venti*), and field madder (*Sherardia arvensis*) appear for the first time as crop weeds in cereals (KÜSTER 1985a).

The number of weed species increased dynamically in the Iron Age. At the beginning of the Bronze Age, there were 120 species. At the beginning of the Iron Age however, already 157 weed species were present. By the end of the Iron Age this number grew to 212. In other words, some 55 new weeds were introduced during the Iron Age, lasting merely 800 years (WILLERDING 1986c). The advancement of cereal and intertilled crops is shown by the appearance of 11 new *Chenopodietea* species and 6 new *Secalietea* weed species. This trend might have something to do with the changes taking place in farming and crop production methods, but it can also be explained by climatic differences. Seeds of several short stalked weed species can be found among the grains of cereals. Certainly, the question remains whether these were present in former ages as well but not detected because of mid-height reaping practices, or are simply newcomer weeds. Frequently found black-bindweed and false cleavers seeds obviously refer to an elevated reaping height from ground level of cereal stalks. It is interesting, however, that characteristic weed species such as korn-rade (*Agrostemma githago*) and cornflower (*Centaurea cyanus*) (c.f. DICK 1989) are missing.

Unsolved phytopathological problems may have been caused in cereal crops by fungal infections. Wintering sclerotium residues of ergot (*Claviceps purpurea*) present a health risk to people whose bread is made of flour contaminated with ergot.

No species characteristic for meadows and pastures were found previously, yet ruderal vegetation is well represented. Subfossil pastures and meadows are suggested by burnt hay fragments beginning in the Iron Age when grazing of livestock in paddocks was initiated. Before the Roman Period, associations like Cynosurion are known.

An increasing number of weed species in the Late Bronze Age can be associated with the a type of harvesting close to the ground. This is obviously connected to livestock husbandry that uses straw as fodder. In this way, biomass production also grew, yet in lack of nutrient replenishment arable lands depleted soil fertility more quickly. Low nitrogen weeds frequently appearing in the Early Iron Age indicating a depletion of cropland. The number of nitrophil species increases only in the Late Bronze Age: alkafoun fumitory (*Fumaria vaillantii*), green field-speedwell (*Veronica agrestis*), and grey field-speedwell (*V. polita*). This apparently reflects different forms of nutrient replenishment such as manure fertilisation.

In the Early Iron Age in Central Europe, the presence of such weeds (19 species) is associated with a boreal ecology as indicated in recent cases of a boreal spread (WILLERDING 1986c). All this supports a conclusion that the climate changed for the worse. The same is supported by the so-called oceanic/continental coefficient (2:1) where the number of boreal species dropped to three while that of humid oceanic species doubled. Weeds in the Late Bronze Age, however, have ecological requirements supporting an improved climate.

7.4.5.3. Dyer's plants and herbs

A number of plants found in nature are used for dying. For instance, the immature husks of walnut colour fabric brown, bedstraw (*Galium* spec.) roots and alder (*Alnus* spec.) bark make red, Scotch heather, common heather (*Calluna vulgaris*) and horsetail (*Equisetum* spec.) green, while dyer's woad (*Isatis tinctoria*) makes the fabric blue, weld mignonette yellow, common madder (*Rubia tinctorum*) red again. Unfortunately, the detection of these plant species is fairly difficult as the husks or seeds are only rarely seen at archaeological sites.

The dying capacity of dyer's woad is well-known. Its leaves, when in a state of initial decay in lack of oxygen, leak blue coloured liquid. The plant was discovered as early as in the Neolithic in France (de l'Adouste/Joursques, Bouche-du-Rhone). Following this, Iron Age ceramic pot fragments in the upper Danube region in Heuneburg/Hundersingen held its imprint (dated to 6th and 5th centuries BC, respectively). There are finds known from the period in the fourth and third centuries BC from the sites near Cuxhaven in Feddersen Wierde and Northern Denmark (Ginderup in Thy) (KÖRBER-GROHNE 1987). Dyer's woad was probably brought to the north from the southern and southeastern regions of Europe. *Isatis tinctoria* var. *praecox* is native to the Pontus East-Mediterranean areas. It is a steppe plant which ranges from Southern Europe to the Caucasus. It may have arrived to the territory of Germany along the river Danube at the beginning of the Celtic period together with hemp (*Cannabis sativa*). The first occurrence of hemp in Germany is the princely grave in Hochdorf (6th–5th century BC). It has become widespread subsequently along the seaside regions of the North Sea.

Seeds of the yellow colouring weld mignonette (*Reseda luteola*) were found in the botanical material unearthed from the Swiss lake dwellings (Robenhausen, Pfäffiker See/Zürich). An especially great amount was found in the Brises-Lames/Auvernier lake dwellings near lake Neunburg (17% of all seeds). Following this, the next occurrence of this species is dated to 125 BC, i.e. the Celtic Period (Fellbach-Schmiden/Stuttgart. The plant is of Mediterranean origin and came here from areas near the Mediterranean Sea. It cannot be determined, based on botanical residues alone, whether they were used as dyes or herbs in those early days.

Common madder (*Rubia tinctorum*) and safflower (*Carthamus tinctorius*) are not detected in the Iron Age, yet their presence and use have to be considered.

We know even less of the medical activities of people living in the Iron Age. It can be assumed that in this period, the health protective and cold-remedial effects of certain herbs, such as common elder (*Sambucus nigra*) were already known. Maybe a pot full of stocked elder seeds shows evidence of this, found in the Feddersen Wierde site (ZIMMERMANN – KÖRBER-GROHNE – KROLL 1984).

Fig. 204. Grape-vine (*Vitis vinifera* subsp. *vinifera*) seeds from the La Tène settlement of Sopron-Krautäcker. After JEREM *et al.* 1985.

Fig. 205. Barley (*Hordeum vulgare*) grains from the Scythian settlement of Rákoskeresztúr-Újmajor. Photograph by the author.

Fig. 206. Common millet (*Panicum miliaceum*) naked grains from the Scythian settlement of Rákoskeresztúr-Újmajor. Photograph by the author.

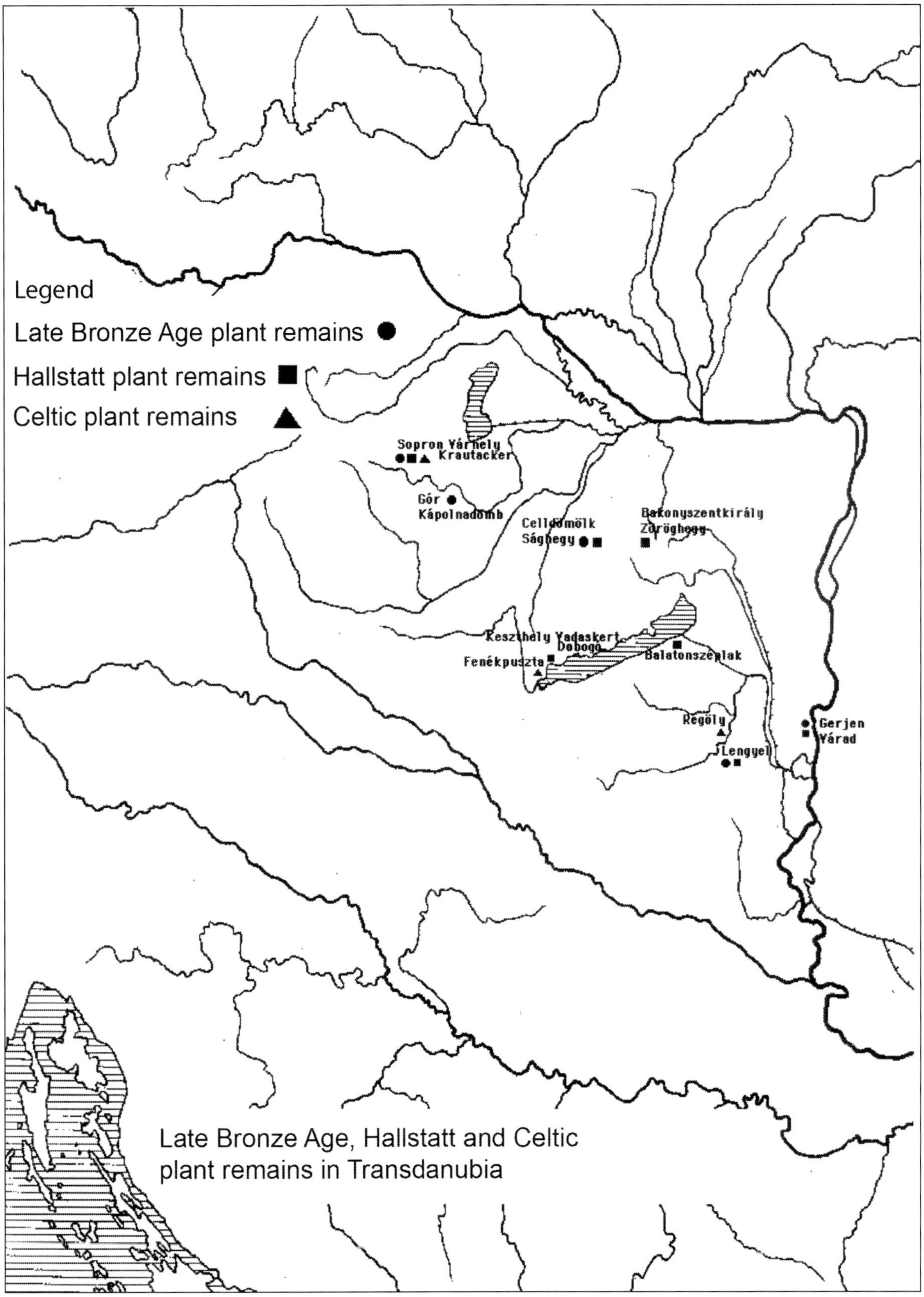

Fig. 207. Late Bronze Age, Early Iron Age and Celtic botanical finds in Transdanubia.

Fig. 208. Barley (*Hordeum vulgare*) grains from the Aggtelek cave (Szultán pamlagja), Hallstatt culture, Inventory of the Hungarian Agricultural Museum, Budapest.

Fig. 209. Grape-vine (*Vitis vinifera* subsp. *vinifera*) seeds from late Urnfield settlement of Sopron-Krautäcker. After JEREM *et al.* 1984.

Fig. 210. Emmer (*Triticum turgidum* subsp. *dicoccum*) naked grains from the Aggtelek cave, Early Iron Age, Inventory of the Hungarian Agricultural Museum, Budapest.

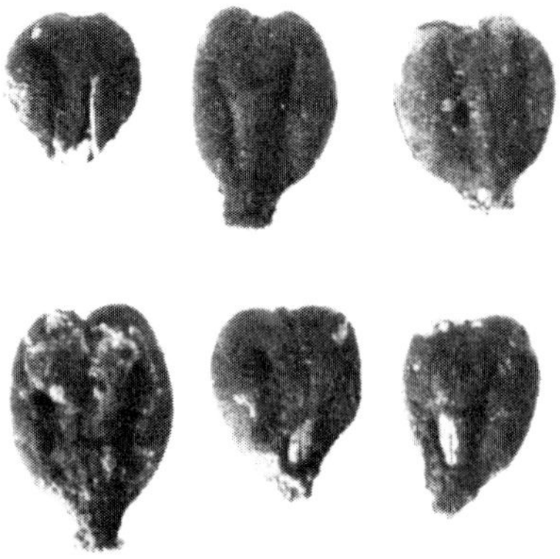

Fig. 211. Grape-vine (*Vitis vinifera* subsp. *vinifera*) seeds from late Hallstatt settlement of Sopron-Krautäcker. After JEREM *et al.* 1985.

7.5. The Roman Period

7.5.1. Crop production archaeology in Pannonia

The Romans expanded their rule up to the Danube in the first decades of the 1st century AD. The entire area of Transdanubia with its centre at Aquincum (today Óbuda) belonged to the Pannonia province. Farming implement finds, archaeobotanical residues, epigraphic data and iconographic portrayals equally confirm the extensive knowledge and high level agricultural skills of the Romans. Production of cereals, legumes, grapes and fruits was described by professional authors (Columella, Cato, Plinius, Varro).

The era in Pannonia extends from the beginning of the 1st century to approximately the middle of the 5th century AD. It is hallmarked by the appearance and spread of new tiller and farming implements and technologies, in addition to the appearance of formerly unknown crops. A whole series of cultivars were introduced to the Carpathian Basin at this time (GYULAI 1998a). Primitive varieties and cultivars existed side by side. During the Roman dominance in Pannonia, a number of fruit pollen occurrences become continuous, e.g. walnut and grape (ZÓLYOMI 1980; ZÓLYOMI – PRÉCSÉNYI 1985). Many fruit species found a new home in Pannonia; such as apricot, peach, pear, plums, walnut and black wine grapes. Their production has continued ever since. All this is witnessed by a wealth of botanical finds (P. HARTYÁNYI – NOVÁKI 1973–74; P. HARTYÁNYI – NOVÁKI – PATAY 1967–68).

Advanced agricultural technologies in Pannonia were mixed with native, indigenous traditions. In this country, the Roman Period saw the emergence of the first urban settlements where people were increasingly divorced from their natural environment. Central food supplies of cities and military camps were procured by traders and trained sales managers. Beside colonia (cities) a widely known type of settlement was vicus (rural settlements) and villa rustica (agricultural production centres).

The achievements of livestock husbandry and animal breeding in the advanced Pannonian Roman agriculture are somewhat overshadowed by the fact that the size of domesticated animals was somewhat reduced compared to earlier periods judged on the basis of bone finds (BARTOSIEWICZ 1999). It would be too simple to assume only a climatic reason (e.g., permanent cooling) behind this, although it might have played a role.

As a result of well-developed Roman trade, imported grains and fruits occur as well: common fig, olive, date. The spread of diverse imported cultivars was accelerated by trade and advanced road networks. Only in very rare cases can it be determined if an individual plant residue was locally grown or imported, maybe from as far as an entirely different continent.

Although the number of Pannonian sites where botanical materials were found exceeds thirty, it is a very low figure compared to the number of Roman sites explored in the last one hundred years. In addition, the great majority of these are made up of dispersed, individual finds, and only a handful of cases (such as Keszthely-Fenékpuszta) were systematically sampled.

It has historically significant for local archaeology that the earliest Hungarian seed and fruit finds between 1835 and 1845 are from a Roman Age site in Szombathely, where a Roman grain was found (P. HARTYÁNYI – NOVÁKI – PATAY 1967–68). Botanical finds come mainly from Aquincum and excavation sites in the inner part of Pannonia (around Keszthely, Tác, Sopron) from the exploration of vicus and villa rustica ruins, the limes along the Danube (watch-towers, castrums), and enclosures in graves.

7.5.2. Grain crops

Food requirements of the population and military stationed in Pannonia were mainly met by cereals. Therefore, not surprisingly, the overwhelming majority of seeds and harvested materials found here are cereal grains (P. HARTYÁNYI – NOVÁKI 1973–74; P. HARTYÁNYI – NOVÁKI – PATAY 1967–68). Unfortunately, many of them are in fragments, unsuitable for determination of the exact species.

The most important changes in Roman crop production are modifications in the composition of cereal crops. Other (earlier) hulled wheat varieties that were resistant, less demanding and/or less exacting with regard to agrotechnical requirements of the soil or nutrient supply, had relatively low yields. These were replaced by naked wheats that were more productive but in many respects more exacting and sensitive. Hulled grain *einkorn* was found at merely three sites: Keszthely-Fenékpuszta castrum, Leányfalu-Móricz Zsigmond Street watchtower, Tác-Gorsium civic city, and only in small quantities (*Fig. 230*). This must have been the case before the late Emperors' Age as well, as the find in Gorsium – as opposed to the others – comes from the 2nd century AD. However, the much desired hulled wheat of earlier periods, *spelt* was continuously grown at the sites of Budapest, Bécsi Street 38–42, Budapest, Corvin Square, and Keszthely-Fenékpuszta castrum.

By the late Emperors' Age, the key staple cereal was common bread wheat and rye. Their grains were found in substantial amounts at Budakalász-Luppa pub, Late Roman watchtower, Kékkút, basilica No. 2., Budapest-Lajos Street, Dunaújváros, Esztergom-Castle, Leányfalu Móricz Zsigmond Street watchtower, Keszthely-Fenékpuszta, Sopron-Beloiannisz Square and Városház Street, Tác-Gorsium,

Tokod (*Fig. 228*). In finds around Keszthely (Keszthely-Fenékpuszta, Keszthely-Mosóház, Keszthely-Vadaskert) and at the site of Budapest, Bécsi Street 44, naked grain club wheat appears in addition to common bread wheat. The significance of rye in provisions is proven by the two litres of carbonised rye grains found in the Late Roman watchtower at Budakalász (*Figs. 219, 221–223*).

Common millet was also eaten, albeit its amount and significance fell behind those of wheat and rye. The majority of carbonised grains come from Late Roman barbicans and other military installations (canabae): Budakalász-Luppa pub, Budapest-Körte Street, Leányfalu-Móricz Zsigmond Street, Óbuda, Bécsi Street 38–42, Óbuda Corvin Square, but it can be found at other places as well: Dunaújváros, Nemesvámos-Balácapuszta, Sopron-Beloiannisz Square and Városház Street, Tokod (*Fig. 220*). The same applies to common oat (*Avena sativa*), albeit it was found only twice: in Tokod and Dunakömlőd.

The decorative villa built in Nemesvámos-Balácapuszta in the last third of the 1st century AD was inhabited as late as the 4th century (B. THOMAS 1964). Gyula Rhé, who carried out excavations several times here between 1904 and 1912, explored a plastered pool made of pure limestone in floorless room 19 in the north-western corner of building No I. The storage pool, still bearing the tracks of former planking was full of carbonised common millet and wheat (LACKÓ – RHÉ 1912). The plaster of the building walls contained bread wheat-, rye- and barley straw.

Much less barley was found compared to earlier ages, both in terms of quantity and locations: Budakalász-Luppa pub, Keszthely-Fenékpuszta, Kékkút, Leányfalu-Móricz Zsigmond Street, Nemesvámos-Balácapuszta, Sopron-Városház Street (*Figs. 214, 218*). This can be explained by changes in eating habits. Barley starts to loose its use as "bread grain" in the Roman Period, and, in addition, the Romans did not get accustomed to the habit of drinking beer. However, advanced six-rowed (fodder) barley, which is suitable for animal feeding, was found at several locations: Óbuda Lajos Street – Nagyszombat Street 38–42, and Bécsi Street 44, Corvin Square No. 1. Two rowed (beer) barley was found only in the Sopron-Beloiannisz Square and Óbuda Corvin Square sites.

From the archaeobotanical point of view, one of the most researched Roman sites is Keszthely-Fenékpuszta. At the area of the fortified Roman settlement (castrum), known since the 18th century, the first scientific archaeological excavation was carried out in 1883 under the leadership of Vilmos Lipp, which lasted with minor interruptions up to 1901. Unfortunately, no botanical finds are known from this time. However, Árpád Csák's excavations between 1899 and 1908 resulted in recovery of approximately 80 cm^3 of seeds (in: BAKAY – KALICZ – SÁGI 1966). These are – according to the determination of Miklós Füzes – einkorn, emmer, club wheat, and barley grains (Miklós Füzes and Károly Sági, in: HARTYÁNYI – NOVÁKI – PATAY 1967–68). During excavations started in 1970 and led by Károly Sági, a substantial amount of seeds (several hundred kilograms!) and carbonised seeds were unearthed in front of the southern gate to the fortress (*Figs. 231–238*). According to the investigations made by Miklós Füzes, cereals account for the overwhelming majority of seed and fruit remains coming from the Late Roman Period in Keszthely-Fenékpuszta, naked barley occupying first place among them. There is somewhat less of common bread wheat and rye followed by the rarely seen common millet with common oat least.

In 1993, at the northern fortress gate of the Keszthely-Fenékpuszta fortress, at the excavation led by Róbert Müller, a substantial amount of Roman botanical material was collected. According to the preliminary investigations, no significant differences are shown between the cultivars found here and the typical species composition found in Transdanubia from the Roman Period: hulled bread wheat species of earlier ages, einkorn and emmer are encountered only in traces in the finds, their role taken entirely by the naked grain common bread wheat and club wheat, both requiring more advanced agrotechniques. Significant amounts of rye and common millet were also found. Barley was not really common here.

Garden plants were represented by pea, common lentil and horsebean. Those living here, in order to supplement their diet, ate wild fruits grown in their neighbourhood.

Miklós Füzes identified the imprints of barley ears (*Hordeum* spec.) and pea husks (*Pisum sativum*) in the grave covering bricks of the Late Roman cemetery in Keszthely-Dobogó (in: HARTYÁNYI – NOVÁKI – PATAY 1967–68).

The cereals coming from the Dunakömlőd (Lussonium) Late Roman fortress explorations differ in species composition from the Pannonian set of grain remains described above (GERÓCS – KOVÁCS – TORMA 1995). In terms of number of grains, hulled emmer is ahead of common wheat. One can also find six-rowed barley, rye and oat as well. All these raise the possibility that we might have discovered the plants of another people with different agriculture, maybe those of the indigenous population. However, commercial activities with the Barbaricum might also play a role.

For similar reasons, the site at Budapest, Bécsi Road No. 38–42 deserves attention. Grain crop species found here are different in composition from those grown in Pannonia in general. Since the number of barley grains exceeds that of common wheat to a great extent and common millet and emmer were also found in relatively high numbers, we believe that the traditions of the Iron Age might have survived here. This is also reflected by the fact that while Roman settlers in Pannonia (e.g. Keszthely-Fenékpuszta) grew already large seed lentil, those living here continued to cultivate a plant typical for the Arching, i.e. small seed lentil and bittervetch.

7.5.3. Garden plants

The general observation that high level farming presupposes an advanced level in garden produces as well seems to be especially confirmed by botanical remains in Pannonia. It is part of the history of legumes of the Roman period in Hungary that the first researcher of botanical troves in Switzerland, Oswald Heer mentions a Roman dwarf horsebean from Hungary in 1865 (in: P. HARTYÁNYI – NOVÁKI – PATAY 1967–68). All the major legumes occurring in this age are found at different Hungarian Roman excavation sites: field pea (Budakalász-Luppa pub, Keszthely-Dobogó), small seed lentil (Óbuda, Bécsi Road No. 38–42.; Óbuda, Corvin Square No. 1; Sopron-Beloiannisz Square), while in Fenékpuszta, in the burnt layer of the castrum (455 AD) small seed varieties were mixed with large seeds (Miklós Füzes, personal communication), horsebean (Óbuda, Corvin Square No. 1; Dunaújváros), and bitter vetch (Budapest, Bécsi Street 38–42) which slightly toxic in its raw state and is entirely vanished today. The imprint of pea (*Pisum sativum*) husks was seen in the grave covering bricks of the Late Roman cemetery in Keszthely-Dobogó (Miklós Füzes and Károly Sági, in: HARTYÁNYI – NOVÁKI – PATAY 1967–68).

The muskmelon, cantaloupe (*Cucumis melo*) seeds found in Budapest Kaszásdűlő-Raktérrét, the earliest melon seed remains known in the Carpathian Basin, are of cultural historical significance (DÁLNOKI 1998). It must have been a plant held in great esteem but one cannot exclude the possibility of its imported origin.

Fabric remains preserved in the sarcophagus found in 1962 at Jablonka Road No. 15 in the Aquincum area can be connected to the production of flax, but not necessarily to local processing (PATAY – PÓCZY 1964).

Thus far, no spices were found, in spite of the fact that based on written and iconographic sources one can assume a high level gastronomic and drinking culture in the territory of Pannonia.

7.5.4. Weeds and gathered plants

In the Roman Period, the number of plant species was increased indicating a strengthening of crop production. A number of new weeds also appeared in the Roman Period (*Figs. 216–217*). Corresponding to a new way of reaping, by iron sickle harvesting close to the ground, these were mainly short stalk weeds. New species also appeared alongside extant crops including field bindweed (*Convolvulus arvensis*), catchweed (*Galium aparine*) and lady's bedstraw (*G. verum*), curled dock (*Rumex crispus*), field cowwheat (*Melampyrum arvense*), annual woundwort (*Stachys annua*), just to mention the most important ones (*Figs. 224, 226–227*). These weeds were apparently not removed completely by winnowing, or by other contemporary cleaning methods, which explains their high number.

Korn-rade and black-bindweed were weeds (*Secalietea* species) of autumn sowing cereals (six-rowed barley, emmer and common bread wheat). Spring sowing cereals (oat and common millet) and legumes (small seed lentil, bitter vetch, horsebean) might have weeds such as smooth finger-grass (*Digitaria ischaemum*) and a goosefoot species. Although they may occur in ploughlands as well, fan-hen or white goosefoot (*Chenopodium album*), maple-leaved goosefoot (*Ch. hybridum*), yellow mignonette (*Reseda lutea*) and creeping buttercup (*Ranunculus repens*) might be interpreted as members of ruderal associations. These species can be found in any area subjected to human impact, such as the margins of roads or houses, or where the nitrogen content of the soil is higher than usual around stables and garbage pits. In addition to the species mentioned here indicating human settlements, those originating from meadows might also occur: e.g. lady's bedstraw (*Galium verum*), large hop clover (*Trifolium campestre*).

During the exploration of basilica No 2 in Kékkút, surprisingly, many weeds and plant seeds from the natural environment were found among grains of cultivated crops: korn-rade (*Agrostemma githago*), soft brome (*Bromus mollis*), fodder vetch (*Vicia villosa*), great burdock (*Arctium lappa*), crosswort (*Galium cruciata*), fiddle dock (*Rumex pulcher*), broad-leaved dock (*Rumex obtusifolius*), curled dock (*Rumex crispus*), water dock (*Rumex hydrolapathum*). The circumstances by which they surfaced suggest that they might have been winnowing remains of cereals which were deliberately burned. Plants attached to twigs and branches taken from the natural environment of the time were used for heating.

Among the cereal remains from the excavations at the Keszthely-Fenékpuszta castrum at the end of 19th century described above, Miklós Füzes found the seeds of several weed species: korn-rade, brome grass, field bindweed, and goosefoot. However, Roman weed seed examinations were most helpful in 1970 in deciding whether the large amounts of cereals found in the burnt layer of the castrum were grown locally or came from imports. Having examined the species composition of the sparse weed seeds, he argued for imported cereals. According to his view, weeds like korn-rade (*Agrostemma githago*), mitre cress (*Myagrum perfoliatum*), ball mustard (*Neslea paniculata*) are basically members of Mediterranean flora and therefore the cereals found in Fenékpuszta very likely imported from that other region. While the loess covered surface of the Keszthely-Fenékpuszta ridge would have been very suitable for crop production once former forest cover was felled, unstable political conditions of the time render this possibility out of the question.

Local residents, in order to make food more diverse and meet vitamin requirements, ate fruits encountered in the natural environment along with their domestic crops (*Fig. 229*). Stones of medlar (*Mespilus germanica*), service tree (*Sorbus domestica*) and mahaleb cherry (*Prunus mahaleb*, syn. *Cerasus mahaleb*) were found in Tác-Fövenypuszta during the exploration of a 2nd-century building; a carbonised crab apple fruit was found here and in Dunaújváros (Intercisa) in a grave at a Late Emperor's Age churchyard (P. Hartyányi – Nováki – Patay 196768) (*Figs. 212, 213, 215*). Carbonised stones of cornelian cherry in a storage pit at the site of Óbuda, Bécsi Road No. 38–of Óbuda, and dewberry seeds at Kaszásdűlő-Raktárrét were unearthed (Dálnoki 1998). Hazelnut shell fragments and pubescent oak (*Quercus pubescens*) oak acorns found in Keszthely-Fenékpuszta castrum were certainly locally collected plants (Füzes unpublished data).

The fruit of danewort is not flavorful, yet it is a very useful plant. These otherwise medical portions of the plant were even used for dying leather at one time. A significant seed cache was unearthed in Tác-Fövenypuszta (P. Hartyányi – Nováki – Patay 1967–68). The berries of common elder, collected and eaten for a very long time, also provides a medical herb. Non-carbonised stone seeds were found in surprisingly great amounts (18 g) at the excavation of Mária Pető, in Budapest, Vörösvári Road No. 20–22 from the Late Roman legion military camp (P. Hartyányi 1975–77).

7.5.5. Orchards at the beginning of the Roman Period

Today's orchards and vineyards trace their ancestry back to Roman foundations in spite of evidence that the Pannonian-Illyrian population of Transdanubia might have engaged in certain fruit growing practices even before the Roman conquest. According to archaeological and archaeobotanical finds, advanced level fructiculture and viticulture were carried out in the Roman villa farm gardens in the 3rd and 4th centuries AD. Fruits domesticated by the Romans did not perish after the fall of the empire in the Pannonian area.

Our walnut production was founded by the Romans as well (Sági – Füzes 1967). Several walnut finds refer to the presence of local (?) walnut farming from the Roman Period. Walnuts "enough to fill a little bag" were found in Aquincum civic city (Budapest third district, Aranyhegyi patak) in a cremation grave from the 4th or 2nd century (P. Hartyányi – Nováki – Patay 1967–68). Walnut shell pieces and a septum in a grave at the Roman cemetery in Balatonberény and shell fragments in Fenékpuszta were found by Miklós Füzes.

The Romans also introduced Peach growing in Hungary. Several sites are known from the Roman Period in Pannonia such as Budapest, third district Kiscelli and Pacsirta Street corner, where a Roman layer was explored, and a grave in the 4th–2nd-century cemetery of Győr-Homokgödör (P. Hartyányi – Nováki 1973–74). Miklós Füzes found carbonised stones during the exploration of Keszthely-Fenékpuszta castrum. In a grave of the Late Roman cemetery at Tác-Margittelep and in a pit refill from the Late Roman Period in Ménfőcsanak intact stones were found. A peach stone found at the Óbuda-Kaszásdűlő site most probably was not imported, but perhaps from a local intercrop tree in a vineyard (Dálnoki 1998).

Apricots arrived in Central Europe by Roman efforts. Apricot in Hungary is known only as a domesticated species, escaping individuals do not survive. Apricot trees have a short expected lifetime and bear fruits only by continuous care. It can be seen as persistently grown in Central Europe since the Roman Period. However, there is no archaeobotanical evidence indicating that this tree had been known in areas north of the Alps before the Romans. Celts could have brought it from the Black Sea area via the Danube River (Willerding 1980). Werneck (1955) assumes that fructiculture in the Eastern Alp regions – including apricots – was developed from native wild varieties and cultivars introduced before the Roman conquest. Supporting evidence is rarely found. Given slow transport means, import of apricots was out of the question. They were necessarily locally grown like the apricot stone found in a stone sarcophagus containing a 4th-century female corpse at Jablonkai Road No. 15 in Budapest third district (Patay – Sz. Póczy 1964). This might have been put into the sarcophagus fresh since if it had been a sun-dried fruit, its sarcocarp would have been preserved in the environment saturated with preservatives. Stones were encountered in the Óbuda-Kaszásdűlő site as well (Dálnoki 1998).

Sweet cherry finds are known from many places in the Roman Empire. Romans spread cherry cultivars to the provinces. Identification of these stones is made more difficult by the overall similarity of wild and sweet cherries. Cherry findings are known from Pannonia. In Aquincum, stones of a sweet cherry cultivar were found in a female mummy grave from the 3rd or 4th century (PATAY – SZ. PÓCZY op.cit.). Service tree berry stones and carbonised cherry fruit stalks found during the exploration of basilica No 2. in Kékkút suggest that people preferred the consumption of improved varieties. Füzes has drawn the conclusion from the service tree stones found at the Keszthely-Fenékpuszta Roman castrum that the fortress must have perished early in autumn somewhere in the mid-5th century AD, as the tree bears the fruits at end of September through mid-October.

Plums were also favored: stones of garden plums (*Prunus domestica* cf. *oeconomica*) were unearthed from a grave in the fourth-century Roman cemetery of Balatonberény. Due to morphological similarities Füzes believed he identified the ancestor of today's Beszterce plums (SÁGI – FÜZES 1967). Imprints of common plums stones were cast from the mortar of a hypogeum in the Roman graveyard in Keszthely-Mosóház that was identified by Füzes as rough-grained plum (in: P. HARTYÁNYI – NOVÁKI – PATAY 1967–68). Further Prunus species finds are known from the second century cremation grave of Budapest III Aranyhegyi patak (Aquincum) and the second-century ash layer of a house in Tác-Fövenypuszta (Gorsium).

According to the descriptions made by Károly Sági and Miklós Füzes, the pear fruit found in the Late Roman grave in Balatonberény (in: SÁGI 1965) is very similar to the "pear ripening with barley, although it is somewhat shorter and blunter at the "cupulate" part. At the same time it resembles pear depictions on Italian frescoes. Along with several other fruit remains, common pear seeds were found at the Óbuda-Kaszásdűlő site (DÁLNOKI 1998).

As a result of the well-developed Roman trade, one has to reckon with the occurrence of imported fruits as well. From the 1st and 2nd-century Roman graves of Győr-Homokgödör, common fig (*Ficus carica*) and date (*Phoenix dactylifera*) remains were found. According to personal communications from by Füzes, the excavations carried out at the area of the Keszthely-Fenékpuszta castrum turned up carbonised Arabian coffee (*Coffea arabica*) beans. Unfortunately, the current whereabouts of the finds are uncertain. Common fig seeds were also found in an early Emperor's Age site in Óbuda-Kaszásdűlő. Local growing of common fig however has to be confirmed by additional finds. According to the opinion of Füzes, those living in Pannonia were affluent in foods of Mediterranean origin. His view is based among others on the stone fragments of an olive tree (Olea europea) floated from the well mud of the Keszthely-Fenékpuszta fortress.

It is known from descriptions and iconographic depictions that Romans grew different kinds and species of fruits in their gardens. Maybe this is supported by the hazelnut shell found in a grave cover brick of a late Roman cemetery in Keszthely-Dobogó.

Almond (*Prunus amygdalus*) remains are also known: shells were encountered in Dunaújváros (P. HARTYÁNYI – NOVÁKI – PATAY op.cit.), while the green, fluffy husk (exocarpium) was unearthed from a grave cover brick of a Late Roman cemetery in Keszthely-Dobogó (SÁGI – FÜZES 1967).

7.5.6. The Pannonian roots of our viticulture

The Romans raised viticulture to a very high level . Grapes were mainly propagated from seeds and cuttings, but grafting skills were also widespread. They developed a number of special vineyard implements. Vineyards were provided with systematic tillage and nutrient replenishment. Low head wine growing was typical. Branches could grow freely and, if the grapes dragged them down, they were supported. In gardens, man-sized square trellises decorated the view and letting grapes grow onto trees was preferred. Many different cultivars were grown. Romans loved wine. The Greek wine cult was carried on in the form of Bacchus feasts. The provinces had prosperous viticulture.

There is considerable archaeological evidence in Pannonia that grapes were grown by the Romans using high-level agrotechniques (MÜLLER 1982). The Balaton-highland, with a submediterranean climate at that time, provided especially favourable conditions. The hills around Buda were likewise suitable. At the end of the 1st century, Emperor Domitianus issued a decree prohibiting wine planting and growing in the provinces. The death of Emperor Probus in Pannonia was caused by an overly strong inclination to wine-growing (RAPAICS 1940). Allegedly, Pannonian wines were too dry and were not held in great esteem.

The advanced level of viticulture in Pannonia, suggested by written and iconographic sources, implement finds and structures was surprisingly not supported by archaeobotanical finds except in a few sporadic cases. A few carbonised grape stones were found in the burnt layer of a 2nd century structure in Tác-Fövenypuszta (P. HARTYÁNYI – NOVÁKI – PATAY 1967–68). Miklós Füzes found very few grape remains in the Keszthely-Fenékpuszta castrum, yet he stated with certainty that they were of local production, since he found fragments along with the stones, stalks and branches (*Fig. 225*). Finally in 1985–86, at the graveyard exploration carried out at Óbuda Kaszásdűlő-Raktérrét, a my grape stones, mixed with soil but well-preserved in a subfossil state due to moist conditions, were found at the bottom of a planked storage pit/well. Seed material from the area around Aquincum is dated to the 1st century AD by Paula Zsidi, the archaeologist responsible for excavations. The overwhelming majority of the stones, weighing half a kilogram, flotated and selected under a stereoscope proved to be the stone of black wine grapes, although seeds and remains of other plant species were also identified among them.

Stones of woodland European grape and black wine grapes are very similar. Distinction is made between them on the basis of a so-called "beak" (TERPÓ 1976). Among the intact and adult stones found, there were also juvenile forms that suggests harvesting. We also found that part of the grape where it joined the stalk. These so called "brushes" confirm the supposition that we are concerned with grape remains left over after pressing the grapes. In addition, bramble and common elder seeds, known by ethnographers as colouring agents for wine making were also mixed among the grape stones (DÁLNOKI 1998).

Further metric measurements (size and shape) were carried out on the subfossil Aquincum grape stones. For this purpose, the best tool proved to be an Apple-MacIntosh based computerised image analysis system. Randomly selected 300 pieces of intact and adult grape stones were subjected to 6000 measurements. The most important of these was "form factor", in other words a value independent from size and without dimension, which is suitable for the description and identification of shapes and forms of different objects – in this case, grape stones. Data clusters were evaluated using a statistical program. They were processed into histograms. According to the examinations made, the 300–piece sample can be divided into 4 different populations (*Fig. 239*). The four groups of different beak forms and stone shapes most probably refer to grape cultivars, which is made more plausible by the fact that in the early Emperor's Age advanced level viticulture blossomed in the environment of Aquincum and several grapevine cultivars were grown.

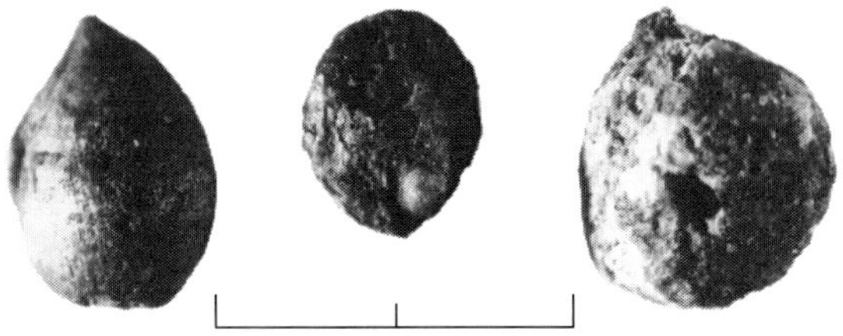

Fig. 212. Mahaleb cherry (*Prunus mahaleb*), plum (*Prunus* spec.) and cherry (*Prunus avium*) stones from a Roman house in Tác, Inventory of the Hungarian Agricultural Museum, Budapest.

Fig. 213. Sorvice tree (*Sorbus domestica*) seeds from a Roman house in Tác, Inventory of the Hungarian Agricultural Museum, Budapest.

Fig. 215. Wild sorvice (*Sorbus torminalis*) seeds, common bread wheat (*Triticum aestivum* subsp. *vulgare*) grains and grape-vine (*Vitis vinifera* subsp. *vinifera*) seeds from a Roman house in Tác, Inventory of the Hungarian Agricultural Museum, Budapest.

Fig. 214. Two-rowed barley (*Hordeum vulgare* subsp. *distichum*) hulled grains from Sopron, Városház Str., Inventory of the Hungarian Agricultural Museum, Budapest.

Fig. 216. Field cowwheat (*Melampyrum arvense*) seeds from Sopron, Városház Str., Inventory of the Hungarian Agricultural Museum, Budapest.

Fig. 217. Spring wild-oat (*Avena fatua*) naked grain from Sopron, Városház Str., Inventory of the Hungarian Agricultural Museum, Budapest.

Fig. 218. Two-rowed barley (*Hordeum vulgare* subsp. *distichum*) hulled grains from Sopron, Beloiannisz sq., Inventory of the Hungarian Agricultural Museum, Budapest.

Fig. 219. Charcoal cereal spikes from a Roman watchtower, Budakalász-Őrtorony, Lupa csárda, Inventory of the Hungarian Agricultural Museum, Budapest.

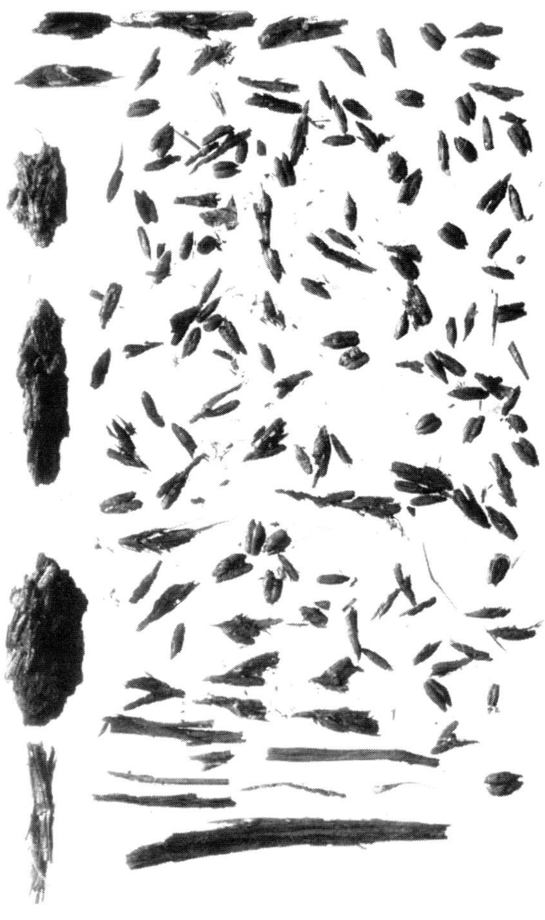

Fig. 221. Cereal remains from a Roman watchtower, Budakalász-Őrtorony, Lupa csárda, Inventory of the Hungarian Agricultural Museum, Budapest.

Fig. 220. Common millet (*Panicum miliaceum*) naked grains found on the peel of an oven in a Late Roman building, Budapest, Körte utca 29., Inventory of the Hungarian Agricultural Museum, Budapest.

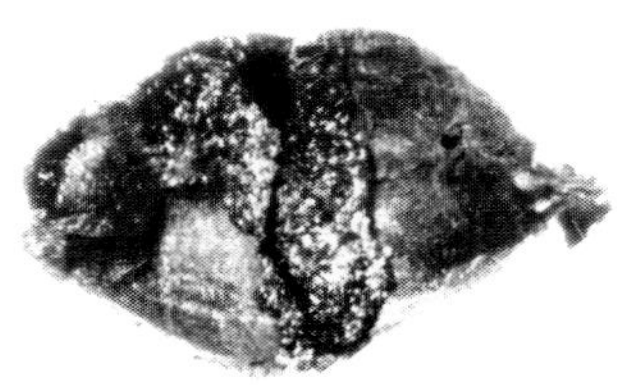

Fig. 222. Barley (*Hordeum vulgare*) grain from a Roman watchtower, Budakalász-Őrtorony, Lupa csárda, Inventory of the Hungarian Agricultural Museum, Budapest.

Fig. 223. Barley (*Hordeum vulgare*) grain from a Roman watchtower, Budakalász-Őrtorony, Lupa csárda, Inventory of the Hungarian Agricultural Museum, Budapest.

Fig. 224. Late Roman common elder (*Sambucus nigra*) seeds from Budapest, Vörösvári utca 20., Inventory of the Hungarian Agricultural Museum, Budapest.

Fig. 225. Late Roman grape (*Vitis vinifera*) seeds from Budapest, Vörösvári utca 20., Inventory of the Hungarian Agricultural Museum, Budapest.

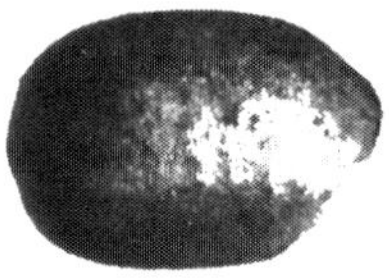

Fig. 226. Leafy spurge (*Euphorbia esula*) seed from a brickgrave found in a 4th-century cemetery in Intercisa-Papsziget, Inventory of the Hungarian Agricultural Museum, Budapest

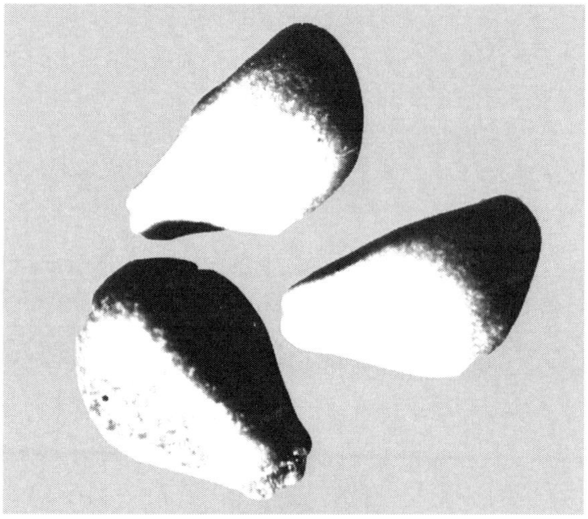

Fig. 227. Field bindweed (*Convolvulus arvensis*) seeds from a brickgrave found in a 4th-century cemetery in Intercisa-Papsziget, Inventory of the Hungarian Agricultural Museum, Budapest.

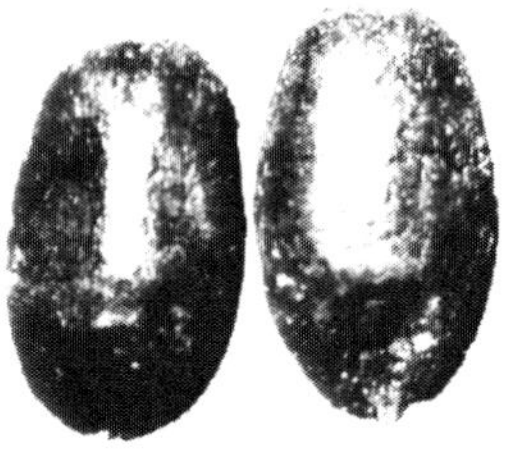

Fig. 228. Common bread wheat (*Triticum aestivum* subsp. *vulgare*) grains from the Roman settlement in Tác, Inventory of the Hungarian Agricultural Museum, Budapest.

Fig. 229. Crab apple (*Malus silvestris*) seeds from a Roman house in Tác, Inventory of the Hungarian Agricultural Museum, Budapest.

Fig. 230. Einkorn (*Triticum monococcum*) grains from a building in roman settlement Tác, Inventory of the Hungarian Agricultural Museum, Budapest.

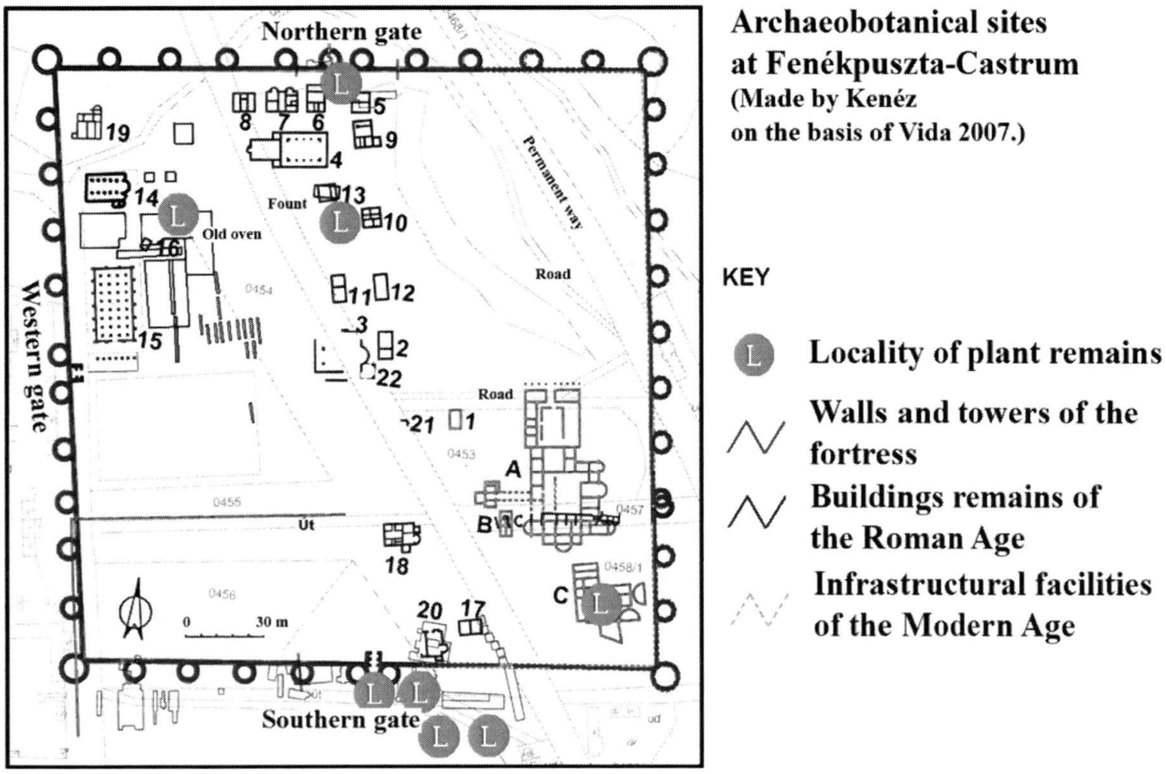

Fig. 231. Roman castrum Fenékpuszta, reconstruction of Tivadar Vida with the location of the archaeobotanical samples.

Fig. 232. Collected wheat grains in a matchbox – handwriting of Árpád Csák, 1905. Photograph by Gábor Gyulai.

Fig. 233. Rye (*Secale cereale*) grains from the Roman fortress Fenékpuszta. Photograph by Miklós Füzes.

Fig. 234. Walnut (*Juglans regia*) stone fragment from the Roman fortress Fenékpuszta. Photograph by Miklós Füzes.

Fig. 235. Ball mustard (*Neslea paniculata*) seed from the Roman fortress Fenékpuszta. Photograph by Miklós Füzes.

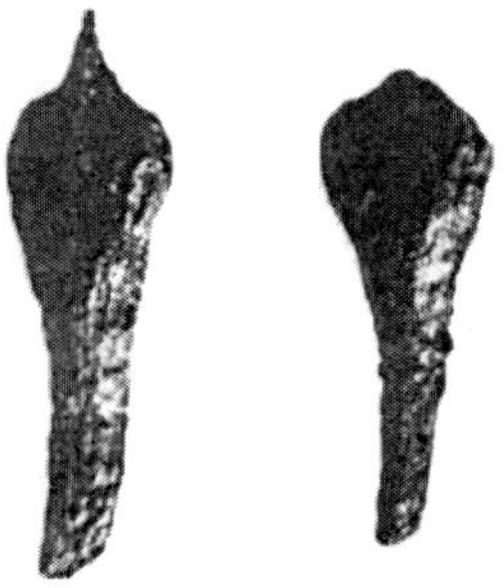

Fig. 236. Mitre cress (*Myagrum perfoliatum*) seeds from the Roman fortress Fenékpuszta. Photograph by Miklós Füzes.

Fig. 237. Olive (*Olea europaea*) stone from the Roman fortress Fenékpuszta. Photograph by Miklós Füzes.

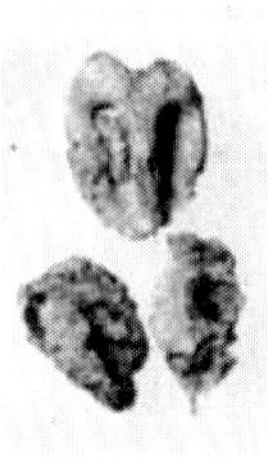

Fig. 238. Grape-vine (*Vitis vinifera* subsp. *vinifera*) seeds from the Roman fortress Fenékpuszta. Photograph by Miklós Füzes.

Fig. 239. Grape-vine (*Vitis vinifera* subsp. *vinifera*) seeds from the Roman layer of the Kaszásdűlő-Raktárrét site, clustered by the computer based on shape. Photograph by the author.

7.6. The Barbaricum in the Roman Period

7.6.1. Botanical finds in the Barbaricum in the Roman Period

While the high level crop production and horticulture practices of the inhabitants in Pannonia are supported by considerable evidence, relatively little is known of the plant growing habits of the "barbarians" who lived in the Great Hungarian Plain. Addressing this issue will substantially improve our awareness of the culture and lifeways of these peoples. Fewer investigations have been conducted among the sites and settlements of nomadic people or people of nomadic origin populating the Great Plain beyond the limes, bounded by the River Danube. Archaeologists working in the area have not devoted enough attention during their excavations to archaeobotanical processing of botanical remains. However, it has been established that people of supposedly Celtic-German origin were living in the eastern part of the country in settlements of late Emperors' Age. Their cropping culture must have differed from that of the Romans to a great extent, particularly in terms of the cultivars as no changes had taken place since the prehistoric ages (P. Hartyányi – Nováki – Patay 1967–68).

Contrary to the common bread wheat that was dominant in Pannonia but required an advanced level of agrotechniques, people in the Barbaricum continued to grow hulled wheat varieties such as einkorn and emmer. We believe that, rather than derived directly from eastern traditions, this reflects the cereal crop production heritage of earlier prehistoric peoples that once lived in the Great Hungarian Plain.

Barley was found only at the sand pit of Szirmabesnyő-Sajóparti. Here, rye was encountered as well. Common millet was a widespread and much favored cereal. Its naked grains were identified in the sites of Földeák, Garadna and Hódmezővásárhely. Fabric tissue imprints found in Isaszeg are thought to have originated from flax, although this does not mean that it was produced locally. Species from cereal weed associations and natural plant associations are also very valuable finds (Arka, Garadna, Szilvásvárad, Szirmabesnyő, Zalkod).

Only the imprint of foxtail millet (*Setaria* cf. *italica*) was identified in a daub fragment on a fireplace of a 2nd–3rd-century AD settlement in Arka, but the grain itself has not yet been recovered (Skoflek – Árendás 1969). Therefore, it cannot be said if it was locally grown.

We only have a very limited number of sites of Sarmatians living in the Great Plain at this time. Contemporary literature offers little information of the Sarmatian Period. Ptolemaeus wrote of their cities in the middle of the 2nd century AD (in: Párducz 1971), while Plinius Secundus (Nat. hist. 18, 100.) reports that their staple food was millet porridge mixed with horse milk and blood.

The Sarmatians started as nomadic people of Oriental traditions. Animal bones found in Sarmatian settlements refer to sophistocated animal husbandry (Bökönyi 1968). The natural environment of the plains at the time (water logged forest-steppe) however was not really favourable for keeping large sized livestock in a migrating, nomadic way and therefore they soon became settled farmers. Sparse farming implements (c.f. sickle fragment from Ózd) support this observation (Párducz – Korek 1958). According to archaeobotanical findings from the Late Sarmatian Period so far (Hódmezővásárhely-Solt Palé, Szalkaszentmárton-Dögtemető) their main crop was common millet, yet hulled emmer wheat was also cultivated (P. Hartyányi – Nováki – Patay 1967–68).

The Sarmatians settled in the area around the river Tisza in the 1st century AD, and by the 4th or 5th century they completely switched to farming and crop systems, retaining only warfare patterns from their nomadic past. The transition must have been a slow process. In its contemporary state, the area around the Tisza was not suitable for nomadic life. Unfortunately, we do not have any archaeobotanical material from the beginning of the Sarmatian Period in the plains. However, it is certain from the third period of the Sarmatian age (260–270 AD – app. 480 AD) that significant changes took place in their settlement patterns, expanding to the east up to the Körös rivers and the Maros and in the north, to Budapest and the source of the river Bodrog (PÁRDUCZ 1971).

Earlier interpretations of Sarmatian agriculture was substantially expanded as a result of excavations carried out between 1987 and 1990 in the Gyomaendrőd area. At the site of Endrőd No. 170 even palynological examinations were conducted in 1987. Within the framework of the research project "Micro-region Békés County" by the Archaeological Institute of the Hungarian Academy of Sciences (MTA), samples were taken for the purpose of subsequent pollen processing from a well and a house, both containing Sarmatian layers at the site Gyoma No. 133. Pollen of several different cereals and weeds, and other species coming from the natural environment of the age were identified (MEDZRIHRADSZKY 1996). The crops of the settlement inhabitants can be determined from the pollen of cereals (barley, bread wheat), autumn and spring sowing grain weeds, and that of ruderal species. A varied environment is suggested by the pollen of natural vegetation. Open waters, forest and grazing land (i.e. steppe) all occurred in the neighbourhood.

Botanical macrofinds, analysed between 1987 and 1990 by Dénes B. Jankovich, gave similar results to the pollen analysis at Endrőd No. 170. Botanical samples came from Sarmatian (4–5th centuries) houses and pits of different uses. Cereal species identified refer to a settled, tiller culture. Ploughland crops included six-rowed barley, emmer, common bread wheat and common millet, while garden produce was pea. Of wheats, the less developed, therefore agrotechnically less demanding hulled emmer wheat was in production. The relatively great number of barley and common millet finds suggest the survival of nomadic dietary patterns.

Further significant results of Sarmatian archaeobotany were obtained by Csaba Szalontai and Katalin Tóth in 1998 and 1999 at the Kiskundorozsma-Nagyszék site dated to the 3rd–4th centuries AD. Almost all of the archaeological structures were soil sampled, sometimes more than once. As a consequence of these representative samples, nearly 30 thousand carbonised items of 90 (!) different species were found from demolished houses and pits of various uses. Surprisingly, many grain species were encountered, however, only a few of them are of any significance. Their most important cereals were six-rowed barley and common millet, reflecting doubtlessly a continuation of their nomadic traditions.. Hulled wheats were also produced, first of all emmer, and next einkorn. Spelt is found only sporadically. Of the naked grain common bread wheat and the characteristically cornered club wheat grains hardly any were found. Other grain types were also known, yet they played only a subordinate role: naked barley, two-rowed barley and rye. Sporadically found common oats were identified only twice as coming from cultivars, the rest are remains from weeds (spring wild oat). On the other hand, the material contained little garden produce. Merely a few carbonised seeds of small seed horsebean were identified.

Many samples contained thrashing remains such as glumellae, ear spindles, fragments, weeds, etc. Weeds encountered together with cereals are tall and, therefore reaping probably was done using hooks or sickles.

Further species indicate the diverse natural environment of the time. The majority of weeds come from winter cereal weed associations (Secalietea) (bread wheat, rye, maybe barley as host): goatgrass, korn-rade, bearded wheat-grass, ground pine, corn snapdragon, spring wild-oat, bare (naked) oat, field brome, soft brome, rye brome, infertile brome grass, musk thistle, smooth finger-grass, field camomile, red fingergrass, barnyard grass, black-bindweed, catchweed, cleavers, henbit dead-nettle, field gromwell, yellow mignonette, sheep's sorrel, annual knawel, field madder, perennial sow-thistle, creeping thistle, annual woundwort, hare's foot clover, narrow-fruited cornsalad, wild turnip, narrow-leaved vetch, summer vetch, field pansy, cotton thistle.

Common orache, yellow goosefoot, spike leaved goosefoot, Scarlet pimpernel, dwarf mallow, sticky and green bristle-grass, common orache, and common chickweed are members of spring sowing grains (common millet, maybe spring barley), in other words a hackfruit weed associations (*Polygeno-Chenopodietalia*).

Given the high number of species from the natural environment, the environment must have been very diverse. Natural processes such as wind, runoff and animals might have contributed to the accumulation of species in the samples. The local environment around the settlement was a waterlogged landscape with common reed. Common and tabernaemontanus bulrush are characteristic species of the common reed zone of sedge marshes. Species from tall sedge associations included tussock sedge/yellow sedge, and common spike-rush. Small water-pepper and celery leaved crowfoot were encountered at the water edges. This was the habitat for reed sweet-grass and fox sedge/prickly sedge as well. Yellow sedge/ pendulous sedge ia a marshland species. Common meadow-rue also thrives in a wet environment. Purple moor-grass might have formed an association in flat areas, but it could have come from forests lining the bog as well. Such a moor forest is also suggested by remote sedge. The other edge of the forest toward the settlement must have been drier, as indicated by the presence of ground-ivy, sloe (blackthorn), wild marjoram remains. A park forest could also be present near the settlement providing buckthorn and woodland (European) grape. Wild strawberry achenes might have originated from a forest clearing. Ground-ivy may have come from a forest edge while a more arid area could sustain summer Michaelmas-daisy.

Many relict species from meadows and pastures suggest livestock grazing in the vicinity of the settlement. The great number of plant species suggests a large and diverse environment consistent with a large area necessary for raising an assumedly great number of livestock. The meadows and pastures had explicitly wet parts, periodically inundated or waterlogged. Basically, these are the habitats preferred by slender tufted-sedge, prickly sedge, pendulous sedge and common spike-rush. Similarly wet parts are preferred by perennial flax, maybe caraway, provided of course that it did not come from local cultivation. In spite of this, the majority of species are those preferring moderate water supply, which simultaneously is an indication of the diversity in such areas: ribwort-plantain, sun spurge, meadow vetchling, black medic, hop clover, yellow trefoil, bur medic, ribwort-plantain, common sorrel, large hop clover, parsley-pier, brown knapweed, red clover, and white clover. This meadow and pasture, having mainly typical habitat characteristics, could be effectively arid in some places, as shown by white melilot, lesser meadow-rue, field and wild sage, wall germander, wild rye.

Ruderal species (downtrodden weed associations) originated from human settlement, human-visited areas (such as roads, stables). Their number is significant. The ruderalia of the settlement refer mainly to a moderate habitat (with average water supply): fan-hen or white goosefoot, wild barley-grass, white horehound, annual meadow-grass, knotgrass, white campion, broad-leaved dock, creeping buttercup, rough cocklebur. A lesser proportion prefer arid parts: thyme-leaved sandwort, and cammon mallow. Their presence gives rise to the conclusion that settlements and roads of significant size might have occurred. It should be noted here that fan-hen or white goosefoot may be regarded either as a weed or as a ruderalia.

Wheat Seed-Gall Nematode (*Anguina tritici*) remains were also found, mainly burnt, together with emmer wheat.

Seed and fruit remains recovered from the Szihalom-Pamlényi Sarmatian graveyard (excavation by Adél Váradi 1997) fit well into the pattern of the crop production system based on botanical remains. This is archaeobotanical material of a population dealing with settled crop production based on nomadic traditions and preserving traditional eating habits. This site is not at all typical. It is not a settlement, but graves where finds might represent purely ritual purposes. There were mainly carbonised remains of cereals, the majority being common millet, along with some emmer and common bread wheat grains.

Fig. 240. Maple-leaved goosefoot (*Chenopodium hybridum*) seeds from a pit in Szirmabesenyő-Sajópart, 2th–4th centuries, Inventory of the Hungarian Agricultural Museum, Budapest.

Fig. 241. False cleavers (*Galium spurium*) achenes from a pit in Szirmabesenyő-Sajópart, 2th–4th centuries, Inventory of the Hungarian Agricultural Museum, Budapest.

Fig. 242. Emmer (*Triticum turgidum* subsp. *dicoccum*) spicelets (*furca bicornis*) from a pit in Szirmabesenyő-Sajópart, 2th–4th centuries, Inventory of the Hungarian Agricultural Museum, Budapest.

Fig. 243. Annual woundwort (*Stachys annua*) nutlets from a pit in Szirmabesenyő-Sajópart, 2th–4th centuries, Inventory of the Hungarian Agricultural Museum, Budapest.

Fig. 244. Vetch (*Vicia* spec.) seed from a pit in Szirmabesenyő-Sajópart, 2th–4th centuries, Inventory of the Hungarian Agricultural Museum, Budapest.

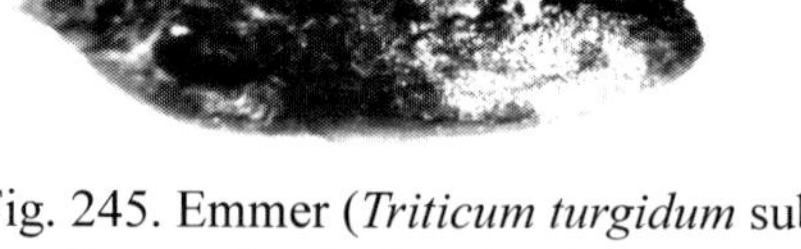

Fig. 245. Emmer (*Triticum turgidum* subsp. *dicoccum*) naked grains from a pit in Szirmabesenyő-Sajópart, 2th–4th centuries, Inventory of the Hungarian Agricultural Museum, Budapest.

Fig. 246. Water-pepper (*Polygonum hydropiper*) nutlets from a pit in Szirmabesenyő-Sajópart, 2th–4th centuries, Inventory of the Hungarian Agricultural Museum, Budapest.

Fig. 247. Spring wild-oat (*Avena fatua*) grains from a pit in Szirmabesenyő-Sajópart, 2th–4th centuries, Inventory of the Hungarian Agricultural Museum, Budapest.

Fig. 248. Spring wild-oat (*Avena fatua*) grains from a pit in Szirmabesenyő-Sajópart, 2th–4th centuries, Inventory of the Hungarian Agricultural Museum, Budapest.

7.7. The Migration Period

Botanical finds from the Carpathian Basin and surrounding countries indicate that in the Migration Period a much more modest crop production system replaced Roman agriculture. At most places, settlements became sporadic and somewhat impoverished. Although all cultivars of former ages are still found, they are very modest in extent and grains are smaller. Common millet is dominant, a characteristic type of cereal for quickly moving nomadic people (WASYLIKOWA *et al.* 1991). Analysis of weed associations also indicates impoverishment (WILLERDING 1986a).

A number of people appear in the Carpathian Basin during the Migration Period. The underlying cause for the migration of these masses of people might have been climatic in nature. Historical climatology, using the achievements of natural sciences and historical sources alike, calls the period between 370 and 1220 AD "the long warming". This climatic change was manifested first of all in reduced precipitation in the inner parts of Eurasia and the Mediterranean shores. Extensive areas in Eastern Europe and Central Asia, especially between 300 and 400, were subjected to prolonged droughts. Due to the dried up lakes, rivers and meadows, pastoral people started to migrate westwards. Yet another wave of drought in the 7th–8th centuries triggered a second large wave of migration. The Mediterranean was afflicted with a similarly severe drought between the 4th and 9th centuries AD.

We know relatively little about the crops of folks living in the Carpathian Basin at this time. However, these finds solve not only a number of burning questions of archaeobotanical research, but provide substantial help to archaeologists, linguists and historians alike. The only reasonablly certain pattern associated with the first stage of the Migration Period is the typical crop system of nomadic peoples.

Pollen diagrams (ZÓLYOMI 1971 1980; ZÓLYOMI – PRÉCSÉNYI 1985) confirm the above stated pattern. Annual (NAP) pollens from the Migration Period show a general decline, while oak (*Quercus*) pollens increased slightly. The number of pollen from cereals and goosefoots (*Chenopodiaceae*) is reduced in the Migration Period compared to the Roman Period. These data indicate a decline in plant cultivation and a drop in settlement numbers. Pollen of cereals and fruits (e.g., walnut and grape), as well as non-tree pollen increased again only after 800 AD. According to a personal communication from Miklós Füzes, the boundary of the forest-steppe ranging along the middle black earth belt could shifted slightly more to the south during the 9th and 10th centuries. The movement of the boundary towards the north is explained by a combination of tree logging and climatic warming.

Any substantial changes can be demonstrated only from the 9th century, when pollen profile values start changing. Reduced oak and increased annuals (NAP) pollen indicate an increasing number of permanent settlements entailing the growth of ploughland cultures. In the 10th centory, oak (*Quercus*) pollen drop further with a slight but transient increase in the numbers of hornbeam (*Carpinus*), European beech (*Fagus*) and hazelnut (*Corylus*) pollen, signalling a possible short cool period. Cereals, fruit (walnut and grapes) and non-tree pollens (NAP) were increasing. All these are direct consequences of forest clearings resulting in the formation of secondary steppes.

7.7.1. A climatic history of the Migration Period

FIRBAS (1949) named the era beginning 800–600 BC to the present Subatlantic European beech phase for the Alpine foreground region. ZÓLYOMI (1952) subdivides Subatlantic phase into two stages: the European beech stage II and the Present. In the European beech stage II, to approximately 1200 AD, a mild, drier climate dominated while in the Present stage, since approximately 1200 AD, European beech recede while conifers spread and a cultural steppe is formed. Palynological examinations of Magda Járai Komlódi (1966) in the plains further refined these categories. She separates two more categories in European beech stage II: the first has increasing temperature and is later replaced by a late Subboreal cooling period. At the start of the present age, she calculates a period of Subatlantic climatic improvement. A general cooling occurs from the middle 8th century BC to the middle 2nd century BC. This is followed a warmer period from the middle 2nd century BC to the end of the 4th century AD, then reverted to cooling from the end of the 4th century AD to the end of the 8th century AD cooling. A long-term warming trend returns from the end of the 9th century AD to the end of the 13th century AD.

In the Present period (100–1950 AD) (Xth pollen zone) the current zonal and extrazonal plant associations (bogs, marches, forests, forest steppe interchanged with meadowland) and soil types (marshland, bog meadow, alkaline and grassland soils) were formed. Due to human impact, the zonal oak-forest steppe of the Great Hungarian Plain has become a culture wasteland, converted to agricultural land. The native vegetation survived only in leftover (relict) areas. Reduced precipitation is suggested by the natural decline of European beech and the development of former forest soils towards black earth.

Historical climatology, summarising the results of interdisciplinary examinations, presumes significant changes in climate during the Holocene (KORDOS 1982). The period called “Greek stage” in climatology (750–150 BC) corresponds to the Subatlantic phase. The “Roman stage” (150 BC to 350 AD) experienced a climate warmer than today. The climatic period after the Roman Period (350–700 AD) was colder and drier in both Northern and Central Europe as well as in Central Asia when compared to the preceding period (KORDOS op.cit.; SCHWARZBACH 1950). The “Little Climatic Optimum” period, lasting until app. 1200 AD, was only interrupted by a short cooling between 700 and 800 AD. This may be reflected in semi-subterranean residential (pit) houses equipped with internal fireplaces appearing in the 7th century (BÓNA 1971). Schnitnikov (in: SOMOGYI 1984) states that climate gradually became humid after 1000 AD, becoming characteristic only after the 13th century.

According to interdisciplinary and historical sources for historical climatology, the period of "the Little Climatic Optimum" (800–1200 AD) was the warmest period in Europe during the last two thousand years (RÁCZ 1993). Precipitation started to increase only after the tenth century. This climate change resulted mainly in increased temperatures in the northern part of Europe. Warming lasted from the 7th century, interrupted by some minor fluctuations, up to the end of the 12th century. All this was favourable for the spread of agricultural plant species.

Results similar to the conclusions above are given for the climatic conditions using the so-called malaco-thermometer method, calibrated with ^{14}C data for the last 30,000 years of history in the Great Plain (HERTELENDI – SÜMEGI – SZÖÖR 1991), and the malacofauna examinations of the Hungarian-hill range (FŰKÖH 1990). Approximate results are also given for the Hungarian Holocene climatic changes by László Kordos (1977) using the so-called "vole thermometer" method. Voles (*Microtus arvalis*, *M. Agrestis*, *Arvicola* spec., *Myodes glareolua*, *M. oeconomus*, *M. gregalis*, *Dicrostonyx torquatua*) are small rodents very sensitive to temperature changes. Their geographic distribution is mainly dependent on climatic conditions. Using bone remains of these small rodents, July mean temperatures over the entire period of Holocene in Hungary were successfully determined. With the help of a "vole thermometer", you can read July temperature means. The July mean temperature at the end of the Roman Period (15.5° C) starts increasing following the Late Migration Period and by the 8th century it rises more than one degree centigrade. After a minor faltering, it rises even more rapidly, reaching 18.5° C and falling back thereafter. These are in fact all conclusions drawn from proxy data of climatic history.

Wet conditions are shown by the frequency of the remains of water vole (*Arvicola* spec.) (arvicola humidity). The minimum value falls between 500 and 1000 AD, which means a relative drought on the Great Plain (SOMOGYI 1984).

The appearance of the Great Plain in the Migration Period was strikingly different from that of today: it was made diverse by open waters, marshlands, bogs, meadows and forests. In the Late Migration Period and Hungarian conquest, approximately one eighth of historical Hungary and one quarter of present-day Hungary was periodically or permanently covered with water (SOMOGYI 1984). This land was continuously inhabited since the Neolithic period. Parts suitable for settlement were mainly those free of floods and inundations. Flora and fauna of the marshes provided a constant supply of food for humans and livestock alike; after the receding floods, grass grew in abundance and water was still plentiful.

The zonal oak-forest steppe of the Great Plain was only gradually transformed into an artificial agricultural landscape from the Late Middle Ages on. The Great Plain of the Subatlantic phase differed from the one today in terms of hydrology as well (Somogyi 1984). The river Tisza and its tributaries were surrounded by extensive flood plains, marshes and swamps. Bogs in Szatmár Bereg were associated with waterlogged parts of both Hortobágy and Nagykunság; water filled depressions of the two Sárréts and the rivers Bodrog, Tisza, Körös and Maros; and the bogs and moors south of the river Maros. The landscape was passable only through elevated grounds free of floods.

The nature transformation activities of humans settled in the Great Plains are not to be underestimated (SOMOGYI 1984). Tatarian maple oak-forests of the loess ridges and lily of the valley, May lily oak forests of the sandhills were noted from the Migration Period on. Forest clearance, agricultural encroachment, grazing and trampling caused soil erosion. Anthropogenic impacts in the Boreal phase resulted in general moving sand phenomena in the sandy landscapes in the Subatlantic phase as a consequence of natural vegetation cover deterioration. Vegetation decay persisted in terms of time and space and causing sporadic quicksand formation on the surface of the Great Plain.

The forested steppe was slowly transformed into culture steppe. The streams of Tarna and Gyöngyös still run partly in the mighty earth works built by the Sarmatians, i.e., the Ditch of Csörsz. Artificial microformations, rising from the surrounding ground surface, such as settlement mounds (tells) and kurgans, possess particular wildlife. The landscaping effects of human activities were most expressed in hydrological-water management transformations carried out in the deeper lying regions of the Eastern Hungarian territory where water was abundant, having lasting impacts on the land profile, soils and wildlife. The alluvial plains started to look more like a human-made landscape.

These changes affected mostly the floodplains and the higher, non-inundated areas of the flat lands silted up by the river Tisza and its tributaries. However, landscaping efforts were made at the basin margins and the areas of alluvial fans as well.

However, former construction of river bank dunes by the Kurgans, and the Ároktő and Hejőkürt ditches, dykes and embankments (Csörsz-dyke) of the Roman Period did not have so great an impact on the contour as to influence landscape development of the area (NAGY 2000). The natural evolution of the landscape was continuous since the Subatlantic phase and it was not before the Tisza river regulation works of recent times – started in 1846 and completed by building embankments along the river in 1938 – that the landscape was radically transformed.

7.7.2. The archaeobotanical heritage of the Early Migration Period

Botanical finds in Hungary from the Early Migration Period are extremely rare (P. HARTYÁNYI – NOVÁKI – PATAY 1967–68). In addition, some of the finds (Cikó, Szegerdő, Táp-Borba puszta), lacking archaeobotanical examination, have been permanently lost. A summary is provided below of all the information obtained from the crop production systems of the age. The discussion is limited to those inhabitants whose agriculture is known to some degree.

It is well known that the Huns, who settled in during the middle of the 5th century, were nomadic, warfaring people. Unfortunately, no botanical material has been recovered to date associated with their settlement in the territory of Hungary. We know from Priscos rhetoris' contemporary description that their food was made of common millet, their drinks of barley (in: HARMATTA 1952). The husbandry of the Asian Huns (hsiung-nu) is discussed by Chinese chroniclers (Yen S-ku, Chan-shu) (in: TASKIN 1968). These sources and the excavations of settlements in the Buriat Republic suggest that both domestic livestock (i.e., cattle, horse and sheep) were kept in the regions conquered by the Huns while land was in tillage. At the site of Ivolga motte (3rd century BC – 1st century AD) a ploughing iron, a spade blade, a sickle-formed iron knife and grain pits were found (DAVYDOVA 1968).

It may be assumed that in addition to the labor of their Chinese captives, Huns also participated in farming. Common millet, barley and bread wheat grains provide evidence for crop production. However, no hulled wheats (einkorn or emmer) were found which are otherwise typical for European sites. A surprising feature is the joint appearance of common bread wheat, club wheat and black wine grapes, all requiring advanced agriculture. Common millet was identified at several Hun burial sites, such as, I'movaya Pad and Noin-Ula. These and the implements found here cast new light on the methods employed by the Huns (BERENOVÁ 1986). Basically, the nomadic Huns, just like the subsequent Avars, must have had a certain degree of farming skills, and they have arrived to the forested steppe region of the Great Plain equipped with these skills.

Caches associated with the Gepids, a people who lived in the area beyond the Tisza, are explicit evidence of a farming lifestyle (MÜLLER 1982). In Kiszombor, in a Gepid graveyard, an iron sickle, the remains of a small spade and non-specified fabric remains were unearthed (CSALLÁNY 1961). A hazelnut shell was also identified in the 3rd–4th-century Gepid grave of Kisvárda-Darusziget (P. HARTYÁNYI –

NOVÁKI – PATAY 1967–68). Cereals that were found in remains in Eperjes-Csikóstábla (excavations by Csanád Bálint 1976–77) in earth samples taken from the floor of a house destroyed by fire must have been grown locally. Most are common millet, the rest are common bread wheat and six-rowed barley grains.

In 1986 excavations were carried out in Devín, near Bratislava, Slovakia, along the limes at the time, at a settlement dated to the 5th century AD, populated by Danube Germans or maybe Kvads. A carbonised grain layer, a completely intact bread and several pieces of bread were found among pot fragments in the demolished remains of a fireplace (PIETA 1988; PIETA – PLACHÁ 1989). The composition of the cereals obviously stocked for kitchen use shows an advanced level of agriculture: 66% rye, 21% common bread wheat, 11% barley, 1.6% common millet.

The Longobards, migrating through Pannonia, were mainly livestock farmers. En route towards Italy other ethnic groups migrating together with them provided their grain supply. Their botanical heritage is restricted to a bread wheat straw fragment and a few fabric imprints identified as flax found in a grave of the Vörs Longobard cemetery (FÜZES 1963).

7.7.3. The archaeobotanical heritage of the Late Migration Period

We do not know much of the lifestyle of the Avars, the people moving in to the Carpathian Basin in 568 AD and again later in 670. Those surviving until the Hungarian conquest were settled in the Great Plain and more or less in the eastern and southern band of Transdanubia. In contrast to many graveyards, only a few settlements are known. Examining the Late Avar settlements in the environs of Endrőd, Béla Miklós SZŐKE (1980) arrived at the conclusion that the Avars in the Körös region lived in semi-subterranean houses and followed a sedimentary lifestyle. Taking into consideration pedological and botanical factors, he stated that Avar sites from the 8–9th centuries form a semi-circle and remain in the alluvial meadows of the southern floodplain of the river Körös. They do not cross the northern edge of the Békés-Csanád-loess ridge (loess-steppe). All this indicates that the former nomadic people settled down and followed a peasant lifestyle.

It is thought that the main crop of the Avars was common millet, yet this does not mean that they continued to be nomads, but rather that they insisted on their historical tradition with respect to eating habits. One can assume for certain that by the 7–8th centuries, the Avars have changed their way of living. They settled down and conducted a farming system mixed with livestock husbandry (KOLLAUTZ – MIYAKAWA 1970). Latest research presumes that the Late Avars interacted with the Saltovo-Mayak culture based on ceramic pot fragments found in the Late Avar settlements of Békés county (Béla Miklós Szőke, personal communication in: MÜLLER 1982).

The bones of domesticated animals found as grave goods in Avar graveyards (many hens and a few pigs) also indicate more or less permanent settlements (BÖKÖNYI 1968; MATOLCSI 1967–68). Settlements are also indicated by underground houses found in Dunaújváros and Kölked (BÓNA 1971). Unfortunately no major grain finds to date preclude confirmation of supposedly intensive ploughland cultivation. However, the archaeological artefacts found in graveyards refer to such cultivation, mainly iron sickles and sporadically iron hoes or spades (FETTICH 1964; KOREK 1943). The Avars first used a delving (muzzling) type of plough equipped with handled plough iron, but lighter that that used previously in the Roman Period. Subsequently, they must have used wooden ploughs in place of very expensive iron implements (MÜLLER 1982). Avar social differentiation must have been influenced by the dominating role of cropping as the number of graves with modest grave goods increases in Late Avar cemeteries (MÜLLER op.cit.).

Botanical caches associated with the Avars are only sporadic in nature (*Figs. 249–252*). Unfortunately the large number of recent graveyard and settlement excavations were only partially supplemented with comprehensive archaeobotanical examinations. It is certain, however, that a major staple food was bread from common millet, as indicated by carbonised grains from the Szirák Avar graveyard (HAMPEL 1897). Barley was also produced. In the excavation by Andrea Vaday in 1995 at Kompolt-Kistéri hamlet, barley grain imprints were found at the bottom of a pot from the Late Avar layer of the site. It might have come from locally grown cultivars as the chaff and tailings of locally threshed grain crops were mixed with clay during the manufacture of ceramics as an attenuating agent.

At the Vörs-Papkert site at the excavation led by László Költő in 1994, in one of the Avar skeleton graves the carbonised carpological findings were identified as common wheat and pea. The remains were most probably put into the grave for ritual purposes, and they also provide evidence that the Avars living here grew the most genetically advanced naked grain hexaploid common wheat.

Gábor Lőrinczi explored several Avar graveyards in Csongrád county: an Early Avar graveyard at Szegvár-Oromdűlő in 1990 and a Late Avar graveyard at Pitvaros-water reservoir in 1993. All botanical residues came from soil samples floated from graves (and grave contents, e.g. ceramics. Such evidence exposes the funeral habits of the time. As these are mostly cultivars and remains of foods made from them, they might have been interred with the deceased as food in a burial rite, and not in fear of sortilege. It remains to be seen why these remains are carbonised, as no traces indicated cremation or burying in any of the cases. According to the present assessments in archaeobotany, these remains cannot be carbonised in soil, so they must have been interred in this state. There are two possible solutions for this: either they are random caches or burnt scraps of well used cooking pots or they were deliberately overheated in order to get them burnt, thus ensuring their durability.

Only a few botanical finds were found in the Early Avar graves of the Szegvár-Oromdűlő graveyard: a few common bread wheat grains, grasses, a non-specified hulled grain fragment and naked grains of green or rough bristle-grass. During the flotation of a sample coming from grave ceramics, two pieces of husked common millet grain was also observed, burnt together with – or rather burnt into – food scraps. They lacked the germ portions that are usually broken away when husked.

The botanical remains in Szegvár-Oromdűlő are typical for a settled population growing tillage crops and garden produce. One of the favourite plants of those living here was doubtlessly common millet. Both their food and drink were made of it. The common bread wheat find, which assumes a higher level of agriculture, might have been the produce of a settled population dealing with crop production.

In spite of expectations, no food remains were found in floated soil samples of the Pitvaros-water reservoir Late Avar graveyard. We do not know the reason for this. It means that no food was put into the pots, or foods that were put in were not previously burned or carbonised and therefore decayed. On the other hand, there were many grain fragments that could not be identified by species as a result of their fragmented appearance. Of these, only barley (most probably six-rowed) carbonised grains were specified with certainty. In this age, barley served for human nutrition. It might have been by accident that a piece of husk from a vexillate plant was put into the pot of a grave.

In three graves in the 6–7th-century graveyard of Keszthely-Fenékpuszta, grape seeds and common bread wheat grains coming from the straw once put under the corpse were found (FÜZES 1963). Unfortunately no ethnic identity is given to the 5–6th-century site in Sopron, yet species-rich crop grains and legumes were encountered here in great numbers (P. HARTYÁNYI – NOVÁKI – PATAY 1967–68).

Field bindweed (*Convolvulus arvensis*) seeds might have been put into Avar graves for mortuary ritual purposes explored in Szőreg and Vác-quarry (P. HARTYÁNYI – NOVÁKI – PATAY op.cit.). In one graves in the Budakalász Avar graveyard, field bindweed seeds were found in great numbers and stocked. The appearance of field bindweed seeds in such a large amount indicates a yet unknown but necessarily special burial habit. Similar finds (c.f. bindweed seeds) are only found with the conquering Hungarians of the Káli graveyard. In another Avar graveyard in Gerjén-Várad, carbonised oak acorns were observed around the skulls in several graves (WOSINSZKY 1896).

Historical literature maintains that in terms of tillage culture, the Slavs were one of the most significant ethnic groups in the Migration Period (ERDÉLYI 1982). According to assumptions, they settled in three distinct waves: first in the 6th century, taking Transdanubia, Transylvania and Bánát, second during the Avar immigration and finally at around 680 AD. The farming and livestock-tending Slavs, having spread over a large region and being different in both culture and tongue, survived the Hungarian conquest first in the framework of the Avar kaganate, then, the latter having collapsed, scattered in a large area. The proportion of immigrants versus the local population as well as the issues related to the knowledge and skills and propagation material remain unanswered down to today.

Since the 830s, especially in the forested, hilly regions of Southern-Somogy, the Zala region and the Balaton highlands, significant amounts of Christian Avars and Slavs were present who had similar cultures (CS. SÓS 1985; VÁNDOR 1986; MÜLLER *et al.* 1989; KÖLTŐ *et al.* 1996). The role of the Slavs in the 9th century is far from understood. Their graveyards and settlements are not clearly separated from the Avar materials in all cases, and often it is not possible to distinguish them.

Several scientists dealing with the period have carried out research on the Hungarian language in order to make indirect conclusions based on Slavic loan words about the cropping culture of the Slavs at the time of the Hungarian conquest (RAPAICS 1943; BALASSA 1965–66; GYŐRFFY 1977). Some words in livestock husbandry are doubtlessly Slavic: akol, jászol, járom, iga (pen, feedbox, yoke, harness). The transfer of words connected to farming is also significant; harrow, hoe (originally meaning digging), eke (plough) and its parts: lemez (blade), csoroszlya (coulter) (eke and csoroszlya are probably even more ancient, coming from South-Ukraine) gerendely, pating (the chain or rope fixing the plough to the fore-carriage), kabola (bed-plough), kakat-szeg, barázda (furrow), láz (clearing), mezsgye (balk). Loanwords of meadow and pasture management coming from Slavic are széna (hay), villa (pitch-fork), kazal (haystack), gereblye (rake), petrence, kasza (scythe), kalangya (storage of cut crops on stalk), szín (barn), lapát (shovel), csép (thrash, displacing the old Hungarian expression "másfélfa"), parlag (fallow), ugar (the old Hungarian equivalent was the word "nyomás", meaning lea-land), kalász (ear), szalma (straw), polyva (chaff). Several plant denominations come from Slavic as well; gabona (crop, grain) (uncertain origin: kása (porridge), or bread plant), mohar, rozs (rye), zab (oat), len (flax), bab (bean) (horsebean), lencse (common lentil), kapor (common dill), retek (reddish), tök (gourd), squash (kobak) ("pumpkin" = Cucurbita pepo is of the New World, before America was discovered only Lagenaria siceraria kabak and lopótök (gourd and squash) cultivars were known), káposzta (wild cabbage), uborka (cucumber), cékla (beet-root), murok (carrot), bükk (European beech), cser (Turkey-oak), jegenye (silver poplar), gomba (mushroom). Many problems might have been caused earlier on by vadmák (field poppy), kertimák (corn poppy) and konkoly (korn-rade), or ragya (blight). They are all Slavic loan words (KNIEZSA 1955).

However, one must not erroneously conclude from the analysis of loan words that the Hungarians learned stabling livestock management, animal feeding and crop rotation from the Slavs (CS. SÓS 1973). The majority of the items above must have been known, used and produced in the east as well. Moreover , they do not mean a production method or implement or plant associated with a particular ethnic group. It seems that many of the loan words coming from the Slavic language were taken over unnecessarily (such as gabona, lopó, tök, csép, mezsgye, ugar), since the Hungarians were already in possession of these skills as early as in the times of the Kazar kaganate (GYULAI 1987, 1989).

No doubt the Slavs, albeit indirectly, were exposed to Antique traditions. Their agriculture advanced further as under the influence of interactions with the Avar and Bulgarian-Turkish population, and/or they were connected to Greek, Frankish and Bavarian tillers or missionaries. As a proof for their tilling skills, a symmetric 16.4-cm-long plough iron belonging to a "raló" (delving plough) was uncovered in Zalavár, and "heavy bed-plough" finds are also known from Moravian areas (MÜLLER 1982). Asymmetric plough irons associated with Slavic population in the late Migration Period were used for furrowing in Transdanubia. These ploughs were already equipped with a coulter, i.e. a part serving for vertical cutting of the earth (MÜLLER 1982).

GYŐRFFY (1977) based his statements on mainly linguistic arguments when he claimed that the Slavs living here could grow autumn wheat and spring barley along with autumn sowing rye and spring sowing common millet. Reaping was done with a sickle. Grain was not trodden as the Hungarians did, but thrashed using a flail. They differed also in growing flax instead of hemp.

Doubts are emerging in respect of spring barley production. As a matter of fact, this crop has lower yields and at the same time requires more favourable conditions of soil and climate than the autumn variety of barley. Also, spring barley is mainly used for beer, while the autumn sowing cultivar is suitable for animal feed (GYULAI 1989).

According to MOLNÁR (1961), linguistic analysis shows that Slavs living in the Carpathian Basin acquainted Hungarians with the art of growing and processing flax. In our opinion, however, the word len (flax) can equally be derived from the ancient Bulgarian "lentja", the Latin "linum" and the German "Lein".

Textiles – as their fabric is decomposed in the humification process – are sustained only very rarely, under extremely special conditions, such as charcoal generation, peat formation or preserved by heavy metal ions. Charcoal-like fabric remains, soaked in heavy metal ions coming from Migration Period Avar graveyards explored by Nándor Fettich and Gyula Rhé in Jutas and Öskü, were identified mostly as flax (*Linum* cf. *usitatissimum*) by Miklós FÜZES (1964), and in one case as cotton (*Gossypium* spec.). This however refers only to the actual raw material of the textile, because the fibre plants might have come either from import or from flax grown and processed locally. These finds suggest that the Hungarian conquerors arrived into the Carpathian Basin with the knowledge of growing and processing flax.

According to 9th century finds in Hungary, the earlier forms of hooked sickles in use before the Hungarian conquest continued. Short scythes with a slightly bent form appeared as well as the main tools for gathering fodder (MÜLLER 1982). Typical finds include straight grubbing hoes with widening blades and the hoes with even wider blades at the neck. The latter were most probably used in viticulture. In the northwestern and western parts of Transdanubia, the spade-shoe assumedly appeared through western influence. Therefore, the appearance of the spade-shoe in the Carpathian Basin cannot be related to the Hungarian invasion. The first accurately dated spade-shoe find does not appear until the Late Migration Period from Mohács. It was used by the Romans, but it was yet not found in Pannonia. Farming implements dated at the 8th century and found in southwestern Slovakia (hoes, spade-shoes, plough irons, vine pruning knifes, sickles) are claimed by BERENOVÁ (1986) to be of uniformly Slavic-Avar origin.

According to the archaeological finds, the general utility tools of the Slavs and Avars living in Transdanubia at the end of the Migration Period became similar to each other, while the botanical finds reveal the image of a settled population possessing a shared agricultural system (GYULAI 1994).

One of the most important botanical findings in Hungarian archaeobotanical research was uncovered from the late Migration Period Fonyód-Bélatelep site, the excavation of Béla Horváth in 1964 (GYULAI – HERTELENDI – SZABÓ 1992) (*Figs. 253–254, 258*). Based on ^{14}C tests, the age of the settlement, made up of lake dwellings, can be dated from the second half of the 7th century to the end of the 9th century. Considering the amount, quality and especially the species composition of the finds, this is the largest quantity of remains of this kind ever found. A rather accurate picture can be gained about botanical knowledge, crop production practices, fruit consumption patterns, horticulture and viniculture of the period directly preceding the Hungarian conquest by analysing all 181,000 seeds of 64 different plant species found here. According to evidence provided by these plant macrofossils, the inhabitants of the lake dwellings pursued extensive farming activities. They did not specialise in one main crop but produced a wide variety of grain crops, kitchen produce, fruits and grapes (*Figs. 255–257, 259–313*). This implies self-sufficiency. The main crops of the inhabitants were barley, common bread wheat, club wheat, common millet, rye and common oat. Of the barleys, naked barley and two-rowed barley were produced at levels equal to six-rowed barley.

The low number of kitchen produce species and seeds (common lentil and pea only) seems surprising, since current opinion holds that productive ploughland cultures could be formed only in places where garden produce at an appropriate level was associated with it. This cannot be the case here, however. It is much more likely that some special kind of production or dietary custom can be detected (cf. abundant grain and poor legume finds of sites from the early Árpádian Period).

Fruit finds (9 species and 1400 pieces of remains) suggest a substantial amount of fruit consumption. The inhabitants of the settlement met their needs partly by gathering and partly by growing fruit trees. Stones and shells of sloe (blackthorn) (*Prunus spinosa*), hazelnut (*Corylus avellana*), crab apple (*Malus silvestris*), certain wild pear species (such as Hungarian-pear), single-seed hawthorn (*Crataegus monogyna*), danewort (*Sambucus ebulus*) might have come from gathering activities, perphaps also wild sour cherry (*Prunus cerasus* subsp. *acida*) and service tree (*Sorbus domestica*). During the life of the settlement, shallow waters and muddy marshland covered the Nagyberek region where water-chestnut (*Trapa natans*) must have been abundant and gathered for supplementing the inhabitants' diet. Walnut, cherry, plum, apricot, peach and grape remains show that these fruit cultivars established by the Romans did not disappear from Pannonia following collapse of the empire. Residents on the western slopes of Fonyód Castle Hill could have established vineyards and orchards. In fact, Nagyberek was not suitable for agricultural use before draining was completed in 1914. Metric data (seed dimensions, length, width, height) indicate that the fruits mentioned above might have had several forms (cultivars?) just like common wheat and rye.

A unique find is a mummified apricot fruit bearing characteristic symptoms of Monilia fungi (*Monilia* cf. *fructigena*) infestation – circular monilia cushions around the peduncles (i.e. concentrically zoned sporodochims) (GYULAI 1988). We have also found pea weevil (*Bruchus pisorum*) and grain beetle (*Calandra granaria*) damages and a number of insect remains (saw-fly, mosquito species, lace-wing fly, velvet wine-dressed beetle) were also found. Even turfed remains of earthworms were uncovered.

A large number of weed species (19 species) suggests cereal production as well: for instance, korn-rade, spring wild-oat, brome grass, field cowwheat (*Melampyrum arvense*), field poppy, corn poppy (*Papaver rhoeas*), redshank (*Polygonum persicaria*), wild radish (*Raphanus raphanistrum*), annual woundwort (*Stachys annua*), lesser stitchwort (*Stellaria graminea*), large yellow-vetch (*Vicia grandiflora*), smooth tare (*V. tetrasperma*), fodder vetch (*V. villosa*). Herbs collected from the vicinity were also used. Numerous fan-hen or white goosefoot seeds raise the possibility of dietary use, for instance, by milling into flour. A number of parallel finds are known from both domestic and international archaeobotanical and ethnobotanical research (cf.. Túrkeve-Terehalom Middle Bronze Age sites). Some species identified as possible natural colouring agents include common agrimony (*Agrimonia eupatoria*), lady's bedstraw (*Galium verum*), walnut, crab apple, cammon mallow, sloe (blackthorn), danewort and common elder.

Seeds and fruit remains of the original flora appear in great numbers (29 species). Analysis of floristic elements and area types showed that vegetation in the area was European and Eurasian in nature, mixed with a substantial Mediterranean influence. Based on the area types Eurasian species dominated the environment of the site at the time along with the presence of a strong submediterranean influence. This suggests a somewhat warmer climate at that time. Radiocarbon dating by Ede Hertelendi determined that the settlement was active from the 7th to the 9th centuries, a warming period, providing paleoclimatic confirmation of the archaeobotanical data.

The excavation was significant from a historical point of view as well, because it resolvess a long lasting debate by demonstrating that the water level in Lake Balaton was lower in the periods preceding and during the conquest than it is today. The settlement layer explored in Fonyód-Bélatelep lay at 103.4 m above Adriatic sea level. As a considerable amount of grain was produced and stored at this level, the ground must have been completely dry at the time. The settlement was established on an island covered with peat, and taking soil capillarity into account, we have to assume a lower water table than that of the settlement: about – 102.4 m Adriatic compared to 104.1 m Adriatic today.

Decades of excavations, led by Ágnes Ritoók and Miklós Béla Szőke, of a parking lot exposing the 9th century site of Zalavár-Vársziget, initiated collection and flotation of soil samples for botanical remains in 1994 (GYULAI 1998b). This was the first time that botanical macro remains, seeds and fruits, were recovered (*Figs. 314–317*). In terms of species composition, the finds provided the richest carbonised set of plant remains (103 species!) found in Hungary yet. According to historical data, the Carolingian court centre was populated by common people of mixed ethnic composition. Most seeds and fruit remains come from the fill of a "well", which was probably a planked storage pit. The carbonised diaspores were possibly deposited during the cleaning process after the burning of the settlement at the time. The botanical material is dominated by cultivated plants and their weeds. The most important crops were common millet, six-rowed barley and common bread wheat while legumes included lentil, pea and chick-pea. Flax was ground as an oil and fibre crop. Garlic (*Allium sativum*) slices found here are among the earliest spice finds. A great number of fruit remains: walnut, peach, garden plums and black wine grapes, confirm the contention that growing of domesticated cultivars introduced into Pannonia during the Roman Period was continued and probably survived until the conquest.

The large number of autumn sowing or grain crop weed association species (*Secalietea*) indicate locally grown autumn sowing host plants. The number of species introduced from the natural environment of the time is also surprisingly high. In part, these consist of fruits and wild berries rich in vitamins: hazelnut, cornelian cherry, dogwood, sloe (blackthorn) and dewberry. According to ecological grouping by habitats, the natural environment around the settlement was diverse. A large area of open water existed quite close to the settlement proving common reed beds at the margins as a source of common bulrush (*Schoenoplectus lacustris*) achenes followed in the succession by an association dominated by tall sedges including tussock sedge/yellow sedge (*Carex elata/gracilis*) and common spike-rush (*Eleocharis palustris*) achenes. Fox sedge (*Carex vulpina*) and lesser spearwort (*Ranunculus flammula*) lived at the water edge. Forests alos existed nearby. Crab apple (*Malus silvestris*) and wild pear (*Pyrus achras*) might have been gathered in the surrounding woods. Livestock was kept on grazing land and ploughland was created by forest clearing. The number of ruderal weed associations is very high; consequently, the dimensions of the settlement must have been considerable.

Although the avalanches of folk migration caused a decline of villa farming built on Roman foundations, some traditions, albeit fragmentarily, survived. Mostly wine growing and fruit production skills survived these chaotic centuries. It is not surprising that several black wine grape finds are known in the Balaton area from the Migration Period: black wine grape stones were found in a grave of the 9th-century cemetery of Balatonszentgyörgy and the 6th–7th-century cemetery in Keszthely-Fenékpuszta, the former together with apricot stones (P. HARTYÁNYI – NOVÁKI – PATAY 1967–68; BAKAY – KALICZ – SÁGI 1966). In Főnyed-Szegerdő, a 9th-century settlement's house wall desdtroyed by fire yielded a piece of vine-branch (SÁGI – FÜZES 1967). Likewise, a walnut find unearthed from the Keszthely-Halászcsárda site was dated to the 9th century (P. HARTYÁNYI – NOVÁKI – PATAY 1967–68).

SÁGI – FÜZES (1967) state that grape and fruit finds from the Migration Period show that the conquering Hungarians conducted a very high level of fructiculture and viticulture around Keszthely. Róbert MÜLLER (1982) in spite of the fact that no fructiculture or viticulture implements or tools are known from Hungary in the Early Migration Period. Inasmuch as wine grape stones retain their morphological traits even when grown wild, the level of viticulture Christian missionaries encountered here in the 9th century must have been primitive. According to KISS (1964), the report found in the Suidas Chronicle describing Avars getting drunk, offered as proof for a continuity theory, does not necessarily mean that they grew the grapes themselves. However, this scenario has been recently revisited (VÁCZY 1974).

We also have botanical finds from Slavs who lived in the Upper-Tisza region of the eastern part of the country. During the exploration of Panyola-Ásottfok, a 9th–10th-century Slavic settlement (István Erdélyi's excavation 1996) samples taken from the inner rooms of earth houses found contained carbonised seeds and fruit remains (GYULAI 1996e). The number of cultivated plant species is considerably low. Of

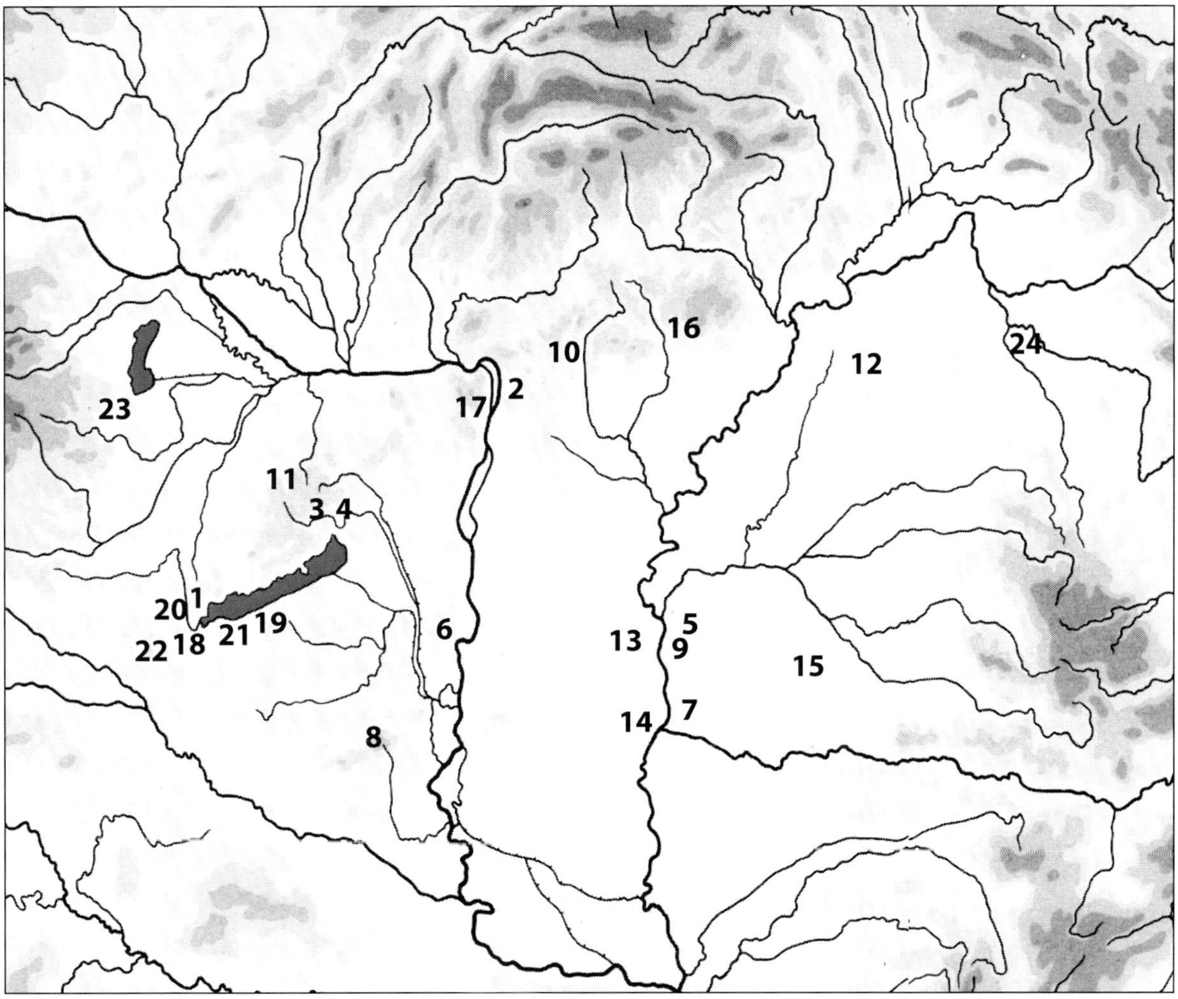

Fig. 249. Archaeobotanical sites from the Late Migration Period.
Early Avar Age: 1. Fenékpuszta; 2. Vác; 3. Jutas; 4. Öskü; 5. Szegvár.
Middle Avar Age: 6. Gerjén; 7. Szőreg; 8. Cikó; 9. Mindszent.
Late Avar Age: 10. Szirák; 11. Táp; 12. Hajdúdorog; 13. Kistelek; 14. Röszek; 15. Pitvaros; 16. Kompolt; 17. Budakalász; 18. Vörs.
Caroling Age: 19. Fonyód-Bélatelep; 20. Zalavár; 21. Balatonszentgyörgy; 22. Főnyed-Szegerdő; 23. Sopron.
Slavic period: 24. Panyola.

the grain crops, only barley, and of legumes, only small seed lentils were found. If this does not reflect inadequate research, such finds suggest that these population practiced a lower level of agricultural skills. It is interesting to note that common bunt, stinking smut sclerotiums, which damage crops, were also found here.

Reviewing the archaeobotanical and farming implement finds, we can state with great confidence that the conquering Hungarians found an impoverished stock-farming and tiller population in the Carpathian Basin with an agriculture of various origins, whose living conditions are very unclear.

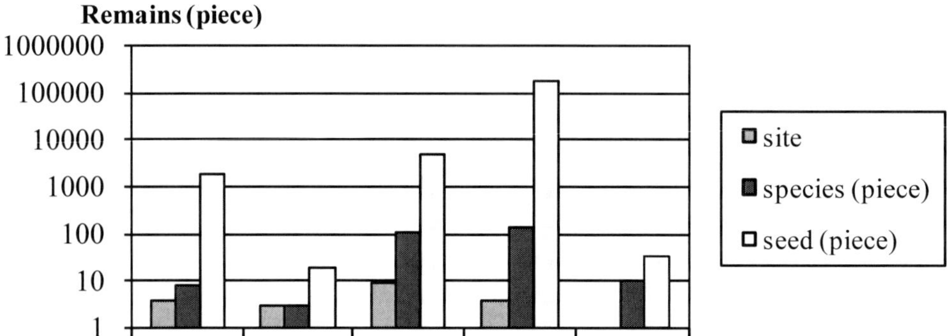

early Avar age	middle Avar age	late Avar age	Caroling age	Slavonic
4	3	9	4	1
8	3	114	132	11
1850	20	5093	184783	32

Fig. 250. Distribution of the archaeobotanical remains in the Late Migration Period.

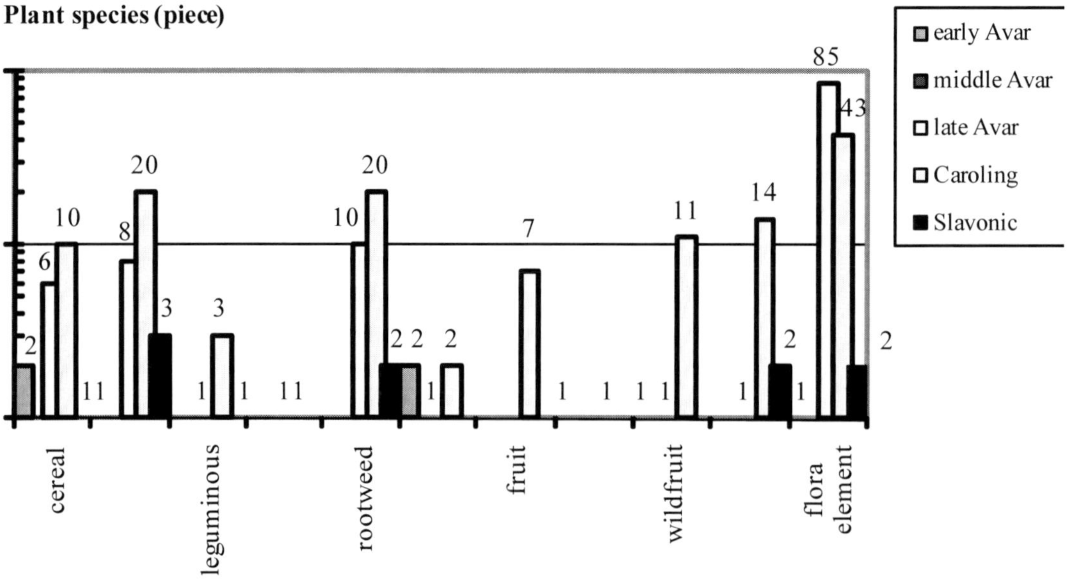

Fig. 251. Anthropological distribution of the plant species in the Late Migration Period.

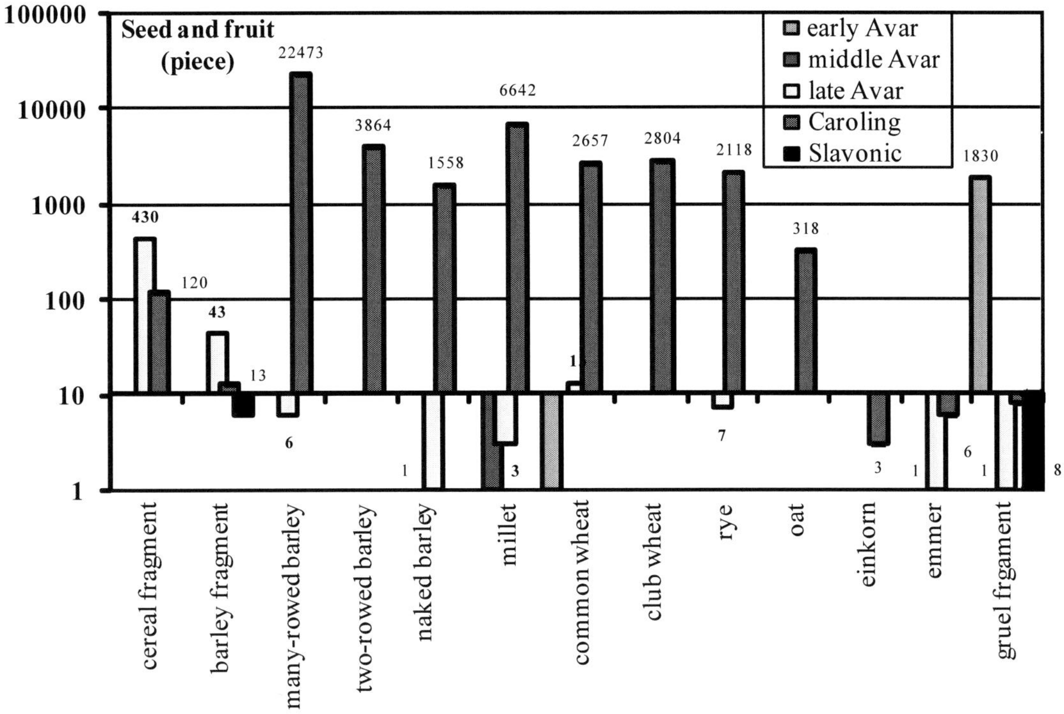

Fig. 252. Distribution of the cereals in the Late Migration Period.

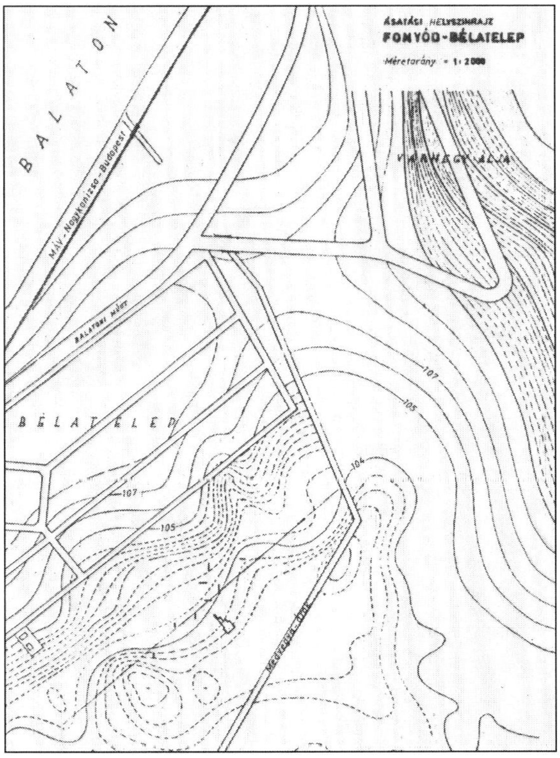

Fig. 253. Location of the lake dwelling settlement in Fonyód-Bélatelep at the Lake Balaton. Photograph at the excavation.

Fig. 254. The excavation of the lake dwelling settlement in Fonyód-Bélatelep. Photograph at the excavation.

Fig. 255. Hazelnut shells (*Corylus avellana*) from Kisvárda-Darusziget, Gepid, Inventory of the Hungarian Agricultural Museum, Budapest.

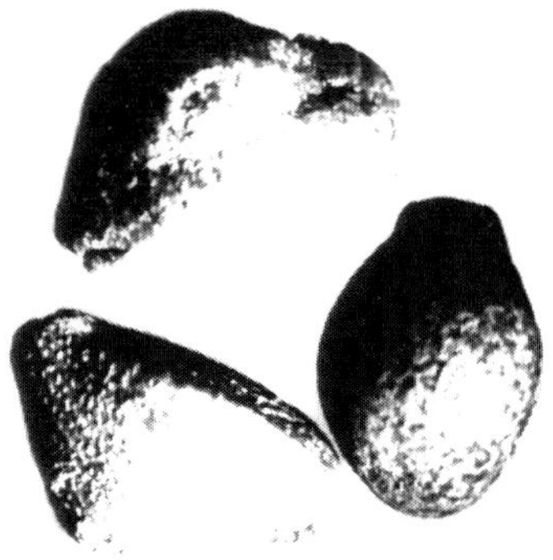

Fig. 256. Black-bindweed (*Fallopia convolvulus*) nutlets from a grave in Szőreg, Avar Period, Inventory of the Hungarian Agricultural Museum, Budapest.

Fig. 257. Two-rowed barley (*Hordeum vulgare* subsp. *distichum*), hulled grains with scars from burn from Fonyód-Bélatelep Late Migration Period lake dwelling. Photograph by the author.

Fig. 258. Wattle-and-daub wall, lake dwelling settlement from Fonyód-Bélatelep Late Migration Period lake dwelling. Photograph at the excavation.

Fig. 259. Six rowed barley (*Hordeum vulgare* subsp. *hexastichum*) hulled grains from the middle with straight axis from Fonyód-Bélatelep Late Migration Period lake dwelling. Photograph by the author.

Fig. 260. Small seed lentil (*Lens culinaris* subsp. *microsperma*) seed from Fonyód-Bélatelep Late Migration Period lake dwelling. Photograph by the author.

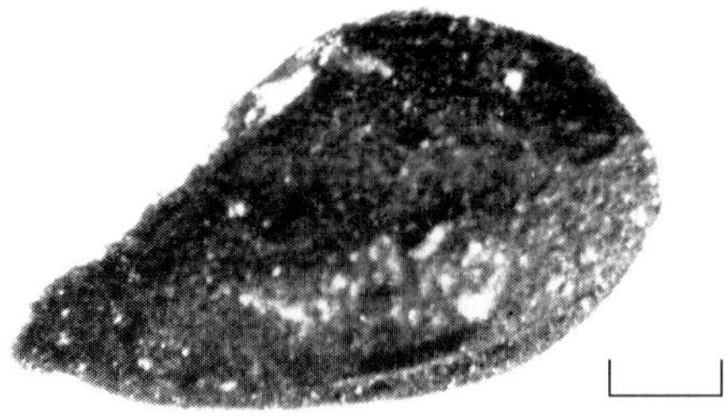

Fig. 261. Einkorn (*Triticum monococcum*) grain from Fonyód-Bélatelep Late Migration Period settlement. Photograph by the author.

Fig. 262. Common oat (*Avena sativa*) naked grain from Fonyód-Bélatelep Late Migration Period site. Photograph by the author.

Fig. 264. Six rowed barley (*Hordeum vulgare* subsp. *hexastichum*) hulled grains from the side with oblique axis from Fonyód-Bélatelep Late Migration Period lake dwelling. Photograph by the author.

Fig. 266. Common oat (*Avena sativa*) hulled grains from Fonyód-Bélatelep Late Migration Period site. Photograph by the author.

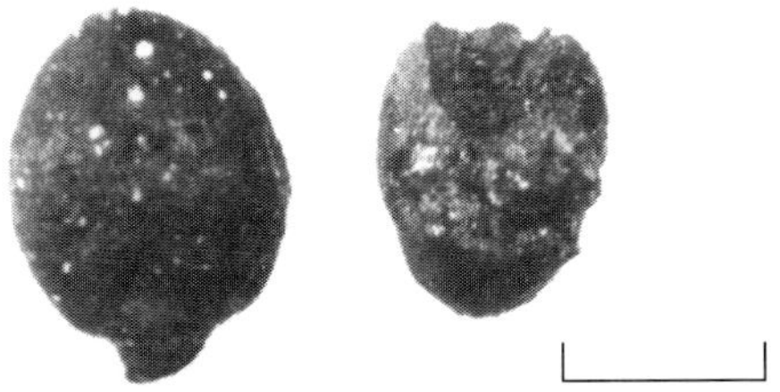

Fig. 268. Common millet (*Panicum miliaceum*) naked (husked) grains (*caryopsis nuda*) with and without the germ disk from Fonyód-Bélatelep Late Migration Period lake dwelling. Photograph by the author.

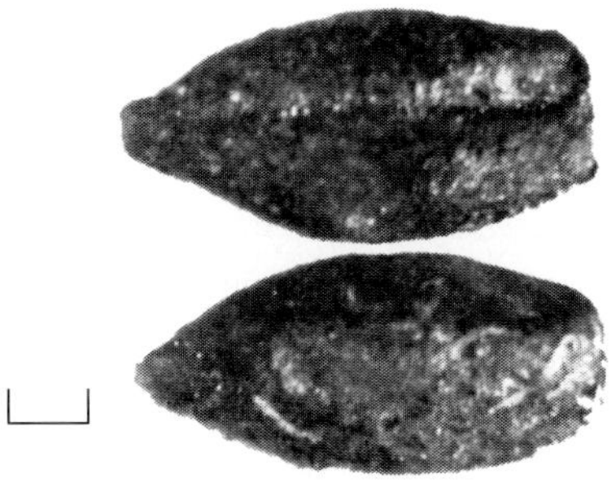

Fig. 263. Two-rowed naked barley (*Hordeum vulgare* subsp. *distichum* var. *nudum*) grains in dorsal and ventral view from Fonyód-Bélatelep Late Migration Period lake dwelling. Photograph by the author.

Fig. 265. Rye (Secale cereale) and barley (*Hordeum vulgare*) burnt grains from Fonyód-Bélatelep Late Migration Period settlement. Photograph by the author.

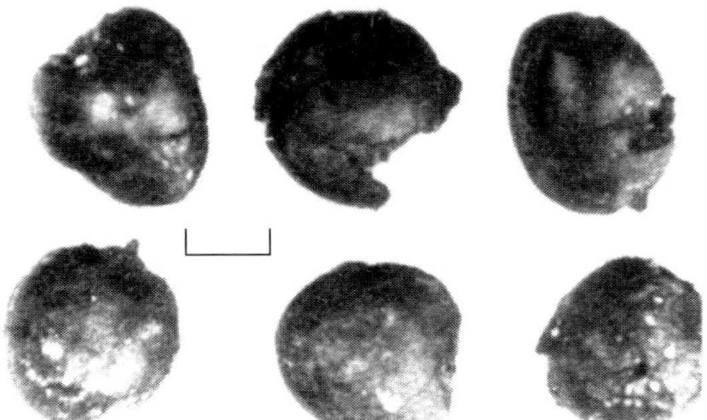

Fig. 267. Common millet (*Panicum miliaceum*) hulled grains (*cum caryopse corticata*) from Fonyód-Bélatelep Late Migration Period lake dwelling. Photograph by the author.

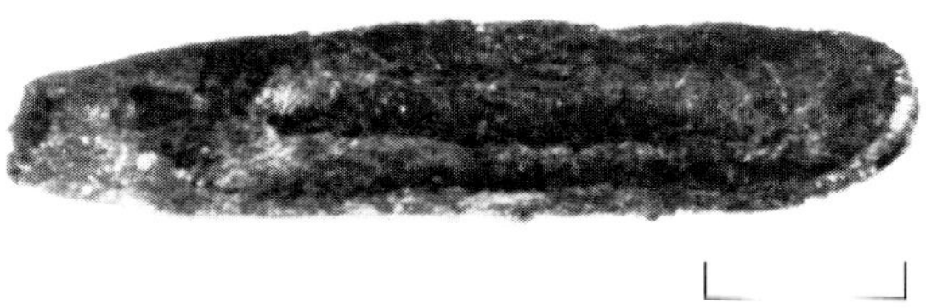

Fig. 269. Rye brome (*Bromus secalinus*) grain from Fonyód-Bélatelep Late Migration Period settlement. Photograph by the author.

Fig. 270. Pea (*Pisum sativum*) seeds from Fonyód-Bélatelep Late Migration Period lake dwelling. Photograph by the author.

Fig. 273. Spring wild-oat (*Avena fatua*) hulled grains from Fonyód-Bélatelep Late Migration Period settlement. Photograph by the author.

Fig. 275. Field cowwheat (*Melampyrum arvense*) seed from Fonyód-Bélatelep Late Migration Period settlement. Photograph by the author.

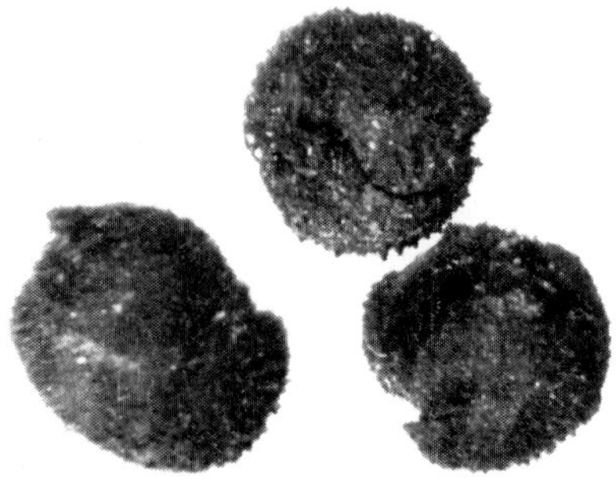

Fig. 271. Korn-rade (*Agrostemma githago*) seeds from Fonyód-Bélatelep Late Migration Period lake dwelling. Photograph by the author.

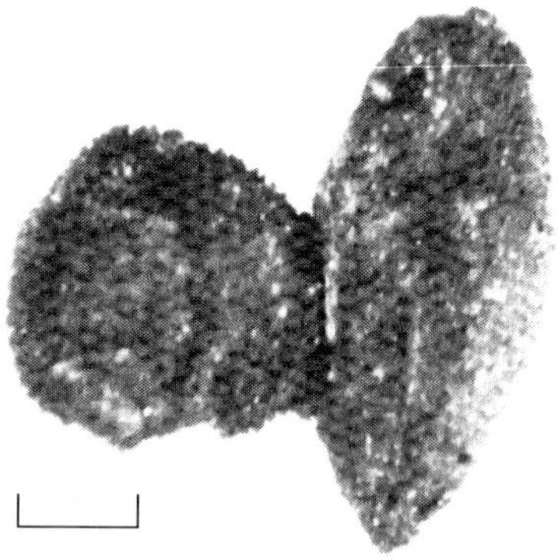

Fig. 272. Korn-rade (*Agrostemma githago*) seed burnt togather with barley (*Hordeum vulgare*), from Fonyód-Bélatelep Late Migration Period settlement. Photograph by the author.

Fig. 274. Field poppy or corn poppy (*Papaver rhoeas*) seed from Fonyód-Bélatelep Late Migration Period settlement. Photograph by the author..

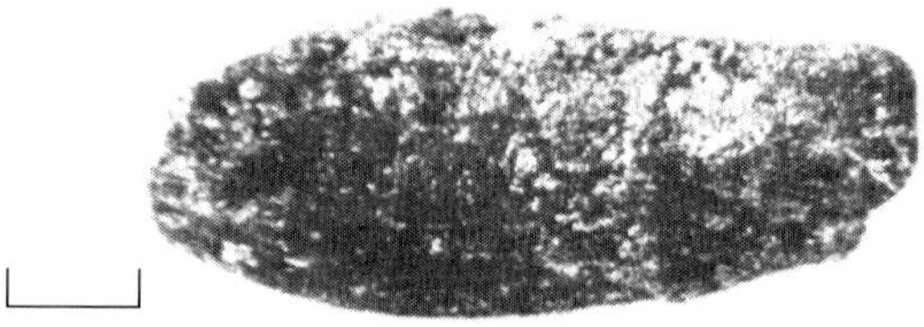

Fig. 276. Wild rye (*Secale silvestre*) naked grain from Fonyód-Bélatelep Late Migration Period settlement. Photograph by the author.

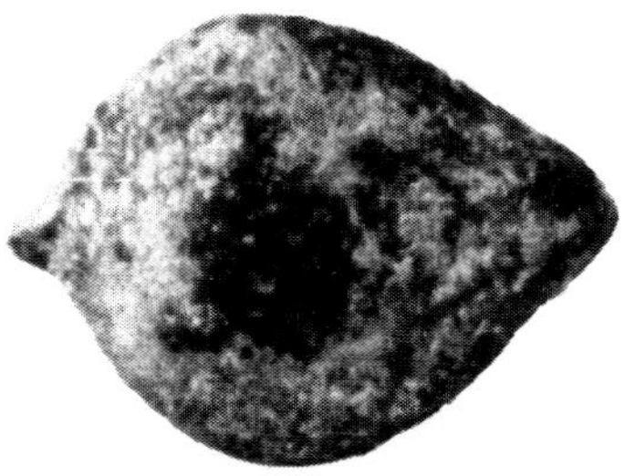

Fig. 277. Redshank (*Polygonum persicaria*) nutlets from Fonyód-Bélatelep Late Migration Period settlement. Photograph by the author.

Fig. 278. Wild radish (*Raphanus raphanistrum*) seeds from Fonyód-Bélatelep Late Migration Period settlement. Photograph by the author.

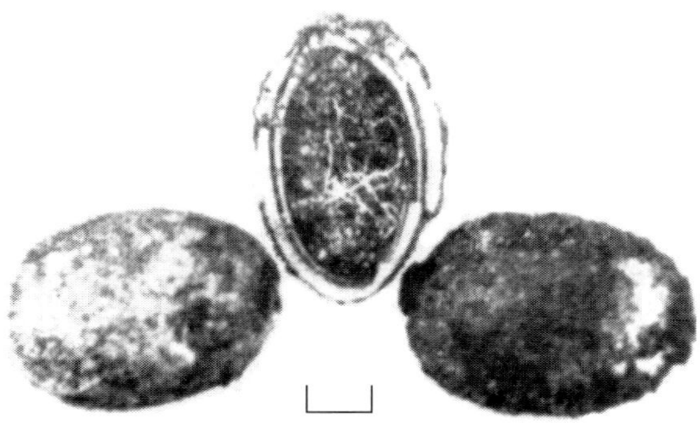

Fig. 279. Hemp (*Cannabis sativa*) achenes from Fonyód-Bélatelep Late Migration Period settlement. Photograph by the author.

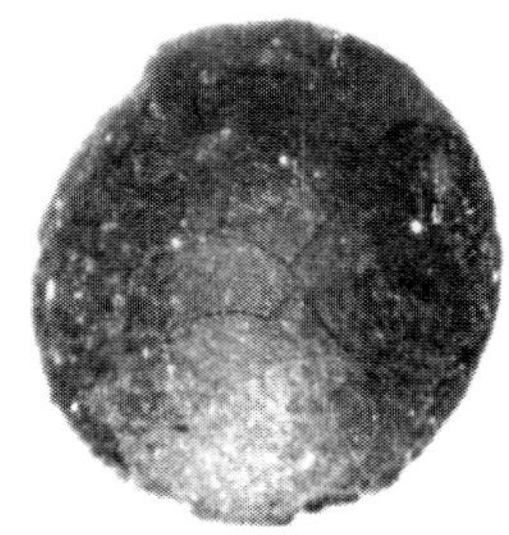

Fig. 280. Hairy vetch (*Vicia villosa*) seed from Fonyód-Bélatelep Late Migration Period settlement. Photograph by the author.

Fig. 281. Common elder (*Sambucus nigra*) seeds from Fonyód-Bélatelep Late Migration Period settlement. Photograph by the author.

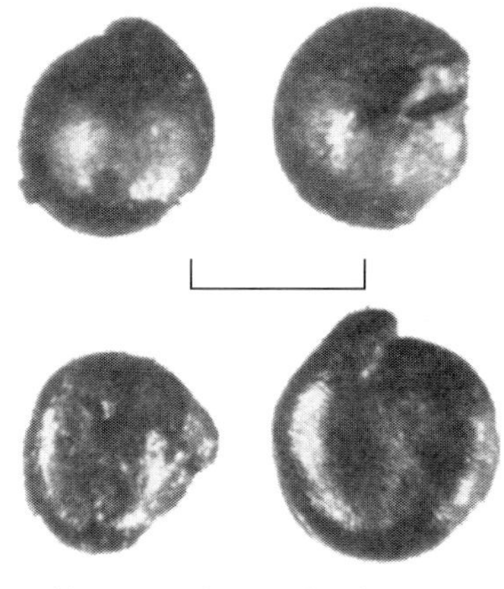

Fig. 282. White goosefoot or fan-hen (*Chenopodium album*) seeds from Fonyód-Bélatelep Late Migration Period settlement. Photograph by the author.

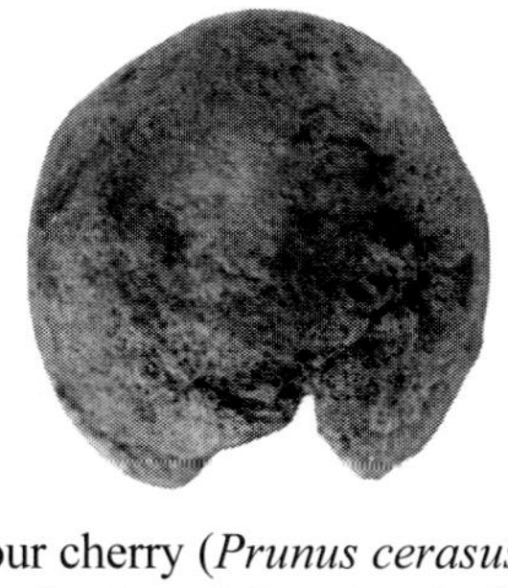

Fig. 283. Sour cherry (*Prunus cerasus*) stone from Fonyód-Bélatelep Late Migration Period settlement. Photograph by the author.

Fig. 284. Cherry (*Prunus avium*) stone from Fonyód-Bélatelep Late Migration Period settlement. Photograph by the author.

Fig. 285. Crab apple (*Malus silvestris*) mummified fruit from Fonyód-Bélatelep Late Migration Period settlement. Photograph by the author.

Fig. 286. Danewort (*Sambucus ebulus*) seeds from Fonyód-Bélatelep Late Migration Period settlement. Photograph by the author.

Fig. 287. Mahaleb cherry (*Prunus mahaleb*) half stone from Fonyód-Bélatelep Late Migration Period settlement. Photograph by the author.

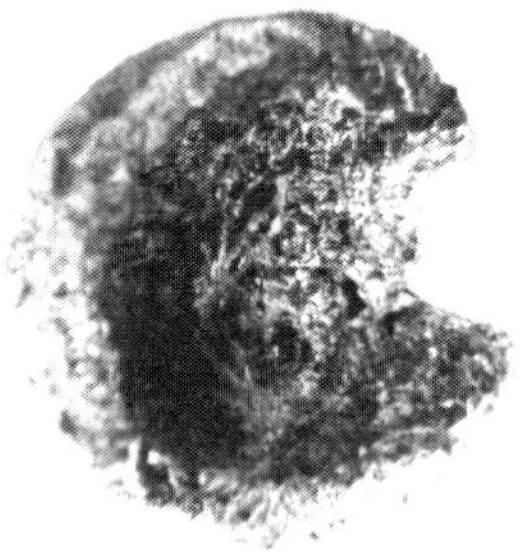

Fig. 288. *Prunus* spec. stone fragment from Fonyód-Bélatelep Late Migration Period settlement. Photograph by the author.

Fig. 289. *Prunus* spec. stone from Fonyód-Bélatelep Late Migration Period settlement. Photograph by the author.

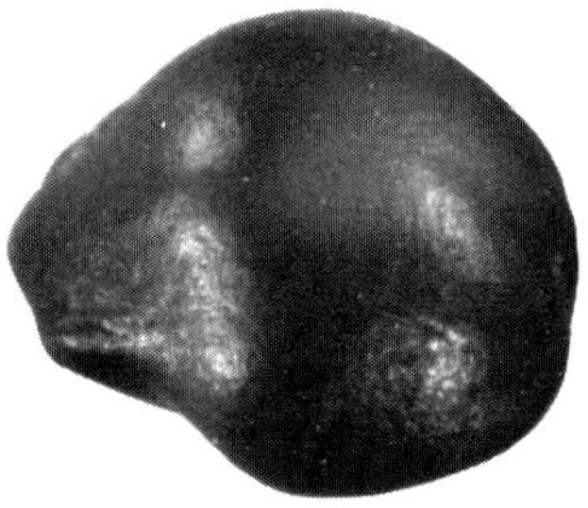

Fig. 290. Single-seed hawthorn (*Crataegus monogyna*) stone from Fonyód-Bélatelep Late Migration Period settlement. Photograph by the author.

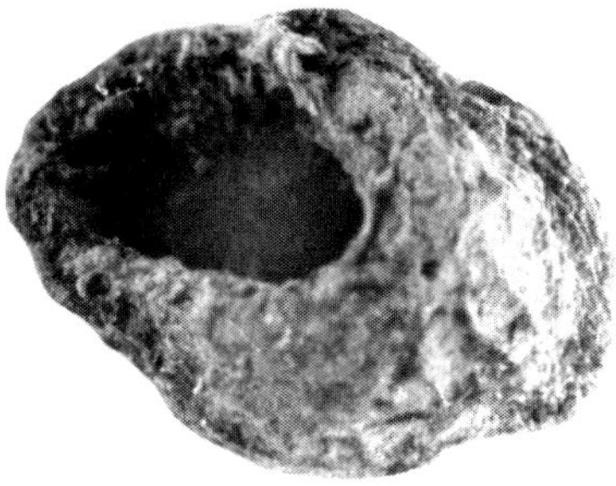

Fig. 291. European bird cherry (*Prunus padus*) stone from Fonyód-Bélatelep Late Migration Period settlement. Photograph by the author.

Fig. 292. Hungarian(?) pear (*Pyrus* cf. *magyarica*) mummified fruit from Fonyód-Bélatelep Late Migration Period settlement. Photograph by the author.

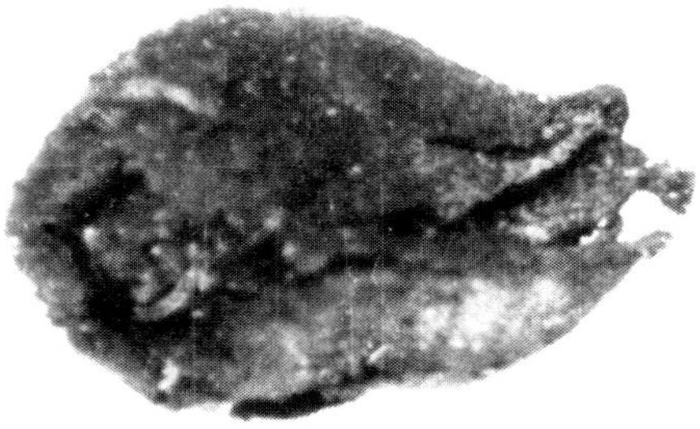

Fig. 293. Service tree (*Sorbus domestica*) seed from Fonyód-Bélatelep Late Migration Period settlement. Photograph by the author.

Fig. 294. Apricot (*Prunus armeniaca*) fruit mummy with Monilia damage from Fonyód-Bélatelep Late Migration Period settlement. Front view. Photograph by the author.

Fig. 295. Pear (*Pyrus* spec.) fruit-stalk (*pomum*) from Fonyód-Bélatelep Late Migration Period settlement. Photograph by the author.

Fig. 296. Peach (*Prunus persica*) stone from Fonyód-Bélatelep Late Migration Period settlement. Photograph by the author.

Fig. 297. Apricot (*Prunus armeniaca*) fruit mummy with Monilia damage from Fonyód-Bélatelep Late Migration Period settlement. Rear view. Photograph by the author.

Fig. 298. Peach (*Prunus persica*) stone from Fonyód-Bélatelep Late Migration Period settlement. Photograph by the author.

Fig. 299. Peach (*Prunus persica*) stone from Fonyód-Bélatelep Late Migration Period settlement. Photograph by the author.

Fig. 300. Peach (*Prunus persica*) stone fragment from Fonyód-Bélatelep Late Migration Period settlement. Photograph by the author.

Fig. 301. Walnut (*Juglans regia*) shell from Fonyód-Bélatelep Late Migration Period settlement. Photograph by the author.

Fig. 302. Garden plum (*Prunus domestica* subsp. *oeconomica*) fruit mummy from Fonyód-Bélatelep Late Migration Period settlement.

Fig. 303. Grape-vine stones (*Vitis vinifera* subsp. *vinifera*) from Fonyód-Bélatelep Late Migration Period settlement. Photograph by the author.

Fig. 304. Walnut (*Juglans regia*) shells from Fonyód-Bélatelep Late Migration Period settlement. Photograph by the author.

Fig. 305. Birthwort (*Aristolochia clematitis*) seed from Fonyód-Bélatelep Late Migration Period settlement. Photograph by the author.

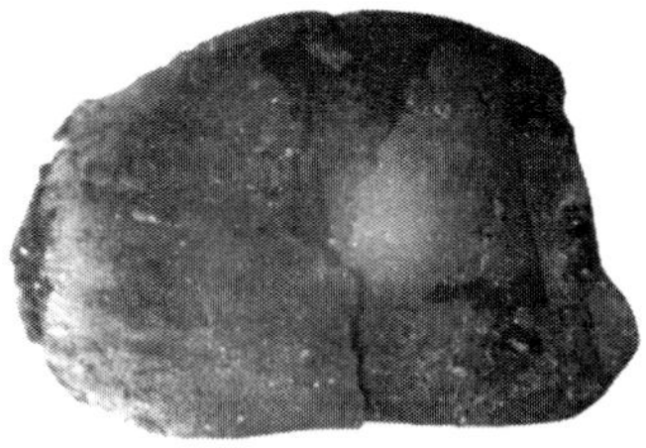

Fig. 306. Hazelnut shells (*Corylus avellana*) from Fonyód-Bélatelep Late Migration Period settlement. Photograph by the author.

Fig. 307. Common viper's-grass (*Scorzonera humilis*) achenium from Fonyód-Bélatelep Late Migration Period settlement. Photograph by the author.

Fig. 308. Sedge (*Carex* spec.) tricarpellat glans from Fonyód-Bélatelep late Migration Period settlement. Photograph by the author.

Fig. 309. Nodding bur marigold (*Bidens cernua*) achenium from Fonyód-Bélatelep late Migration Period settlement. Photograph by the author.

Fig. 310. Water-chestnut (*Trapa natans*) juvenil glans from Fonyód-Bélatelep late Migration Period settlement. Photograph by the author.

Fig. 311. Common reed (*Phragmites australis*) fragment with shoot from Fonyód-Bélatelep Late Migration Period settlement. Photograph by the author.

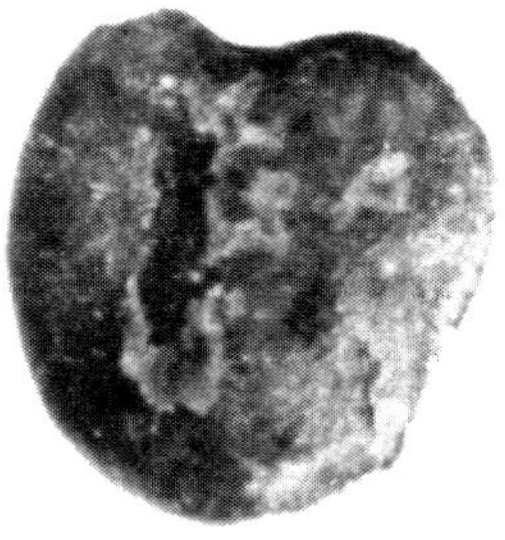

Fig. 312. Alsike clover (*Trifolium hybridum*) seed from Fonyód-Bélatelep Late Migration Period settlement. Photograph by the author.

Fig. 313. Common buckthorn (*Rhamnus catharticus*) seeds from Fonyód-Bélatelep late Migration Period settlement. Photograph by the author.

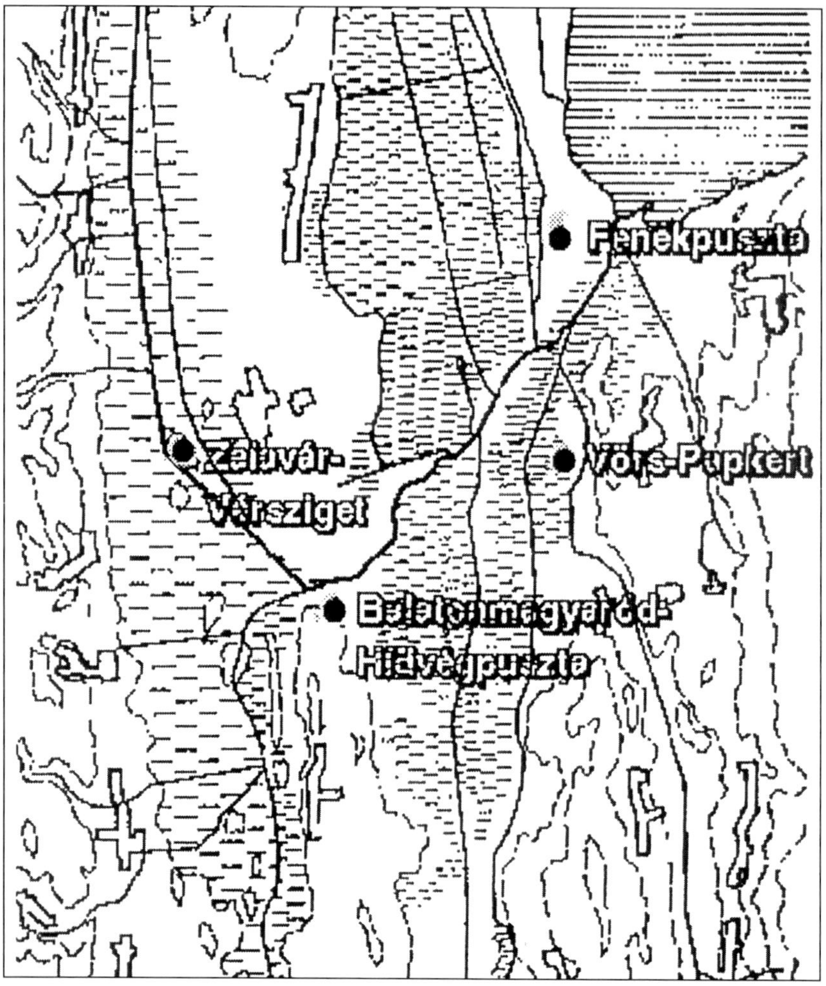

Fig. 314. Archaeobotanical finds from the Kis-Balaton area.

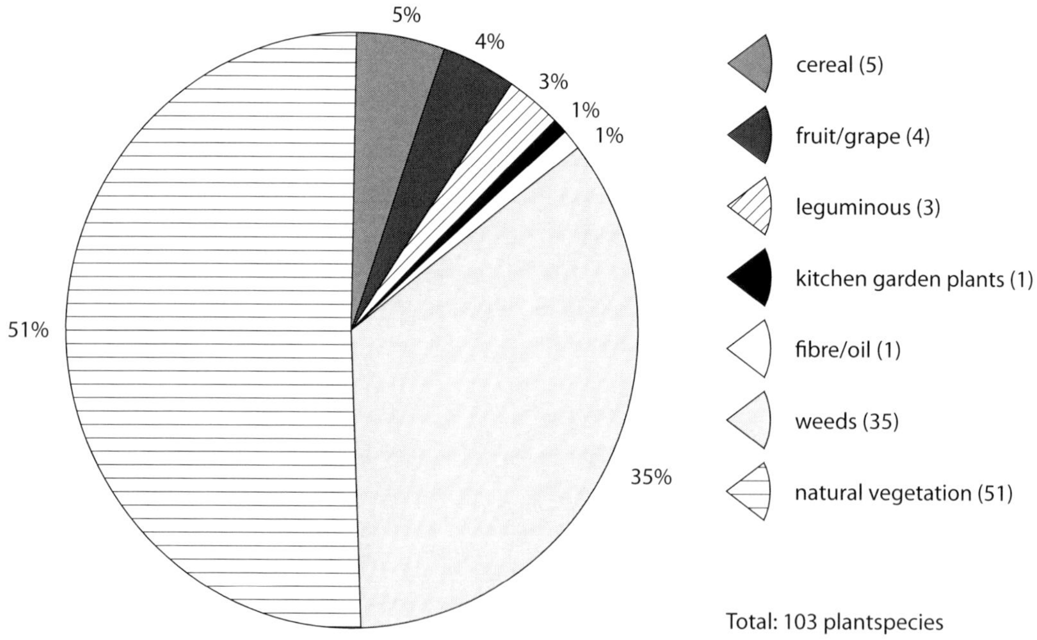

Fig. 315. Anthropological distribution of the species from Zalavár Late Migration Period fortress.

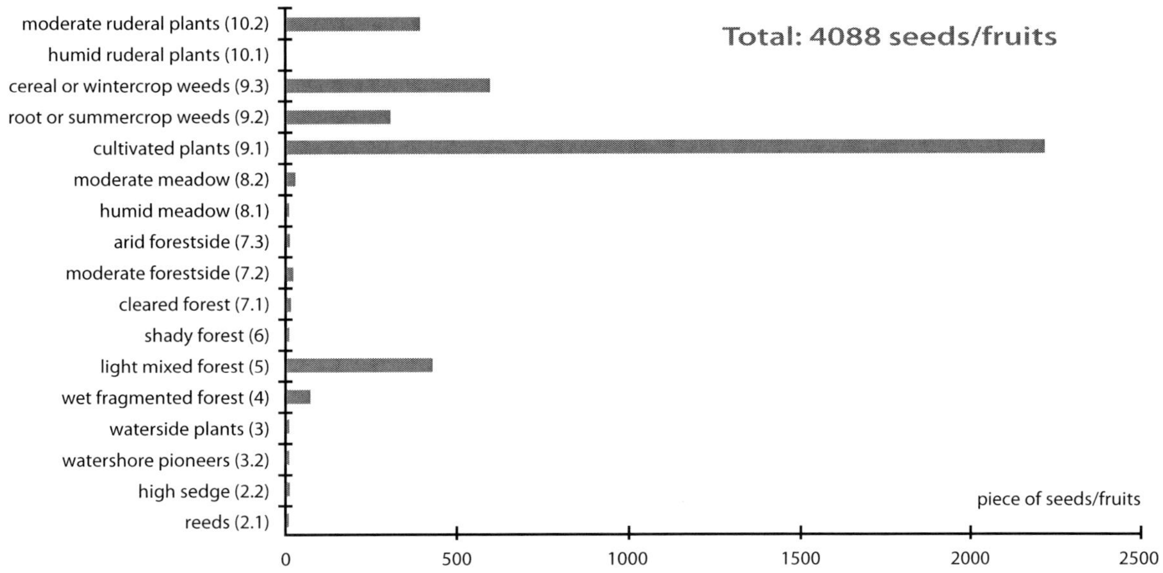

Fig. 316. Anthropological distribution of the seeds from Zalavár Late Migration Period fortress.

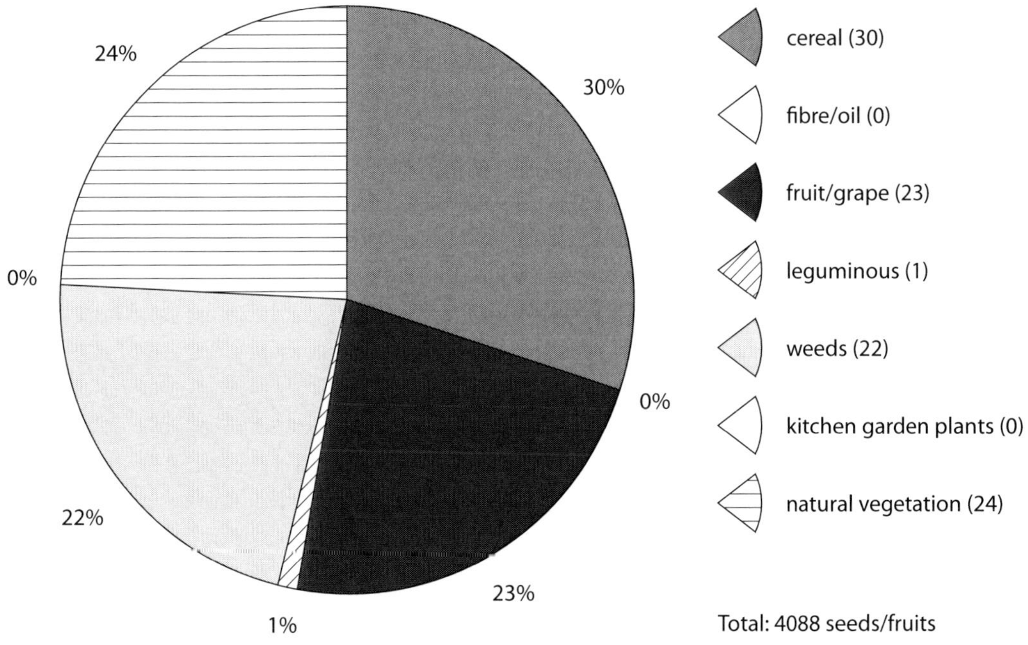

Fig. 317. Ecological distribution of the seeds from Zalavár Late Migration Period fortress.

7.8. The Hungarian Conquest and the Árpádian Period

7.8.1. Farming skills of the Hungarians before the conquest

The most significant event of the 10th century in the Carpathian Basin was the Hungarian conquest. This is the most fiercely debated period of Hungarian history, sometimes involving extreme views. However, our information on the way of life of these Hungarians is very sparse, especially as regards farming and crop production skills.

Opinions of the conquering Hungarians have been much altered as a result of research carried out in recent decades. The concept that the Hungarians were livestock keepers prevailed for a long time, which would have entailed that farming be the obligation of the conquered people, as is typical for most nomadic people. The ancestors of Hungarians are sometimes considered as nomads, horseback nomads, or semi-nomads, even a half settled people, depending on how much the sedentary lifestyle of the Hungarians is accepted. Unfortunately, we do not have any botanical material associated with the Hungarians before the conquest.

However, re-evaluation of contemporary Arabic and Byzantine sources combined with the analysis of Finno-Ugric, Turkish and Slavic denizens has slowly changed our perceptions of the lifestyle of the conquering Hungarians. We make inferences about their way of life from the finds of people living in the south Russian steppes in the 9th century and the records of contemporary chroniclers.

Arabic-Persian and Byzantine sources mention the Hungarians living in Levedia as steppe horseback nomads (in: Kovács – Veszprémy 1995). Djayhani, a scholar from Bokhara around 920 wrote a large volume on the people along the Volga R iver and those around the Black Sea, among them the Hungarians. His text was lost but had been used by Ibn Rusta when writing on the Hungarians around 930: "They have tents, and they migrate following fresh grass and green vegetation". Between 1050 and 1053 Gardízi supplemented this description as follows: "Anywhere they go, they travel together with all their loads and storages, the escort of the ruler and the throne, the tents, and the livestock." Leo the Wise, Byzantine Emperor (886–912) in his opus written after 904 entitled "War tactics" noted, mentioning Hungarians as "Turks": "They are followed by a large number of horses, stallions and mares, partly for eating and drinking their milk, partly in order to pretend to be a crowd. They do not camp out in camps surrounded with ditches like the Romans, but they spread out by clans and kins up to the very day of war, grazing the livestock permanently; in times of war, the horses necessary are kept beside their Turk tents fetterlocked up to the point of deployment for action…"

These descriptions appear similar to those of all contemporary nomadic people of the age, since nomads migrated in the wake of fresh grass and pastures. Grassland vegetation – depending on its exposure in the landscape – grows in different times during spring or early summer, but it provides food supply only for a limited time. Therefore, stock-farmers need to migrate continuously over a large area while living in tents or other types of easily dismantled dwellings.

As was customary, the Hungarians also had their winter quarters. These quarters ranged up to the Black Sea and included the two rivers as well. According to the records of Ibn Rusta: "When winter days come, each of them withdraws to the river which is closer to him. They stay there for the winter and fish. Winter dwelling is more suitable for them here." Ploughlands, which he found worth mentioning, should have been close to the winter quarters: "There is plenty of trees and water in the country of the Hungarians. Its soil is wet. They have much arable land. They keep on conquering the Slavs living next to them. Heavy food contribution duties are levied on them and they are treated as captives."

As tillage requires a sedentary lifestyle and constant work, some had to stay at the winter quarters for this purpose permanently. This is not surprising, as a nomadic lifestyle does not exclude a certain lower level of farming. Even nomadic peoples customarily ate food of vegetable origin.

The ancestors of the Hungarians in Levedia were not nomads in the classical sense, but stayed within a certain district, engaged in the so-called grazing-ground rotation livestock farming. The point is that herders graze the land with their livestock in a certain sequence, returning to the same place after a certain period. Herds were followed only by shepherds while other family members who stayed home farmed surrounding fields.

Hungarians, originally fishers and hunters, became stock-farmers and crop producers during the time spent in the Mid-Volga, Káma, Beleja regions. BALASSA (1973) also believes that nomadic migration must have been only a transient feature generated by external pressure, such as moving to Levedia and later on into Etelköz. In fact, the Hungarians stayed within the area of Levedia from the 6th to the 9th centuries. Abandonment of the migrating lifestyle is associated with an abundance of grazing land.

As it was mentioned earlier, this was a time of permanent warming, and the fight for pastures and drinking water started a migration surge from Central Asia. In our view, this is the underlying cause behind the Pecheneg attack in the mid-9th century, which resulted in forcing the Hungarians to leave Levedia and move to the area between the rivers Dnester and Seret, in other words to Etelköz.

The traditions of the Saltovo-Mayak culture had an impact on the Hungarians in Levedia for three centuries (VÁCZY 1958). This huge culture cannot be associated with a single ethnic group; rather, it is a cultural historical configuration extending to a large area bordered by the upper ranges of the river Don in the north, the Caspian Sea and the river Volga in the east, the Crimean and Cuban to the south and Donec and the Azovi-sea in the west (PLETNJOWA 1978). Animal bone remains of the north (shrubland) and south (grassy steppe) are different due to the distinct ecosystems and divided into two distinct ethnic groups as well. The park-forest steppe might have been populated by the Alans and the grassy steppe by the Kazars or Bulgarians. The melting pot of several cultures in the Saltovo-mayak culture constituted the Kazar Kaganate.

People in the Saltovo-Mayak culture were not stock-farmers, but were settled crop farmers. Through the examination of the archaeological heritage of the 8th–9th-century people in Eastern Europe, we can derive indirect conclusions about the lifestyle of the conquering Hungarians.

The agriculture of peoples living adjacent to the Hungarians in the area of the Saltovo-mayak culture is reflected by plant remains unearthed from the excavation site of the Majackoje castle beside the river Don. Majackoje castle was built in the 8th–9th centuries to guarding its borders. Its inhabitants were mainly the Alans (Ace, Yace), but Kazars and Permian Ugric (probably Mordvin) elements also occurred. From a biogeographical point of view, the area belongs to the grassland flats zone and lies at the boundary of the forested and tall grass steppe in the forested steppe zone. During his excavations in Majackoje castle, FÜZES (1987) found all the cereal grains known from earlier ages (emmer, spelt, field and club wheat, barley, common millet), legumes (common lentil, pea) seeds and fruit remains (wild sour cherry, pear, apple). Linen and hemp fabric remains also appeared. Providing there was enough communication, this is indirect evidence for crop production of both the Late Avars and the Hungarians. It is very likely that club wheat, which disappeared from Pannonia after Roman domination, was reintroduced into the Carpathian Basin in this era.

In addition to botanical finds from Majackoje and Rogalik, iconographic and implement remains prove that the immediate neighbours of the Hungarians, the inhabitants of the Saltovo-mayak culture, were settled farmers. They produced all the cereals common in this age, conducted forest-clearing husbandry and cultivated their lands with ploughs. They also knew and gathered forest fruits.

The Hungarians of Levedia engaged in economic and political alliances with the Kazar Kaganate for a considerable time. Bulgarian-Turkish peoples in the Kazar empire exerted strong influence on the culture of the Hungarians. This was the time when words connected to husbandry were introduced into Hungarian, supplementing and sometimes replacing the set of Finno-Ugric words related to farming and agriculture.

The most important Finno-Ugric words connected to farming before the conquest were as follows: (köles) common millet, ed (grain), kenyér (bread) (porridge), fiu (boundary), csegely (wedge-shaped ploughland), fürt (bunch), meggy (sour cherry), fű (grass), ág (twig), falu (village), ház (house), nyomat (track), vág (cut), tér (land?) (Moór 1943; Gombocz 1960; Lakó 1967–78; Makkai 1980).

The most important Bulgarian-Turkish words connected to farming before the conquest were as follows: eke (plough), árpa (barley), búza (bread wheat), arat (reaping), sarló (sickle) (?), boglya (hay stack), gügyü (a handful of common reed, or grain), tarló (stubble-field), őröl (grind), kölyű (mortar for crushing grains), szór (cleaning grains), dara (groats), ocsú (tailing), gyümölcs (fruit), alma (apple), körte (pear), mogyoró (hazelnut), dió (walnut), kökény (sloe), galagonya (blackthorn), som (cornel-berry), grapes (grape), karó (prop), kocsány (stalk), szűr (filter), bor (wine), borsó (pea), feketebors (black pepper), kender (hemp), kóró (dry stalk of a weed), tiló (swingle), csepű (hurds), orsó (spindle), torma (horseradish), üröm (wormwood), kabak (bottle-gourd), komló (common hop), csalán (nettle), gyom (weed), gyertyán (hornbeam), gyűrűfa (ringtree), kőris (ash-tree), tátorján (a kind of flower), bojtorján (burdock), kalokány, katáng (wild chicory), gyopár (cat's foot), gyékény (bulrush), cötkény, kökörcs, kökörcsin (anemone), kikirics (colchicum), káka (club-rush) (Zichy 1923; Moór 1943; Gombocz 1960; Ligeti 1986).

It is very useful to compare the climate, soil type and natural vegetation of the forested steppe zone in Levedia with those of the Great Plain first invaded by the conquerors. We will find a striking similarity. Forested steppes constitute an intermediate zone between flat grasslands. Steppe and forest interchange here in a mosaic pattern. The forests, rich in steppe elements, take on the character of parklands. A typical forest association is steppe oak-wood. Continuous forests are established only along rivers. This expansive zone, stretching beyond the Ural, always provided a secure livelihood for stock-farming populations living here. Folk migration took off from here as well. The grassland steppe soil, which is rich in humus (such as in the Ukraine), support production of a very good quality bread wheat. Bread wheat from the Tisza region belongs to the variety named "proles hungarica" (Mándy 1972). The Carpathian Basin (mainly the Great Plain, which has a continental climate) and its adjacent regions (Galicia, Bukovina, Bessarabia, Western-Ukraine) belong to the most significant bread wheat producing ranges. Local varieties called "ancient Hungarian bread wheat" show morphological and physiological similarities to the wheat grown in Galicia and Western-Ukraine. They may even be identical.

The Great Hungarian Plain and a large part of the Transylvanian Basin, similarly to the Southern-Russian steppe, belong to a forested steppe zone (Hortobágyi – Simon 1981). Climate in the forested steppe is continental; summer is dry and hot, fall is short and winter is cold. Annual precipitation falls short of 400 mm. Steppe lots are rich in fescue (*Festuca sulcata, F. pseudovina, F. valesiaca*) species, their most common sedge species being the dwarf sedge (*Carex humilis*), a number of onions (*Allium* spec.), anemone (*Colchicum* spec.) and peony (*Paeonia* spec.). Adonis (*Adonis* spec.) is encountered as well. The parkland steppe formed on black earth soil contains oak-groves adjacent to tall grass zones. Typical species is common oak (*Quercus robur*). Oak-woods are mixed with Tatarian maple (*Acer tataricum*), field elm (*Ulmus minor*), and sometimes linden (*Tilia cordata*). Their margins are invaded by crab apple (*Malus silvestris*), wild pear (*Pyrus pyraster*) and hawthorn (*Crataegus* spec.). Wet localities contain Padus avium as well. Undergrowth is characterised by lily of the valley and May lily (*Convallaria majalis*). Common pine (*Pinus silvestris*) is also found here. Typical species in the riparian parkland forests are ash trees (*Fraxinus excelsior*), common oak (*Quercus robur*), common alder (*Alnus glutinosa*), and white willow (*Salix alba*).

The largest sandy regions in the Carpathian Basin are between the Danube and the Tisza, Nyírség and Deliblát. Botanical and historical research show that these areas became "human wastelands" not earlier than the Late Middle Ages. Previously, the Great Plain was still covered by woods, parkland forests and sandy forests. These have survived at the edges of these sandy regions in the form of relict forests. The largest river in the Great Plain is the Tisza, which meandered through the whole Tiszántúl (area beyond the river Tisza) before the regulation works in the 19th century (1846–67) (BODNÁR 1999). Riverbed changes in the Tisza were followed by regulation works in all the tributaries as well (Bodrog, Sajó, Hernád, Takta). This does not mean however that swamps had been everywhere. Alkali grasslands had existed in the Great Plain for a long time. Water regime and vegetation in these ancient alkali regions did not change following regulation works in the 19–20th centuries. These are in contrast to the secondary alkali grasslands, which were formed as a result of drainage and inland dewatering works replacing formerly alkali-free or only slightly alkali grasslands or marshy floodplain areas (MOLNÁR 1999).

The lake district of the area between the Danube and the Tisza, which was less alkali at the time supported grazing and provided good quality hay. Open waters covered the entire area of Southern Hungary and the Great Plain while water periodically covered ranges, marshlands and swamps. Floodplains ensured an excellent and versatile livelihood for the population settled here. Livestock husbandry, such as horse, cattle and pig raising, fishing and different activities, such as ploughland cultivation, fructiculture, grassland and pasture management can also be inferred based on ethnographic parallels (ANDRÁSFALVY 1973). Water-edges, rich in nutrients, provided an obviously rich source of dietary resources. In shallow stagnant waters, in areas covered by floodwater and in marshes and bogs, plankton propagated supporting an abundance of crayfish and fish.

Locations suitable for human settlement were on the ridges free from floods, elevated above the waterlogged environment. The climate also was favourable, as the period called the "little climatic optimum" ranging from 800 to 1200 was the warmest period in the last two thousand years. The climate began to shift to increased humidity after 1000 but the process became definitive only after the 13th century (RÁCZ 1993).

The Hungarians, arriving in the Carpathian Basin, found several other Celtic populations along with the Avars. GYŐRFFY (1977) summarises as follows: "... The remains of the Gepid-Avar-Slavic society, amalgamated for centuries, must have been inhomogeneous in itself. Semi-nomadic lifestyle can be assumed in the Great Plain flatlands, while on the forested hill and mountain ranges small livestock farming was practised together with slash-and-burn farming". Although the observation of ethnic composition is not the task for a historian or a botanist, the description cited must have been valid only for a part of the Great Plain and even then not exclusively. The Great Plain was not barren land without any forests at this time. According to the latest results of microregion examinations made in the area around Endrőd, asteppe boundary might have existed here in the Late Migration Period (personal communication from Dénes B. Jankovich). This assumption seems to be confirmed by the phytolith finds from the settlement (personal communication from Irwin Rovner).

MÜLLER (1982) assumes that the farming skills of Hungarians were no less advanced than those of the people living here. Otherwise it would difficult to explain how the Hungarian people could create a strong feudal state in only a century (ERDÉLYI 1982).

Opinions about the lifestyle of the conquering Hungarians differ among various souces. According to an earlier perception, they were nomadic people who had their lands tilled by Slavic slaves. Archaeological heritage suggests however that reality was more complex. The number of Hungarians conquering the land exceeded that of the resident populations. Analysis of the Bjelo-brodo-culture graveyards, which are identified as an anthropologically less homogenous common people of the conquerors, suggests that in contrast to the dominating elite, they conducted a settled life form and engaged in farming (SZŐKE 1962). Graveyards of the common people, which were more numerous, were always found on the fertile loamy, loess grassland and rubiginous forest soils of the Great Plain and Transdanubia, and in each case close to the rivers swarming with fish. The sandy grasslands of the Great Plain were not occupied, people settled only in areas of fertile soils. This can be explained by crop production (NÉMETH 1973). In contrast the burial sites of the elite with horses and rich enclosures can be found in the sandy soil steppe regions (Nyírség, Duna-Tisza region, Mezőföld, Small Plain) (BÁLINT 1980). The insistence on steppe surroundings and lifestyle explains why the first settlements of the Hungarians did not cross the boundaries of the European beech forests (KNIEZSA 1938). This region is the westernmost occurrence of their former quarters, the forested steppe. They insisted on husbandry methods based on long traditions and the typical areas suportding them, which represented preferred and secure livelihood. All these allow for the conclusion that during the conquest, the elite had a Turkish-type, nomadic lifestyle with changing environs. Winter and summer pastures were switched at definite intervals. Nyírség and the area between the Danube and the Tisza were still appropriate for this pattern at the time.

BÁLINT (1980) found a number of similarities between the Eastern-European flatlands, which gave home to large surges of folk migration and the Great Plain steppe regions. Summarizing earlier research, he stated that sites from the period of the conquest and the Árpádian Period are found mostly in the forested steppe regions of the Carpathian Basin. He refers to GLASER (1937) and MOLNÁR (1954) who insisted that "the climate in Hungary was identical with the zone of the forested steppe in the Ukraine (along the Voroniesh-Kuybishev-Novghorod line)". However, this statement holds true only for the area of the Nyírség. The area around the river Körös resembles more the Pontus steppe regions. The issue is further complicated by the fact that forested steppes and steppe meadows are distinguished by their formation on sandy, loess and cohesive soils. These were utilised in different ways according to the finds from the Migration Period and the Conquest Period.

Natural conditions at the time of the conquest are reflected in settlement names (GYŐRFFY 1963, 1987). György GYŐRFFY – Bálint ZÓLYOMI (1994) even made a vegetation cover map of the Etelköz and the Carpathian Basin as it existed a thousand years ago. Tamás GRYNAEUS (1996) argued that an equally important task would be to complete the data of place names and lands derived from plant names with botanical finds from the period and as an example, he prepared a map of oak, European beech, pear and grape occurrences in Hont county during the Árpádian Period.

As a matter of fact, the natural geographical conditions in the Carpathian Basin in the 10th–11th centuries did not favour livestock husbandry based on shifting pasturage. Yet, a more restricted version of this nomadic lifestyle, especially in the Great Plain, is very likely. The excavation at Felgyő proves that the conquerors living here had a settled way of life and lived in log-walled houses or permanent yurts.

Tillage by ploughs could not be unfamiliar to the conquering Hungarians. Ploughland might have been tilled according to the agrotechniques of the time, using a delving plough equipped with a handled plough iron, but through association with the eastern Slavs, they must have known the bed plough as well (BALASSA 1973). According to grave goods, crops was reaped with sickles. Since the 10th century, a new implement for collecting fodder appeared: the straight edge, widening blade *scythe* (MÜLLER 1982).

If one has to classify the husbandry of the conquering Hungarians, the most suitable word would be "semi-nomadic". This includes migrating livestock just as well as a limited extent of cropping and tillage. Even though sheep and cattle bones dominate animal remains, quite typical for the nomadic lifestyle, this does not contradict the predominantly stock-farming lifestyle of the conquering Hungarians who arrived with farming and crop producing skills into the Carpathian Basin.

7.8.2. Botanical finds from the age of the conquest

Only a very few seed finds are available from the conquest age which relates to crop production (P. HARTYÁNYI – NOVÁKI – PATAY 1967–68). The archaeological emphasis for this period has focused on graveyards where botanical finds are usually very rare. An exception is a conqueror belonging to the elite, whose grave was located within the current country boundaries in Zemplén, which contained common millet grains. Knowing the eating habits of nomadic and semi-nomadic people, this find is expected, as common millet is a favoured cereal with a short growing season and rapid development requiring relatively little tending.

According to research scientists dealing with the history of crop production in Hungary, the most important cereals in the Migration Period, the age of the conquest and the Árpádian Period were common millet and barley. Consequently the most important food was *gruel* (RAPAICS 1934; GAÁL 1978). Archaeobotanical research has modified this concept.

The only Hungarian botanical find from the age of the conquest (beginning of the 10th century) comes from Lébény-Billedomb, the 1993 excavation of Miklós Takács. Several soil samples were collected from settlements of the conquering Hungarians. A great number of seeds and fruit remains of 30 different plant species were found (GYULAI 1997b). Hulled wheat types, typical in prehistoric ages, were not grown at all, only the more advanced naked grain common wheat are found. In addition to six-rowed barley, albeit to a smaller extent, two-rowed barley and naked barley grains were also found. An important gruel plant was common millet, as unearthed carbonised common millet gruel pieces show. It has to be noted in connection with recovered rye grains that they can be sown separately or mixed with wheat (abajdoc, maslin). Common millet might be sown as an aftercrop. It is very likely that among other crops, six-rowed barley, which is encountered in the greatest numbers, was also consumed, as it was customary in this age. Common millet, common bread wheat and rye were necessarily used for human consumption. Specific eating habits with respect to garden produce was minor given the fact that only pea has been recovered as opposed to the many different cereals (c.f. similar finds from the Avar Period).

Weeds relate to the presence of specific cultivars. Members of spring cereal or hackfruits weed associations (*Polygeno-Chenopodietalia*) are present to a greater extent while members of winter cereal weed associations (*Secalietea*) occur to a smaller extent. Crab apple, identified by carbonised seeds, came from gathering. Species gathered from the natural environment suggest a diverse environment: reed-beds, forests, meadows. Ruderalia such as yellow mignonette (*Reseda lutea*), white melilot (*Melilotus albus*), knotgrass (*Polygonum aviculare*) and creeping buttercup (*Ranunculus repens*) indicate permanent settlements.

7.8.3. Grain finds from the early Árpádian Period

Relatively more evidence has been recovered from the period (10th–11th centuries) after the conquest (P. HARTYÁNYI, NOVÁKI – PATAY 1967–68; P. HARTYÁNYI – NOVÁKI 1973–74) (*Figs. 318, 321*). In the early period, common millet continued as the primary grain crop. This is especially true for the Great Plain. In Kardoskút, during excavation of a 10th–13th-century village, cereal grains were found among burnt straw under an oven. Numerically, the most important grain was common millet followed by bread wheat and rye. In Tiszaörvény, during excavation of an 11th–13th-century village, common millet awn remains were observed, apparently stored in considerable amounts in a corner of a house (*Figs. 323–326*). Weeds of cultivated common millet might have been redshank (*Polygonum persicaria*), fan-hen or white goosefoot (*Chenopodium album*), field pepperwort (*Lepidium campestre*), knotgrass

(*Polygonum aviculare*), flower-of-an-hour (*Hibiscus trionum*), all picked from among the awn remains. This weed flora indicates spring sowing crops. The cropland of this period might have been situated in an environment of ruderal areas with typical habitats. Based on height of the weed species found among the grain crops, reaping might have been done at a low height, apparently with a scythe.

Dominance of common millet remains supports the contention that, in the Great Plain after the conquest, nomadic lifestyle and nomadic pasture rotating large livestock husbandry continued to a significant degree. Common millet, to be grown more easily and ripening more quickly when compared to wheat, was the typical crop of nomadic and semi-nomadic husbandry.

Periodic (i.e. short-term) pasture rotation livestock production in steppe areas and the associated lifestyle is confirmed by archaeozoological evidence. István VÖRÖS (2000) found significant differences in species composition by regions when he projected animal bone data from 37 Árpádian Period sites onto a map. A common feature of animal bone finds in the Árpádian Period from the Mid-Tisza-region and southern Hungary is a donimance of small ruminants (mainly sheep) and horse bones. However, a considerable amount of cattle bones were also encountered. Pig bones were fewest in number. This is clear evidence for the survival of pasture rotating large livestock farming. In contrast to this, in Transdanubia along the river Danube and northern Hungary, cattle finds are most abundant, followed by pigs, indicating sedentary, permanent settlements. Small ruminants are fewer, followed by horse.

When examining finds from the 10th–11th centuries, common bread wheat, a crop assuming much more advanced production skills than common millet, appears, although only sporadically. In Biharkeresztes-Ártánd-Nagyfarkasdomb, common wheat grains were found in a grave in the 11th-century cemetery.

Settling down and switching to crop production seem to have occurred more rapidly in Transdanubia and in the northern parts of the country. Botanical finds from the western and northern parts of the country are characterised by the presence of more valuable grains requiring longer growing seasons. These cereals require a higher level of agrotechniques and continuous tending. Grains found are identical in terms of species composition with finds in western countries from the same age. It is not by accident that the finds are encountered in and around castles and seats of counts, and monasteries dated to the early Árpádian Period, which served as centres to the slowly consolidating and uniform feudal order and production method.

The basis for husbandry was established in the village system by the 11th century. Settlements and permanent quarters became customary. The village system was consolidated by the 12th century (GYŐRFFY 1977). In Transdanubia, such settlements were traditionally formed, which doubtlessly facilitated the spread of the village system.

Crop production in the western part of the country is characterised at this time by grains with longer growing seasons. In a grave of a 10th–11th-century cemetery explored by Gyula Török in Halimba, a bread wheat grain fell out from among chaff imprints at the bottom of a pot (PARÁDI 1959). Carbonised grains were found in a house of the 10th–11th-century Keszthely-Halászcsárda settlement (BAKAY – KALICZ – SÁGI 1966). Barley was also grown, although only sporadic finds are known from this period. When the ancient Christian cell trichora of Pécs was renovated in the 11th century, the plaster under the paint contained plant parts, providing calcified barley awn remains (P. HARTYÁNYI – NOVÁKI – PATAY 1967–68). In 1995–1996, at the excavation led by Gábor Ilon, houses dated to the 10th–11th centuries were found in the Pápa-Hantai Street site. Plant residues were found in daub and soil samples. (TORMA 1996a). Daub contained mainly imprints of glumaceae (*Poaceae*) leaves and straw stalks, with enclosed barley grains.

Common bread wheat proved to be the most important crop in the Sopron-Városház Street 9th–10th-century layer (P. HARTYÁNYI – NOVÁKI – PATAY 1967–68). This was followed by rye and common millet was present only sporadically. Separation of crops into autumn sowing (bread wheat, rye hosts) and spring sowing (common millet host) crop weed associations is indicated by species of weeds recovered: korn-rade (*Agrostemma githago*), field cowwheat (*Melampyrum arvense*), knotgrass (*Polygonum aviculare*), redshank (*P. persicaria*), bindweed (*Convolvulus arvensis*), hemlock (*Conium maculatum*), annual woundwort (*Stachys annua*), charlock (*Sinapis arvensis*), spring wild-oat (*Avena fatua*) and goosefoot (*Chenopodium* spec.). Based on the ecological needs of these weed species, ploughlands must have had a moderate level of water supply and been rich in nutrients. One can infer the height of reaping from the height of the weed species: it was done close to the ground, about one third of the height of the crop, probably using a scythe.

The same can be said about the remains coming from the northern part of the country: more valuable kinds of crops were grown here as well. Adjacent to the Roman church at Esztergom-Kovácsi, 11th-century graves provided botanical materials characterised by common bread wheat and rye, both having longer growing seasons.

A carbonised grain layer was found during the excavation at the 10th–11th-century castle of a count at Hont (P. HARTYÁNYI 1981–83) (*Figs. 319–320, 322*). The sample from this layer contained a small amount of rye and common bread wheat, together with a very high level of weed infestation. Korn-rade seeds were especially numerous. Somewhat less, but still significant are catchweed, cleavers (*Galium* cf. *aparine*) and brome grass (*Bromus secalinus*).

When one compares the earliest finds of the period after the conquest, the conclusion is that common millet played an important role mainly in the Great Plain, while common bread wheat and rye did the same in Transdanubia. The finds from the Plain support the notion of limited nomadic patterns in the period after the conquest. The finds from Transdanubia and from the northern part of the country suggest a sedentary lifestyle and a more advanced level of agriculture. The Danube River, which is a historical as well as floristic boundary, also divided the country into two major areas of different crop production: the Great Plain, producing more archaic plants (see production of emmer below) and Transdanubia, a more advanced region integrating the traditions of Roman agriculture.

Grain finds from later excavations in the 12th–13th centuries in the Plain start to show similarities with those found in Transdanubia both in terms of species composition and their relative importance (P. HARTYÁNYI – NOVÁKI 1973–74). Apparently, the population in the Great Plain became settled only a century later, by the 12th–13th centuries. The alteration of sowing seeds, representing a quality change in crop production, was completed by this time. Growing high nutrient common bread wheat and rye became customary. The significance of common millet declined but it was retained in production up to the Modern Period as an aftercrop providing gruel dishes. Considerable amounts of carbonised common bread wheat grains and somewhat fewer rye grains were found in the Early Árpádian Period graveyard of Cegléd-Madarászhalom. Common millet was not found at all. In the purified stock, korn-rade and cress (*Lepidium* spec.) seeds are only sporadically found. Unspecified cereal awn fragments were found in the daub unearthed from the Árpádian Period pit of Doboz-Hajdúirtás. The bedstraw (*Galium* spec.) and brome grass (*Bromus* spec.) diaspores identified indicate autumn sowing cereals.

Although no oat is represented among the botanical finds after the conquest, it cannot be excluded as linguistic evidence indicates its presence. However, even if present, it must have been of very low significance. Of the wheats, mainly common wheat and naked club-wheat were grown. Unless the conquerors brought sowing seeds with them, these must have been the more advanced common wheat and club wheat.

In 1995–1996, in the northern sector of the M0 highway under construction at Rákospalota-Újmajor site No 1, an Early Árpádian Period village, the long forgotten former Sikátor, was explored under the leadership of Zoltán Bencze. The large majority of the carbonised diaspores unearthed from the waste pits and external or internal ovens dated to the 12–13th centuries were cereals (GYULAI 1999a).

Corresponding to other sites from the Árpádian Period, six-rowed barley, common and club wheat, rye and certainly the inevitable gruel-plant, common millet, were also found here. The dominance of six-rowed barley is striking. It has to be noted here that the dominance of barley and common millet has always been a feature of quickly moving livestock farming people. Thus, one cannot exclude that they preserved a certain amount of steppe traditions as regards their crop production and eating habits.

It is very surprising that grains and furca bicornis remains coming from chaffing of a hulled bread wheat, namely emmer, are also encountered. This is a strange and uncommon, yet not unique feature in this period. The majority of archaeobotanical finds show that hulled wheats (einkorn, emmer) were not grown after the Migration Period.

In the archaeobotanical finds of Early Árpádian Period houses and pits explored at 170 sites in Endrőd, a late occurrence of emmer is found again (excavation by Dénes B. Jankovich, in press). Notwithstanding this, the production of cereals requiring a more advanced level agriculture was also started. Barley remained an important kind of grain. Rye production was insignificant. Naked grain common wheat, more demanding from an agrotechnical point of view, was also grown. Hulled emmer, staple grain crop in earlier times, was somewhat less important, yet it occurs sporadically.

Emmer persisted for a short time after the conquest as a typical representative of earlier (prehistoric ages and Migration Period) cultures of the Great Plain, but its production became insignificant.

In 1998, Anna Endrődi and Attila Horváth explored a Late Árpádian Period (12th–13th centuries) settlement segment in Csepel, Rákóczi Ferenc Road (ditches, open fireplaces/ovens, waste pits). Although the botanical material is poor, it indicates cultivation. Three quarters of the species are grain crops: six-rowed barley, rye and common bread wheat. The finds are dominated by six-rowed barley. A third of it is rye, and less common bread wheat.

No hulled wheat was found here or in Szigetszentmiklós-Waterworks, where Anna Endrődi made a rescue excavation in 1999 before the construction of a MOL gas pipeline and explored some Árpádian Period structures. Grain remains coming from here correspond to the species composition found at other sites of 12th-century Hungary and the data of written sources from the time: the main crops were six-rowed barley, common bread wheat and rye.

In 1996 at the Rákospalota-Újmajor site, most probably another segment of Sikátor village, was excavated by Anna Gyuricza. This time, a well, dated to the 13th–14th centuries, rich in plant residues was also identified. According to the archaeobotanical finds recovered, the number of crops in cultivation in the Late Árpádian Period was further increased. Expansion means, when compared to the Early Árpádian Period, club wheat and common oat. The number of club wheat grains, like that of common millet grains, is low, both of them being typical for the Middle Ages in Hungary. There is somewhat more rye. The ratio between the grains of common bread wheat and rye is 3:1. This raises the possibility of their joint production (triticum mixtum, abajdoc, maslin). The number and composition of weed species is similar to those at other Early Árpádian Period sites. Of the legumes, pea and big seed lentil production is confirmed. This site is the earliest occurrence of garden radish (*Raphanus sativus*) in the Carpathian Basin. Some other plant species found here are also edible: e.g. dwarf mallow (*Malva neglecta*) as a bread substitute, and fan-hen or white goosefoot leaves as salad. The environment of the time is characterised by species from the former forest edge with average habitat potential and from meadows.

The consolidation of the feudal Hungarian kingdom resulted in a dynamically developing agriculture. As time went by, only cultivars requiring a high level agriculture but ensuring safe yields were grown.

The gradual penetration of crop production is indicated by palynological results as well. Cereal pollen increased at the time of the conquest around Lake Balaton (Zólyomi 1980). From the 8th century until the end of the 11th–12th centuries, cereal pollen numbers doubled. Goosefoot (*Chenopodiaceae*) pollen increase can be associated with an increase in the numbers of settlements and the expansion of existing ones. Species belonging here occur where human activities and livestock farming increase the nitrogen content of soils. The doubling of Non Arboreal Pollen can be connected to slash farming. The increase of birch pollen numbers can be associated with slow degradation, triggered by forest clearing. All these are direct consequences of deforestation and the formation of secondary steppes. The gradual penetration of steppe vegetation is indicated by the fact that in the period from the 800s until 1100, wormwood (*Artemisia* spec.) pollen doubled when compared to the previous period.

Garden produce was also present, although as a result of the specific eating habits (meat and gruel based diet) kitchen vegetables did not play a significant role. The fragment of a tray full of garden lentil (*Lens culinaris*) and common lentil vetch (*Vicia sativa* var. *lentisperma*) seeds was unearthed during the excavation of the Visegrád-Várkert dűlő 9th–11th-century settlement (P. HARTYÁNYI – NOVÁKI – PATAY 1967–68). The Early Árpádian Period common lentil and pea (*Pisum sativum*) seeds might have come from local production as well.

There are archaeological evidences as well to support the concept that the conquerors were familiar with fibre crops and their use. Fibre tests demonstrated that the fabric residues of the 10th-century graveyard in Szabadkígyós-Pálliget were made of flax, hemp and cotton. The production sites of both the plants and the textiles are unknown. Of fibre crops, only hemp (*Cannabis sativa*) was confirmed as locally grown. Substantial amounts of stocked hemp achenes were encountered both in the 9th–11th-century Visegrád-Várkert dűlő and the Early Árpádian Period Endrőd settlements.

The early Hungarians got their cereal seeds either by arriving with them, or receiving them from the people found here (or some combination) or maybe through the Western "hospes" who settled in later. The farmer's life might have been aggravated by the many weeds presumably introduced by the last. Even though grain finds from houses and pits explored by the archaeologists were already cleaned and stocked, in other words prepared for use, they still contained a great number of weed seeds: catchweed, cleavers, bromegrass, field cowwheat, knotgrass, redshank, bindweed, annual woundwort, spring wild-oat, fan-hen or white goosefoot. Obviously, using the cleaning procedures, which practically meant scattering and husking, it was not possible to remove these. Many seeds of korn-rade are found, which might have caused poisoning when mixed in flour. These grain weed diaspores confirm local growing of autumn sowing grains, which is important proof of stable settlem and it also provides evidence that reaping was made at a low level, probably using a scythe.

We have finds from the conquest period regarding burial habits as well. Bindweed (Convolvulus arvensis) was found in one of the graves at the Káli cemetery and, in a sample coming from a grave in the cemetery at Hajdúdorog-Kati dűlő (excavation by István Fodor), a substantial amount of caper spurge (*Euphorbia lathyris*) seeds were found, along with sporadic fan-hemp and Aaron's beard (*Hypericum perforatum*). These seeds were most probably put beside the corpse and around the skull as a sign of grief. Caper spurge has an effect of facilitating digestion, Aaron's beard is a sedative that facilitates digestion and, as an herb, heals wounds. Thus, these might represent herbological knowledge of the early Hungarians.

7.8.4. Remains of fructi- and viticulture

Walnut and sour cherry grow in forests without any human intervention, just as do crab apple, wild pear, sloe (blackthorn), hawthorn, wild strawberry, cornel-berry and hazelnut. Early production, before the 13th century, was mainly restricted to gathering and preserving these species. Early names of places are indicative. It is noted that, since the Árpádian Period forest clearing included cutting down all trees except walnut. Thus, smaller or larger "walnut groves" were formed. Walnut was also deliberately planted beside wells or at the ends of vineyards. Walnut pollen has consistently increasing since the turn of the millennium, making a connection with the grape plantations, which are clearly indicated by data from the time (ZÓLYOMI 1980).

A large majority of Hungarian words in connection with grape-growing and fruit production are of Bulgarian-Turkish origin (GOMBOCZ 1960). They were exposed to the language during their co-existence with the Kazar Kaganate. According to a more recent opinion, our viticulture has double roots: the fructicultural and viticultural skills of the Hungarians brought in from the east were supplemented and completed by the vine-growing skills of Roman origin Pannonians in the Carpathian Basin, and it was here that they were amalgamated into one integrated whole (SÁGI – FÜZES 1967; FÜZES – SÁGI 1968; FÜZES 1971). Even if Roman viticulture and fructiculture survived up to this point, they must have been melted into the knowledge base of the Hungarians arriving into the Carpathian Basin.

Adopting Christianity facilitated the spread of fruit and grape production. Hardly one century after the conquest, at the beginning of Latin literacy, the donation charters and documents tell us about orchards, vineyards and viticulturists (PÁKAY – SÁGI 1971).

The first Hungarian kings made an effort to populate the sparsely populated regions in the country by giving special patents and privileges to German, French and Italian settlers. Settlers and religious orders arriving in this country must have brought locally native fruits and selected grape varieties from their respective homelands.

Forests covered hill and mountain ranges in the country at this time and, therefore, new vineyards were mostly planted in clearings. These vineyards and orchards were attached to monasteries and early manors. It was not before the 13th century that inlots with adjacent orchards were formed in peasant farms.

Sporadic fruit finds from the Árpádian Period include a carbonised peach stone from a grave beside the Romanesque church in Esztergom-Kovácsi; walnut shell pieces found in a house of an 11th-century settlement in Keszthely-Halászcsárda, which indicate the production of more advance cultivars, albeit in a limited manner (P. HARTYÁNYI – NOVÁKI – PATAY 1967–68). In the exploration of the Pápa-Hantai Street site of 10th–11th-century houses, peach stone fragments were again encountered (TORMA 1996b). The metric data of these stones are substantially different from those in recent varieties, but they are similar contemporary ones.

Evidence of early viticulture are provided by black wine grape seeds found at Early Árpádian Period settlements. These seeds were found at the already mentioned settlement in Rákospalota-Újmajor.

Compared to other sites, the culture layer explored in the cellar under Hunyadi Street 22 in Budapest, first district in 1973, is unusually rich in cultivated plant remains. Here, a series of locally grown and gathered plants were obtained from a pot dated to the 13th century, but a number of common millet remains and several weed species were also encountered. Seeds were identified by István Skoflek and Mrs. Hortobágyi. The grape seeds were analysed by Géza Facsar (in: P. HARTYÁNYI – NOVÁKI 1973–74). A large part of the seeds are black wine grape, but sour cherry, hazelnut, sugary melon and watermelon, walnut, apple, rose, bramble, black mulberry, sloe (blackthorn), alkekengi and danewort are also found.

The language at the beginning of Hungarian literacy was Latin. Latin texts of diplomas preserved a number of Hungarian words. These words could not be replaced by Latin words, or it was not the scribes' intention to do so. It is interesting to examine these earliest Hungarian words in terms of regional use (TERPÓ 2000). Many words refer to the natural environment already in the earliest diplomas from the 11th century and since; such as, hegy (mountain), szikla (rock), szurdok (ravine), erdő (forest) and liget (grove). The occurrence of words related to original vegetation in documents from the 13th–14th centuries refer to the utilisation of the different plant stocks; for instance, fenyves (conifer-wood), bükkös (European beech-wood), tölgyes (oak-wood), égeres (alder-grove), füzes (willow-grove), nyáras (poplar-wood), cserjés (shrubbery), nádas (reed-bed) and füves-gyepes (grassland-pasture).

The natural environment in the Carpathian Basin at the time was abundant in waters and marshes. This wealth can be seen in the Hungarian words of charters. Names of trees and bushes were given readily to boundary marks such as service-tree, willow, maple, Scotch pine, cornelian cherry, dogwood, sloe (blackthorn) and European chestnut. These obviously reflect ancestral ownership in the ancient homeland, yet they are suitable for the characterisation of the landscape as well. Words of land use might have connections with slash husbandry, since a part of these boundary marks could have been trees left over from former forests. The earliest records referring to a cultural landscape (meadow, arable land) come from the mid-11th century. Words of horticulture (orchard, garden) originate at least that early.

The Great Plain was always characterised by sandy areas and a limited amount of sand movement. The triggers were erosion due to precipitation and wind movements, for the main period of sand movement ended in the southern Nyírség at the end of the Pleistocene. Dunes formation, however, continued, eroded away and created again. The pace of this natural evolution was changed by humans who settled in the region. The area was already populated in the Neolithic. New settlers lived on the top of the dunes. Dune movement triggered by human activities such as vegetation stripping and ploughing mainly affected the higher dune tops. Wind eroded the dunes and changed their original shapes.

Several low dunes can be found in the area of contemporary southern Nyírség settlements (Vámospércs, Bagamér, Álmosd, Bánk) but even more can be identified around the villages formed between the 11th and 14th centuries that perished during the Turkish invasion (Kiss – Bódis 2000). All these relate to the larger extent of human impact on shaping landforms in the Middle Ages. On the one hand, the environmental transformation in the Árpádian Period (400 years) lasted longer. On the other hand human activities in the Middle Ages were much more local in nature than today, however, at an actual locality, it had a more intensive impact than in the last one and a half centuries.

Crop production shows a uniform process and continuous development from the conquest to the 12th–13th centuries. It cannot be exactly estimated, yet the Moravian-Frankish crop production skills must have had a strong influence on the Hungarians. At the same time, one has to reckon with the integration of late Avar practices. The consolidation of the feudal state, skills and knowledge paralleled the amalgamation of residential peoples into a single unit. As services, trade and commerce, information flowing in from abroad, the centralised royal power and the formation of the landed class fostered the spread of religious customs and literacy and a dynamically evolving set of information, implements and seeds that made their way to all parts of the country.

7.9. The Late Middle Ages

Climate started to change in Europe from 1000 AD. (Rácz 1993). The weather was becoming cool and humid from the beginning of the 13th century. Cool, wet summers interchanged with extremely cold winters. This was the "Early Medieval Cool Period" which lasted until 1450. Agricultural production declined as a result of the already permanently cool and humid weather by the mid-14th century. The history of these centuries is accompanied by famine and pandemic.

This "Little Ice Age", which lasted for several centuries, was felt in the Carpathian Basin as well. After the warm and dry summers of the mid-16th century, weather became humid and cool by the end of the century. This period is characterised by harsh and cold winters. Weather changed again for the better at the beginning of the 18th century (Rácz 1993). This global climate change however did not cause any decline in the development of Hungarian crop production. Moreover, it provided an explicitly favourable period between the 14th and 16th centuries. Due to the supply needs of developing Hungarian

Fig. 318. Spring wild-oat (*Avena fatua*) naked grains from Visegrád-Várkert, 9th–11th centuries, Inventory of the Hungarian Agricultural Museum, Budapest.

Fig. 320. Rye (*Secale cereale*) grains from the *ispán*'s castle at Hont, Early Árpádian Period, Inventory of the Hungarian Agricultural Museum, Budapest.

Fig. 319. Common bread wheat (*Triticum aestivum* subsp. *vulgare*) grains from the *ispán*'s castle at Hont, Early Árpádian Period, Inventory of the Hungarian Agricultural Museum, Budapest.

Fig. 321. Common bread wheat (*Triticum aestivum* subsp. *vulgare*) grains from Cegléd-Madarászhalom, Late Árpádian Period, Inventory of the Hungarian Agricultural Museum, Budapest.

Fig. 322. Spring wild-oat (*Avena fatua*) grains from the *ispán*'s castle at Hont, Early Árpádian Period, Inventory of the Hungarian Agricultural Museum, Budapest.

Fig. 323. Knotgrass (*Polygonum aviculare*) nutlets from the floor of a house in Tiszaörvény (11th–13th centuries), Inventory of the Hungarian Agricultural Museum, Budapest.

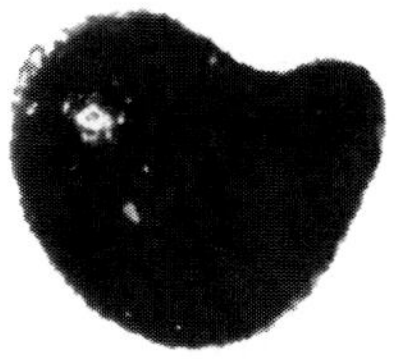

Fig. 324. Flower-of-an-hour (*Hibiscus trionum*) seed from the floor of a house in Tiszaörvény (11th–13th centuries), Inventory of the Hungarian Agricultural Museum, Budapest.

Fig. 325. Redshank (*Polygonum persicaria*) nutlets from the floor of a house in Tiszaörvény (11th–13th centuries), Inventory of the Hungarian Agricultural Museum, Budapest.

Fig. 326. White goosefoot or fan-hen (*Chenopodium album*) seeds from the floor of a house in Tiszaörvény (11th–13th centuries), Inventory of the Hungarian Agricultural Museum, Budapest.

townships, grain crops, vegetables and fruits were produced in larger amounts. At this time the export production of meat and wine commenced.

Thanks to the agrotechnical innovations introduced in the period called "High Middle Ages" (bed-plough, horse-collar, three-field rotation in farming) it was possible to apply secondary use of several domesticated species as draught animals. Extensive arable lands and grasslands supplied the produce to ever increasing, sometimes urbanised settlements.

László GAÁL (1978) called the period in Hungarian crop production between 1350 and 1750 "developing phase", which is also interpreted as the "age switching to cash crops". The population of Hungary was less affected by the cooling of Western and Northern Europe at the beginning of the 14th century (RÁCZ 1997). In this period, the Carpathian Basin was less densely populated and therefore the first wave of the Central Asian pest did not cause any serious crisis. At the same time the agricultural production of the country turned for the better during the cooler years. No supply disturbance and famine on a Western scale were experienced in Hungary. Moreover, the crisis there made Hungarian products more appealing. This was the time when Hungarian cattle export was initiated towards Northern Italy and the German areas.

Most probably, the cooling of the "Little Ice Age" in the second half of the 16th century could have been endured by the country had it not become a battlefield. The population of the country, weakened and dispersed by the expansion of the Turkish Empire and later by continuous warfare of the fifteen years war, was directly and adversely affected by the extremely cool and humid weather of the first decade of the 17th century (RÁCZ 1997). Entire regions were abandoned as a result of military, political and ecological consequences.

Written sources (see the Diploma-Archives from the Sigismund-period) mention wheat six times, common millet three times, oat and hemp once between 1387–1399. Likewise, between 1400-1410 references occur to millet three times, oat eight times, while hemp occurs once. Tithe lists in the 16th century mention lentil, pea, flax, buckwheat, cabbage, carrot, onion, garlic, and fruits (apple, pear) or grape in addition to the cereals.

"Turkish wheat" (maize) first appears at the end of the 17th century and in the 18th century we encounter common tobacco and potato (GYULAI 1999b). In Transylvania, einkorn, spelt and most probably emmer continued to be grown and a new plant, bean (Phaseolus vulgaris) appeared. The list of plants grown in Upper Hungary are known from a Murány (Gömör county) instruction issued in 1662: bread wheat, spelt, rye, barley, oat, pea, common lentil, vetchling, buckwheat, flax, hemp.

As ploughlands surrounding the settlements were already formed by the 12th–13th centuries, new land and meadow could only be obtained by breaking new ground. Ploughland was usually formed by clearing of forests on hillsides or in the mountains. Although landlords owned these plots, peasants were still encouraged by a number of benefits, such as easement and exemption from tithe, to engage in hard work. The first stage of clearing lands can be put to the 14th–15th centuries, the period during which population doubled, and the second stage at the 18th century, when an economic boom started after the expulsion of the Turks.

Although we have no ploughing tracks, the research of old ploughlands has a tradition in Hungary. Gyula NOVÁKI (1975–77) found some remains (terraces, stone rows, balks) associated with old ploughlands in two different parts of the Börzsöny mountain (Nagybörzsöny, Bernecebaráti). These were lands cleared before the 19th century, possibly medieval in origin. Field-remains were found in the surrounding area of contemporary Tamási beside Kosba or Kospa, a village abandoned in the Turkish era and Sarvaly, beside Sümeg (NOVÁKI 1984–85). Both perished villages were overgrown with forests and former ploughlands were identified only by superimposed soil terraces and stone rows between them.

7.9.1. Botanical finds as sources of diet in the Middle Ages and the Early Modern Period

Botanical finds at medieval sites provide great assistance in increasing our knowledge of the period. The botanical material unearthed in excavations is an accurate and faithful mirror of crop production and botanical knowledge of medieval people. With its help, eating habits, plant use and the former natural environment are investigated (KARG – JACOMET 1991). They are also suitable for comparison with local written and iconographic sources (WILLERDING 1984). Archaeobotanical finds mostly confirm written sources (c.f. SZAMOTA – ZOLNAI 1902–1906).

Medieval seeds and fruit remains come usually from privies, wells, waste pits, i.e., places where the waste of households goes. Roads, cisterns and sewers, in particular, preserved seeds and fruit remains in good condition. Diaspores coming from here were conserved as a result of anaerobic conditions, although in certain instances surface corrosion can be significant.

The consolidated feudal order formed a uniform crop production system in Hungary. More or less the same plants were grown everywhere. This can be demonstrated by a number of archaeobotanical finds. The seeds processed by Imre Deininger, originating from Torna-Szádelő Valley (today Slovakia) represent the heroic stage of archaeobotanical research (in: LEHOCZKY 1883). An approximately three-centimetre-thick cereal layer, consisting of mainly rye and to a smaller extent common wheat and barley, is dated to the Mongol invasion.

Important evidence for a uniform medieval crop production culture is seen in the late medieval botanical material of Muhi. Although the samples collected and locally floated in 1995 at the Muhi medieval excavation site led by József Laszlovszky and Tamás Pusztai are still not completely processed, the cereals identified so far (common bread wheat, club wheat, rye, six-rowed barley, common millet) confirm the level of farming typical for the age.

All cereals typical for the Hungarian Middle Ages were found in the samples taken from the medieval manor house excavation site in the forest beside Baj Öregkovács-mountain (excavation by Sándor Petényi 1998): common bread wheat, six-rowed barley, rye, common millet.

In 1998 at Szarvasgede, at the excavation led by Miklós Takács and István Paszternák, similar kinds of cereals were found. Archaeobotanical examinations indicated that intensive agricultural activities were carried out here in the mid-15th century. A number of carbonised cereals grains were floated from the soil samples taken at the archaeological structures. Along with common bread wheat, rye and six-rowed barley were also grown.

Led by András Horváth Pálóczi, archaeologist at the Agricultural Museum, several wells were explored at the late medieval Cuman settlement in 1984–87 at Lászlófalva-Szentkirály. Wells were floated in situ; seeds and fruit remains were separated from other organic residues in a work which lasted for years. It is not known whether six-rowed barley, a grain crop that occurred most frequently in the finds, was grown as fodder or was intended for human consumption. However, common millet, common bread wheat and rye necessarily must have served for human consumption. Common bread *wheat* and *rye* grain ratio is close to one to one. This raises the possibility of their mixed cropping.

Mixed growing of these two cereals was typical in the Hungarian Middle Ages. As early as in the Árpádian Period, joint production of bread wheat and rye was widespread. Abajdoc (abenác, abajdos) means a mixed crop. It was also called "maslin", "triticum mixtum", or "cerealiam promiscuam". King Ladislaus I provided for the taking of the tithe separately and not in the maslin: "In annona vero commixtum non accipiat, sed separatim". Wheat and rye were not mixed subsequently but sown together. Allegedly, this was done for security purposes. This way, even if one of the species would not ripen, the other still might bear a yield. An interesting observation is that maslin was still grown at the beginning of the 19th century.

In our view, joint production of bread wheat and rye provided more advantages than simple harvest safety. Stalks of maslin are more resistant to being blown over. It is also possible that people realised at that time, also demonstrated in recent East-German crop production experiments, that mixed production of different cereal species results in more uniform stock, higher yields (stimulating effect), and more resistant to plants pests (host hiatus). (Ákos Mesterházy, personal communication).

Half of the archaeobotanical material found at Dunaföldvár-Öregtorony site (ruins of a 17th-century house) consist of rye, somewhat less of common wheat and even less of common millet (P. HARTYÁNYI – PATAY 1970) (*Figs. 417–424*). In the 16th–17th-century layer of Hollókő castle, bread wheat and rye occurred in almost identical amounts (P. HARTYÁNYI – NOVÁKI – PATAY 1973–74) (*Figs. 425–428*). Sometimes rye surpassed the amount of wheat (devastation layer of Szolnok castle from the beginning of the 18th century), and in other places (a burnt house from the 16th–17th centuries, found at Pécs, Surgery Clinic) there was more bread wheat (*Fig. 362*). In these sites, common oat occurred as well.

A significant amount of seeds and fruit remains unearthed from excavations carried out in downtown Vác during the seventies and eighties represents well crop production in the Middle Ages (GYULAI

1995b) (*Figs. 411–416*). The archaeobotanical material collected at excavations carried by Zsuzsa Miklós between 1986 and 1992 originated from different centuries of the Middle Ages, thus ann opportunity to record changes in crop production spanning centuries at one special location. In the 13th century, albeit in small amounts, barley, common wheat and rye were still produced in nearly equal amounts. These species were accompanied by club wheat, six-rowed barley and common oat in the 15th–16th centuries.

A favourite gruel plant was common millet in the Hungarian Middle Ages. It can be found at almost all medieval sites: for instance, in the destruction layer of a 15th–16th-century housing estate in Nagyvázsony-Csepely (in a 40 cm^3 mass of mostly hulled aggregate) and in the 17th-century dwelling-house at Székesfehérvár-Palotai Street 5 (P. HARTYÁNYI – NOVÁKI – PATAY 1967–68) (*Figs. 327–328, 348–361*).

We do not know the average yields of medieval cereals. Yet, one can draw some conclusions by assessing the evolution of ploughing methods in the Middle Ages combined with average yields of the 16th century as reconstructed from documented data. Accordingly, in the interval ranging over thirty years of duty registers, the following ratio can be applied to peasant farming: three times the amount sown in wheat, twice that of barley, five times as much in rye and three times as much in oats (KIRILLY 1968). In our view, these values are very low and it cannot be taken literally or uncritically –because producers had a conflicting interest – that duty requirements would reflect the actual situation.

BALASSA (1973) notes the following: "Growing cereals can be considered as a general practice in the lowlands in the 12th–13th centuries. As to production efficiency, it can be stated that in 1219 in Pelőke village, beside Derecske, when two ploughs of land were confiscated 200 stooks of cereals were taken away, meaning that one plough of land resulted in 100 stooks (200 butts of grain)". MOLNÁR (1949) also dealt with the issue: "The average yield, when using a regulated crop rotation system, could be taken as 2–3 times, in other words, taken an average of 2.5 (and assuming that 1 yoke of land was sown with 125 litres of grain), this meant 312.5 litres per acre, i.e. a total of 6250 litres. In a three-course rotation the average yield in Hungary was 3–4 times the amount of the seeds and thus taken the 3.5 mean value, the amount of grain produced on one acre was 437.5 litres average".

Crop yields were not any higher in other parts of Europe. Valuable data have survived from Carolingian times. According to these, the yield was twice the amount sown in spelt, 1.6 times in rye, and 2.2 times in barley. Would it be possible that village people constantly lived at the verge of famine? According to the statement made by FÜZES (1977), production yields of grains increased from the Neolithic to the mid-19th century not more than 15–20 percent in weight (acceleration percent), which is due to metric increase in grain size (*Figs. 439, 441–447*).

Farming skills accumulated knowledge and tradition through millennia. However, soil fertility is limited. In the prehistoric ages, depleted arable land was abandoned and a new areas waere ploughed. Since the Roman Period, fertility was replenished by fertilisers.

MAKKAI (1968) has shown that in contrast to the slash and burn method practiced earlier on, which depleted soil quite quickly, the heavy plough used in the Árpádian Period conserved the fertility of the soil. "Crop yields were grown by the possibility to grow plants with a higher nutrient content. The typical grain of slash and burn farming, common millet, which prefers freshly cleared land, was pushed back gradually in favour of the more valuable but more demanding wheat, rye and barley".

In this period of the Middle Ages, a two-course rotation system was practiced. One part (campus) remained fallow, the other was ploughed three times using a plough draught by six oxen. Mainly wheat, maslin (wheat and rye), barley and oat were produced. Twice as much autumn sowing grain was grown than spring sown. Sometimes only autumn sown grain crop was produced. Harvesting and threshing were done manually. The expansion of grain crops met the needs of newly settled families perfectly. In a later stage of the Middle Ages, three-course rotation took over. "In a land divided into two fields, the cultivated half was sown with autumn and spring crops separately... In a three course system, one part was always sown with autumn crop, however as spring crops required less space, in certain cases the part reserved for spring crops was also partly (sometimes in whole) sown with autumn crops" (BALASSA op.cit.).

A number of ways were known for the cleaning of threshed grain. First of all, husking and dispersing were employed, but bolting, washing and picking as well (BALASSA (1963–64). After cleaning, grain was stored in pits dug in the ground. Grains were smashed by breaking or milling. Grinding stones are common in settlements and even in grave enclosures. The greater legend of St. Gellért preserved the story of woman singing while grinding.

7.9.2. Plant remains in the Budavár medieval wells

In the course of the exploration and reconstruction of Buda castle over the past fifty years, archaeologists found a number of wells. However, botanical examinations were made only in a few cases. The first archaeobotanical examinations were made at the well of Dísz Square No. 10, led by Imre Holl (*Figs. 329–339, 343*). Seeds from the well, dated to the 14th century, were identified by Zoltán Zsák. A species list was prepared, but the number of seeds for each species was not published (HOLL 1966). Most of the botanical finds were made up of grown and gathered fruits, melons and grape stones. In 1955 Győző Gerő explored a 13th–14th-century well at Úri Street 40 (*Figs. 340–342, 344–347*). Weed seeds were recovered along with grown and gathered plants (P. HARTYÁNYI – NOVÁKI – PATAY 1967–68). In 1966, at Dísz Square No. 8, yet another well with seeds was found, dated by Katalin I. Melis to the 13th–14th centuries based on associated ceramics. The majority of the seeds originated from fruits, garden produce and spices (P. HARTYÁNYI – NOVÁKI 1973–74). Compared to other contemporary sites, an unusually large quantity of garden produce remains were identified in a medieval well filling explored in a cellar of Hunyadi Street 22 in 1971. A wide variety of produced and gathered plants were unearthed from a pot dated to the 13th century surrounded by a burnt layer. Common millet and a number of weeds also occurred (Seeds were indentified by István Skoflek and Mrs. Hortobágyi (SKOFLEK – HORTOBÁGYI 1973). Grape stones were analysed by Géza Facsar (FACSAR 1973). Many of the seeds belong to black wine grape, but sour cherry, hazelnut, sugary melon and watermelon, walnut, apple, rose, bramble, black mulberry, sloe (blackthorn), alkekengid and danewort were also present. In 1975, Katalin H. Gyürky explored waste pits and a privy cut in rocks beside the ruins of a 14th–15th-century house at Hess András Square No. 1. Seeds were processed by Borbála P. Hartyányi (P. HARTYÁNYI 1975–77). Based on seed size, she noted considerable similarity between the identified species (common fig, cherry, sour cherry, melons) and those coming from the Budavár wells earlier on, confirming local production. In addition, imported seeds are also found among them. Sloe (blackthorn) had several varieties. It is possible that they were sold in the marketplace close to the former Lady in Harvest church (today Hess András Square).

Finds from medieval wells explored by Zoltán Bencze in 1998–1999 at the site of the former Military Headquarters are described in detail below (*Figs. 366–381*). It is noted that at the site of the former Teleki palace on the western side of Saint George Square, the floated filling of well No. 8 from the 14th–15th centuries, still under investigation, is about to yield even more seeds and fruit remains (excavation by Dóra B. Nyékhelyi 1999–2000) (*Figs. 382–386*).

Fruit remains probably originated from orchards and vineyards of the Budai mountain range, where they still exist today: apple, pear, peach, walnut, medlar, almond, cherry, cherry plum or myrabolana, sour cherry, garden plum, bullace plum, greengage, service tree or garden service, and quince. This is species rich material even when compared to contemporary fructiculture, clearly indicating versatile consumption and processing habits; such as juices, preserves, wine, brandy, etc., symptomatic of high level horticulture. Morphological analysis of grape stones demonstrates that several varieties were grown. The number of gathered, wild fruit species is also high as are weeds and plants obtained from the natural environment of the time.

A research project, recently intiiated, promises to yield significant results. A great number of diaspores are being recovered from the fill of a 15th-century well in Budapest I., Kapucinusok Street (excavation by András Végh 2000) (*Figs. 387–403*).

Several earth and mud samples were taken for archaeobotanical examinations from two wells explored in the yard of the former Military Headquarters (Dísz Square No. 17) of Buda Castle in 1999. These were stratigraphic samples from a well, code named "I/2." With only one layer dated to 13th–14th centuries (8. layer), and from the fill of a rock-carved well named "27/2." During the explorations, proceeding from top to bottom, the first samples were taken from the brick debris layer (No. 5) dated to the 15th century. No samples were taken from the layer above (No. 4), a strongly mixed, grey, clayey, trash-containing level. The overwhelming majority of samples come from a watery-muddy layer, No. 6. When sampled, upper and lower parts were unnecessarily subdivided as later both parts proved to date to the 14th century.

These well findings are of key importance. They come from the first stage of Buda castle construction, from a period of the emerging and expanding capital of a strong feudal state. At the same time, they bear information concerning growing and eating habits.

Carpological remains (seeds and fruits) from the two Buda castle wells are all direct botanical finds. No indirect plant residues (imprints) occurred among them. These originated from kitchen garbage and other household waste, which were put into the wells either deliberately (well filling), or by accident (wind drift). Thanks to immersion in water, remains of fruits, weeds and contemporary floristic elements were preserved in an excellent state. Water created anaerobic conditions that in turn prevented microbes from decomposing the diaspores.

Most of the seeds and fruits, close to one hundred thousand pieces, come from the 14th-century layer No. 6. Macro remains concentration (the number of seeds per 1 kg mass of unfloated soil) is the largest here. This might be explained by the well filling being much slower at the beginning with only household waste and kitchen garbage being thrown into it (layer No. 6). Later, this became much more rapid and intensive. Remains of food based on plants are fewer and more of the fill comes from construction debris and levelling. The final intention to fill in the well was more decisive. With respect to species inventory, no real difference can be seen between layers No. 5 (15th century) and No. 6 (14th century) , but in the samples taken from layer No. 8, this value has doubled. Most species (82) come from the 13th–14th-century samples of well No. I/2.

In general, cultivated plants, weeds and species of contemporary flora occurred in the well layers with different frequency. While grains were all carbonised, associated weed seeds, fruit remains and reproductive organs of natural vegetation elements did not carbonise. Grains occurred in the well layers with various frequencies. They were all carbonised, meaning kitchen refuse. They were burnt during food preparation (cooking or frying) or some preparatory work that made their further use undesirable. A large part was obviously successfully processed and eaten as the number of grains is very low.

In the 13th–14th-century well No. I/2 samples, common oat. naked (*caryopsis nuda*) and hulled grains (*cum caryopse corticata*), occurred in roughly the same ratio, all in a carbonised state. Some carbonised hulled grains of six-rowed barley (*cum caryopse corticata*) and carbonised grains of common bread wheat are also encountered. Although no common millet grains were found, husking refuse in forms of one thousand husks occurred in a non-carbonised form.

On the other hand, the number of grain species dropped strongly in the layers of the rock-carved well No. "27/2": *rye* was found individually in layer 5 (15th century) and layer No. 6 (14th century) while one hundred fiftyd annaked (husked) carbonised common millet grains were found in layer No 5 and seven thousand five hundred glumellae in layer No. 6 from the 14th century. Naked or "husked" common millet grains swelled as a result of burning, yet their elongated shape can be distinguished. The endospore fell out of the grains. This is evidence of cleaning as the eye often breaks out during threshing. Glumellae coming from the cleaning procedure did not carbonise. This is expected as common millet was a key crop for Hungarians in the Middle Ages. A significant amount was grown traditionally by Hungarians through time until the appearance of maize, the new gruel plant. The sparse occurrence of rye grains can be explained by the fact that it was rarely sown in the Middle Ages as a separate crop, more often in mixture with common wheat. It was grown as a single crop in Northern and Western Europe, as it is less sensitive to cold and dry winters, survives on acidic soils and it is the grain of sandy soils. In other words, it endures conditions that are not suitable for wheat. The protein content of bread grains is high, therefore suitable for making dough. The Hungarian word is of Slavic origin occurring first in a document dated to 1292 as a place name (MOLNÁR 1961). In the Hungarian Diploma Dictionary, a different date can be found: "Roswago" – 1478 (SZAMOTA – ZOLNAI 1902–1906).

It is very likely however that after the conquest, Slavic influence encouraged Hungarians to grow oats, as indicated by its name (SZABÓ L. 1982; BALASSA 1973). Even if this was so, that does not mean that they had not known it previously. According to personal communication by Füzes, oat was found at the Majackoje gorodišče exploration beside the river Don and also much further to the south, in layers coming from the Saltovo-mayak culture. This cultural configuration included Hungarians as well before the conquest. Thus, the word zab (oat) is a superfluous loanword.

According to documentary evidence, the significance of oat grows after the 13th century (GAÁL 1978). It occurs as a person's name in the form of "Zob" in 1211 for the first time. Later on it turns up as "zobozauth," a product in a charter from 1271. Its production might have been significant since the 12th century, as it is noted by Ansberg, a German priest, in connection with his travels in Hungary (1189): King Béla III presented Emperor Frederick I "with boats and wagons burdened with the oat destined for the horses" (SZAMOTA 1891).

Due to its resistance to low temperatures, relatively short vegetation period and extraordinary adaptability, barley was one of the most widespread cereals in the past. Given its undemanding nature, it was grown under arid conditions, on poor soils or even slightly alkali soils, and it used to be one of the most important staple food plants. It has been produced in Europe since the Neolithic and used for making mush, bread, beer, or as fodder.

Weeds occurring together with cereals, to be described later, germinate in autumn and mature at the beginning of summer. Consequently, it is probable that the wheat was sown in autumn

Under climatic conditions prevailing in the Carpathian Basin, excellent quality bread wheat is grown. Hungarian wheat is praised by Pierre Choque, who served Ladislaus II: "Buda township and the castle stands on top of a low mountain, with further mountains on the side, disappearing in a distance, which are very fertile. Here, excellent wine grapes were produced and a kind of bread wheat which I have never seen before…" (HARASZTI – PETHŐ 1963). In connection with this, MÁNDY (1972) notes: "During the Middle Ages, wheat was produced regularly in this country. According to its ancient name: "gilice" wheat, "iglice" bread wheat, most probably this meant the reddish grain (var. erythrospermum) bread wheat variety…"

It could not be determined whether sporadically occurring common wheat and rye grains were grown separately or together. In fact, joint cultivation of the two plants, "abajdoc", was typical for the Middle Ages in Hungary. Wheat and rye were not mixed subsequently, but sown and harvested, milled and used this way. It is also possible that abajdoc was grown for security purposes. This way, even if one of the species would not ripen, the other still might bear a yield. A further advantage is seen in the stalks being more resistant in maslin to being blown over.

In summary, it can be concluded that in the 8th layer of the earliest well No. "I/2" more cereal species, associated with kitchen processing, were found than in other places. This shows the picture of an archaic household striving for self-sufficiency. Grain was rather grown than purchased at the marketplace, processed at home and for this purpose, stored. In a wide variety, all the key grain species were eaten: barley, bread wheat, rye, oat, common millet. In the layer numbered six (14th century), only common millet glumellae (from cleaning) and sporadic rye grains were found. We do not know for certain, but it is possible that bread was no longer baked at home but purchased from the baker. Six-rowed barley was also used for human consumption at this time. The relative proportion of the sparse grains does not suggest any of the cereals in a dominating position.

In layers from the 14th and 15th centuries, the number of species of cereals decreased. Maybe a civic lifestyle, which has become general, did not reckon with storing and using grains at home. Bread was probably bought from bakers. Therefore no more grains were included in household waste.

7.9.3. Husbandry in the areas under Turkish occupation

The grain crop production boom was stopped by the Turks. The country was torn into three parts and the constant warfare did not favour crop production. The size of land left fallow increaased and production yields fluctuated. A number of settlements were abandoned.

The part of the country dominated by the Turks suffered from the imposition of heavy taxes. The census of settlements in smaller administrative units (*náhije*) belonging to the Buda Sandjiak (major administrative centre) between 1546 and 1590 was preserved (KÁLDY-NAGY 1985). Turkish taxes were based – similarly to the Hungarian taxing system of the time – on cereals and vineyards income. In the process, even kitchen produce and orchards were on record. Turkish tax officers distinguished between bostan (field) and bágche (orchard) gardens. Field gardens were situated at riversides, at the edges of towns and villages, but they could also be in the ploughland of the neighbourhood. Mainly cabbage and melons were grown in them (KÁLDY-NAGY 1970). Turkish taxes included wheat tithe, maslin ("mahlut" = mixed crop) tithe, must tithe, pasture benefit, wild cabbage tithe, barrel levy, fruit tithe. These all show that the population – if they were allowed – continued to deal with a variety of crops. Based on the amounts of tithes, the volumes of wine and cereals produced by peasant farms can also be established (VASS 1983). Archaeobotanical findings support the economic data of settlements in the Vác administrative district: Vác (GYULAI 1995b), and Szarvasgede (MOLNÁR 1997) during the Turkish era.

Devastation layers from the Turkish Period are relatively well researched from a botanical point of view (*Figs. 363–365*). Plant remains of the Pogányszentpéteri monastery, destroyed in the 16th century (excavation by Róbert Müller 1967), included many common bread wheat and rye grains, in other word "maslin" (FÜZES 1972). Between 1969 and1974 Imre Holl and Nándor Parádi led the excavation of a village razed during the Turkish era in the 16th century in Sümeg-Sarvaly. Botanical finds from six houses devastated and burned during the Turkish Period and adjacent debris were processed by István SKOFLEK (1984–85) and Borbála P. Hartyányi (in: NOVÁKI 1984–85). Beside grains and seeds of carbonised common bread wheat, rye, common millet and weeds, fruit remains were also encountered.

In Színház Street, which is situated in an area where the Buda castle existed in the Turkish Period (excavation by István Feld 1995) and in Óbuda, Medve street (excavation by András Végh 1995) common bread wheat and six-rowed barley grains were also found, but the numbers indicated that they must have not been very significant. Much more important were common millet and common oat.

Quite often common wheat and rye production was abandoned altogether in settlements dominated by the Turks, and only oat was grown, albeit in a limited manner. All this means spring crops, that is, less work and safer yields. This was obviously encouraged by the fact that autumn sowing cereals had a higher tax levied on them. The same conclusion could be drawn from examination of grain residues from the Turkish Period found in downtown Vác (GYULAI 1995b). Regression of crop production stopped only after the Turks were driven out in the 18th century.

The examinations carried out by Bálint ZÓLYOMI (1980) at Lake Balaton also confirm the decline of grain production during the Turkish occupation. This can be explained by a reduced amount of ploughland. As an opposite process, intertilled cultures started to develop since the 17th century included maize. An increase of goosefoot (*Chenopodiaceae*) pollen indicates expansion of intertilled crops as well.

During excavation of the Pilisszentkereszt medieval monastery, pollen analytic tests were carried out (ZÓLYOMI – PRÉCSÉNYI 1985). As this is so far the only pollen analysis connected to an archaeological exploration project, and the operation of the monastery can be put to the period between the end of the 12th century and the middle of the 16th century, a number of important conclusions can be drawn with respect of the life in that period. Non-Arboreal Pollen (NAP) numbers are high. Goosefoot (*Chenopodiaceae*) and wormwood (*Artemisia* spec.) ratios are high. Gardens and orchards are suggested by a high frequnecy of walnut pollen, which is necessarily connected to an advanced level of fructiculture. Low grain pollen numbers can be explained by reduced ploughland areas. Expansive forests must have existed in the neighbourhood. The lack of birch pollen suggests that these forests were not cleared.

7.9.4. Kitchen produce and spices

In order to diversify their diet, late medieval people grew kitchen produce and spices. Common lentil and pea were the most common vegetables. The seeds of these legumes were found in the Lászlófalva-Szentkirály late medieval Cuman settlements, just as in the finds of Muhi, a deserted medieval village. In Lászlófalva-Szentkirály cucumber was also encountered. Garden parsnip can be used both as a root plant and as a spice. Opium poppy seeds probably come from locally grown plants. Fibre plants were hemp and flax. (Flax seeds can be prepared as a dish.) Many hemp seeds were found. Written documents are confirmed by the archaeobotanical finds.

According to Hungarian medieval seed finds, watermelon (*Citrullus lanatus*) and muskmelon, cantaloupe were especially favoured garden produces (P. HARTYÁNYI – NOVÁKI 1975). Melon is mentioned for the first time by the Besztercei-dictionary as: “gereg dyne”. Up to the 16th century, the name of the melon did not differ from that of the cucumber and the two kinds of melons were also distinguished later. Balázs Szikszai Fabriczius in his Nomenclatura (1561) also refers to them in this manner. Delicious kinds of melons must have been produced in the territory of the Hungarian Kingdom. János Lippay (1664) maybe refers to this when he mentions that King Albert I died in 1439 as a consequence of the extreme amount of melons eaten. Melons grown in the Middle Ages were mostly yellow on the inside. Red melons spread only after the Turks were gone. Allegedly, up to the 18th century, two thirds of all the melons grown in this country were yellow inside.

Seeds of both species occur in Szécsény (TORMA 1994). They were present in Lászlófalva-Szentkirály as well. Both sugary and water melon seeds were conserved in excellent condition and large number in wells explored in Budavár: Dísz Square No. 10 (15th century) (excavation by Imre Holl 1954), Úri Street 40 (13th–14th centuries) (Győző Gerő 1955), Dísz Square No. 8 (Katalin Melis 1966) (HARTYÁNYI – NOVÁKI – PATAY 1967–68; P. HARTYÁNYI – NOVÁKI 1973–74).

In the 13th–15th-century cisterns and wells explored at the Teleki palace in Buda castle (well I/2. and 27/2. rock carved well mentioned earlier when describing grains) (excavated by Dóra B. Nyékhelyi 1998–99) seeds and fruits of all key products, mentioned frequently in medieval cook-books were found: common dill (*Anethum graveolens*), wild celery (Apium graveolens), wild cabbage (*Brassica oleracea*), turnip/naven (*Brassica rapa* var. *rapa/ B. r.* var. *oleifera*), hemp (*Cannabis sativa*), muskmelon, cantaloupe, watermelon, common hop (*Humulus lupulus*), grass pea wine (*Lathyrus sativus*), big seed lentil (*Lens culinaris* subsp. *macrosperma*), common flax, and opium poppy. Based on the shape and form of seeds, several cultivars of sugary and watermelon were produced. The same place was the site where an important and formerly very expensive import spice, black pepper (*Peper nigrum*) seeds, were found. The very diverse supply of vegetables and herbs indicates an advanced level of garden produce culture.

More than the usual amount of spices, cereals and garden produce species associated with everyday meals and kitchen processing were found in the earliest layer (13th–14th centuries) of the wells at the former Military Headquarters in Buda Castle. Interestingly, no garden legume was encountered in any of the layers. According to the lessons learned from plant remains at other sites, consumption of legumes (pea, lentil) was part of everyday life in this age. Maybe a special eating habit lies behind this. At any rate, the image of an archaic household striving to self-sufficiency emerges from the finds.

Most of the fibre plants were present here. Flax and hemp refer to self-sufficiency in themselves. It cannot be said if opium poppy was used as an oil plant or for flavouring dishes.

Cultivated and domesticated plant remains from layers dated to the 14th and 15th centuries suggest that the species number did not only decline with regard to cereals and their associated weeds, but also in relation to garden produces, oil and fibre plants.

It can be imagined that common fig came from locally growing bushes, although, similarly to pomegranate, its importation is also an option. Black pepper represents a special role among the finds as it was very rare and considered to be exotic, consequently expensive in the Middle Ages. On the basis of sugary and water melon seeds found in all the layers it can be concluded that these must have been a favourite delicacy. Based on the shape and form of the seeds, several cultivars (varieties) were grown, which means different origin at the same time.

Of garden produce, wild celery (*Apium graveolens*) achenes are found only in the 15th-century layer No. 5. It is noted here that garden celery (*Apium graveolens* convar. *rapaceum*) has its primary gene centre from the Caucasus to the Mediterranean. In ancient times, only the wild form was eaten. Very little is known of the history of its production. Wild celery was probably grown in Hungary first in Monostorapáti (Hajdúság). It can be assumed that it has been continuously grown here since the 15th century. Beside this site, no other well contained any of its remains.

Watermelon (*Citrullus lanatus)* seeds came from the earliest layer *No.* 8 (13th–14th centuries), as well as from layer No. 6 (14th century) in great abundance. Non-carbonised seeds are very similar, yet distinct varieties with slightly smaller and greater seeds, respectively, can be distinguished among them. (P. HARTYÁNYI (1975–77) reached a similar conclusion earlier, when comparing watermelon seed sizes found in Buda castle.) Watermelon production was started in Hungary in the Middle Ages. As the name shows, it might have resulted from Byzantine influence.

Similar statements can be made with regard to the conditions under which muskmelon, cantaloupe (*Cucumus melo*) was found. The nearly one thousand non- carbonised seeds from layer 8 can be divided into two distinct types: one is shorter and broader, and the other is longer and slimmer. (The majority of the seeds were put into the latter group.) This indicates at least two varieties and consequently more than one place of origin. Compared with recent melon seeds the first group shows similarities with the Turkesthan ("Togo") varieties, while the latter are closer to the rock-melons and pineapple melons of today. Turkesthan melons are smooth, non-segmented, their skin is thin and tanned. Pineapple melons are slightly segmented, their skin is thin and smooth. Rock melons with their ribbed and rough skin come from Italy and appeared possibly as early as in the Middle Ages in Hungary. (Cantaloupe is a general name for early medieval varieties with Italian origin.) Muskmelon, cantaloupe is a plant cultivated much earlier than water melons. The earliest Hungarian occurrence of the name melon is in an 11th-century charter: "predium quod vocatur dinna" (SZAMOTA – ZOLNAI ibid.) Its seeds were found several times in the fillcccccc of the 13th–14th-century wells at the excavations carried out in the Buda castle area.

Cucumber (*Cucumis sativus*) seeds were found only in layer No. 8, albeit half a hundred of them. Due to the great similarity, thin and elongated cucumber seeds are very difficult to distinguish from sugary melon, but the work was facilitated by the fact that all of the seeds were found in good condition. The earliest occurrence of cucumber is known from Zalavár at a Late Migration Period site (9th century) (GYULAI 1998b). It cannot be said which species The Cucumis seeds found in Bronze Age Szilhalom could not be identified to species (P. HARTYÁNYI – NOVÁKI – PATAY 1967–68). The earliest mention from Hungary comes from 1405 (RAPAICS 1940).

Both wells contained remains of spices. Common dill (*Anethum graveolens*) was a favourite spice in medieval Buda occurring in the earliest layer No. 8. Extensive use is indicated by three hundred achenes found in layer 6. It was identified at other sites in Budavár, such as the well under Dísz Square No. 8 (P. HARTYÁNYI – NOVÁKI 1973–74).

The few common hops (*Humulus lupulus*) *achenes* found in layer 8 might have been "beer spices" from brewing production. No riverside groves or wet forests could be found on Buda Castle Hill, yet under natural conditions these are habitats suitable for hops. Common hops from the Middle Ages is also common in German town wells (c.f. WIETHOLD 1995).

Wild marjoram (Origanum vulgare) achene found in layer 8 might have come from local production as well. It was a kitchen spice and did not get into the well from a forested, shrubby region. Wells were rich in fibre and oil plants as well. Common flax (*Linum usitatissimum*) seed found in layer 8 was a plant grown for fibre and oil. Opium poppy seeds (*Papaver somniferum*) came only sporadically in the earliest layer 8. However, five hundred seeds from layer 5 show its popularity. As for its use, one can only say that no imprints can be seen on the intact seeds.

Several dozens of hemp (*Cannabis sativa*) achenes were found in layers 8 and 6, and it cannot be said if they were grown, spontaneously introduced or weeds. Provided they were grown, the fibre might have been used for making ropes, strings and accessories such as belts or cords. Hemp is a Eurasian (Mediterranean) floral element with flatland-mountain distribution which survives on sandy, clayey, loess or turf soils alike. It is a dioecious nitrophil species with three-metre-long stalks. It bears fruit only in intertilled cultures or as an aftercrop of cereals following harvest (HUNYADI 1988).

In both Buda castle wells, remains of plants from import were found. The great abundance of fig *(Ficus carica)* was classified as an imported crop with certain reservations, as it might well be that the seeds came from fruits grown locally on the Mediterranean climate southern slopes of the hills around Buda. Its seeds were found in all three layers. Most of them, about three thousand, come from layer 6, more than five hundred from layer 8, and the least amount, half a hundred from layer 5. *Common fig* is of Mediterranean origin. It was a favourite, planted decorative and domesticated crop in the Middle Ages. Its area spread far up to the north. Through a possible Italian connection, a number of fig shrubs arrived in Hungary. The name was first mentioned at the beginning of the 14th century in the Schlägli dictionary as a tropical fruit (JESZENSZKY – KÁRPÁTI 1963). They argue that the fig shrubs on the Gellérthegy are several hundred years old and are of Italian origin. It is probable that the great number of common fig seeds coming from several wells of medieval Buda also come from local production and are not imported.

As a result of continuous connections with the Mediterranean, the decorative fruit pomegranate (*Punica granatum*) also frequently occurred on the table of wealthier lords and consequently a dozen seeds were found in layer 6. (According to the opinion of Dezső Surányi, the seeds, due to the sub Mediterranean climate of the period, might easily have come from local production.)

A real specialty is black pepper (*Piper nigrum*), found in each of the three layers in the form of a few round, striated, sometimes rived blackish stones. Seeds of black pepper, an expensive exotic spice in the Middle Ages, were found earlier in a few medieval sites in Germany (WIETHOLD 1995). All this confirms that these seeds came from kitchen waste.

Turkish landlord taxes, collected in the part of the country under Turkish occupation, were as follows: wheat tithe, maslin ("mahlut" = mixed crop) tithe, must tithe, pasture benefit, wild cabbage tithe, barrel levy, and fruit tithe. The population – albeit suffering from the taxes – continued agricultural production. In the 16th century, tithe censuses mention lentil, pea, flax, buckwheat, cabbage, beetroot, onion and garlic, also fruits (apple, pear, grape).

Cereals (common bread wheat, rye, common millet) and kitchen produce (flax, vetchling, horsebean, common lentil, garlic) mentioned in the tithe census were found without exception in the houses of Nagyvázsony-Csepely dated to the 15th–17th centuries (Júlia KOVALOVSZKY 1957–58). Carbonised rye grains found here can be divided into two groups by shape, which in our view is an evidence for the early use of multiple varieties (HARTYÁNYI – NOVÁKI – PATAY 1967–68).

The gem of garden culture were melons. The Turks assisted the Hungarian population in acquiring skills necessary for growing melons. Turkish-Hungarian correspondence by Hassan, Aga of Buják in his letter written in 1670 to István Koháry, Chief Captain of Szécsény, includes references to two beautiful "körg" (Greek) melons added as a sign of his appreciation (PÁLMÁNY 1984). The value of the melons was emphasised by the fact that when fleeing impending danger, the population took melon seeds as part of their valuables (TAKÁCS 1917).

Other cultivated plants also surfaced from waste pits of the Turkish era: cucumber (Budapest, I. district Színház Street, Budapest Medve Street) opium poppy and garden parsnip (Budaújlak). Although one could wonder about the reason for its appearance, buckwheat (*Fagopyrum esculentum*), of basically Central Asian origin, is known so far only from Szentgotthárd (TORMA 1996b). It has never been significant in Hungary in spite of its high nutrient value and medical properties as it contains depressor and arteriosclerosis solving rutin and other flavono-glycosids. It was rather eaten in regions populated by Germans or Slavs as gruel or the flour was used for baking pastry, cookies, or the seed crop was given to animals as fodder and the straw as bedding. However, feeding it as a single fodder leads to fagopyrism (rashes appear on the skin of achromatic animals) (KÁRPÁTI – BÁNYAI 1980). It was introduced in Central Europe by the Mongolian and Turkish peoples in the late Middle Ages. Beside mentions in charters and documents, achenes were found in the 13th–14th-century layer of Opava, Bohemia (SCHULTZE-MOTEL 1972).

I would like to mention here a research project that has just started, yet it already has promising results. A great number of diaspores, more than ever before, were found in the year 2000 in the fill of a 15th-century well in Budapest Teleki Palace. Several centuries of water cover preserved the seeds and fruits in excellent condition; so much so, that their condition was tested using biotechnological methods. Genetic material of common millet, sugary and water melon was successfully extracted from remains recovered from the well in the yard of the Teleki palace (Gyulai G., Törjék O., Heszky L. – Gyulai F. unpublished; and I. Nagy personal communications).

Similar result were obtained through examination of melon seeds found in a 15th-century well explored in 2000 in Budapest I district, Kapucinusok Street (GYULAI *et al.* 2001). The research has two distinct lines: plant regeneration and DNA isolation. Unfortunately, plant regeneration experiments through germinating tissue development were not successful so far due to endogen mycoplasma and phytoplasma infestation. PCR reactions (DNA polimerase chain reaction, in other words the propagation of DNA segments) however presented well reproducible DNA stocks that can be used for species analysis. Future molecular experiments might demonstrate the origin of the plants mentioned and their genetic relations with contemporary varieties.

7.9.5. Late medieval horticulture in Hungary

A significant expansion of fruit species occurs in the Late Middle Ages compared to previous periods (SURÁNYI 1985). Fruit trees grew first in protected places such as monastery gardens, or between external and internal walls of towns (WILLERDING 1983b). However, since the 14th century, conscious fructiculture can be assumed in monastery gardens, in castles, in fenced gardens surrounding town walls and manors. According to the archaeobotanical findings, many species were grown.

According to botanical finds unearthed in excavations of German, Swiss and Polish towns, an advanced level and nearly homogeneous horticulture was present. In addition to growing of domesticated fruit trees, wild fruit was also deliberately gathered. Likewise, this pattern occurred in medieval Hungary (P. ERMÉNYI 1975–77). HAJNALOVÁ (2001) arrived at a similar conclusion when she examined fruit remains at Slovak sites.

The clergy played an essential role in creating orchards and vineyards throughout central Europe, and likewise in establishing Hungarian horticulture. Thanks to the operation of Benedictine and Pauline orders, the church and secular orchards blossomed both in terms of species and area. The orchard (*pomarium*) was always established close to human dwellings, while vineyards were located in a more distant place, e.g., on the vineyard hill, a slope having good and fertile soil (*promontorium*). Ploughlands were at a distance. This division still can be found today in the life of the hamlets in the Great Plain.

Peasant orchards were mainly established by the Hungarian kingdom. As early as the 14th century, wine-growers and horticulturists appeared. Their growing numbers were facilitated during the rule of Louis the Great when they could sell their products in well-established market places. In contrast to real estate, fruit and wine was readily marketable. Citizens of country towns could also possess such areas. Vineyards were planted with fruit trees as well such as stone fruits, apple, pear and service tree.

Since the 15th century, the demand for fruits has increased and fruit production became a profitable business. Beside fresh fruit, dried fruits were also in demand. Inhabitants of country towns, like in the vineyard hills, created common orchards. Fruit trees and vine-plants were planted irregularly without any arrangement. Sowing of seeds and propagation by slipping were the simplest and preferred methods. Parallel to the peasant economy, manorial farms were also formed, which depended on peasant labour (FRISNYÁK 1990). After the 16th century, stately gardens were established one after the other (SURÁNYI 1985–86).

Fruit remains from different parts of the country indicate that fructiculture was gradually expanded and, similarly to crop production, became increasingly uniform in terms of species grown as verified by many finds of seed and fruit remains. On the 15th–17th-century housing estate of Nagyvázsony-Csepely, at the excavation site of Júlia Kovalovszky between 1957 and1958, dried apple fruits (sun-dried apples?) and black wine grape stones were found (HARTYÁNYI – NOVÁKI – PATAY 1967–68). During the exploration of the former Bátaszék Abbey (excavation by Ilona Walter, Attila Gaál, Tamás Pusztai 1994), a carbonised apple was found in the moat.

In 1992, during the excavation by Andrea Pölös, a large amount of grape and peach stones were found in the filling of a waste pit from the Turkish Period in Budaújlak, Bécsi Road. In 1966, at the site explored by Katalin Melis in Buda Castle, the 13th–14th-century well filling yielded an almond shell (P. HARTYÁNYI – NOVÁKI 1973–74).

Based on excavated findings, the late medieval population in Vác knew and grew walnut, peach and the typical plums of Middle Ages, myrobalan plums. In the late medieval layer of Budapest I. Színház Street, peach stones and common pear seeds were found. The fruit remains coming from the Lászlófalva-Szentkirály late medieval (15th–16th centuries) Cuman settlement indicate high level and diverse fructi- and viticulture. All major fruit types known in the era were found here: walnut, apricot, sweet cherry, sour cherry, garden plum, bullace plum, peach, and black wine grapes. The 15th-century plum pits and grape stones found in 1998 at Szarvasgede are also the reminders of an expanding horticulture.

It is known from written and iconographic sources that medieval Hungary was rich in native (endemic) fruit varieties (GYULAI 1999b). Carpological finds from recently found medieval wells seem to verify the sources perfectly. Due to continuous water cover of deposits in wells, more and more fruit remains are found. However, it would be too early to conclude as to fruit varieties actually grown, although personal communication by Dezső Surányi indicated such analysis is underway.

During the excavations made in the parish-church in Szécsény (by Katalin F. Bodnár 1988–93) several pits, dated to the second half of the 16th century and the first half of the 17th century were explored. Remains from all fruit types typical for late medieval Hungary appeared here (apple, common pear, cherry, sour cherry, walnut, plums, and grape) (TORMA 1994). The find is especially rich in plums: myrobalan plum (*Prunus cerasifera*), *domestica* subsp. *oeconomica*). According to comparative studies, many of them are like Dobzó plum which is identical to the medieval mud plum variety and the yellow plum, mentioned by Balázs SZIKSZAI FABRICZIUS (1590), respectively.

During the 1990s, an environmental archaeological excavation was made in Visegrád, where a former royal palace is situated. The excavation was preceded by historical-ecological research and a horticultural-historical study (UBRIZSY-SAVOIA – GYULAI 1993). It is known from contemporary written sources that there were several gardens in the Visegrád Palace in the time of King Matthias: within the northern walls, a terraced decorative garden with lime trees and buxus, and outside the walls, to the north a vineyard and an orchard (Bonfini, Aelius Lampridius Cervinus, Gaspar Ursinus Velius, Miklós Oláh, in: BALOGH 1966). In order to locate the latter, excavations were carried out under the leadership of András Pálóczi Horváth in 1993–2000 (PÁLÓCZI HORVÁTH 1999). A map edited using ground penetrating radar assisted in the research. As a result, it became obvious that the garden was built at the end of the 14th and the beginning of the 15th centuries. A fountain dated to the Sigismund-age, found during the excavations, designed to stand in the middle axis of the whole garden, must have had a significant role.

During the work designed to clarify the history of the garden, several garden wells were identified which contained a number of implements connected to garden work (sticks, hoe, water jug, etc.). Yet, accurate reconstruction of the garden and its contemporary environment is best derived from analysis of the large amount of seeds floated from well mud (PÁLÓCZI HORVÁTH – TORMA 1999). During the reign of Louis the Great and later King Sigismund, remains of many different fruits (apple, pear, cherry, sour cherry, apricot, peach, plums, walnut, hazelnut, muskmelon, cantaloupe and grapes) made their way into the wells where the wet mud conserved them in an excellent state. The number of plant species diaspores coming from the contemporary natural environment is also relatively high.

At the rescue excavation made by Gábor Ilon in 1995–1996 in Pápa, Deák Ferenc Street, remains of a plank built in the 16th century were found. In the samples taken from the straw filling material of the beam level of the plank, seeds and fruit remains were identified associated with twig pieces (Torma 1996a). The continuously wet environment preserved them in an excellent, so-called subfossil state.

A great part of the remains are cultivated plants (walnut, apple, pear, plums, grape), but wild fruits (hazelnut, wild forest strawberry and so on) are also found in a considerable number.

Never before were so many fruit remains found at a Hungarian excavation as in the recent Buda castle wells: Teleki palace (Dorottya B. Nyékhelyi 1998–99), former Military Headquarters (Zoltán Bencze 1988–99). More than two hundred thousand seeds and fruit remains were identified from 13th–15th-century wells and cisterns.

Most probably, the fruit remains recovered by flotation from a very considerable volume of earthen fill come from the orchards and vineyards still extant today in the Buda hills. This material, which is rich in species even when compared to recent fructiculture, infers a diverse consumption and processing tradition (such as juice, preserves, wine, brandy, and so on), referring necessarily to a high level horticulture.

In the two Buda castle wells examined by us (well I/2. and the rock carved 27/2 well) fruit remains were found indicating a diverse consumption and processing tradition indicative as a whole to a high level horticulture. Remains of all major fruit species known in the age were found here. The material is especially rich in plums: myrobalan plum, bullace plum, garden plums. Sloe (blackthorn) plums, just as sloe (blackthorn) had several varieties. The finds in the earliest layer No. 8 of the I/2 well (13th–14th centuries): apple, pear, cherry, sour cherry and plums were present along with further species such as apricot, almond, walnut and medlar. In other words, the amount of fruit purchased at the marketplace had grown. This correlates with an increase of the population.

Fruit was mainly consumed fresh. No burnt remains were found. The diversity of domesticated and wild fruit or fruit-like species is surprisingly high. A lot more species were eaten than today! Resistant, indigestible parts of the fruits were discarded and scattered. Most fruits were mostly harvested or collected in autumn. It can be taken for granted that these were purchased at the market. A large area would be required for the production of such a great number and different kinds of fruits. The surroundings of Buda were filled at this time with orchards and vineyards. People gathering wild fruits knew the local forests, meadows and grasslands well. They knew what and when to collect. Fruits could not only be eaten fresh but processed into lictarium, juice, wine, and maybe also brandy. Certainly, these processes leave no traces on the seeds. Some of the plants found here have medical properties as well. It cannot be said if they were used for this purpose, yet they attribute a medical knowledge to the population in an indirect way. However, the distribution of fruit species is uneven. Their number and species diversity in the earliest layer No. 8 is less than in the more recent layer 6.

The sweet cherry (*Prunus avium*) pits found in all three layers but mostly in layer 6 (27/2 well, 14th century) in great abundance resemble *Prunus avium* L. convar. *duracina* in their shapes, that is, crispy cherries. The fruit of this plant is larger than that of conv. *juliana* (L.) Janchen (heart-cherry) – though such are found as well. It has hard pulp, blackish colour and is crispy. It ripens around June (Soó 1966; Pór – Faluba 1982). Wild cherry (mazzard cherry), wild sour cherry, mahaleb and dwarf cherry can be considered native in the Carpathian Basin (Soó 1970). These species spread through forests west along the Danube between 6500–2500 BC in conjunction with the mixed-oak woods (P. Erményi 1975–77).

It can be assumed that the domesticated black and soft heart-cherries and the equally domesticated crispy cherries were formed by mutation from the fruit of the hardly 10 mm diameter wild cherry. (The domestication of the wild cherry can mainly be seen by the increase of pulp mass and volume.) According to rich finds from the Middle Ages, sweet cherry and sour cherry were important in feeding the population. Dishes could be made of cherries without expensive sweeteners, which however was required in the case of sour cherries. Wine and brandy was made out of them (Willerding 1978). Folk medicine also took advantage of them. Herbal tincture was cooked from the dried stalks of sour cherry and sweet cherry. The reduced extract is widely known as an expectorant, diuretic, heart-stimulating and obesity-preventive drug (RÁPÓTI – ROMVÁRY 1983).

In addition to the deliberate growing of sweet cherry and sour cherry, the fruits of escaped varieties and wild fruits were probably also gathered. (Both sweet cherry and sour cherry grow readily in nutrient rich alluvial soils in floodplains without any human intervention.)

The monastic orders, especially Cistercian and Benedictine orders, substantially contributed to the spread of sweet cherry and fruit species in general. Fruit trees grew first in protected places, such as monastery gardens, and between external and internal walls of towns (WILLERDING 1983b, 1986a). It is a frequent find in the medieval Buda castle wells.

Sweet cherry (cseresznye) is a word of Slavic origin (KNIEZSA 1955). It occurs in several 13th-century documents in a written form: "Cheresna" (1256), "ad arborem Cheresne" (1265), "Venit ad arborem Cheresna" (1268) (SZAMOTA – ZOLNAI ibid.).

Sour cherry (*Prunus cerasus* L.) occurs in all the layers in a quantity similar to that of the sweet cherry. However, compared to layer 8 (13th–14th centuries), layer 6 (14th century) contained one hundred times more. The subfossil pits mostly resemble *Prunus cerasus* L. subsp. *acida* (Dum.) Dostál (wild sour cherry) pits. The wild sour cherry is a little shrub-like tree. Its fruit is 12–20 mm in diameter, roughly round, blackish-brown coloured, sour. It ripens in mid-June (SOÓ – KÁRPÁTI op.cit.; SCHERMANN 1966).

Domesticated sour cherry varieties were created using the verified influence of ground cherry and dwarf cherry. Its distribution area is identical to that of sweet cherry. Today's sour cherry cultivars are considered a hydrid of the two species (PÓR – FALUBA 1982). Sour cherry was domesticated much later than cherries. It does not appear in Central and Eastern European finds before the Early Middle Ages. The earliest occurrence of sour cherry in Central Europe is in the 7th–9th-century settlements of Mikulčice, Bohemia (OPRAVIL 1966). It is a frequent find in the medieval wells unearthed at Buda Castle.

According to various sources, vineyards were used for growing fruits as well (P. ERMÉNYI 1975–77). Orchards were protected by hedges made of thorny plants, wild sour cherries and sloe (blackthorn) (LIPPAY 1664).

The word for sour cherry (meggy) is of Finno-Ugric origin (LAKÓ 1971). The original meaning suggests that it was the edible fruit of some kind of shrub, maybe dwarf sour cherry along the river Volga (MOHÁCSY – MALIGA 1959). Sour cherry appears in documents around the same time as the sweet one: "caput vallis Medyuozou vocate" (1296). In 1544 there is a sentence written in Hungarian: "vöttem uramnak megget" (I bought sour cherries for my master). The word "meggy" was a favourite name for places: "Mediesteluk" (1233), "Medyesmal" (1275), "Mydies" (1297), "Medyesmezeu" (1304) (SZAMOTA – ZOLNAI op.cit.).

Although in small numbers, myrobalan plum (Myrobalana) (*Prunus cerasifera*) was also found in the layers. However, myrobalan plum is not to be understood as the red leaved shrub but rather a green leaved larger bush that is still found in Hungary at many places. The fruit is yellow or yellowish red. The pits are not "S" shaped but almost symmetric, coned as both ends. Pits of the same species were found earlier in the medieval Buda Castle area (P. HARTYÁNYI 1975–77).

Plums were also found in great abundance in each of the layers. However, the ratio of different species fluctuated wildly. Garden plum (*Prunus domestica* L. subsp. *domestica*) occurred in almost identical amounts in both ages (half a hundred). There were more bullace plums (*Prunus domestica* subsp. *insititia*). In contrast to layer 8, half a thousand were counted in layer 6.

Bullace plum pits can be divided into three varieties by shape (c.f. KNÖRZER 1987):

- *Prunus domestica* subsp. *insititia* var. *Juliana* – Bullace plum. Relatively small and flat pits, with sharp piked tips on both ends. Most of the pits are these.

- *Prunus domestica* subsp. *insititia* var. *oxicarpa* – Bullace plum. Large, oval and flat pits, symmetric and evenly piked tips. The number of pits in this group is half as many.
- *Prunus domestica* subsp. *insititia Jusl.* var. *subrotunda – Bullace plum.* Large and flat, almost round pits, hunched toward the ventral edges, therefore symmetric, with rounded tips at both ends. The smallest amount was classified here.

The difficulties surrounding bullace plums can be illustrated by the statement of PÉNZES (1950): "the exact description of bullace plum... is still uncertain today, the Linneaean description concerns only the flower, while on the other hand, the species might have quite different fruits. Usually it is also of the ancient type: white, roundish petals, hanging stalks, round fruit, somewhat elongated, egg-shaped pits, with irregular depressions on the surface; soft, sweet pulp... *Prunus insititia* appears in Hungary at many places around old arable land in forests, such as the Buda hills, and it must be much older than Prunus domestica..."

The wild (escaped) recent sloe and blackthorn plums of South-Western Germany are divided by KÖRBER-GROHNE (1983) into two distinct varieties: one has a fruit with a diameter of 18–20 mm, yellowish-green colour, tastes sweet-sour, and the other is around 20 mm in diameter, and its colour and taste are more like that of sloe and blackthorn. Both kinds have an "S" shaped pit. According to his opinion, the ancestors of these were introduced from the Caspian Sea region during the Neolithic.

Plum (szilva) is a word of Slavic origin (KNIEZSA 1955). The earliest mention in documents: "Predium Sciluas"(1231), "Lewstahsylvaia" (1263–1326), "In loco Zylvavelg" (1314), "Silwafa" (1337), "Kukynzylwa (1334) (SZAMOTA – ZOLNAI op.cit.).

Plum must have become known, widespread and preferred in the Middle Ages as a result of the selection work started in the Neolithic. Garden plums (*Prunus domestica* subsp. *oeconomica*), bullace plum (*P. domestica* subsp. *insititia*), myrobalan plum or myrabolan (*P. cerasifera*) and greengage (*P. domestica* subsp. *italica*) have uncountable numbers of different local varieties, confirmed by plum finds in Germany, Bohemia and Poland. Processing of the Buda Castle wells, previously and more recently indicates that Hungarian plum growing skills were also at an advanced level.

In the excavation of Szécsény 16th–17th-century settlement all the above-mentioned plums subspecies nutlets were encountered (TORMA 1994). Perhaps the most famous Hungarian fruit in the Middle Ages was Besztercei or Hungarian plum. Many 14th–17th-century Hungarian sites (Gyöngyöspata, Kereki-Fehérkő, Hollókő and Szécsény) contained its stone-fruit (P. HARTYÁNYI – NOVÁKI – PATAY 1967–68; TORMA 1994).

A pple (*Malus* spec.) was one of the favourite fruits of those living in medieval Buda. It was found in each of the layers and the number of remains exceeds one thousand. A longitudinal streakiness can clearly be seen on the brownish black seeds and their fragments. A number of core fragments were also found (pericarpium) which appear in the field of vision of the microscope as a shiny structure, yellow on one side. It is assumed that these mainly come from domesticated fruits.

It is not known whether the apples from which seeds were found in Buda Castle (13th–15th centuries.) three times were domesticated or wild (P. HARTYÁNYI – NOVÁKI – PATAY 1967–68; P. HARTYÁNYI – NOVÁKI 1975; P. HARTYÁNYI 1975–77), but they were found here is well. At any rate, the number of seeds suggests that contemporary people were fond of this fruit.

Apple has one of the largest growing areas among fruits. Crab apple (*Malus silvestris*) can be regarded as native in the whole of Europe with the exception of northern Scandinavia. Crab apple is often found in dry and hornbeam oak forests, elm-ash clearings, pastures – here as a hold-over tree. It is a flatland-mountain European (-Mediterranean) species: *Querco-Fagea* subsp. *desyphylla Carpinion, Alno-Padion* (SOÓ – JÁVORKA 1951; SOÓ – KÁRPÁTI 1968; SOÓ 1980). Remains pertaining to the consumption of crab apple can be found from the Middle Ages (13th–14th centuries).

According to documentary evidence, cultivars in addition to crab apple started to dominate in an ever-widening area. The 9th-century work entitled Capitulare de villis lists sweet and sour, winter and summer apples. WILLERDING (1983a) concludes from the analysis of 15th century charters that several varieties of apples existed in Germany during the Middle Ages.

Apple (*Malus domestica*) was a much favored and eaten fruit in the Middle Ages. Apple seeds are particularly frequent in the medieval layers of German towns. It has not been determined if these seed are from locally grown cultivars or from wild varieties. Malus domestica finds are known in Hungary from Nagyvázsony, Kereki and Hollókő castles (P. HARTYÁNYI – NOVÁKI – PATAY 1967–68; TORMA 1994).

The word apple (alma) is Bulgarian-Turkish (LIGETI 1986). A charter from 1093 mentions “apples” (SZAMOTA – ZOLNAI op.cit.). Crab apple frequently occurs in land descriptions: “arborem mali” (1217), “inde currit ad almafa” (1225), “arborem pomi Pirosalmafa” (1422) (in: SURÁNYI 1985).

RAPAICS (1943) writes. “Medieval documents preserved the names of a number of fruit varieties… these were primitive cultivars, which can be seen from the adverbs like gypsy, jaw and wild… Pear: ripening with the barley, honey, mud, red, winter. Mud pear must have been a yellow coloured pear, and winter pear was one of the old winter pears. Apple: red, willow, winter. Red apple in those times was Simonffy red, which is also known as blue apple or gypsy apple, and Kenézi red, which is also an ancient cultivar. Plums: bullace plum, mud plum. The medievals knew only sloe (blackthorn) plums, blue and yellow, the latter later named dobzó plums. (NB: the garden plum was created by the crossing of sloe (blackthorn)- and myrobalan plum through amfidiploidia.) Sour cherry: wild sour cherry. Beside these, the special Hungarian walnut cultivars and – provided they can be distinguished – Hungarian service tree berries also originated in the Middle Ages” –

Pear (*Pyrus* spec.) appears in a more modest amount when compared to apples, yet its seeds were found in each of the layers. Similarly to apples, it cannot be determined if the dark brown, thin and elongated pear seeds come from cultivated or wild fruits. Beside domesticated pear varieties, a number of wild forms (*P. commumis* var. *achras*, *P. amygdaliformis*, *P. salicifolia*, *P. elaeagrifolia*, *P. nivalis*) are also edible.

According to the evidence given by iconographic, written and archaeobotanical sources, pear was grown in large parts of Europe in the later stages of the Middle Ages. *P. domestica* finds are known from the following places: Leewarden (ZEIST – NEEF 1983) and Voorst (13th century) in the Netherlands (Vilsteren 1984), Münsterhof (12th–13th centuries.) beside Zurich, Switzerland (JACQUAT – PAWLIK – SCHOCH 1982), Saint-Pierre-Lentin (16th century), France (PETIT 1983), Lübeck (PAAP 1984), Duisburg (KNÖRZER 1983), Heidelberg and Ladenburg (15th–16th centuries) (MAIER 1983), Göttingen (WILLERDING 1984), Höxter (WILLERDING 1986b) Ahren beside the Rhein in Germany (KNÖRZER 1984), Olomouc (14th and 16th–17th centuries) in Slovakia and Bohemia (OPRAVIL 1984), Slapanice (9th century) (KÜHN 1979) beside Brno, Most (CULIKOVA 1981) and Tábor (16th century) (OPRAVIL 1985), Crakow (9th–15th century) in Poland (WASYLIKOWA 1973), Worcester in England (15th century) (GRIEG 1981). Buda Castle yielded two 14th-century well fillings where its seeds were found (P. HARTYÁNYI – NOVÁKI – PATAY op.cit.; P. HARTYÁNYI 1975–77).

The word pear (körte) is of Bulgarian-Turkish origin, it got into the language during the late ancient Hungarian period (LIGETI 1986). The first written occurrence (“kurtuel”) was found in the charter of the Tihany Abbey (1055) (LIGETI 1986). Elsewhere, it can be identified in the form of “kertvel fa” (1158), “piri silvestris” (1237), “arbor vadkörtvél (1274), “arpaly reukuruel” (1338), “Hintouskurthuel” (1342) (CSŐRE 1980).

Two dozen brown-black seeds found in layer 8, are thought to be quince (cf. *Cydonia oblonga*). due to their oval, slanted wide shape which is flatly protruded. Possible difficulties in identification also contributed to the fact that no seeds were found so far in medieval sites. However, they are found at German medieval sites (WIETHOLD 1995).

Black mulberry (*Morus nigra*), which is a frequent find in general, was found in the Buda Castle medieval wells as well (c.f. P. HARTYÁNYI 1975–77), in both layers (Nos. 8 and 6.).

The remains found in layer 6 indicate a wider scope of fruit varieties both in terms of volumes and species diversity. Walnut (*Juglans regia*) was encountered only in layer No. 6, from the era of King Sigismund. Three dozen yellowish shell fragments were found which are of typical shape and identifying traits can be clearly distinguished indicating they were domesticated cultivars. Walnut remains are classified as paper shelled and elongated walnut.

Pollen analytic research carried out in Lake Balaton region (ZÓLYOMI 1953, 1980) and in Pilisszentkereszt (ZÓLYOMI – PRÉCSÉNYI 1985) suggests that walnut was not native in the Carpathian Basin. Its pollen appeared as a sporadic find in the Bronze Age for the first time, and it was only continuously present in greater numbers since the Roman Age. It is very likely that the plant, which comes from Asia Minor and the Balkans, was introduced to the Carpathian Basin by people from those regions who lived along the river Danube.

Hungarians acquired walnuts through Alan or perhaps Turkish influence, probably as long ago as in the ancient homeland between the Volga River and the Azovi Sea. Walnut (dió) is a Bulgarian-Turkish loanword in Hungarian (LIGETI 1986).

Since the Árpádian Period, trees were felled with the exception of walnut trees. In this way, smaller or larger walnut groves were formed and fruit of the trees collected. Walnut was readily planted beside wells, or at the end of grape rows. Later on, it was introduced into gardens as well.

Walnut is found as "dyo, geo, gyia, gyio, gyo, dia, gia, gyofa, gyabukar" in medieval charters. The first occurrence in a document is dated to 1015, the foundation charter of Pécsvárad. It was frequently used as a place name: "dias, gyias, gyios, gyos" (SZAMOTA – ZOLNAI op.cit.). Walnut was a favourite dry fruit in the Middle Ages. Several Buda Castle well finds confirms this.

A tea prepared from the tree's dried green leaves is still drunk to treat stomach and gut ailments, high blood pressure and worms. It also has a purifying and appetiser effect. Externally, it can be used for treating rashes, abscesses, frostbites, sore throat, ophtalmia, and cataracts. The seed (walnut meat) is used to prepare cold pressed oil (*Oleum junglanctis*) against arteriosclerosis. Green parts are used for making suntan oil, hair dye, and liquors (RÁCZ – RÁCZ-KOTILLA – LAZA 1984; RÁPÓTI – ROMVÁRY 1983).

Peach (*Prunus persica*) stones were only found in layer 6. Intact and light coloured pits are brownish in colour, with rough and deeply cored outer parts. According to their shapes, stones found in layer 6 can be classified as convar. *scleropersica* (duránci, clingstone). They are much smaller than recent cultivars.

The word apricot (barack) is of Slavic origin (SURÁNYI op.cit.). This however does not necessarily mean that early Hungarians learned about the fruit in the Carpathian Basin. In all probability, it was introduced in Levedia.

Almost all Central European countries have apricot remains from later periods of the Middle Ages in sewers, graveyards, wells, pits. Apricot was a fruit of medieval orchards. Its stone was encountered in the Buda Castle wells several times.

Almond (*Prunus amygdalus*) was found in layer 6. The shell is dotted with small holes, yellowish, slightly fragmented and typically oval in shape. It was grown. Compared to apricot, almond is much less frequent. In spite of this, it was encountered twice in medieval Buda.

In layers 5 and 6, one and a half hundred rust-coloured, spongy, rough and striated medlar (*Mespilus germanica*) seeds were found. Medlar is a frequent find in sewers and wells of Central European towns (c.f. KNÖRZER 1987). Its earliest find was in medieval Buda wells. The relatively high number of seeds shows that the mushy autumn fruit was in high esteem.

A special feature is the dark brown, typically curved, wide oval service tree (*Sorbus domestica*) seed found in layer 5. It must have belonged to cultivated varieties. The service tree easily escapes from gardens. It is a common native species in large parts of Hungarian forests. It appears in dry oak woods, Turkey oak-oak woods, karstic shrubberies, rock forests, loess oak woods, hornbeam-oak woods, bushes, vineyards and forest edges. Today it is no longer grown and sometimes it escapes (SOÓ 1966). Previously, the service tree was favored for its fruit and was easily grown. Service berries were collected half-ripen from the trees and matured (cured) suspended. The Capitulare de villis, written under the rule of Charles the Great, recommended the planting of service trees in gardens. The service tree was the stock for medlar. It was readily planted among vine-plants. Service tree berries are still used in a variety of ways by people of countries to north of Hungary: brandy and marmalade is prepared from them (SURÁNYI op.cit.). A S. domestica find from the Sümeg-Sarvaly 16th-century site was probably gathered (SKOFLEK 1982; KASSAI 1991). Berkenye (service tree or berry) is a Slavic word (KNIEZSA 1955). Its first occurrence is in the foundation charter of the Tihanyi Abbey (1055) as "brokinarea" (SZAMOTA – ZOLNAI op.cit.).

The much beloved common fig (*Ficus carica*), a Mediterranean plant, was planted as decoration as well as crop in the Middle Ages. Its range extended far to the north. It was grown in Central Europe since the 15th century and in England since the 16th century. In Hungary, a number of fig bushes were probably introduced during the Renassaince through Italian connections. JESZENSZKY – KÁRPÁTI (1963) consider that the fig trees in Gellérthegy are several hundred years old and of Italian origin. The Turks also ate this fruit readily and probably planted them as well. Entz maintains that fig shrubs in Mecsek are of Turkish origin (in: RAPAICS 1940). It is likely that the great number of common fig seeds coming from medieval Buda, Lászlófalva-Szentkirály and Szécsény were of local production and not imported.

As a result of continuous relations with the Mediterranean, the decorative pomegranate (*Punica granatum*) also appeared on the table of the higher classes. As a proof, Budavár 1999 well/cistern explorations yielded 15th-century remains including pomegranate seeds. Dezső Surányi argues that the seeds must have belonged to fruits grown locally under the submediterranean climate of the age.

In spite of all the devastation wrought by the Turks, horticulture is considered to have been continuous (ÁRENDÁS 1982). Each of the former species were continued to be grown and the number of species in horticulture even increased. A number of Balkanic kitchen products (Smirna melon), fruit (Makaria pear, Kadarka grapes) and decorative plants (tulips) were introduced (RAPAICS 1940).

According to descriptions, 17th-century orchards had a square shape. Grafts were planted in arranged rows in the manor orchards and vineyards. Fields were bordered by roads. In turn, roads were edged by plum, sour cherry, hazelnut and ligustrum hedges. János LIPPAY (1664) recommends, among others, thorny plants like sloe (blackthorn) and wild sour cherry be used as protective hedges. According to various sources, fruit trees were also grown in vineyards (P. ERMÉNYI 1975–77). Of the propagation methods, mostly grafting was widespread. It was fashionable among the nobility to exchange seeds. Many data are available in baronial correspondence on sending and exchanging seeds or grafts. In a letter by Tamás Nádasdy, for instance, he asks for "öregszemű" (old type) cherries. Emperor Maximilian also wanted to get fruit cultivars from Hungary for gardens in the Neugebaude in Vienna. (RAPAICS 1940).

7.9.6. Wild fruits

Master and servant both were fond of wild fruits. Wild fruit was included in the income of an estate (RAPAICS op.cit.). It was consumed fresh, dried, in the form of wine, brandy, vinegar, syrup, marmalade, and preserves. Medical effects were also known. It was known that a decoct made of sloe (blackthorn) flowers purifies the blood and the kidneys, or that wild strawberry leaves are good for indigestion (AUGUSTIN *et al.* 1948).

Sloe (blackthorn) has special significance among gathered wild plants. Sloe (blackthorn) fruits are strikingly bluish-bloomed in sunny, warm places (due to the vax layer on them) and almost black in cooler places. One of the German names for sloe, "Schwarzdorn" might easily refer to this feature, while of the Hungarian "kökény" sloe (blackthorn), according to PÉNZES (1950), Rapaics demonstrated that it comes from the word "kek" = "blue". RAPAICS (1940) argues it was originally Bulgarian-Turkish in origin.

A number of medieval cookbooks discuss uses of sloe (blackthorn) for vinegar, and marmalades, but it was also known as a drug. Probably, it was also planted, rather than wild, in many places. Sloe is known from several locations at Buda Castle (cf. P. HARTYÁNYI – NOVÁKI 1975; P. HARTYÁNYI 1977). The same is confirmed by several thousand sloe pits found in excavations of the Szécsény late medieval parish-church (Torma 1994). It was identified at other Hungarian medieval settlements as well: Hollókő castle and Gyöngyöspata (P. HARTYÁNYI – NOVÁKI – PATAY 1967–68). According to pit remains, sloe (blackthorn) had both large and small, round and elongated seed varieties.

Hazelnut, rich in oils, was also preferred and used in several ways. A number of fragments were encountered from a well explored in the garden of the Visegrád Palace and the Buda Castle area. Hazelnut also has several shapes.

The sour crab apple was also readily eaten as shown by the seeds found in the Lászlófalva-Szentkirály late medieval site.

Field strawberry was readily eaten. Its seeds are frequently found in Buda Castle medieval wells, and they were encountered during exploration of the Szécsény parish-church as well.

Dewberry was a popular gathered plant. Major amounts of stones were found in Szécsény and Lászlófalva-Szentkirály, in the Buda Castle wells and the Dunaföldvár-Öregtorony medieval layers (P. HARTYÁNYI – NOVÁKI – PATAY 1967–68).

Cornelian cherry and dogwood shells are also common in medieval layers at Budapest, Hollókő and Szécsény. The Hungarian area of the cornelian cherry includes towns in mountain ranges as mentioned above (PRISZTER 1990). In Vác, for instance, single-seed hawthorn pits were also found, confirming that the tasty and vitamin-rich fruits of this plant were also consumed (GYULAI 1995b).

All the above-mentioned wild fruit trees were represented in the Vác 16th–17th-century layer (GYULAI 1995b) and the material of Lászlófalva-Szentkirály Cuman settlement.

Elderberry and danewort seeds are often found at medieval sites. Common elder is a typical shrub in European forests. Its berries have been collected and eaten for a long time. Due to its colouring effect, it is still used as a colouring additive in drinks and dishes. The flowers can be used for making juice or fermenting wine.

Danewort (*Sambucus ebulus*) grows in clusters at forest edges, in clearings, alongside roads and in places with trash. Annual trumpery stalks spring forth from the rhizomes creeping away underground. It blossoms in June and July while the fruit ripens in August or September. The berries were once used for dying leather and it is a medical herb. However, the berries of danewort are not tasty, therefore they can be eaten only conditionally. An interesting observation is that the berries are eaten by ringworms. Sometimes it cannot be determined if the pits were scattered by people or taken by ringworm species

(*Lumbricidae)* into the cultivated layer, as it was observed at other sites (personal communication by Helmut Schlichtherle).

At Hungarian sites, danewort is more common than common elder. Larger amounts were found in the Sály-Latorpuszta 10th–15th-century graveyard (10 cm^3!) and in the Nagyvázsony-Csepely 15th–16th-century settlement (P. HARTYÁNYI – NOVÁKI – PATAY 1967–68; P. HARTYÁNYI – NOVÁKI 1975).

Common elder stock is known from Buda Castle buildings (13th–15th centuries), Tihany brethren dwellings (13th–14th centuries) and Hollókő castle (16th–17th centuries) (0.5 !) (P. HARTYÁNYI – NOVÁKI – PATAY 1968; P. HARTYÁNYI – NOVÁKI 1975; P. HARTYÁNYI 1977). Seeds found in the Budaújlak Turkish Period wastepits and in Hollókő castle prove that this "wild fruit," which is rich in vitamins and can be used in many different ways, preserved its significance in the Turkish era.

Black mulberry (sycamine) (*Morus nigra*), or as it was called in the Middle Ages, "bramble tree", a native in Hungary, has been definitively grown since the 15th century. The fruit was a favourite delicacy. Hungarian archaeobotanical research has data of its occurrence since the Middle Ages. The tiny seeds are frequent finds in medieval wells (Budavár, Szécsény), and they were present in the Turkish waste pit of Budapest II district, Medve Street. White mulberry (*Morus alba*) was domesticated only later, in the 18th century, when silkworm production started (RAPAICS 1940; SURÁNYI 1985).

All the wild fruits were found in 1998–99 in the Buda Castle area (Teleki palace, Szent György Street) in the mud-fill of 15th-century cisterns and wells. Wild fruits and fruit seeds were encountered mixed with domesticated varieties. In our view, some of these were produced locally. Wild fruits played at least an equal role in food supply when compared to cultivars.

Seeds of cornelian cherry appear in all the layers, proving that this fruit was very much favoured. The dogwood ranges from Western Asia to Southern Europe. It grows into shrubs or short trees and it is common in Hungarian forests or forest edges. It is often planted. The wood is dense, easily workable and much in demand. According to archaeobotanical finds, the berries were eaten since the prehistoric ages. The scarlet red, sour, but vitamin rich fruits are eaten fresh or processed into marmalade or juice (DANERT *et al.* 1981). In Russia, it is sun-dried; in France, it is stored in honey or sugar syrup or processed into a gel. The juice is much admired. The Romans fermented it for wine. In Greece and Italy, it is still used for making wine. To make marmalade, it requires a lot of cooking as the fruit contains very little pectin. The green, unripe berries in vinegar and spiced with fennel or bay leaves taste like common olive (HELM 1982). Cornelian cherry in pits is common at Hungarian medieval sites: Budapest, Hollókő, Szécsény. The range of its habitat includes the towns in mountain ranges as well as the Buda surrounding (PRISZTER 1990).

Hazelnut shells, from layer 6 in the 27/2 rock-carved well (14th century), might have been the remains of the giant filbert, Lambert nut (*Corylus avellana* f. *oblonga*). One cannot decide if it was locally grown or wild. Hazelnut is a flatland-mountainous, Central European (-Mediterranean) *Querco-Fagea* species. It is common in forests and shrubbery, sometimes dominating. Several varieties are planted (SOÓ – JÁVORKA op.cit.; SOÓ 1980; SCHERMANN op.cit.). In the Middle Ages, together with other wild fruits, hazelnut played a greater role in food supply. It is also known as a drug. The bark is used for a decoct good for diuretic purposes. The leaves (*Coryli folium*) contain flavonoids and tanning agents, while compounds built up with certain metals have a deodorant effect (RÁCZ – RÁCZ-KOTILLA – LAZA op.cit.). The name is a Bulgarian-Turkish loanword, coming from the time of hunting-gathering before the conquest, an eponym for medieval places (RAPAICS 1934, 1940). In all probability, this was known by Hungarians in their ancient homeland. It was one of the key oil sources for Central European gatherers beside European beech achenes and walnut. Its first written reference is in the foundation charter of the Tihany Abbey (1055) ("Munarau bukurea") (SURÁNYI 1985). Hazelnut (mogyoró) was common in place names as well: "Monorosheg" (1212–1328), "Munurouspotok" (1234–1243), "Monoruswelge" (1287–1328) (SZAMOTA – ZOLNAI op.cit.).

Field strawberry (*Fragaria vesca*) was held in high esteem, as proven by the presence of thousands of seeds. It was found in all layers. This tasty perennial plant is common in forest clearings. It is one of the fruits with a very long history as a gathered plant. The fruit is eaten fresh, as a juice, marmalade, and a filling in dough. It is also a medical herb. Dried leaves (*Fragariae folium*) were used to make tea against colitis, or as a diaphoretic (RÁPÓTI – ROMVÁRY 1983). It is also an ingredient in a tea blend against diabetes mellitus.

Ground cherry, dwarf cherry (*Prunus fruticosus*) pits found in layer 8 (I/2. well, 13th–14th centuries) are also classified as wild fruits. Although not very tasty, the fruit of these little trees growing in groves and shrubberies or forest edges can be eaten. Mahaleb or Turkish berries, often present in considerable numbers, is sometimes a preferred forest fruit (*Prunus mahaleb*). Pits were found in all the three layers. Mahaleb is a hillside-mountain species from the Pontus-Mediterranean, often found in Transdanubia. It is a shrub, but sometimes grows to be a middle-height tree with beautiful canopy. Major stock for sweet cherry and sour cherry, it is planted and sometimes escapes (PÓR – FALUBA 1982; SOÓ 1966; SOÓ – KÁRPÁTI op.cit.). It is a *Quercetea* (*Tilio-Acerion*) element, found in rocky forests, linden-ash rock forests, calcinophil oak woods and rock grass slopes (SOÓ – JÁVORKA op.cit.; SOÓ 1980). The fruit is 8–10 mm in diameter, poor in flesh and juice, somewhat bitter. The pit is found in lake dwellings in Switzerland, where it was possibly eaten sometimes as a fruit. Assumedly it is found in a number of medieval place-names (RAPAICS 1940). The leaves contain kumarin and the wood used is for making implements.

Sloe (blackthorn) (*Prunus spinosa* agg.) pits can be divided into the following groups by shape and form:

- *Prunus spinosa* agg. (round) – typical of the majority of the pits,
- *Prunus spinosa* agg. (elongated) – fewer pits,
- *Prunus spinosa* agg. subsp. *macrocarpa* (= subsp. *fruticans*) (elongated, large) – only a few pieces.

Round and elongated pits attest to the fact that several varieties were known at this time. The unusually large pits coming from layer 8 are especially interesting. Pénzes notes that the *fruticans* stock found in Farkasvölgy, which has round fruits as big as 2 cm across, used to be grown in cultures (PÉNZES 1950).

The sloe (blackthorn) is a flatland-hillside, European (Mediterranean) species; Prunetalia (Querco-Fagea): it grows at forest edges, bushy places, on pastures (SOÓ – JÁVORKA op.cit.; SCHERMANN op.cit.; SOÓ 1980). It is common from the plains to the subalpine landscape (SOÓ 1966).

Sloe supply played an important role from the Mesolithic to the Modern Period. Medieval sloe (blackthorn) finds include Krakow (KLICHOWSKA 1964), Göttingen, Leisenberg, Northeim (WILLERDING 1978), Lübeck (PAAP 1984). Its use was discussed in charters and cookbooks (WILLERDING 1983b). According to German medieval written documents, a great number of sloe was used for making vinegar or marmalade (WILLERDING op.cit.) It is very likely that it was planted at several places. At several locations in Buda Castle we found sloe (blackthorn) pits that belonged to more than one variety (c.f. P. HARTYÁNYI – NOVÁKI 1975; P. HARTYÁNYI 1975–77). The Buda medieval law book listed sloe (blackthorn) as "common fruit" (Relkovič 1905 in: P. HARTYÁNYI 1975–77). It was regularly seen on medieval market places in Buda. It is known as a medical herb. Flowers (*Pruni spinosae flos*) were used to make diuretic and laxans tea. Fruits (*Pruni spinosae fructus*) were cooked to decoct medicine against colitis. The fruits could be processed into marmalade, once the frost had bitten them, which was used as a binding medicine (SOÓ – JÁVORKA op.cit.; RÁCZ – RÁCZ-KOTILLA – LAZA op.cit.)

The word kökény (sloe) was originally Bulgarian-Turkish in origin (LIGETI 1986). Documents contain it in the form "Cucen" (1211), and as a place name "Chucinus" for the first time in 1251 (SZAMOTA – ZOLNAI op.cit.).

More than a thousand rose (*Rosa* spec.) seeds suggest that this plant was popular. Seeds of different *Rosa* species are very similar to each other and therefore cannot be distinguished. The genus *Rosa* contains a great number of species. Many of them are planted. "Wild roses" can be found in shrubberies, clearings, forest edges and roadsides, and also in dry forests and shrub forests as well. "Bramble rose" is a common name for those wild roses that have edible fruits ("rose-hips"), a natural source of vitamin C. Rose hips grow red from the end of August and become sweet after frost arrives. They can be eaten in several different ways: fresh, dried, tea, juice, marmalade, wine, etc. The most valuable part is the sarcocarp, once cleaned of seeds and hairs. The distillate made of it has a restorative effect and is used for treating atrophic conditions (RÁPÓTI – ROMVÁRY 1983).

Bramble was a popular fruit as well. Its seeds were present in large amounts in all the layers (mainly *Rubus caesius*, to a smaller extent *Rubus fruticosus* agg.). Bramble is a thorny bush found in many places: groves and marshland forests, floodplains, ditches, or even stubble fields. The bloomy blue fruits have been gathered for a long time and used in many different ways. It is known as a drug. Its dried leaves (Rubi fruticosi) make laxative or sometimes diaphoretic tea (RÁPÓTI – ROMVÁRY op.cit.).

Interestingly, wild raspberry (*Rubus idaeus*) is a rare find. Seeds are found only in layer 6. It is possible that wild raspberry was not part of the gardens around Buda at the time. This shrub grows in cuts, clearings, wet and moist forest parts and edges. However, the true homeland of sweet and good smelling wild raspberry is mountains. According to archaeobotanical finds, it has been gathered since the earliest times and used in many different ways. Sometimes it was even used for making wine. Dried leaves (*Rubi idaei folium*) served for making herbal tea against diabetes, kidney and bladder ailments or as a diaphoretic (RÁPÓTI – ROMVÁRY op.cit.).

Common elder (*Sambucus nigra*) is a fruit that was presumably used and eaten in spite of the fact that it was found only sporadically in layer 6. Common elder (Sambucus nigra) is a typical bush of forests in Europe and Asia Minor). Birds contribute to its propagation. The fruit has been gathered and consumed for a long time. Due to its strong colouring effect, it is used as an additive in foods and drinks even today. Flowers can be used for making drinks or wine. Common elder was first found in Budapest in a Late Roman Period legion camp (18 g) (excavation by Mária Pető), and in buildings in Buda Castle (13th–15th centuries), Tihany frater dwellings (13th–14th centuries) and Hollókő castle (16th–17th centuries, 0.5 l!) (HARTYÁNYI – NOVÁKI – PATAY op.cit.; HARTYÁNYI – NOVÁKI op.cit.; HARTYÁNYI 1975–77). Elderberry (bodza) is a Slavic word (KNIEZSA 1955). In a written form it was first encountered in a charter dated to 1211 as "boz" (SZAMOTA – ZOLNAI op.cit.). Sambucus species are relatively short lived and therefore they were not suitable for boundary posts, as opposed to other fruit trees. Therefore, they show up in the documents relatively infrequently. Csőre collected elderberry names in the period 1231–1337: "buzbukur, sumbukfa, bukfa, bozyabukur, felbozfa" (CSŐRE op.cit.).

7.9.7. Medieval viticulture

According to the written documents, grapes were generally planted widely in terms of both varieties and area (*Figs. 404–410*). Archaeobotanical finds confirm this perfectly. Late medieval layers contain large amounts of black wine grape stones everywhere: Buda Castle area, Óbuda, Vác, Szécsény.

As opposed to farming, viticulture requires higher level skills and was therefore easier for the peasants. Having paid their relatively small duties, they were free to sell their grapes. Where natural conditions allowed the growing of grapes, people were inclined to purchase their staple grain needs from the proceeds of sales of the wine they produced. In other words, wine production contributed to the development of the production of marketable goods to a great extent.

The grape stones unearthed from the late medieval waste pit of Külsővat accompanied by oven tiles were not subfossil, or even carbonised but calcinated (excavation by Gábor Ilon 1986–87). A carbonised ash stain coming from the fireplace of an oven produced *oak* charcoal pieces (determination by Ernő Horváth) as well as oat-grass and barley grains (in: ILON – SABJÁN 1989).

Substantial amounts of grape stone finds surfaced from the Buda Castle 14th–15th-century dwelling house pit (P. HARTYÁNYI 1977), 15th century wells and cisterns (P. HARTYÁNYI – NOVÁKI – PATAY 1967–68; P. HARTYÁNYI – NOVÁKI 1973–74), Teleki palace 15th-century wells and Szécsény 16th–17th-century church exploration site suggesting that vine growing was widespread in the Middle Ages (TORMA 1994).

However, these remains give little help in identifying the varieties grown. FACSAR (1970) carried out metric measurements on medieval grapes and reportedly identified Kecskecsöcsű and Makkszőlő table grape cultivars. In his view, Gohér, Furmint, Fehérszlanka, Vékonyhéjú, Betyárgrapes, Fügér, Mézes, Lisztes, Balafánt, Dinka, Berkenyelevelű, Szagos bajnár, Rózsagrapes varieties were highly probable cultivars in Buda in the 13th century (in: CSOMA 1994–95). We have also observed different kinds of stones in Budapest II district Medve Street, Budavár, Szécsény, Szentkirály, Vác medieval structures.

Wine grape stones were unearthed in unexpectedly great abundance from two wells explored in Buda Castle in the yard of former Military Headquarters (Dísz Square No. 17) in 1999. This is not at all surprising if one considers the large number of vineyards that would have filled the Buda hills and mountains up to the wine-pest epidemic in the 19th century. Tens of thousand of stones were found, most of them yellowish brown or dark brown. Most of the stones had long "beaks" and round, pear-shaped forms, but some of them were short with definite beaks. They likely represent a mixture of several different kinds.

Help can be obtained for identification of the earliest grape varieties from the processing of archaeological seed finds and the analysis of written sources. In order to stress the importance of this research, a Task Force for Agrarian History was formed in 1956 for the purposes of studying historical grape varieties and the clarification of strain names.

The Hungarian names of the oldest strains were preserved (in: RAPAICS 1940). The name Hasártó grapes is encountered in a charter from 1410: "vinee Hasártógrapes dicte". This must have been a soft-pulped grape. Another document from 1454 mentions Cserszőlő: "vineam suam Cserszőlő". This must have been identical with the very early ripening table grape around Lake Balaton called Gohér. The long, yellowish-white sweet grapes ripened in August. Also from the 15th century comes the first mention of Kecskecsöcsű. This is a late ripening strain, which can be traced back to Antiquity, with large clusters, medium sized and typically elongated uvae and thick, tough pulp. Fabriczius mentions Vadfekete grape. It appears later under other names as well: Cigányszőlő, Csókaszőlő. This late ripening grape has mid-sized clusters and round berries. The wine was austere, sour and dark red. The very little red wine which existed in Hungary in the Middle Ages was made of this strain. Fabriczius' dictionary contains Bolgár grapes. RAPAICS (op.cit.) says it was an elongated, large table grape something like Afuz Ali.

As to medieval wines, Hungary was characterised by white wines and hence by white grapes (in: CSOMA op.cit.). Although some varieties providing red wines existed, these could not have been very important. It was not before the 15th century that they were distinguished by the attributes: "vadfekete" and "csóka." The Beszterce dictionary from 1395 mentions "labrusca vas zelew" which must have been a bluish grape.

European medieval sources discuss two kinds of wine, one with higher alcohol content, "*vinum frannicum*" and another cheaper and lower quality "*vinum hunnicum*". The two names might hide different grape strains (Csoma 1996). For those who deal with ampelology, the latter name, which might refer to an assumedly Hungarian and very early grape variety, causes much trouble (Csoma 1994–95). According to Égető this might have been a long selected version of woodland European grapes (in: Csoma 1994–95). Ferenc Schams claims to have found its progeny in Fejér grapes, while Márton Németh thought the same of Hajnos. In Western Europe, this strain was in production for a long time. This might be indicated by the fact that a certain kind of grape was known in 1827 in German territories as "Heinisch." Probably it was the same as the Hajnos green included in the grape strains collection of Demeter Görög established in 1819 in Grinzing (Rapaics 1940).

Wine trade did not stop among the different areas of the country when torn into three parts during the 16th–17th centuries. Many data in documents indicate this (Feyér 1981). For instance, Gyöngyös was engaged in unlimited wine trade by a royal decree in spite of its being occupied by the Turks. By the way, wine production induced tax reductions.

Beside grain, wine provided the most important tax. For instance, according to the analysis of botanical finds from downtown Vác, the number of different grain and fruit species and cultivars still produced in a great variety in the 15th–16th centuries was reduced by the Turkish era. The only exception is grapes with a significant increase and expansion (Gyulai 1995b). Wine drinking was a widespread habit. It obviously had a protective role as well against epidemics breaking out due to lack of clean drinking water. During the constant warfare in the 16th–17th centuries, both the soldiers and merchants drank a lot of wine.

The grape was a popular fruit as well. In the excavation works of medieval Sümeg Castle, led by Károly Kozák in 1957–1963, black wine grape stones were found in the privy pit in the eastern end of the palace, accompanied by 17th-century ceramics (Füzes Miklós and Sági Károly, in: Hartyányi – Nováki – Patay 1967–68).

It might be strange to talk about the evolution of viticulture during the Turkish occupation. Yet the production of table grapes had a boom at this time. Grape berries dried for raisins were very popular. A new black wine grape was also introduced. Serbs fleeing the Turks brought in Kadarka (Katona – Dömötör 1963). Szerémség used to be one of the most famous wine-growing regions in the country before the Turkish occupation. Grapes were grown here since the Roman Period. A famous 16th-century strain was Szerémi green grapes (Rapaics 1940). The first Hungarian area for Kadarka occurred here as well. Kadarka reached the Northern mountain range during the Turkish occupation (Andrásfalvy 1957). Wine growers of the southern slopes here still call Kadarka "Turkish grapes" and "Turkish black".

Thus, the Turkish occupation did not restrict the spread and development of viticulture (Égető 1993). The apparent contradiction comes from the Koran's prohibition of drinking wine. However, taxes on the must and wine produced represented a considerable amount of income. Also, grapes were used to make "pekmez", a kind of boiled, choking, condensate and very sweet must, which was consumed by the Turks very frequently and became a popular drink among the Hungarians as "petymez, peszmeg" (Andrásfalvy 1961). In addition, the prohibition was not strictly applied to "boiled wines", i.e. brandy (Feyér 1981). It is known from the mid-16th-century description of Dernschwam (1984) that in the northern part of Buda Castle (around Maria Magdalene Church) Hungarians were served by a pub and cellar during the Turkish years. Beside Hungarians, Turks grew grapes as well. This is indicated by the "muslim vineyards' dönüm tax", paid to the Vác nahije (Káldy-Nagy 1985). According to a will, a spahi soldier left marc worth one and a half foals (Hegyi 1976). The significance of grapes in the age can be judged by the one-metre-thick layer full of grape stones found in Vác, in a Turkish Period cellar. Beside the stones, pressed berry skin remains were also identified. According to the preliminary findings, several kinds of grapes could be distinguished. Grape marc was buried just as wine growers still do in Zala.

7.9.8. Remains of medieval weed associations

At medieval sites where grains were found some weeds were also identified. Weed remains, mostly the leftover from thrashing or cleaning procedures, were unearthed mainly from fireplaces or waste pits.

Grain was cleaned by husking, sieving or "scattering." Leftover tailing, which also contained lighter "idle" grains damaged by insects, glumellae and other kinds of offal, was fed to animals or burned. Cleaned and stored crops contain fewer weeds as care was taken at that time that no husks, which could make the flour bitter, or toxic weed seeds were mixed into the grain destined for milling.

Weeds made the life of medieval peasants hard. Árpádian Period houses explored by archaeologists revealed that although the grains in them were mostly cleaned and stocked, a number of weed seeds were found in them. The weed seeds were accompanied by the host plant when they were burnted. The removal of these seeds with contemporary means of cleaning, such as husking, winnowing or maybe scattering, would have been very difficult or even impossible. Especially toxic korn-rade seeds were encountered frequently, which, when mixed into the flour, could have caused major poisoning. To a smaller extent, black-bindweed, catchweed, cleavers, mildew, field cowwheat, knotgrass, redshank, bindweed, annual woundwort, oat grass and goosefoot species were also found. Their presence confirms the local production of autumn sowing grains and it can also be concluded that harvesting was made by scythe at a low height.

The number of weed species damaging grain crops increased since the 13th century in connection with expanding crop production up to the 15th–16th centuries. The species number of intertillage crops in production (legumes, pumpkins) grew, which implies the growth of corresponding weeds in spring sowing grain and intertilled crops.

Vác grains were very heavily infested by korn-rade in the 13th century and especially in the 15th–16th centuries. This dangerous species was very difficult to get rid of using the contemporary methods of cleaning, while the others did not pose such a problem. This is why the number of field cowwheat and rye brome seeds is so little. Autumn sowing (such as rye, bread wheat) hosts are implied in korn-rade and annual woundwort finds, while spring wild-oat indicates spring cereals (such as common millet).

In places where herbological studies were made such as Rákospalota-Újmajor and Vác, it was found that segregation of these species is typical. A large part is winter cereal weeds (*Secalietea*): field brome, wild turnip, black-bindweed, false cleavers, stinking nightshade or henbane, white melilot, narrow-leaved vetch and summer vetch. These species indicate autumn sowing hosts (wheat, rye, and maybe autumn barley). A smaller proportion of the weeds come from spring cereal weed associations (spring barley and common millet), or in other words hackfruits associations (*Polygeno-Chenopodietalia*): smooth finger-grass (*Digitaria ischaemum*) and common stork's-bill (*Erodium cicutarium*).

Weed flora has reached an unprecedented richness in the Middle Ages as a result of the expansion of agriculture, associated fertilising practices and other soil replenishment procedures, amelioration and other interventions. Nitrophilous, creeping, shade tolerant species started to dominate.

A great number of weeds were found among the large amount of diaspores coming from the medieval Buda Castle wells (Teleki palace, former Military headquarters) in recent years. Those living here stocked or purchased only as much grain as was necessary for the preparation of their food. Before kitchen use, grain was controlled once again in order to avoid weeds getting into the dishes. Obviously their toxic and distastful properties were known. Korn-rade was especially regarded as a dangerous weed. Carefully selected korn-rade seeds made their way into household waste together with other weeds and from there into the wells.

Korn-rade is as old as crop production in Asia Minor, which was introduced into Hungary with the grains of the Neolithic (Bakay – Kalicz – Sági 1966). It was the first Secalitea species to arrive to

Central and Eastern Europe where there is still a great number of Pontus elements in the vegetation. At the end of the Neolithic, its range reached as far as the Bodensee. It moved on towards England and Northern Germany only in the Roman Period, corresponding to the limes area. It also reached Scandinavia. It reached its largest distribution in the Middle Ages, when it was a widespread autumn grain weed in the whole of Europe (KÜSTER 1985a). Korn-rade can be removed from grain only with modern technologies. Chemical pesticides have finally pushed korn-rade back and it has become an endangered species. Hungary also included it in the list of endangered species (UDVARDY 2000). As an annual ornamental plant, sowing seeds of this decorative archaeopyton pink family species are even marketed.

In terms of numbers of seeds, black-bindweed (*Fallopia convolvulus*) follows after korn-rade among those found in wells I/2 and 27/2. It must have been an important weed, difficult to contain. The only seed encountered in a carbonised state was bedstraw (*Galium spurium*). It was a typical weed of medieval grains, however it did not belong to the major ones in Hungary. Maybe it avoided the thorough and careful cleaning of grains before use and was burned together with them. Relatively widespread species could be charlock (*Sinapis arvensis*) and annual woundwort (*Stachys annua*), which is a spring weed, and blossomed in grain stubble fields mainly during rainy summers. It was a significant source of nectar, the so-called "stubble honey" (ÖRÖSI 1968).

Some other grain weeds were also found in the Buda Castle wells sporadically: ground pine (*Ajuga chamaepitys*), wild turnip (*Brassica campestris*), thorow-wax (*Bupleurum rotundifolium*), red hemp-nettle (*Galeopsis ladanum*), ball mustard (*Neslea paniculata*), annual knawel (*Scleranthus annuus*), field pennycress (*Thlaspi arvense*) andnkeeled-fruited cornsalad (*Valerianella* cf. *carinata*). As wild carrot (*Daucus carota*) seeds were smaller than today's carrots, they might have been weeds.

Spring cereal weed or hackfruit weeds occurred less frequently. Based on the number of remains, bristle-grasses were important weeds: yellow bristle-grass (*Setaria lutescens*) and green/rough bristle-grass (*Setaria viridis/verticillata*). In addition to bristle-grasses, common orache (*Atriplex patula*) and annual mercury (*Mercurialis annua*) were also found.

The presence of the weeds described above allows for the following conclusions. First of all, in the earliest times, in the 13th–14th centuries, in other words, during the time when Buda Castle was built, residents might have produced their own grain (not on the Castle Hill, though), because the many weeds present in the samples, could not have been present in grain purchased at the marketplace of the time. The number of korn-rade seeds is worryingly large and in the Middle Ages people were quite aware of its toxic effects. Among the grains, mildew was also observed.

The majority of weeds belonged to winter cereals hosts (winter wheat, rye): korn-rade, black-bindweed, bedstraw, ball mustard and so on and to a smaller extent to spring sowing grains (spring barley, oat, common millet) or to intertilled crops (like melons): annual mercury, common orache and bristle-grasses. Both tall and short weeds were present, therefore reaping must have been made at low level, maybe using a scythe. Compared to the 13th-century layers of the wells, the numbers of both grain and weed species were reduced in layers from the 14th and 15th centuries. Consequently, Buda dwellers of the time abandoned self-sufficient farming and purchased their food at the marketplace.

7.9.9. Remains of the medieval environment

We consider any attempt to reconstruct the contemporary environment essential. Therefore, whenever it is possible, determinations must be supplemented with environmental and ecological studies.

During the processing work of archaeological seed finds, remains of plant species from the contemporary natural environment were encountered many times. They carry important information on the flora and climate of the time.

Most species coming from the contemporary natural environment are ruderal plants. Namely, species of impacted, devastated soils are always found in nitrogen rich environments (such as roads, houses, stables). Fewer vegetation elements come from more natural environments around the settlements, which is subjected to human impacts to a smaller extent or not at all.

Members of ruderal plant associations imply areas subjected to human impact and refer to the size of the settlement. At the Lászlófalva-Szentkirály site, the number of ruderal species is very high (41 species!), consequently the settlement must have been large. Most of them come from a habitat with average fertility: white, maple-leaved and wall goosefoot (*Chenopodium album*, *Ch. hybridum*, *Ch. murale*), creeping thistle (*Cirsium arvense*), chicory (*Cichorium intybus*), henbit and red dead-nettle (*Lamium amplexicaule*, *L. purpureum*), dwarf mallow (*Malva neglecta*), smooth sow-thistle (*Sonchus oleraceus*), rough cocklebur (*Xanthium strumarium*), and woolly burdock (*Arctium tomentosum*) (the last one has especially many seeds). This means that the settlement, although established on an elevation with average water supply, had drier parts indicated by: hedge bindweed (*Calystegia sepium*), curly welted-thistle (*Carduus crispus*), common fig-leaved goosefoot (*Chenopodium ficifolium*), catchweed, cleavers (*Galium aparine*), alsike clover (*Trifolium hybridum*), and more wet parts as well, as shown by prickly lettuce (*Lactuca serriola*), cammon mallow (*Malva silvestris*), white horehound (*Marrubium vulgare*) and cotton thistle (*Onopordum acanthium*).

In the 15th–16th-century botanical material of Vác there are also many ruderal weeds, indicating a settlement: fan-hen or white goosefoot, soapwort, creeping buttercup (*Ranunculus repens*), knotgrass, wild carrot and common nettle (*Urtica dioica*). The number of ruderal species is low in the Turkish Period, which can be explained by reduction of settlement size (GYULAI 1995b).

The number of ruderalia found in well I/2 and the rock carved well 27/2 explored in the Teleki palace and the former Military Headquarters is high, their presence indicates an area subjected to significant human impact. The species come from a habitat with average fertility for the most part but those living in dry or even wet environments also occur. When the layers are compared on the basis of the number and habitat needs of species found in them, serious environmental changes are revealed. The number of the earliest ruderals from the 13th–14th centuries is three times that found in the layer dated to the 14th century. In addition, these differ substantially in terms of ecological needs as well. Species with wet habitat needs came only from the 13th–14th-century layer (livid amaranth and curly welted-thistle). This disturbed but undeveloped area could be of considerable size. About a dozen species coming from average habitats with impacted areas signal the constant presence of humans. Especially many danewort and fan-hen or white goosefoot seeds originated from this area. Later, wet areas might have been filled in as, in the 14th-century layers, only species with average habitat needs were found (fan-hen or white goosefoot and small-flowered fumitory), while henbane, a species with dry habitat needs also appeared.

As shown by many remains, fan-hen or white goosefoot and livid amaranth (Amaranthus lividus), must have been important weeds, the latter being one of the very few Amaranthus species of the Old World. Maple-leaved goosefoot (*Chenopodium hybridum*) was encountered only in the earliest, 13th–14th-century layer.

Other plants were also found, such as curly welted thistle (*Carduus crispus*), small-flowered furmitory (*Fumaria schleicheri*), creeping buttercup (*Ranunculus repens*), yellow mignonette (*Reseda lutea*), whorled clary (*Salvia verticillata*), narrow-fruited cornsalad (*Valerianella* cf. *dentata*) and rough cocklebur (*Xanthium* cf. *strumarium*).

Some of the ruderal species were also used as herbs. Great burdock (*Arctium lappa*) was probably well known for its medical properties. The drug of its root (Barbanae radix) contained insulin, tanning agents, raisin, vegetable fat, sugar, bitters (RÁPÓTI – ROMVÁRY op.cit.). The decoct of the leaves is a diuretic, diaphoric, gall-stone and nephrolith dissolver, and blood purifier, but it is also suitable for healing various skin diseases and to cure the hair. Natural healers used its seeds were to ease bile and kidney ailments.

Catchweed, cleavers (*Galium aparine*) is also considered as a drug by natural healers (RÁPÓTI – ROMVÁRY op.cit.). Tea made of its dried flowering sprouts has a sedative, hypotonic and narcotic effect. Cammon mallow (*Malva silvestris*) is also a popular drug (RÁPÓTI – ROMVÁRY op.cit.). Its tea is a cough mixture and expectorant. Its infusion was used against sore throat. Cammon mallow occurs in impoverished associations among cultivated plants as a weed. It is very probable that one of the seeds was among the grains of a spring crop (common oat, common millet) and burned with them. Knotgrass (*Polygonum aviculare* agg.) has a number of eco-types. It is common in weed associations, along roads, stubble fields and similarly impacted areas. The flowering sprout is a hemostatic drug (*Polygoni avicularis herba*). The tea is used for curing kidney and bladder ailments, colitis, tracheitis, and associated with other drugs, ulcer of the stomach (RÁPÓTI – ROMVÁRY op.cit.; RÁCZ – RÁCZ-KOTILLA – LAZA op.cit.).

Danewort grows in clusters at forest edges, in openings, disturbed and trash-filled areas. It propagates by underground spread of rhizomes from which new plants emerge. It blossoms in June and July while its fruit ripens in August or September. It has many non-food functions, especially as a medical herb. A drug made of its root (*Ebuli radix*) contains bitter substances, saponin, glycosides and tanning agents. The decoct is diuretic, diaphoric and slightly laxant. A drug made of the fruits (*Ebuli fructus seu bacca*) contains essential oils, bitters, sugar, anthocyan and hydrocyanic acid glycosides. It has similar effects. Danewort fruit marmalade is a vernacular medicine used to treat consumption (RÁPÓTI – ROMVÁRY op.cit.).

The seed of henbane (*Hyoscyamus niger*) is a strong poison and therefore a dangerous weed. It is also known as a drug (RÁPÓTI – ROMVÁRY 1983). The active substance is hioscyamin that is transformed into atropine when dried. It has a narcotic and pain-killing effect. The decoct made of its leaves was drunk for alleviating asthmatic spasms, but it was also useful for easing rheumatic and other kind of pains, such as a tooth-ache. It was customary that horse-dealers made the animal eat henbane seeds so that the horse intended for sale would seem healthier. In the Middle Ages, potions were also made of it.

When making the ecological classification of plant species originating from the contemporary natural environment and found at late medieval excavations, one can infer that the environment was very diverse. This is not at all an accident. It is very useful to compare site selection in historical times and today. While in the case of historical settlements it is usually a place close to nature, exploiting all the advantages, in our day nature is an indifferent factor. The settlements in the Neolithic, Copper and Bronze Ages, but also in the Middle Ages were established at higher grounds in the riparian areas, while in other places in basins of mountain and hill ranges, riversides and alongside the traffic roads. The preferred choice of farming and stock-farming people was areas wedged between marshes and swamps. Marshlands and wetland pastures had great significance in terms of food supply and fodder.

From the point of view of environmental reconstruction, it is always important to know how many culture layers are at the site. In the case of only one culture layer, the tanatocoenological examinations speak for themselves. The recovered evidence can be best compared with contemporary floristic conditions. However, if there are more than one layer, that is a different story. Here, there is an opportunity to monitor all the changes that occurred over time due to human activities.

A great number of floristic elements were encountered in the Lászlófalva-Szentkirály late medieval Cuman settlement wells, but this contained only one culture layer. According to the ecological classification of these, the environment around the settlement might have been very diverse. Waterlogged areas were proximate as the evidence was especially rich in marshland species. The existence of reed-beds was confirmed by four species: water dock (*Rumex aquaticus*, *R. hydrolapathum*), lake and roughseed club-rush (*Schoenoplectus lacustris*, *Sch. mucronatus*). A number of species originated from tall sedges around the margins, tussock sedge/yellow sedge (*Carex elata*/*C. gracilis*), stream-bank sedge (*C. riparia*), common spike-rush (*Eleocharis palustris*), least bur reed (*Sparganium minimum*) and water germander (*Teucrium scordium*). Water-edge pioneers included black pepperos and tasteless water pepper (*Polygonum hydropiper*, *P. mite*). Green sedge (*Carex serotina*), yellow sedge/pendulous sedge (*Carex flava*/*C. pendula*), Bog-rush (*Schoenus nigricans*) and marsh ragwort (*Senecio aquaticus*) might have come from wet grasslands.

A similarly wet, marshy environment is inferred by perennial marsh woundwort (*Stachys palustris*) and the all-pervasive common meadow-rue (*Thalictrum flavum*). Parkland forests decorated water edges. Species from these groves included: alpine grass (*Carex brizoides*), common hop (*Humulus lupulus*), crab apple (*Malus silvestris*) and English ivy leaved speedwell (*Veronica hederifolia*). The existence of the forest is confirmed by remote sedge (*Carex remota*), oak (*Quercus* spec.), wall germander (*Teucrium chamaedrys*) and European cranberry bush (*Viburnum opulus*). Hazelnut (*Corylus avellana*) is usually grown in light average habitat mixed forests and also at average or dry forest edges. In more shadowy places of the forest, wood sedge (*Carex silvatica*) and Sylvan/Rivini violet (*Viola reichenbachiana*/*V. riviniana*) could occur.

Most probably, a part of the land was broken by clearings, since fox sedge/prickly sedge (*Carex vulpina*/*C. muricata*), field strawberry (*Fragaria vesca*), black nightshade (*Solanum nigrum*), danewort and common elder (*Sambucus ebulus*, *S. nigra*) are common in clearings. Forest edge species included also crownvetch (*Coronilla varia*), single-seed hawthorn (*Crataegus monogyna*), sloe (blackthorn) (*Prunus spinosa*), red campion (*Silene dioica*) and sheep's sorrel (*Rumex acetosella*). The forest edge habitat was mostly of average water supply, but it might have had drier parts as well, as indicated by wild basil (*Clinopodium vulgare*), which favors such places.

Beside marshland elements, most of the spontaneous plants come from a typical meadow. The meadow of the time must have been large and expansive because it produced a great variety of species: soft brome (*Bromus mollis*), oxeye daisy or woundwort (*Chrysanthemum leucanthamum*), perennial ryegrass (*Lolium perenne*), common sorrel (*Rumex acetosa*), meadow clary (*Salvia pratensis*), common dandelion (*Taraxacum officinale*) and red clover (*Trifolium pratense*). The species rich area must have served as an excellent pasture for the livestock of the people living there. Creeping cinquefoil (*Potentilla reptans*) and marshmallow (*Althaea officinalis*) come from wetter parts, while lady's bedstraw (*Galium verum*), cypress spurge (*Euphorbia cyparissias*), black medic, hop clover, yellow trefoil (*Medicago lupulina*), silver cinquefoil (*Potentilla argentea*), tormentilla cinquefoil (Potentilla erecta) were at home in drier habitats.

The wells explored in the yard of Buda Castle former Military Headquarters (Dísz Square No. 17) in 1999 contained three culture layers. In terms of plants originating from the contemporary natural environment, these layers differed markedly. Sedges (*Cyperaceae*) were not found before the layer from the 13th–14th centuries: yellow sedge (*Carex flava*) is common on wetlands, hairy sedge (*C. hirta*) in wet, disturbed places, pale sedge (*C.* cf. *pallescens*) in groves, greater Tussock-sedge (*C.* cf.

paniculata) occurs in marshes, bogs, sedge-associations, fox sedge/prickly sedge (*C. vulpina/muricata*) grows in wet meadows, reed-beds and groves, while common spike-rush (*Eleocharis palustris* agg.) lives in marshes and wetland meadows. The majority of sedges are *hydato-helophyton* (watery, marshy) plants: they are typical members of high sedges, common reed-beds, wetland and marsh associations. Wet environment is suggested by gipsywort (*Lycopus europaeus*), which occurs in wetland pastures, cabbage thistle (*Cirsium oleraceum*), which lives at streamlet banks, wetland meadows and groves, carpet cinquefoil (*Potentilla supina*), which lives in muddy, wet habitats, sometimes in ploughland, and creeping cinquefoil (*P. reptans*), which occurs in wet pastures, sometimes in ploughland. Woodland European grape (*Vitis vinifera* subsp. *silvestrist*) stones found here are characterised by an especially short beak, a distinctive trait. The ventricular side is rounded. The dorsal plate is about one quarter of the width of the stone. Under natural conditions, it clings to trees like a liana in wet forests and groves.

In the layer dated the 13th–14th centuries, there were other species as well, which did not come from wet environments. Woundwort (*Chrysanthemum leucanthemum*) lives in meadows and bushy places. The decorative pale globe thistle (*Echinops sphaerocephalus*) is more common in shrubberies, clearings, weed-infested places, while white laceflower (*Orlaya grandiflora*) is a species of bushes and sometimes vineyards. Cypress spurge (*Euphorbia cyparissias*) lives mainly in grassy, weedy places and sometimes in ploughlands. Hoary plantain (*Plantago* cf. *media*) is usually found in meadows and stubble-fields. Due to the deterioration of the naked grain, fescue (*Festuca* spec.) was not identified by species. It must have come from grasslands (meadows, pastures), as a stock-forming grass. There were sheep's sorrel (*Rumex acetosella* agg.) achenes as well. This plant grows in grassy places, stubble-fields, but can be very difficult when it gets into the sowing. Black nightshade (*Solanum nigrum*) seeds were also found. This species occurs in weedy places, fallow land, ploughland and among intertilled crops.

The species unearthed from the 13th–14th-century layer were common only in terms of black nightshade. Here, fewer species came from wet habitats. This included water dock (*Rumex hydrolapathum*) that occurs in reed-beds, high sedge or tussock sedge associations. Woody nightshade (*Solanum dulcamara*) is at home in wet groves, water edges, cuts and marshland forests. Vernacular healers once used it. Otherwise, it is strongly poisonous; yet, the oval red berry was soaked in brandy to cure rheumatism and arthritis, stomatitis, colitis, shiver, or just as a sedative (RÁPÓTI – ROMVÁRY op.cit.). To a certain extent it is hydrophilous. It grows on floodplains, ploughlands, stubble-fields and in intertilled crops, but otherwise indifferent to associations. Curltop lady's thumb (*Polygonum lapathifolium*) is an association-neutral species. . Wayfaring tree (*Viburnum lantana*) is common in moist forests, forest edges and shrubberies. European cranberry bush (*V. opulus*) grows in groves, parkland forests, wet forests, and in cuts as well. Its garden variety is snowball bush (conv. *roseum*). Other species included meadow clary (*Salvia* cf. *pratensis*), which is common in meadows and grassy-shrubby places, and common Saint John's wort (*Hypericum perforatum*), which occurs in grassy, shrubby places and forest edges. Its tea is used against stomatitis, colitis, ingestion, kidney and gall ailments, hypertension, internal and external bleeding. It is a gargle against sore throat and receding gums (RÁPÓTI – ROMVÁRY op.cit.).

In the upper layer dated to the 15th century, only one shrub was found, blood-twig dogwood, red dogwood (*Cornus sanguinea*), which is common in groves and marshland forests or forested, shrubby places.

When the natural flora is examined by layers, hence by ages, a substantial change can be recorded. At the turn of the 13th–14th centuries the environment was very diverse. Waterlogged, wet parts were around. The existence of the reed-beds and sedge associations is confirmed by several sedges (panicled sedge, fox sedge/prickly sedge, common spike-rush). Species like gipsywort, carpet cinquefoil, yellow sedge come from places with permanent water cover. Marshes slowly transformed into groves. Remains of such woody areas are common hop and European wild grapes. Possibly bramble was collected at the edges whose seeds were found in great numbers. Beside the marshland, there were grasslands. The number of species coming from meadows and pastures is relatively high. These meadows must have been very diverse. Certain sedges (hairy sedge, pale sedge) indicate waterlogged parts. However, in places where woundwort, cypress spurge, meadow clary, hoary plantain and Orlay chervil once grew, the soil must have contained less moisture. And these species consitute the majority.

In sum, the environment at the end of the 13th century and the beginning of the 14th century, showing originally a very rural aspect, changed in a few decades. The marshy, waterlogged grasslands that had existed previously in the Buda castle quarter all but vanished and gave way to a developed environment. This corresponds with the building of the royal castle and its surrounding.

At Vác, several periods of the Late Middle Ages are present. This allows documentation of environmental changes. Accordingly, large forest ranges were present in the proximity of the settlement in the 15th–16th centuries. Wild fruits already mentioned must have come from such locations (sloe (blackthorn), hazelnut, crab apple, forest wild strawberry, dewberry, cornelian cherry, dogwood and single-seed hawthorn). Maybe, together with these, some elements of the vegetation made their way into the area such as strawberry ground cherry (*Physalis alkekengi*) and black nightshade (*Solanum nigrum*). People took their land by clearing. The occurrence of common birch's (*Betula pendula*) winged acorns is associated with this. No more plants were introduced from the forest edges in the Turkish Period. Meadow plant species can only be found after the Turkish Period.

The liberation wars at the end of the Turkish Period all but wiped out the great majority of rural Hungarian population engaged in farming. Grain and fruit production did not gain momentum again until the 18th century, after the Turkish occupation (*Figs. 429–438, 440*). Plants of the New World gained ground: maize (corn), gourds and squash appear at the end of the 17th century, while in the 18th century common tobacco, potato and bean are introduced. This era, however, falls outside the scope of research in archaeobotany.

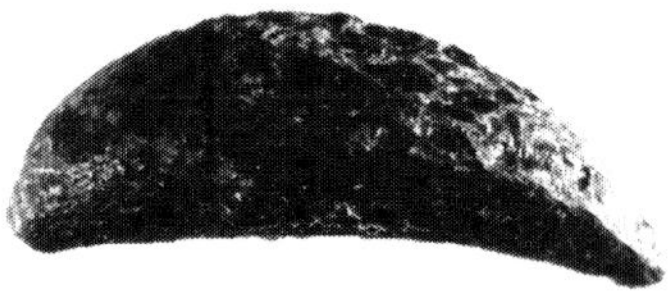

Fig. 327. Garlic (*Allium sativum*) shoot bulb clove from Nagyvázsony-Csepely (15th–16th centuries), Inventory of the Hungarian Agricultural Museum, Budapest.

Fig. 329. Almond (*Prunus amygdalus*) stone fragments from well No 1., Budapest I., Dísz square 10. (Buda castle) (13th–14th centuries), Inventory of the Hungarian Agricultural Museum, Budapest.

Fig. 331. Sour cherry (*Prunus cerasus*) stones from well No 1., Budapest I., Dísz square 10. (Buda castle) (13th–14th centuries), Inventory of the Hungarian Agricultural Museum, Budapest.

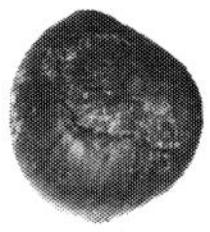

Fig. 333. Hazelnut (*Corylus avellana*) shell from well No 1., Budapest I., Dísz square 10. (Buda castle) (13th–14th centuries), Inventory of the Hungarian Agricultural Museum, Budapest

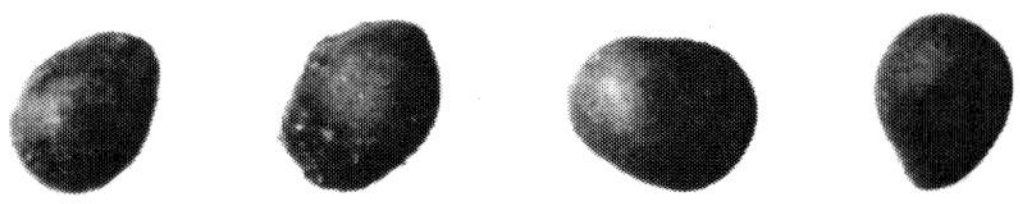

Fig. 334. Common fig (*Ficus carica*) seeds from well No 1., Budapest I., Dísz square 10. (Buda castle) (13th–14th centuries), Inventory of the Hungarian Agricultural Museum, Budapest.

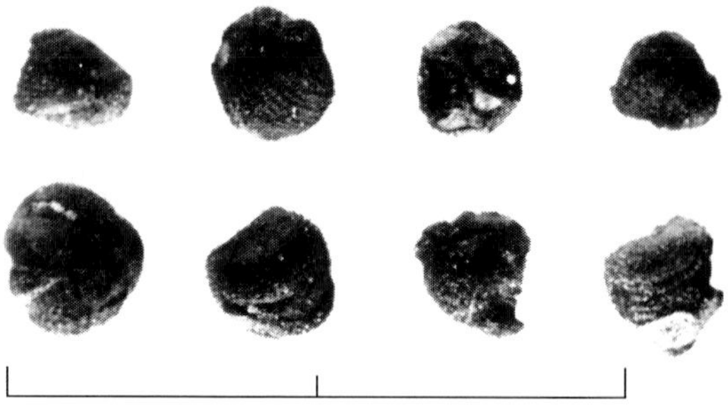

Fig. 328. Korn-rade (*Agrostemma githago*) seeds from Nagyvázsony-Csepely (15th–16th centuries), Inventory of the Hungarian Agricultural Museum, Budapest.

Fig. 330. Sweet cherry (*Prunus avium*) stones from well No 1., Budapest I., Dísz square 10. (Buda castle) (13th–14th centuries), Inventory of the Hungarian Agricultural Museum, Budapest.

Fig. 332. Watermelon (*Citrullus lanatus*) seeds from well No 1., Budapest I., Dísz square 10. (Buda castle) (13th–14th centuries), Inventory of the Hungarian Agricultural Museum, Budapest.

Fig. 335. Walnut (*Juglans regia*) stones from well No 1., Budapest I., Dísz square 10. (Buda castle) (13th–14th centuries), Inventory of the Hungarian Agricultural Museum, Budapest.

Fig. 336. Cherry plum (*Prunus* cf. *cerasifera*) stones from well No 1., Budapest I., Dísz square 10. (Buda castle) (13th–14th centuries), Inventory of the Hungarian Agricultural Museum, Budapest.

Fig. 337. Garden plum (*Prunus domestica* subsp. *oeconomica*) stones from well No 1., Budapest I., Dísz square 10. (Buda castle) (13th–14th centuries), Inventory of the Hungarian Agricultural Museum, Budapest.

Fig. 338. Grape-vine (*Vitis vinifera* subsp. *vinifera*) seeds from well No 1., Budapest I., Dísz square 10. (Buda castle) (13th–14th centuries), Inventory of the Hungarian Agricultural Museum, Budapest.

Fig. 339. Danewort (*Sambucus ebulus*) seeds from well No 1., Budapest I., Dísz square 10. (Buda castle) (13th–14th centuries), Inventory of the Hungarian Agricultural Museum, Budapest.

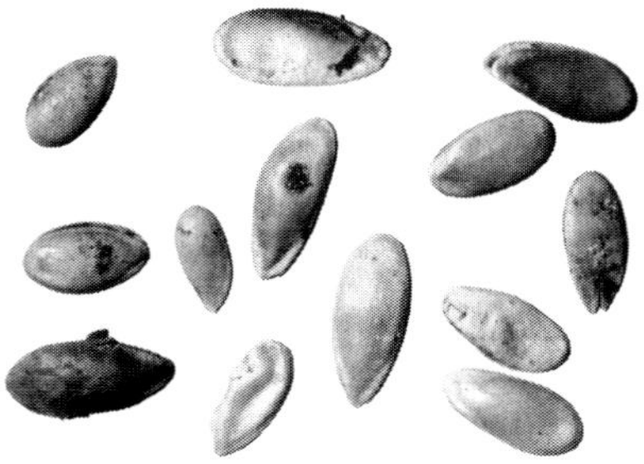

Fig. 340. Muskmelon (*Cucumis melo*) seeds from a well, Budapest I., Úri Str. 40. (Buda castle) (13th–14th centuries), Inventory of the Hungarian Agricultural Museum, Budapest.

Fig. 341. Watermelon (*Citrullus lanatus*) seeds from a well, Budapest I., Úri Str. 40. (Buda castle) (13th–14th centuries), Inventory of the Hungarian Agricultural Museum, Budapest.

Fig. 342. Cornelian cherry or dogwood (*Cornus mas*) stone from a well, Budapest I., Úri Str. 40. (Buda castle) (13th–14th centuries), Inventory of the Hungarian Agricultural Museum, Budapest.

Fig. 343. Sloe, blackthorn (*Prunus spinosa*) stones from well No 1., Budapest I., Dísz square 10. (Buda castle) (13th–14th centuries), Inventory of the Hungarian Agricultural Museum, Budapest.

Fig. 344. Common fig (*Ficus carica*) seeds from a well, Budapest I., Úri Str. 40. (Buda castle) (13th–14th centuries), Inventory of the Hungarian Agricultural Museum, Budapest.

Fig. 345. Walnut (*Juglans regia*) stones from a well, Budapest I., Úri Str. 40. (Buda castle) (13th–14th centuries), Inventory of the Hungarian Agricultural Museum, Budapest.

Fig. 346. Sloe, blackthorn (*Prunus spinosa*) stones from a well, Budapest I., Úri Str. 40. (Buda castle) (13th–14th centuries), Inventory of the Hungarian Agricultural Museum, Budapest.

Fig. 347. Grape-vine (*Vitis vinifera* subsp. *vinifera*) seeds from a well, Budapest I., Úri Str. 40. (Buda castle) (13th–14th centuries), Inventory of the Hungarian Agricultural Museum, Budapest.

Fig. 348. Common millet (*Panicum miliaceum*) naked grains from Nagyvázsony-Csepely (15th–16th centuries), Inventory of the Hungarian Agricultural Museum, Budapest.

Fig. 349. Barley (*Hordeum vulgare*) grains from Nagyvázsony-Csepely (15th–16th centuries), Inventory of the Hungarian Agricultural Museum, Budapest.

Fig. 350. Common millet (*Panicum miliaceum*) naked grains from Nagyvázsony-Csepely (15th–16th centuries), Inventory of the Hungarian Agricultural Museum, Budapest.

Fig. 351. Rye (*Secale cereale*) grains from Nagyvázsony-Csepely (15th–16th centuries), Inventory of the Hungarian Agricultural Museum, Budapest.

Fig. 352. Common bread wheat (*Triticum aestivum* subsp. *vulgare*) grains from Nagyvázsony-Csepely (15th–16th centuries), Inventory of the Hungarian Agricultural Museum, Budapest.

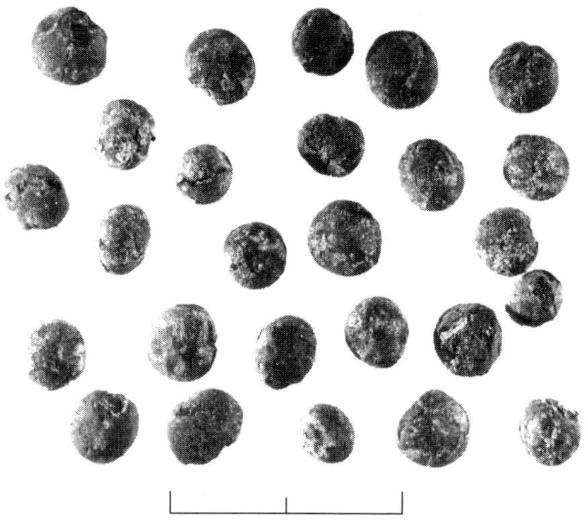

Fig. 353. Big seed lentil (*Lens culinaris* subsp. *macrosperma*) from Nagyvázsony-Csepely (15th–16th centuries), Inventory of the Hungarian Agricultural Museum, Budapest.

Fig. 354. Horsebean (*Vicia faba*) seeds from Nagyvázsony-Csepely (15th–16th centuries), Inventory of the Hungarian Agricultural Museum, Budapest.

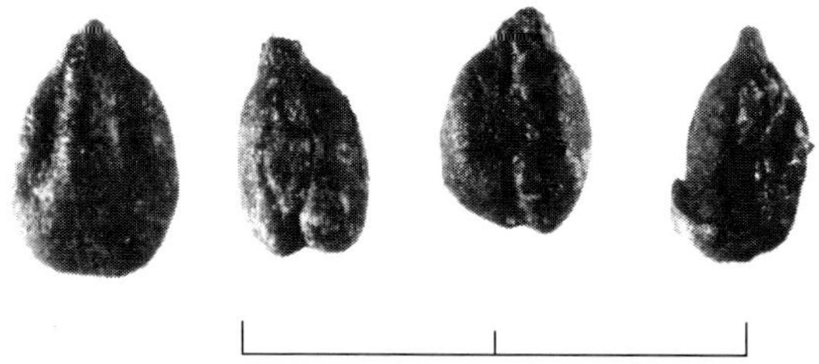

Fig. 355. Grape-vine (*Vitis vinifera* subsp. *vinifera*) seeds from Nagyvázsony-Csepely (15th–16th centuries), Inventory of the Hungarian Agricultural Museum, Budapest.

Fig. 356. Common flax (*Linum usitatissimum*) seeds from Nagyvázsony-Csepely (15th–16th centuries), Inventory of the Hungarian Agricultural Museum, Budapest.

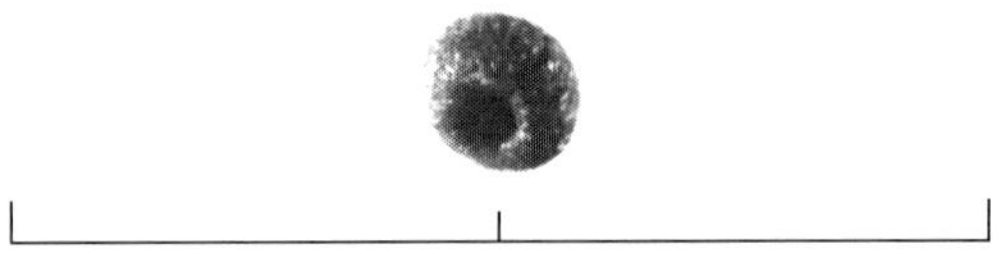

Fig. 357. Catchweed or cleavers (*Galium aparine*) achenes from Nagyvázsony-Csepely (15th–16th centuries), Inventory of the Hungarian Agricultural Museum, Budapest.

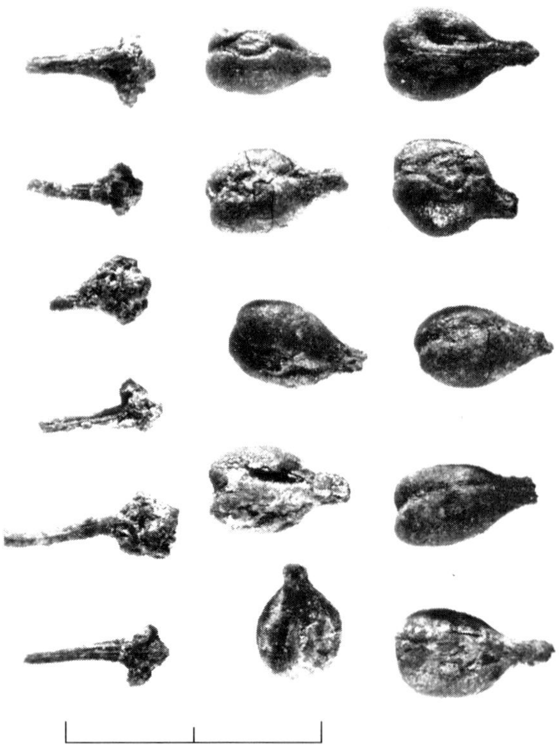

Fig. 358. Grape-vine (*Vitis vinifera* subsp. *vinifera*) seeds and stalk fragments from Nagyvázsony-Csepely (15th–16th centuries), Inventory of the Hungarian Agricultural Museum, Budapest.

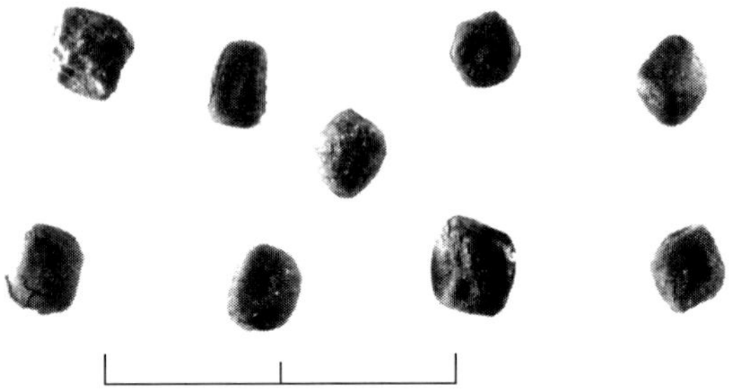

Fig. 360. Vetchling (*Lathyrus* spec.) seeds from Nagyvázsony-Csepely (15th–16th centuries), Inventory of the Hungarian Agricultural Museum, Budapest.

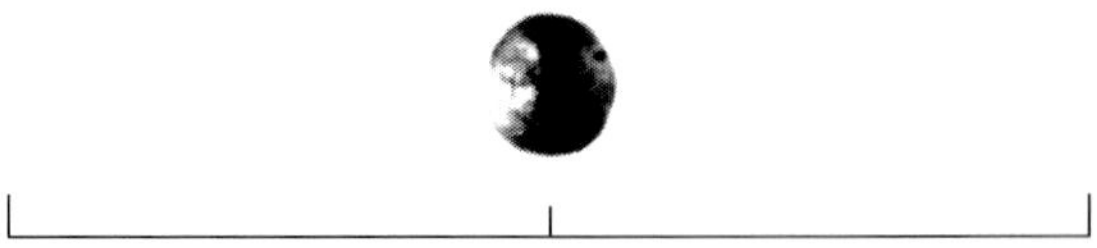

Fig. 359. Charlock (*Sinapis arvensis*) seed from Nagyvázsony-Csepely (15th–16th centuries), Inventory of the Hungarian Agricultural Museum, Budapest.

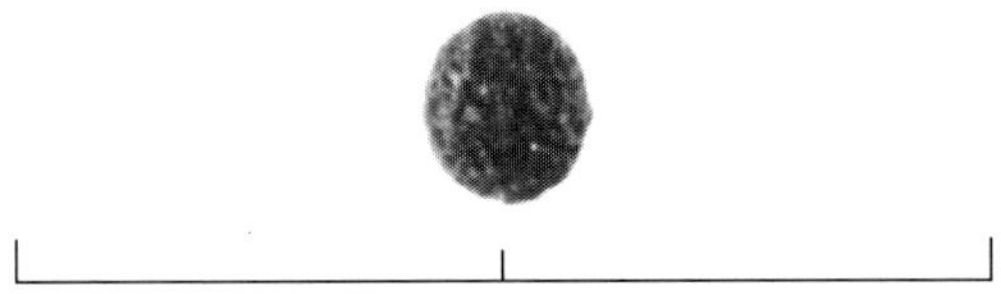

Fig. 361. Ball mustard (*Neslea paniculata*) siliqua from Nagyvázsony-Csepely (15th–16th centuries), Inventory of the Hungarian Agricultural Museum, Budapest.

Fig. 362. Common bread wheat (*Triticum aestivum* subsp. *vulgare*) grains from Pécs-Kórház Str. (16th–17th centuries), Inventory of the Hungarian Agricultural Museum, Budapest.

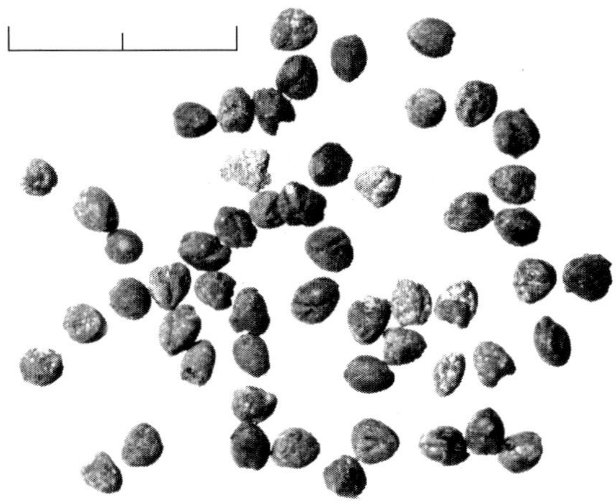

Fig. 363. Common millet (*Panicum miliaceum*) naked grains from the Benedictin monastery, Kaposvár-Kaposszentjakab, Turkish Period (16th century), Inventory of the Hungarian Agricultural Museum, Budapest.

Fig. 364. Common bread wheat (*Triticum aestivum* subsp. *vulgare*) grains fromthe Benedictin monastery, Kaposvár-Kaposszentjakab, Turkish Period (16th century), Inventory of the Hungarian Agricultural Museum, Budapest.

Fig. 365. Horsebean (*Vicia faba*) seeds from the Benedictin monastery, Kaposvár-Kaposszentjakab, Turkish Period (16th century), Inventory of the Hungarian Agricultural Museum, Budapest.

Fig. 366. Sweet cherry (*Prunus avium*) stones from well N° 15 of the former Military Headquarters in the Buda castle. Photograph by Ferenc Hegyi.

Fig. 367. Sour cherry (*Prunus cerasus*) stones from well N° 15 of the former Military Headquarters in the Buda castle. Photograph by Ferenc Hegyi.

Fig. 368. Bullace plum (*Prunus domestica* subsp. *insititia*) stones from well N° 15 of the former Military Headquarters in the Buda castle. Photograph by Ferenc Hegyi.

Fig. 369. Peach (*Prunus persica*) stone from well N° 15 of the former Military Headquarters in the Buda castle. Photograph by Ferenc Hegyi.

Fig. 370. Walnut (*Juglans regia*) stone from well N° 15 of the former Military Headquarters in the Buda castle. Photograph by Ferenc Hegyi.

Fig. 371. Garden plum (*Prunus domestica* subsp. *oeconomica*) stones from well N° 15 of the former Military Headquarters in the Buda castle. Photograph by Ferenc Hegyi.

Fig. 372. Oblong hazelnut (*Corylus avellana*) shell fragment from well N° 15 of the former Military Headquarters in the Buda castle. Photograph by Ferenc Hegyi.

Fig. 373. Almond (*Prunus amygdalus*) stone fragment from well N° 15 of the former Military Headquarters in the Buda castle. Photograph by Ferenc Hegyi.

Fig. 374. Sloe or blackthorn (*Prunus spinosa*) stones from well N° 15 of the former Military Headquarters in the Buda castle. Photograph by Ferenc Hegyi.

Fig. 375. Grape-vine (*Vitis vinifera* subsp. *vinifera*) seeds from well N° 15 of the former Military Headquarters in the Buda castle. Photograph by Ferenc Hegyi.

Fig. 376. Cornelian cherry or dogwood (*Cornus mas*) stones from well N° 15 of the former Military Headquarters in the Buda castle. Photograph by Ferenc Hegyi.

Fig. 377. Common fig (*Ficus carica*) seeds from well N° 15 of the former Military Headquarters in the Buda castle. Photograph by Ferenc Hegyi.

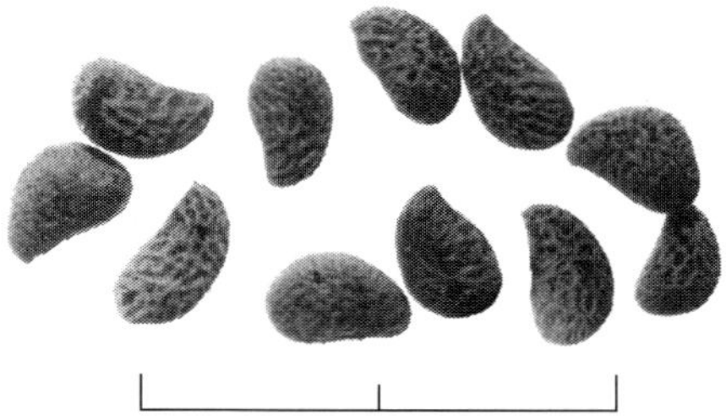

Fig. 378. Bramble (*Rubus caesius*) seeds from well N° 15 of the former Military Headquarters in the Buda castle. Photograph by Ferenc Hegyi.

Fig. 379. Muskmelon or cantaloupe (*Cucumis melo*) seeds from well N° 15 of the former Military Headquarters in the Buda castle. Photograph by Ferenc Hegyi. Type "A": oblong, large seeds (Cantaloupe type).

Fig. 380. Muskmelon (*Cucumis melo*) seeds from well N° 15 of the former Military Headquarters in the Buda castle. Photograph by Ferenc Hegyi. Type "B": short seeds (Turkestan type).

Fig. 381. Watermelon (Citrullus lanatus) seeds from well N° 15 of the former Military Headquarters in the Buda castle. Photograph by Ferenc Hegyi.

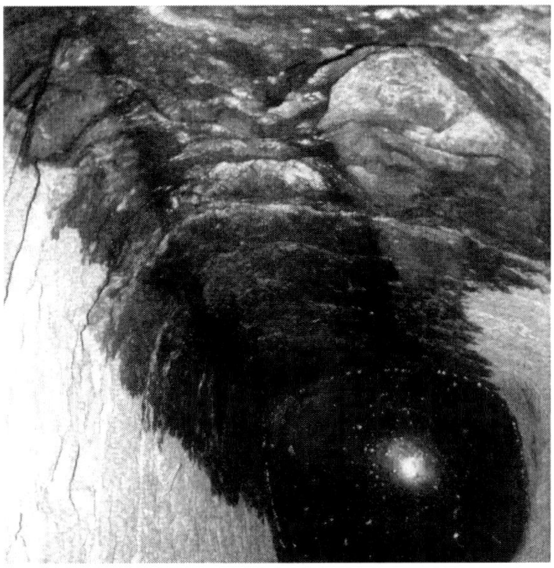

Fig. 382. Rock-well N° 8 of the Teleki mansion in the Buda castle, the beginning of the 15th century. Photograph Dorottya B. Nyékhelyi.

Fig. 383. Seed remains from rock-well N° 8 of the Teleki mansion in the Buda castle, the beginning of the 15th century. Photograph by Gábor Gyulai.

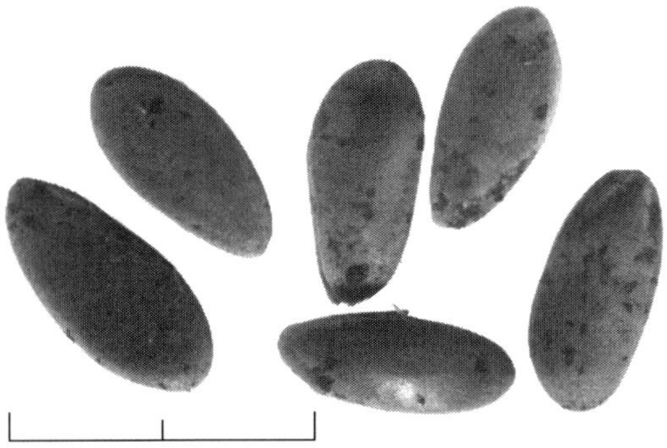

Fig. 384. Muskmelon (*Cucumis melo*) seeds from rock-well N° 8 of the Teleki mansion in the Buda castle, the beginning of the 15th century. Photograph by the author.

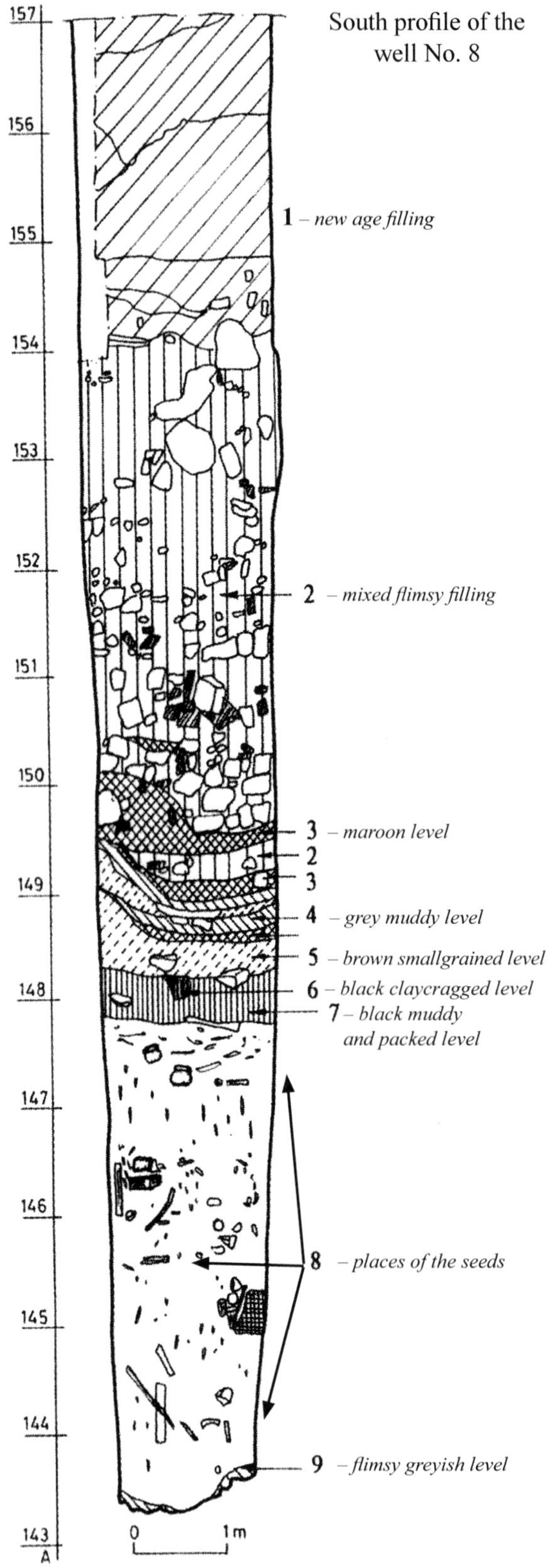

Fig. 385. Stratigraphy of rock-well N° 8 of the Teleki mansion in the Buda castle, the beginning of the 15th century. After Dorottya B. Nyékhelyi.

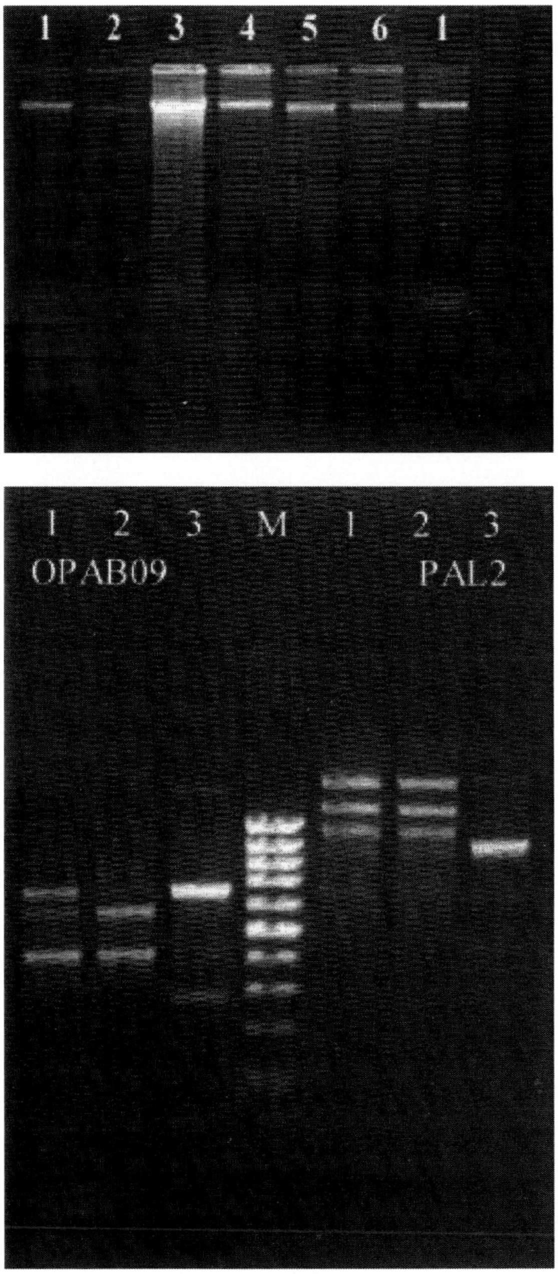

Fig. 386. Findings of genetic examinations of medieval seeds coming from the excavations ın the Buda castle. Photographs by Gábor Gyulai.
1: Total DNA isolates of plant seeds from the Sigismund Period
2: Results of PCR (polymeraze chain reaction) of plant seeds from the Sigismund Period

Fig. 387. Well or sink from Budapest I., Kapucinusok Str. (Buda castle), the end of the 15th century. Photograph by András Végh.

Fig. 388. Well or sink from Budapest I., Kapucinusok Str. (Buda castle), the end of the 15th century. Photograph by András Végh

Fig. 389. Separation of flotated organic remains from inorganic material in Budapest I., Kapucinusok Str. (Buda castle), the end of the 15th century. Photograph by Renáta Bacsó.

Fig. 390. Separation of remains to taxa in Budapest I., Kapucinusok Str. (Buda castle), the end of the 15th century. Photograph Renáta Bacsó.

Fig. 391. Determination of the diaspores using stereo-binocular microscope in Budapest I., Kapucinusok Str. (Buda castle), the end of the 15th century. Photograph Renáta Bacsó.

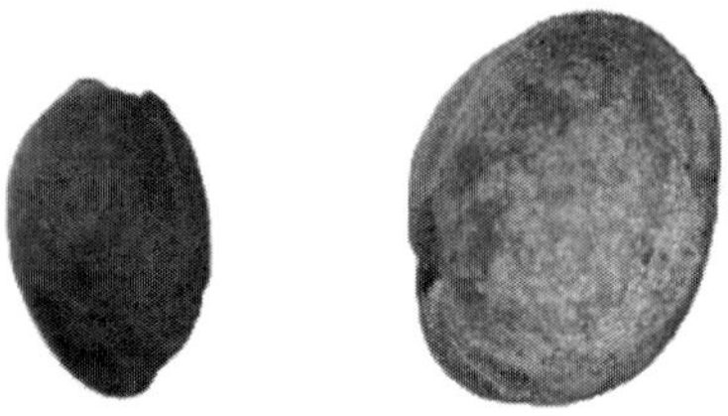

Fig. 392. Sweet cherry (*Prunus avium*) stones from Budapest I., Kapucinusok Str. (Buda castle), the end of the 15th century. Photograph Renáta Bacsó.

Fig. 393. Sour cherry (*Prunus cerasus*) stones from Budapest I., Kapucinusok Str. (Buda castle), the end of the 15th century. Photograph Renáta Bacsó.

Fig. 394. Garden plum (*Prunus domestica* subsp. *oeconomica*) stone from Budapest I., Kapucinusok Str. (Buda castle), the end of the 15th century. Photograph Renáta Bacsó.

Fig. 395. Apple (*Malus domestica*) seed from Budapest I., Kapucinusok Str. (Buda castle), the end of the 15th century. Photograph Renáta Bacsó.

Fig. 396. Watermelon (*Citrullus lanatus*) seed from Budapest I., Kapucinusok Str. (Buda castle), the end of the 15th century. Photograph Renáta Bacsó.

Fig. 397. Sloe or blackthorn (*Prunus spinosa*) stone from Budapest I., Kapucinusok Str. (Buda castle), the end of the 15th century. Photograph Renáta Bacsó.

Fig. 398. Almond (*Amygdalus communis*) stone from Budapest I., Kapucinusok Str. (Buda castle), the end of the 15th century. Photograph Renáta Bacsó.

Fig. 399. Walnut (*Juglans regia*) stone fragment from Budapest I., Kapucinusok Str. (Buda castle), the end of the 15th century. Photograph Renáta Bacsó.

Fig. 400. Haselnut (*Corylus avellana*) shell fragment from Budapest I., Kapucinusok Str. (Buda castle), the end of the 15th century. Photograph Renáta Bacsó.

Fig. 401. Haselnut (*Corylus avellana*) shell fragment from Budapest I., Kapucinusok Str. (Buda castle), the end of the 15th century. Photograph Renáta Bacsó.

Fig. 402. Endospermium of cornelian cherry or dogwood (*Cornus mas*) stone from Budapest I., Kapucinusok Str. (Buda castle), the end of the 15th century. Photograph Renáta Bacsó.

Fig. 403. Cornelian cherry or dogwood (*Cornus mas*) stones from Budapest I., Kapucinusok Str. (Buda castle), the end of the 15th century. Photograph Renáta Bacsó.

Fig. 404. „Spider" shaped grapevine (*Vitis vinifera* subsp. *vinifera*) seed from Budapest I., Kapucinusok Str. (Buda castle), the end of the 15th century. Photograph Renáta Bacsó.

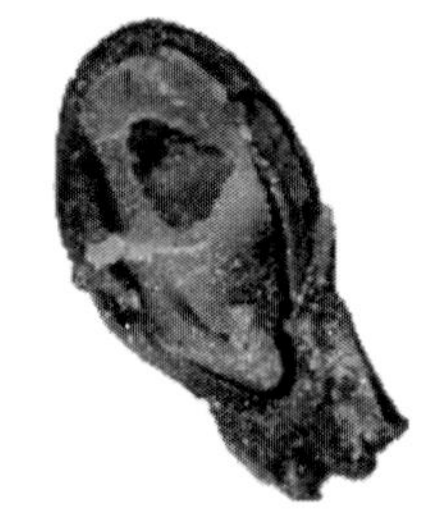

Fig. 405. Endospermium of grapevine (*Vitis vinifera* subsp. *vinifera*) seed from Budapest I., Kapucinusok Str. (Buda castle), the end of the 15th century. Photograph Renáta Bacsó.

Fig. 406. Balkan type grapevine (*Vitis vinifera* subsp. *vinifera*) seed from Budapest I., Kapucinusok Str. (Buda castle), the end of the 15th century. Leg. Géza Facsar. Photograph Renáta Bacsó.

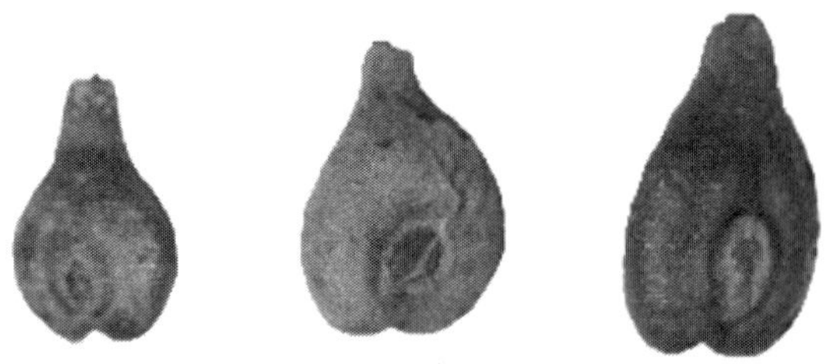

Fig. 407. Balkan type grapevine (*Vitis vinifera* subsp. *vinifera*) seeds from Budapest I., Kapucinusok Str. (Buda castle), the end of the 15th century. Leg. Géza Facsar. Photograph Renáta Bacsó.

Fig. 408. Goher type grapevine (*Vitis vinifera* subsp. *vinifera*) seeds from Budapest I., Kapucinusok Str. (Buda castle), the end of the 15th century. Leg. Géza Facsar. Photograph Renáta Bacsó.

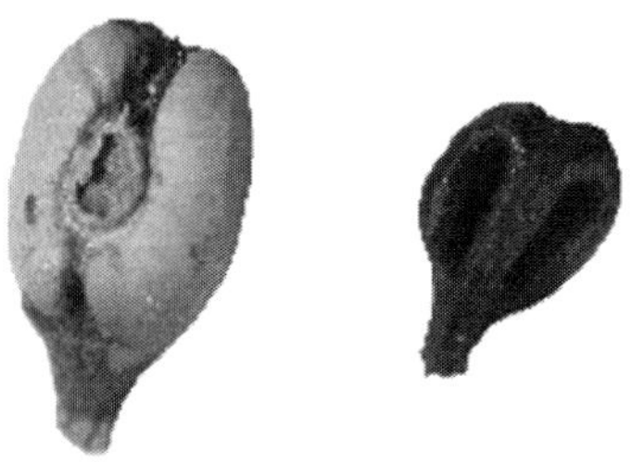

Fig. 409. Goher-Doblerab types grapevine (*Vitis vinifera* subsp. *vinifera*) seeds from Budapest I., Kapucinusok Str. (Buda castle), the end of the 15th century. Leg. Géza Facsar. Photograph Renáta Bacsó.

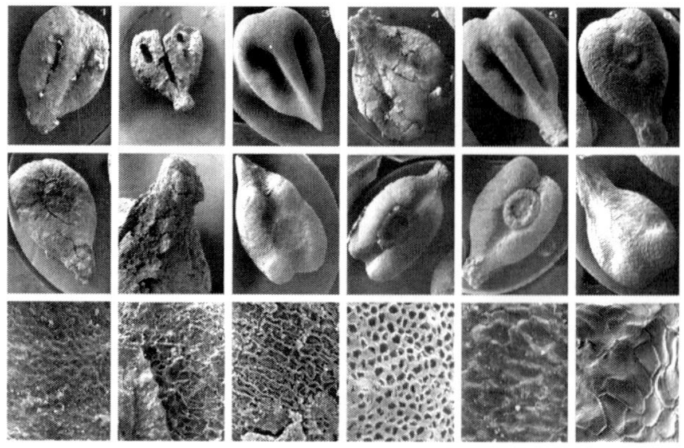

Fig. 410. Morphology of ancient *Vitis* seeds excavated in Hungary. SEM micrographs of seeds excavated at a Roman villa at Aquincum, Budapest, (2nd–4th cent. AD) (1) at Fenékpuszta (2); at a vineyard site near Győr, Ece (11th–12th cent.) (3); at Debrecen (15th cent.) (4); at the King's Palace in the Buda castle, Budapest, (15th cent.) (5). The SEM micrograph of seeds of the contemporary *Vitis vinifera* cultivar 'Kék Bakator' is also shown (6). Upper and middle rows show seed morphology at 20x magnification. Bottom row shows seed coat detail at 500x magnification. GYULAI *et. al.* 2009.

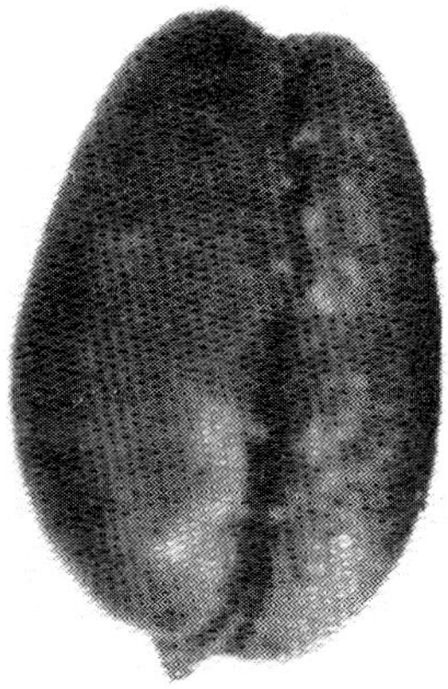

Fig. 411. Common bread wheat (*Triticum aestivum* subsp. *vulgare*) grains from the 16th-century settlement layer of Vác. Photograph by Tibor Kádas.

Fig. 412. Club wheat (*Triticum aestivum* subsp. *compactum*) grains from the 16th-century settlement layer of Vác. Photograph by Tibor Kádas.

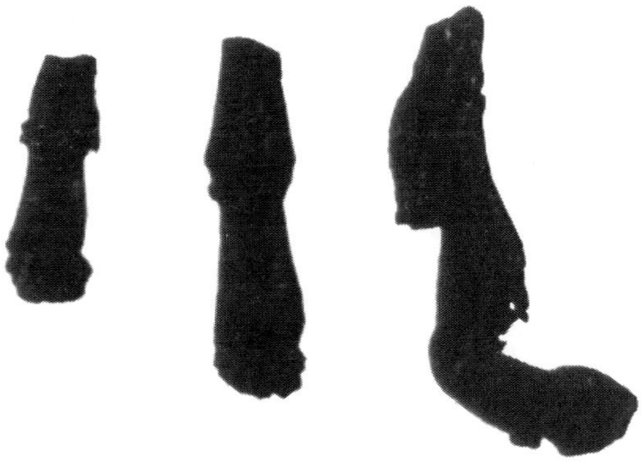

Fig. 413. Bread wheat (*Triticum aestivum*) ear spindle fragments from the 16th-century settlement layer of Vác. Photograph by Tibor Kádas.

Fig. 414. Rye (*Secale cereale*) grains from the 16th-century settlement layer of Vác. Photograph by Tibor Kádas.

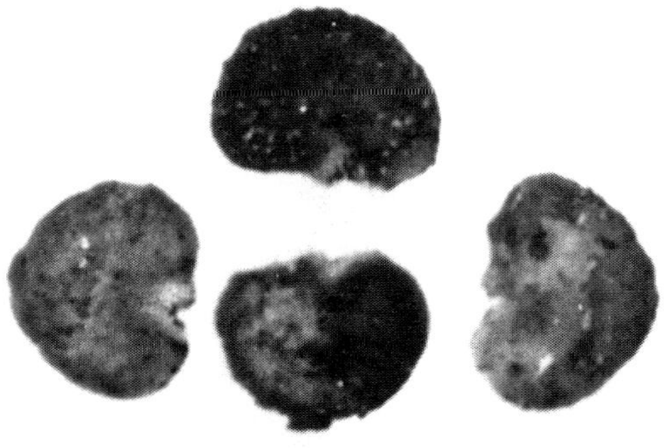

Fig. 415. Jimsonweed (*Datura stramonium*) seeds from the 16th-century settlement layer of Vác. Photograph by Tibor Kádas.

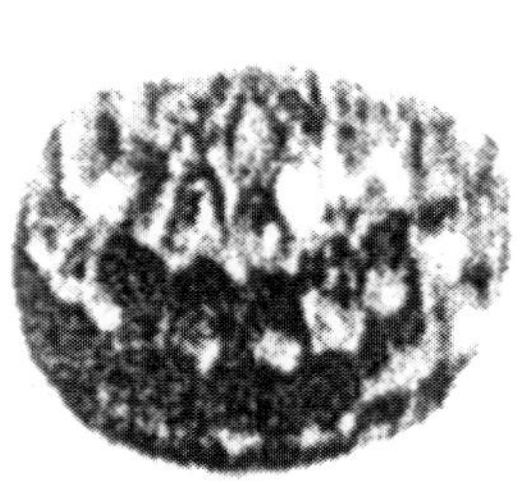

Fig. 416. Opium poppy (*Papaver somniferum*) seed from the 16th-century settlement layer of Vác. Photograph by Tibor Kádas.

Fig. 417. Barley (*Hordeum vulgare*) hulled grain from Dunaföldvár-Öregtorony, the beginning of the 17th century, Inventory of the Hungarian Agricultural Museum, Budapest.

Fig. 418. Germinated rye (*Secale cereale*) grains from Dunaföldvár-Öregtorony, the beginning of the 17th century, Inventory of the Hungarian Agricultural Museum, Budapest.

Fig. 419. Common millet (*Panicum miliaceum*) naked grains from Dunaföldvár-Öregtorony, the beginning of the 17th century, Inventory of the Hungarian Agricultural Museum, Budapest.

Fig. 420. Hare's-ear mustard (*Conringia orientalis*) seeds from Dunaföldvár-Öregtorony, the beginning of the 17th century, Inventory of the Hungarian Agricultural Museum, Budapest.

Fig. 421. Steppen-sesel (*Seseli annuum*) achenes from Dunaföldvár-Öregtorony, Turkish Period, Inventory of the Hungarian Agricultural Museum, Budapest.

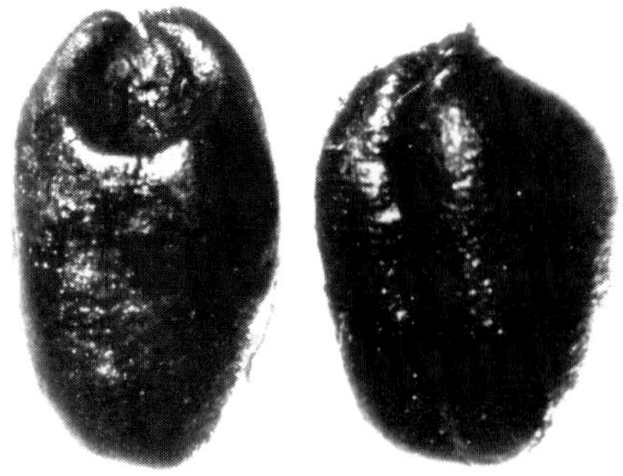

Fig. 422. Cammon wheat (*Triticum aestivum* subsp. *vulgare*) grains from Dunaföldvár-Öregtorony, the beginning of the 17th century, Inventory of the Hungarian Agricultural Museum, Budapest.

Fig. 423. Common wheat (*Triticum aestivum* subsp. *vulgare*) grains from Dunaföldvár-Öregtorony, the beginning of the 17th century, Inventory of the Hungarian Agricultural Museum, Budapest.

Fig. 424. Charlock (*Sinapis arvensis*) seeds from Dunaföldvár-Öregtorony, the beginning of the 17th century, Inventory of the Hungarian Agricultural Museum, Budapest.

Fig. 425. Rye (*Secale cereale*) grains from the gate of Hollókő castle, 17th century, Inventory of the Hungarian Agricultural Museum, Budapest.

Fig. 426. Rye brome (*Bromus secalinus*) naked grains from the gate of Hollókő castle, 17th century, Inventory of the Hungarian Agricultural Museum, Budapest.

Fig. 427. Hoary cress or hoary pepperwort (*Lepidium draba*) seed from the gate of Hollókő castle, 17th century, Inventory of the Hungarian Agricultural Museum, Budapest.

Fig. 428. Thorn-apple (*Datura stramonium*) seeds from the ditch of Hollókő castle, 17th century, Inventory of the Hungarian Agricultural Museum, Budapest.

Fig. 429. White goosefoot or fan-hen (*Chenopodium album*) seeds from the cellar of the barrack-room in Esztergom castle, 16th–18th centuries, Inventory of the Hungarian Agricultural Museum, Budapest.

Fig. 430. Yellow bristle-grass (*Setaria lutescens*) hulled grains from the cellar of a barrack-room in Esztergom castle, 16th–18th centuries, Inventory of the Hungarian Agricultural Museum, Budapest.

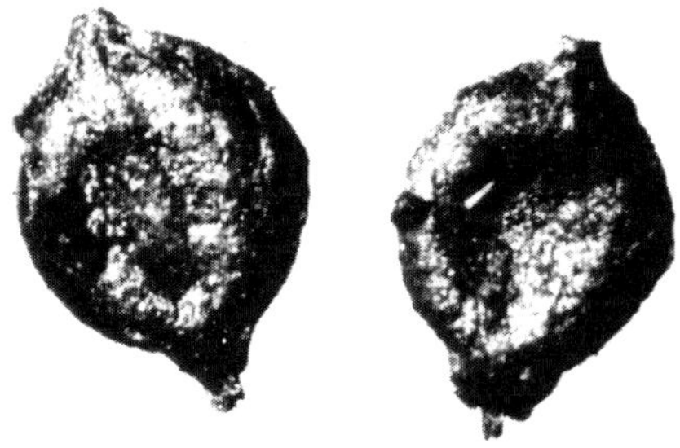

Fig. 431. Curltop lady's-thumb (*Polygonum lapathifolium*) seeds from the cellar of the barrack-room in Esztergom castle, 16th–18th centuries, Inventory of the Hungarian Agricultural Museum, Budapest.

Fig. 432. Annual woundwort (*Stachys annua*) nutlets from the cellar of a barrack-room in Esztergom castle, 16th–18th centuries, Inventory of the Hungarian Agricultural Museum, Budapest.

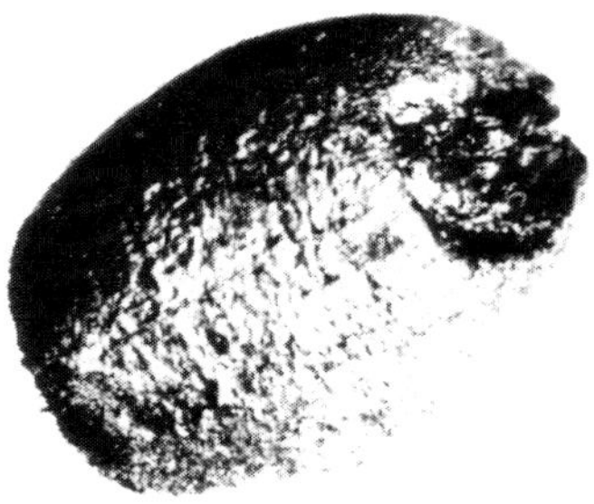

Fig. 433. Common wheat (*Triticum aestivum* subsp. *vulgare*) grain from Feldebrő Roman Catholic church, 18th-century, Inventory of the Hungarian Agricultural Museum, Budapest.

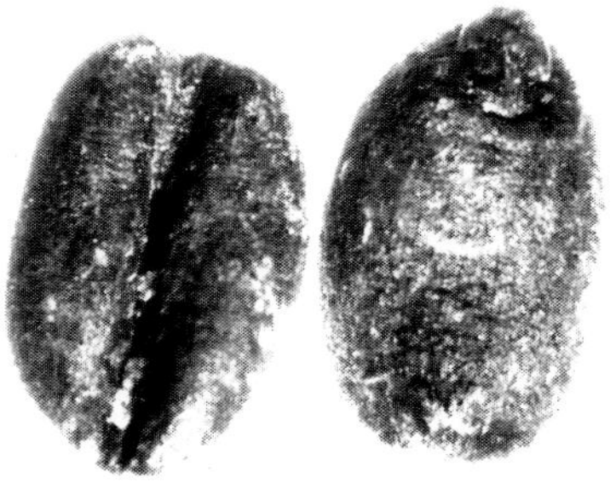

Fig. 434. Common wheat (*Triticum aestivum* subsp. *vulgare*) grains from Feldebrő Roman Catholic church, 18th-century cereal storige pit, Inventory of the Hungarian Agricultural Museum, Budapest.

Fig. 435. Spring wild-oat (*Avena fatua*) naked grains from Feldebrő Roman Catholic church, 18th-century cereal storige pit, Inventory of the Hungarian Agricultural Museum, Budapest.

Fig. 436. Common millet (*Panicum miliaceum*) naked grains from Ebes, 18th-century pit, Inventory of the Hungarian Agricultural Museum, Budapest

Fig. 437. Yellow bristle-grass (*Setaria lutescens*) grains from Ebes, 18th-century pit, Inventory of the Hungarian Agricultural Museum, Budapest.

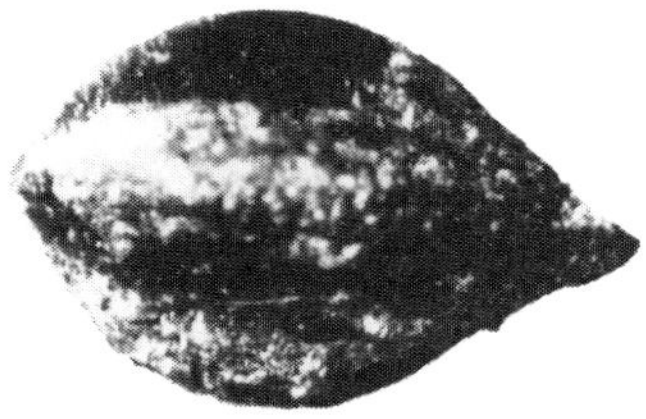

Fig. 438. Black-bindweed (*Fallopia convolvulus*) nutlets from Ebes, 18th-century pit, Inventory of the Hungarian Agricultural Museum, Budapest.

Fig. 439. Spring wild-oat (*Avena fatua*) hulled grain from Tószeg, prehistoric site. Inventory of the Hungarian Agricultural Museum, Budapest.

Fig. 440. Field pennycress (*Thlaspi arvense*) seed from Ebes, 18th-century pit, Inventory of the Hungarian Agricultural Museum, Budapest.

Fig. 441. Two-rowed barley (*Hordeum vulgare* subsp. *distichum*) hulled grain from Tószeg, prehistoric site. Inventory of the Hungarian Agricultural Museum, Budapest.

Fig. 442. Club wheat (*Triticum aestivum* subsp. *compactum*) grains from Szentes Tűzköves, Neolithic or Bronze Age. Inventory of the Hungarian Agricultural Museum, Budapest.

Fig. 443. Emmer (*Triticum turgidum* subsp. *dicoccum*) naked grains from Szob-Ipolypart, prehistoric site. Inventory of the Hungarian Agricultural Museum, Budapest.

Fig. 444. Einkorn (*Triticum monococcum*) spicelets (furca bicornis) from Szob-Ipolypart, prehistoric site. Inventory of the Hungarian Agricultural Museum, Budapest.

Fig. 445. Einkorn (*Triticum monococcum*) naked grains from Szob-Ipolypart, prehistoric site. Inventory of the Hungarian Agricultural Museum, Budapest.

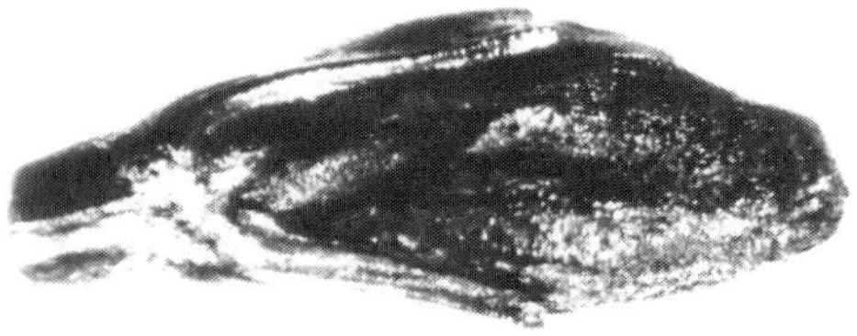

Fig. 446. Einkorn (*Triticum monococcum*) hulled grain from Szob-Ipolypart, prehistoric site. Inventory of the Hungarian Agricultural Museum, Budapest.

Fig. 447. Barley (*Hordeum vulgare*) hulled grain from Tószeg, Roman or medieval site. Inventory of the Hungarian Agricultural Museum, Budapest.

8. Food remains of archaeological cultures in the Carpathian Basin

8.1. Purpose and methods of analysis of food remains

Dishes and bowls found in graves during excavation of graveyards suggest that the deceased was provided with food and drink before embarking on the long road to the other world. The question emerges whether this assumption is true and what kind of food was put beside the dead? Where and under which conditions were such remains preserved and where can they be found now? Is it possible to determine their composition and can we find out how they were prepared? Research on the history of food confronts such questions.

Food remains emerge only very seldom from archaeological sites. The reason is that, under natural conditions, organic matter is decomposed by microbiota very quickly. However, there are certain special conditions; such as, carbonisation by the impact of fire, water coverage or an extremely dry microclimate, under which the potential for their preservation exists.

Ceramics is often the most frequent and most abundant find at excavations. Pottery sometimes occurs in large amounts and its production technology, decoration and form provides an important, if not the most important dating element for classical archaeology. In a number of cases – even when archaeologists find a whole pot – one cannot tell the purpose for its creation and what was cooked or stored in it. The analysis of the food remains coming from such containers is the tool for deciding such issues. Therefore, it is of key importance that pots and pottery fragments during the exploration works be examined thoroughly in situ before washing. Recognition of food remains requires attention and a certain level of experience. Failing to do so means that they are lost forever.

Macroscopic and microscopic examination of food remains does not preclude the need for carpological tests. Yet, with the help of these methods a great deal of precious information can be obtained on the lifestyle and eating habits of ancient cultures that cannot be obtained any other way. Examination of food remains contributes to the compilation of relevant knowledge about the long legacy of development of cake and leavened bread, eating habits of past eras, and gastronomic aspects of past cultures.

Food remains come in separate forms (meal, flour, gruel or bread) or attached to different objects (such as on the surface of metal objects, stuck to the inner wall, the edges or the bottom of pottery and fragments of pottery). However, one also has to include the analysis of stomach contents and faeces remains of the extremely rare marshland- and glacier-corpses here (RICHTER 1988).

Organic remains from archaeological periods are usually preserved in a carbonised state. Burning, which can cause structural changes and permanent deterioration to organic matter, can also conserve it at the same time. For the purposes of macroscopic examinations, sometimes only parts of glumellae – which are rich in phytoliths – survive (NETOLITZKY 1926). If the material to be examined is not yet completely carbonised, the carbonised part can be removed with the help of different chemical procedures. Thus the remaining botanical tissue remains can be subjected to microscopic examinations. Exploration must be determined by the state of the actual sample. In order to make phytolith-rich plant residues visible under the light microscope, special embedding material with high refraction coefficient is used (PIPERNO 1987).

As a result of microscopic preparations requiring special chemical procedures, organic elements (such as tissue residues, fascicula fragments, phytoliths, pollen, spores, hair, colouring agents and dyes, cocoon, and so on) can be distinguished even after several thousand years (NETOLITZKY 1926; LOCHTE 1951, 1954; FEINDT 1989; RICHTER 1987; GASSNER 1989; MEHLHORN – PIEKARSKI 1989). Beside food remains, one can also reckon with the presence of dyes, drugs and poisonous substances. The microscopic examination of cooking remains (soups, vegetable dishes, prepared dishes) baked (burnt) on the inner wall of pottery resembles forensic investigations (GASSNER 1989).

The examination of food remains is a great challenge. During analysis, one is required to have botanical, chemical and gastronomical knowledge.

The field of microscopic processing of food remains was pioneered by NETOLITZKY (1927). WÄHREN (1988, 1989b) developed a macroscopic procedure combined with microscopic techniques for the purposes of processing prehistoric flour, mush, cake and other cooking remains originating from South German and Swiss lake dwellings. In Northern Europe it was HJELMQVIST (1984, 1990), in Germany Schlichtherle (1983), in Switzerland RICHTER (1987, 1988) and WÄHREN (1984, 1987, 1988, 1989b), in Slovakia HAJNALOVÁ (1989b) PIETA – PLACHÁ (1989) who dealt with similar investigations.

MAURIZIO (1916a) carried out ash examinations in order to clarify the origin of cake bread pieces coming from prehistoric lake dwellings. However, only advanced microscopic techniques and instrumental analytical chemical tests made it possible to analyse these sporadic findings appropriately. The very fact that the findings are so sporadic, showing different traits and features, can be the reason why no uniform methodology to be followed has been laid down so far in connection with the comprehensive set of examinations called generally known as the analysis of food remains.

Gas chromatography and atomic absorption spectrophotometry facilitated the study of the macro and microelement composition of seed and fruit finds. Carbonization does not hinder these methods of analysis. Investigations show that a portion of the elements is still present in the residues. The lack of mobile, easily leached elements does not necessarily mean that they were not there when the food was prepared. No sodium, for instance, has been detected thus far from prehistoric food remains. This can be attributed to dissolution and not the fact that no salt was used for the preparation of the foodstuff.

Compared to contemporary food remains, the number of compound clusters to be identified in historical remains is reduced, usually caused by the impact of heat and the age of the find (ROTTLÄNDER 1983a). As a result of carbonisation due to moderated heat impact under anaerobic conditions, the number of compound groups preserved is very limited. No starch, sugar or protein is found in them. However, free amino acids, cholesterol, fatty acids and oleic acids can be identified. Starch can be interpreted as a sugar polymer as it can be transformed into sugar by hydrolysis. At temperatures as low as 190°C it loses water and becomes brownish. Therefore, no starch or sugar can be detected in food remains exposed to heat. Proteins are easy to denaturise. Peptide bounds are disrupted at 200–210°C and become brownish-black. The only exception is free amino acids, which are resistant to both time and heat.

Free amino acids were found in seemingly carbonised cereal grains from various archaeological periods around Lake Balaton (examinations by János Csapó in: GYULAI 1997c). As amino acids deteriorate at different rates when compared to each other, the concentration values are not suitable for determining their original ratios. Their presence does not indicate the type of protein in question, their original concentration or the proportion of each amino acid in the compound. However, based on the method of the so-called amino acid racemization, there is a chance to determine the age of the organic matter (ROTTLÄNDER 1983a; CSAPÓ – TÓTH-PÓSFAI – CSAPÓ-KISS 1986). Initial results indicate that the amino acid racemization used for age determination of bone remains can also be used for the interpretation and evaluation of archaeobotanical finds. Having obtained a sufficient amount of data, it can also become a new method of dating.

Sterane skeleton compounds are also relatively stable against heat impact. Plant and animal hormones are not easy to decompose. Cholesterol for instance endures a heat impact of 250°C for three hours. Cholesterol detectability is very important since it can help us to decide whether the food remains are of plant or animal origin. In terms of detectability fats are most stable. Fatty acids preserve their chemical structure up to 400°C. According to fatty acid tests developed in the Archaeochemical Laboratory of the Institute of Prehistory, Tübingen, fatty acids identified in food remains also indicate the type of food (ROTTLÄNDER 1983b). Gas chromatography of fatty acids less sensitive to heat impacts proved to be a suitable method for the purposes of clarifying whether organic matter is of plant or animal origin. (ROTTLÄNDER – SCHLICHTHERLE 1980).

8.2. Preparation of prehistoric food

Prehistoric housewives must have devoted a considerable of time in food preparation before cereals were served in the form of dishes. Certainly it all depended on what and when they wanted to serve. There are several questions that remain to be answered. Were cereals eaten regularly on a daily basis and how much were they considered a staple food? To what extent were meat, vegetables, fruits and gathered plants represented in the daily food intake and how much were they independent from the dishes made of cereal species?

If one considers that the everyday obligations of a housewife in those times included carrying water, making fire, processing hides and fibre plants, weaving, sewing garments, gathering herbs and other plants, growing and tending crops, cooking and baking, one can say that in spite of the considerable amount of time required for the cleaning of prehistoric wheat varieties, they could possible not have much time to do it.

In prehistory, thrashing and glumellae removal were separate operations in the case of hulled wheats grown.

Advanced type naked wheats such as tetraploid durum wheat, hexaploid common bread wheat and club wheat grains can be deprived of their glumallae by simple threshing. Thus the ear spindle would be torn into uneven parts. Earlier on, threshing was a separate action after harvesting, done on the barn-floor, using a flapper or treading by horse or ox. It is only today that the two operations are completed in one single step using a harvester-thresher combine.

When threshed, ears of hulled wheat cultivars such as einkorn or emmer, which represent the more primitive and less developed varieties, are broken up into spicules. The ear spindle is broken under the spicule (coccoid break). When threshed, spelt also comes apart into spicules but this time a fragment of the next spindle piece is left attached to it (speltoid break).

Thus, the ear of hulled wheats (*spica*) fell apart only into spicules (*spiculum*), and most of the time the glumellae were left on the grains. In order to obtain the grains from among the glumellae, they had to be removed first.

The removal of the glumellae could be done in several ways:

- abrasion and wearing. Only a lactescent, in other words not fully ripen grain can be fretted from the ear. Cereals in waxy maturation can only be disencumbered from the glumellae with dry abrasion between grinding stones. However, this process produces too many broken grains. In our view, grinding mills made of lower and upper stones were mainly used for grinding already husked grains to make meal or flour and they were not used to get rid of the glumellae. In fact, ground chaff makes the flour taste bitter and hence unpalatable. Wooden mortars and "kölyű" (hammer mortar) were used for chaffing buckwheat and millet.
- boiling with lye. Hot lye ash was poured upon cereals in the ears, thus softening chaff, then removing it from the grains by fretting through a sieve (SÁGI 1972). FÜZES (1990) raises the possibility that Neolithic sifter pans with smaller holes might have been used for this purpose. According to ethnographic analogies, millet in the past was mixed with lye-ash and then chaffed half-wet in a mortar. In Asia Minor, in the Taurus mountains, hulled bread wheat, barley, and oat spicules were habitually boiled first, dried, then ground in a hand mill (TAGÁN 1941). Meal on

the surface was separated from groats by sifting and thus glumellae and awn fragments became visible.

- drying and browning. When sundried and eventually browned afterwards, the glumellae open up at the tips and become fragile. It is very likely that grain was burnt when browned in browning dishes and was therefore discarded. According to experimental archaeological examinations, browning in browning dishes proved to be highly efficient, yet in spite of all the efforts some of the chaffs remained. However, when softened and heated for several hours, the glumellae were so much loosened that they fell off by themselves and thus could be easily removed by winnowing. More resistant glumellae could be separated by manual grinding. If spelt was heated for two hours at 50°C, the removal of glumellae took one hour (MEURERS-BALKE 1985). In our opinion, only smaller amounts were chaffed in browning dishes, and only if the condition of the wheats destined for grinding required it.
- treading. In our view, larger amounts of grain were chaffed not in browning dishes but on a daub barn-floor, just as it was done on the southern slopes of the Taurus mountains (TAGÁN 1941 in: FÜZES 1990) and in Anatolia (personal communication by Paolo Biaggi) up to recent times. This was done using a wooden sleigh-like implement with massive sleds, drawn by oxen, donkeys or horses, called "threshing sleigh", of which the sled was hobnailed in an extremely dense manner with silex blades. As wood putrefies, this might be an explanation why silex blades with so called "hook shine" were found in prehistoric sites in such a great number. Experiments also confirm the efficiency of such a tool (ANDERSON 1998).

FÜZES (1990) examined grain imprints found in Neolithic clay daub fragments, wattle-and-daub structures and in the wall of ceramic pots, arriving at the conclusion that grains of hulled cereals – after apparently mild browning, – must have been rubbed out from among the glumellae when subjected to some strong mechanic influence.

Chaffing operations made on the barn floor and in the browning dishes were apparently not mutually exclusive phases, but complementary, strongly related processes. Larger amounts of bread wheat must have been chaffed on the daub barn floor after harvest, storage and drying. Maybe the wattle-and-daub structures, found in great numbers at prehistoric exploration sites and filled with burnt out remains of glumellae are traces of such daub barn-floors and not always of walls.

According to our own observations, removal of chaffs from hulled wheat varieties by rubbing were easiest in the case of spelt, while emmer was more difficult. Einkorn chaffs were the most difficult to remove as they were firmly attached to the grain.

It is noted here that there is a so called "naked grain" *einkorn* as well, but we do not believe that this cultivar, described by ZHUKOVSKY (1964) as Triticum sinskajae, discovered only recently and still found very rarely, would have been present in Hungarian prehistory.

Similarly to the chaffing process, very little is known of sowing in the prehistory. According to demonstration experiments made in the Archaeological Park in Százhalombatta, wheat yields depended very much on the type of sowing (KOVÁCS 2002). According to these tests, carried out twice on 10 m^2 lots, the yield in the case of sowing with digging sticks was 0.2 kg, dispersed sowing yielded 0.6 kg, and sowing in rows 1.4 kg. The same value reached 6 kg when current sowing methods and technology were used. The Sumerians practised furrow-sowing as well and on the Iberian peninsula, a long time ago so-called island sowing was found which resulted in bundles of grain stalks. However, even the most primitive method of digging stick sowing proved to be four times more efficient than ordinary gathering.

The action of grinding must have required considerable time in the prehistory. Single-handed grinding using manual grinder stones required two hours of work for producing 1 kg of good quality meal (GUNDA 1961). Using an oblong shaped grinding stone requiring both hands, the time used for grinding was somewhat less (HENNIG 1966). The form of the grinding stone, consisting of two parts, remained practically unchanged up to the Iron Age. Grinding was made in several steps. The separation of the grist by fractions assumedly took place by using a sieve made of hairs. Certainly, the goal was not always to produce bakery-grade flour. Coarse meal was better suited for making soups or groats-gruel anyway. According to the tests, if grains were first roasted at 200°C, coarse grinding was made much more simple (MEURERS-BALKE 1985). According to the experiments made using Neolithic grinding stones, cereals could be milled to make coarse groats by grinding five times and fine quality bakery-grade flour products by grinding and sieving fifteen times (WÄHREN 1985a).

The former concept, assuming that in prehistory only coarse grist could be made, is now obsolete. HEER (1865) still maintained that the grinding stones used at the time were suitable only for bursting out the grains and not even the germ could be removed, and according to STOKAR (1951) grain could be milled only to coarse grade meals in this period. However, according to the analytical results of the food remains found in the Portalban site, fine quality bakery-grade flour could be created as early as in 4900 BC (WÄHREN 1988). Bread found in Muntelier (3150 BC) was made from fine grade flour in 70% of finds and, moreover, germs were removed from the groats.

8.3. Long forgotten farinaceous plants

One cannot imagine flour nowadays to be made of anything else but grain. However, this is not at all that natural. People in prehistoric cultures did not insist so much on cultivated plants as we do today but tried to co-exist even with weeds. At the dawn of domestication, people had repeatedly selected from the surrounding flora those species which suited them and in many different areas. The relationship between tiller and plants was dynamic, especially at the beginning. The farmer did not strive to eradicate at any price the weeds appearing in his environment and among the cereal plants but he tried to choose from among them those useful to him. For instance, originally rye and oat were both weeds of wheat cereals, and they only became cultivars over time by careful selection. Thus food became more diverse for humans. It would be unreasonable to think that people were forced to try new plants because of famine. Doubtlessly, there were periods of need as well. These species, just like useful kinds of cereals, usually lived in proximity with humans, either around houses and stables or in the local environment. They generally occur in larger numbers at prehistoric sites, rich in grain finds.

There are still a great number of edible fruits and seeds of grasses to be found in the natural environment. In Poland and Lithuania there is for instance, metliza (*Poa lithuanica*), a plant similar to oats. It is found in wet habitats en masse. Grains were gathered and cleaned, mixed to cereal flour and baked to make bread (SZUTÓRISZ 1905). In Brazil, Indian or tuscarora rice, in other words wild rice (*Zizania aquatica* L.) was cultivated, which is now grown in Hungary as well. Love-lies-bleeding species (*Amaranthus caudatus*, *A. hypochondriacus*, *A. cruentus*) were once grown by the Incas in South America. Nowadays they also proliferate in Hungary. In Poland, gruel was made of red fingergrass (*Digitaria sanguinalis* (L.) Scop.) as well. Grains of native wild Hungarian Glumaceae are mostly edible; such as, *Aegilops*, *Avena*, *Bromus*, *Haynaldia*, *Hordeum*, *Secale* species.

There are a great number of plant species in the world that once were used for making mush, bread or other kinds of food (KUNKEL 1984). In the Carpathian Basin, the ratio of wild plant species which have at least some parts (root, stalk, leaves, flowers, fruits or seeds) that are edible raw or as a kind of prepared food, is very high (PRÓNAI 1999). The high number of plant species eaten as food might have something to do with the fact that the historical landscape of the Carpathian Basin is different from the contemporary one.

The natural landscape of the Carpathian Basin and especially that of the Great Hungarian Plain was markedly different from the one today: it was made diverse by open waters, marshes, moors, forests and meadows. In the Late Migration Period and the age of the Conquest, approximately one eighth of historical Hungary and one quarter of contemporary Hungary was covered by water continuously or periodically.

This land has been inhabited continuously since the Neolithic. Places suitable for settlement were mostly the edges of areas free from the floods. Flora and fauna in the marshes provided food for both humans and livestock, grass grew in abundance after the floods were gone and there was no scarcity of drinking water. It was not before the later periods of the Middle Ages that the zonal oak-forest steppe of the Great Plain was transformed into a cultural landscape, an agricultural land.

Marshes and waterlogged areas have always been rich in species. Especially live glumaceae were held in high esteem. The use or consumption of these is long forgotten by now. Only archaeological-botanical finds, written data and ethnological records refer to it. Without setting the objective of completeness, a number of these will be listed below.

8.3.1. White goosefoot

White goosefoot or fan-hen (*Chenopodium album* L.) is an annual weed that grows to 50–120 cm tall. Small, undistinguished green flowers bear many shiny black seeds, covered with perianth. It is permanently associated with human settlements, as the nitrogen and phosphorus content of the soil is higher in these specific contexts. It prefers loose humus-filled clay and loam soils that have a dynamic water balance, but it also survives on flood-plains and alkali soils. It is abundant in any crop, mostly in row crops, sometimes in large numbers (ÚJVÁROSI 1957; SCHERMANN 1966; SOÓ 1970, 1980; HUNYADI 1988). A number of different races have split from it in the ancient Himalayan homeland. The plant is still grown at places in the Western Himalayas and India: leaves and young sprouts are prepared for cooking, husked grains milled for meal or groats used for gruel and pulp (RAPAICS 1934). In times of famine, such as in Russia in 1891–1892, it was mixed with wheat flour to make bread (DANERT *et al.* 1980).

It is one of the most ancient and most common "spinach" foods, but the meal prepared from the older leaves impairs vision (RAPAICS 1934.) and is slightly laxative (Danert *et al.* 1980). Its popular name suggests that the leaves might have been eaten once: "'östörparé', or '-paraj' (blite), or 'paréj',… in other words: esterparé. Its vernacular name is 'fejér libatopp'" (CZUCZOR – FOGARASI 1868). It is also called "lúdlábfű", or "libaláb" (goosefoot). The name of the genus (*Chenopodium*) refers to the goosefoot shape of the species belonging to it. In the time of neology, the composite word of the Greek "khēn" – goose and "podion" – small foot was translated into Hungarian, hence the Hungarian name "top" (small foot) (SZABÓ – PÉNTEK 1996). Only blite (*Ch. Bonus-henricus* L.) has become a cultivar from among the goosefoot species. It was a favourite leafy vegetable of medieval kitchen gardens.

The seeds were also identified in the Scandinavian Neolithic (HJELMQVIST 1955). It is known in Hungary sporadically, likewise since the Neolithic (HARTYÁNYI – NOVÁKI – PATAY 1967–68). Thus, as the finds show, fan-hen or white goosefoot is a Neolithic weed introduced with cultivated plant species. The seeds were always found together with cereals. Based on the large amount of fan-hen or white goosefoot seeds found in the Neolithic lake dwelling in Robenhausen, Switzerland, it can be assumed that it was already used for alimentative purposes at the time (MESSIKOMMER 1913).

Cultures living in the Carpathian Basin once gathered, consumed and eventually grew goosefoot species as a replacement for grain. The considerable amount of fan-hen or white goosefoot seeds found in the two dwelling houses excavated at the Ottomány culture tell settlement of Túrkeve-Terehalom from the Middle Bronze Age (ca. 1600 BC) was stored for consumption at the time (GYULAI in print). This is also confirmed by the cleaned seed stock of many-seeded goosefoot (*Chenopodium polyspermum* L.) found in the remnants of a Late Bronze Age (Vatya culture) house at Tiszaian Feudvar/Vojvodina (KROLL 1990a). Fan-hen or white goosefoot seeds were found in extremely large amounts in the Late Migration Period lake dwelling of Fonyód-Bélatelep (GYULAI – HERTELENDI – SZABÓ 1992). Achenes lacked perigonium, albeit the five-toothed perianth closes off the seed completely, therefore this sign indicates a cleaning operation. More than 70% of the 181 thousand seeds/bearings consisted of this plant species. It was doubtlessly consumed. However, concluded from the circumstances of their occurance, goosefoot finds from the Carpathian Basin were not emergency food. A number of other grain remains indicate abundance. These were rather the heritage of cultures experimenting with a variety of foods.

The findings above show that this plant species, as a "semi-wild" cultivar, was on the way to domestication. The fact that it did not become a cultivated crop can be explained by the different eating habits of later populations, not the changes in the plant's potential.

8.3.2. Water-chestnut

Water-chestnut (*Trapa natans* L.) still occurs in stagnant waters and in oxbows of major rivers (SCHERMANN 1966). The stalk of the annual plant is cylindrical and its length can reach 2 metres. The typical round leaves appear on the surface of the water in June. The spiky, horny fruits start ripening in August and September.

Water-chestnut is a member of the water-lily-pondweed community (*Nymphaeion*, *Hydrocharition*, *Potamion*), it has dominant stands (*Trapaetum*). Flowers blossom from July to August. It can still be found sporadically as a member of the community in the Bükk mountain range, Nyírség, Tisza region, Balaton-highland, Belső-Somogy, Hanság and along the river Dráva in oxbows and stagnant waters (SOÓ – KÁRPÁTI 1968). Since river channelizations and draining of marshes, water-chestnut mostly found a habitat in oxbow lakes in eutrophic and hypertrophic waters.

Water-chestnut, according to its 2–4-cm-long fruit and its shape is usually divided into two subspecies:

- subsp. verbanensis (De Not.) Schinz, which has four short or long horns and usually a crown at the top,
- subsp. muzzanensis Jaeggi, which has only two horns and it is warted between the horns.

It is a thermophilous plant. It does not occur north of the Alps. In the postglacial warm era its area spread somewhat more to the north than today, when it is a Mediterranean and Central European species (SOÓ – JÁVORKA 1951; SOÓ 1980; RAPAICS 1934). ENTZ – SEBESTYÉN (1942) mentioned that BORBÁS (1900) had not yet recorded it and they quote Boros, who in 1927 stated that its presence in Lake Balaton "is doubtlessly due to new colonisation, maybe deliberate introduction". Clusius found water-chestnut in Transdanubia, but Lake Balaton is not mentioned in his works (in: ENTZ – SEBESTYÉN 1942). Would it be possible that the cool climate in the Middle Ages did not favour this termophilous plant and therefore, albeit temporarily, it withdrew and sometimes even vanished?

Water-chestnut is practically extinct in Germany and Switzerland. In Yugoslavia however it is common. In Hungary, it can be considered as an endangered species. One can find a number of water-chestnuts in oxbows of the river Tisza. Recently, it began to spread in still bays of Lake Balaton. In addition to habitat destruction, the impact of muskrats introduced from North America several decades ago contributes a great deal to the scarcity of water-chestnut.

In Hungarian, just in many other languages, this plant was called "water chestnut". It was a delicacy, but it also served as an emergency food in times of famine. Meal and bread, sometimes even cakes, were made of the fruit which the plant bears in August and September. When cooked, it tastes like sweet chestnuts and it also has similar nutritional value (BALASSA 1985). The fruit of water-chestnut, favored for its taste and nutritional value, was not only food for the poor, it was also paid as a tax to landlords. Floating water-chestnut fruits were collected using baskets and rakes, or angled with a piece of fur or sheep tail, trailed from a boat trapping the spiky water chestnuts. In some cases, a piece of fleece was nailed to a stick and the water stirred with it. Entangled chestnuts were removed using a wooden blade. Spikes were cut from the fruits on a so- called "water-chestnut cutting table", where a knife was drawn under hooks nailed into the table board. Earlier, it was collected in great numbers along the Tisza, and it made its way to the marketplaces of surrounding cities, such as Szolnok, Debrecen, Miskolc or Nyíregyháza. It was called "suly" in Ormánság, "susa" in the Siklós range (Hungarian Ethnographic Encyclopaedia 1979). Other names included: "istennyila" (thunderbolt), vad mandola (wild almond), vízi dió (water walnut), vízi gesztenye (water chestnut). A couple of place names in the Plains such as Sulymos ér, Sulymos tó still reflect its former area.

BALASSA Iván (1985) writes in connection with water-chestnut as follows: "The name is in all probability of Finno-Ugric origin, deep-vowel equivalent for sül-, sün(disznó) (hedgehog). The comparison is confirmed by the fact that both have spikes... Water-chestnut was not eaten raw but boiled in water. First, however, it was put in boiling water for a short time ("kövesztés"), so that the spiky shell could be removed more easily. It is worth noting that the procedure has the same word in both the Finno-Ugric and the Hungarian languages. Thus Finnish "kiehua "means" to cook", and Estonian "keema – to cook, to boil". The inner parts of water-chestnut are dried, broken and ground so that the resulting white flour is suitable not only for making bread but also for different kind of cakes."

It was rarely eaten raw, more often cooked and fried, but also dried. In villages of the Plains, it replaced sweet chestnuts. Boiled water-chestnut, mixed with rye flour and fat, was used for making turf-cake. Even the shells were not discarded: cooked, they were poured onto the floor of houses in order to control fleas. "War water-chestnut", an implement forged from one piece of iron mimicking the fruit, was a very potent and dangerous medieval defence tool. Scattered around camps and dwelling places, it served as a protection.

The custom of eating water-chestnuts goes back to prehistoric times. Its remains can be found everywhere in Central European Neolithic and Bronze Age cultural layers. It is a similarly frequent find in South German and Swiss lake dwellings. It was not only eaten, but also processed into necklaces (JACOMET – BROMBACHER – DICK 1989). An important food since the Neolithic, it was a source of meal and flour, "the potato of the age" at Bronze Age sites (RAPAICS 1934). Greek and Roman writers (Dioskorides, Galenus) mention it as well. Even Péter Melius Juhász mentions it in his Herbaria, published in 1578. Thus, it is very likely that our ancestors took note of the usefulness of this plant as long ago as at the time of the Finno-Ugric co-existence. In spite of all this it is very surprising that written records appeared only quite lately. It was mentioned first in a charter in 1319–1342 in the form of "sulmoshordoka" (SZAMOTA – ZOLNAI 1906).

However, it can be found among botanical finds from the prehistoric Swiss and South German lake dwellings. It was collected for the taste – reminiscent of sweet chestnuts – of the inner core. Sometimes it was stringed as a necklace. Albeit it is native in Hungary and it covered large areas in the Carpathian Basin earlier on, only two sites are known from historical times. The Mošorin-Feudvár tell settlement along the Tisza was well researched from an archaeobotanical point of view as well. Among the seeds and fruits of mostly cultivated plants unearthed from the middle urnfield culture layer (12th to 9th centuries BC), water-chestnut was also present (KROLL 1990a). Water-chestnut was gathered from oxbows of rivers for the purposes of consumption and decoration. The second water-chestnut find was discovered in the Late Migration Period lake dwelling remains of Fonyód-Bélatelep, dated between the second half of the 7th century and the end of the 9th century (GYULAI – HERTELENDI – SZABÓ 1992). When the settlement was populated, the Nagyberek region was covered with shallow waters and muddy marshes, where water-chestnut must have grown in large numbers and gathered for supplementing the diet of the population.

8.3.3. Oak acorn

SUTORIS (1905), writing about the indigenous people of Greece, the Pelasgians, states that they ate the acorn of oak. At that time, Southern-Europe was covered with extensive forests. Acorns of trees living there were gathered, dried and subsequently ground to make bread. Later on, the acorn meal was mixed with cereal flour. The Romans protected oak acorns by law. Oak was utilised in a versatile way (wood, leaves, fruits) and the Greek consecrated it to Zeus, the Romans to Jupiter and the Germans to Donar. It was honoured by all Indo-European peoples almost without exception (SUTORIS 1905).

Pliny writes about thirteen different kinds of oak, mentioning also that the acorn of the salt-oak (*haliphloes*) was bitter. KUNKEL (1984) lists 55 oak species from Eurasia, Africa and America, which have edible acorns. A list set up by Rudof Mansfeld, categorised holm-oak (*Qercus ilex* L.) and manna-oak (*Q. brantii* Lindl.) as those that have the most tasty acorns. He notes in relation to holm-oak that the very tasty acorn of this thermophilous tree species occurs everywhere from the western parts of the Mediterranean to the Caspian sea and has bristly lobular, spiky leaves, is eaten in a manner similar to the chestnut. Manna oak, found from north-east Turkey to Northern Iran and from Kurdistan to Iraq, is actively planted in many places for its tasty acorns. Coffee substitute is made of the acorns but they are also used in natural healing.

István PRÓNAI (1999) lists 25 different oak species, all with edible acorns and notes that the acorn of moss-capped oak (*Q. cerris* L.) used to be fed to hogs, but it can also be assumed that smaller amounts were also eaten by people. The acorn of the hazelnut oak (*Q. ballota* L.) was used for milling meal and making bread when mixed with cereal flour. The acorn of the Hungarian or Italian oak (*Q. farnetto* Ten.) is also edible as its tannin content is low like that of the evergreen oak (*Q. ilex* L.), durmast (*Q. petraea* (Mattuschka) Lieblein) and pubescent oak (*Q. pubescens* Willd.), but it is advisable to roast them before eating. The acorns of common (British or American) oak (*Q. robur* L.) were an emergency food and roasted acorns were used to make substitute coffee. Allegedly, it is harmful to your health when consumed in larger quantities. The author mentioned was successful in cooking an excellent dish by peeling the acorns and soaking them in water in order to rid of their bitter taste. They were then ground and used to flavour dough, pasta and cakes as a filling, like a kind of nut. When exposed to frost, or dug permanently in soil, oak acorns lose their bitter taste and become sweeter. It might be interesting to note that *Q. vallonea* was an important staple food for North American Indians. Having no cereals, it was used among others for these purposes: the meal or groats were cooked to make soups or mush, the flour was used to bake bread.

As opposed to the larger acorns of the Mediterranean oaks, those of native Hungarian oak species are smaller, yet edible in the same way (RAPAICS 1940). Their consumption can be traced back in Hungary to ancient times. We have a number of data from the past with Hungarian references to the consumption of oak acorns.

However, oak acorns can also be eaten raw. They have a sweet taste. However, there is always some bitter tasting tannin in them, which makes them unpalatable. The amount of tannin depends on the species, habitat or even the individual tree. The amount can be diminished by roasting, soaking in lime-water, or covering in soil for a certain amount of time. According to ethnological analogies, forest rangers and herdsmen ate roasted acorns as a supplementary kind of food. Otherwise, it was a well-known emergency food.

When communities were short of grain, it was replaced by acorns: they were dried and ground to bake bread. Master Rogerius provides a written record for this in his "Carmen miserabile" when he describes that – fleeing from the Mongols – he arrived into the rock fortress of Fráta (Magyarfráta) beside Gyulafehérvár, where people who gathered there offered him black bread baked from ground oak bark and meal, whose taste "seemed sweeter than French rolls to me." In our view, leftover acorn grist must have been in it as well.

Oak trees have been used in vernacular healing for a long time (RÁPÓTI – ROMVÁRY 1983). For their contractive effects, tinctures of barks (*Quercus cortex*) and leaves (*Quercus folium*) were applied to mitigate external and internal bleeding, to heal inflammation and frost bites. Raw acorns contain quercitin, 7% tannin, 6% protein, 7% sugar, 37% starch and 5–15% fatty oil. The composition can be substantially improved by roasting: the substance will lose its tannin content, while starch is transformed into dextrin (RÁPÓTI – ROMVÁRY 1983). The coffee product made of roasted acorn bearings (*Quercus glandes*) has a roborating effect. The decoct was regularly given to children with rachitis. For the elderly, chronic colitis was treated with it.

According to archaeobotanical finds in the more austere regions of Central Europe (such as Zurich) oak- (*Quercus* spec.) and European beech- (*Fagus silvatica* L.) acorns and masts were eaten as early as in the Neolithic, while roasted European chestnut (*Castanea sativa* Mill.) was eaten in the forests of the protected, soft climate areas (such as Tessin). Fruits were preferred on the table (JACOMET – BROMBACHER – DICK 1989). It can be taken for granted that collected acorns were used for feeding domestic animals during winter since ancient times (HOPF 1991). However, the circumstances under which the finds were uncovered suggest that the major part was still used for human nutrition. It can be found alone, sporadically, or in waste pits mixed with grain, sometimes in great amounts in fireplaces, and sometimes associated with grinding stones. Large amounts of carbonised oak acorns probably originating from leftovers of human consumption are known from Greek Early Bronze Age sites. Of all the seeds and fruits, the frequency of oak acorn findings reaches 50% in Kastanas (KROLL 1983), and 30% in Sitagroi (RENFREW 1966).

Oak acorn finds of Czech and Slovak sites are analysed in a separate publication (HAJNALOVÁ – FURMÁNEK – MARKOVÁ 1999). These finds occurs in all archaeological periods. It can be found in Bohemia from Copper Age Prague-Lysonaje (TEMPÍR 1966), Middle Bronze Age Plžen (TEMPÍR 1966), Late Bronze Age Prague-Bechovice (TEMPÍR 1973), Late Bronze Age Uherský Brod (TEMPÍR 1966), and Hallstatt Period Kyjovice (KÜHN 1979) sites. Carbonised acorns were found in a clay pit of the Kocin Lengyeli culture (HAJNALOVÁ 1989a) as well as in the Early Bronze Age layer of Nyitra castle (HAJNALOVÁ 2001). They were found in bowls of the Middle Bronze Age Ottomany culture site of Vcelince (HAJNALOVÁ – FURMÁNEK – MARKOVÁ 1999), together with cereals (barley, common bread wheat, club wheat and common millet) and legumes (pea) at the Hallstatt Period Zilina-Stráze site, and also in one of the fireplaces of the similarly Hallstatt Smolenice-Molpír site, mixed with einkorn, common wheat, pea and beans (HOPF 1989). Acorns were also found in a waste pit of the medieval monastery of Hronský Benadik (Garamszentbenedek) (HAJNALOVÁ – FURMÁNEK – MARKOVÁ 1999).

Acorns continued to be favourite wild fruits in the Hallstatt Period (KÜHN 1977). The Hallstatt sites of Romania (1200–450/300 BC) also contain oak acorn finds (WASYLIKOWA *et al.* 1991).

The detection of beech mast is much less frequent. In Western Europe it is known from the Neolithic lake dwellings near Zurich (JACOMET – BROMBACHER – DICK 1989) and the early medieval Haithabu (BEHRE 1983).

Former consumption of oak acorns is confirmed by a number of Hungarian archaeological botanical findings. When exploring the rectangle shaped pit house from the Late Neolithic or Copper Age (Lengyel or Pécel culture) in Moha-Homokbánya, Fejér county, approximately 10 cm^3 of carbonised oak acorns were found. The find was very fragmented and it was determined as pubescent oak (*Q. pubescens*) (HARTYÁNYI – NOVÁKI – PATAY 1967–68).

Acorns were also found in the botanical material, otherwise rich on other species as well, which came to light from houses and pits of the former settlement surrounding the Neolithic tell explored at the Tiszapolgár-Csőszhalom site (*Quercus* sp.). In the Csepel-Waterworks area (excavation by Anna Endrődi 1999), more than one thousand carbonised acorns were found at the bottom of a fireplace from the Baden culture (ENDRŐDI – GYULAI 1998–2000). Concluding from the biogeographical conditions of the area, these might have been the fruits of common or American (British) oak (*Quercus robur* L.). The acorns were very fragmented, broken, containing no shells or cups. Further carbonised acorns, probably originating from human consumption, were found in the find material of the Middle Bronze Age motte of Bölcske-Vörösgyír at the River Danube (K. BERZSÉNYI – GYULAI 1998).

Even the advanced Roman culture could not live without gathering wild berries and fruits, where oak acorns doubtlessly belong. This is supported by evidence from the southern fortress gate from the 5th century in Keszthely-Fenékpuszta, where pubescent oak (*Quercus pubescens*) acorns, mixed with a large amount of cereals, were found in the deterioration layer "C" at the exploration works carried out in 1970 (Füzes, unpublished data).

Oak acorn was sometimes an emergency food in the Middle Ages, but at rtother times it counted as a favourite delicacy. Roasted acorns were found associated with a large number of other improved and wild fruit varieties in the material of the 15th-century well explored in the cellar of the former Military Headquarters in the Buda Castle in 1999.

8.3.4. Reed sweet-grass

Of the marshland grass vegetation, the gathering of reed sweet-grass (float grass or glyceria) (*Glyceria maxima* (Hartm.) Holmbg.) had a long tradition. It is a plant from the temperate climate of the northern hemisphere. It grows in large numbers at the edges of marshlands, open waters. The bean shaped, reddish grains with a diameter of 2 mm were eaten. Grists were used for mush, meal for soup and pudding (KUNKEL 1984). The Belarus mixed it with rye flour to make bread (Hungarian Encyclopaedia of Ethnography 1979). It was so valued at a time that MAURIZIO (1916b) called it "the only non-crop grain in Europe." It was abundant in the once waterlogged areas of the Great Hungarian Plain. It still occurs in marshes today, where it provides dominant stands. It ripens from July to August. The truss is mostly jug-eared, sometimes lifted, the surface of the leaves is smooth and the backside is coarse, sometimes elevating.

The name "manna mush" (glyceria) refers to a miracle in the Bible. When Moses led the Jews out of Egypt, manna falling from the sky in a miraculous manner in the desert saved them from starvation: "And the manna was as coriander seed, and the colour thereof as the colour of bdellium. And the people went about, and gathered it, and ground it in mills, or beat it in a mortar, and baked it in pans, and made cakes of it: and the taste of it was as the taste of fresh oil. And when the dew fell upon the camp in the night, the manna fell upon it." (Numbers 11:7–9, King James Version) However, the phenomenon of falling manna, which has caused a lot of trouble and hard thinking to scientists ever since over the centuries, cannot be explained unanimously only with the presence of reed sweet-grass, although the similarity is really surprising.

In season, the easily shattered grains were collected early in the morning, before the morning dew would have disappeared, shaking them in a (horse hair) sieve drawn among the trusses and clapping on them with a stick. This is where the Hungarian name ("due mush") comes from. This usually happened after St. John's Day, but mainly in July. As the grains in the trusses kept on ripening, the operation had to be repeated every now and then, but only at dawn, when shattering was strongest. At times it was gathered from boats with movements mimicking the use of a butterfly-net, but actually using linen bags. The grains, put into bags, were first dried in the sun, then rubbed between the palms, trodden with feet, or in larger amounts clapped in wooden mortars. They were husked (chaffed), then ventilated.

Reed sweet-grass was a commercial item at the end of the 19th century in Germany, named "the manna of Frankfurt" in marketplaces. It made its way up to the Baltic states as well. It was still sold in St. Petersburg at the beginning of the 20th century. In Eastern Europe, it was sold everywhere.

Ethnographical information in relation to reed sweet-grass was summarised by Iván BALASSA (1985b). The first written data comes from around 1500 AD in the Glossary of Timişoara, which indicates the regions around Košice as the primary place of occurrence. There are mentions from 1603 in Szeged, in 1683 from Transylvania. A German-language magazine, "Ungarisches Magazin," published at the end of the 18th century, notes that is was abundant at the river Tisza and beyond. It was called "manna" (around 1500), "grass mush" (1590), "gruel grass flower" (1592), "manna mush" (1783). It was once called "witches' mush." Maybe this designation comes from the 18th century when the grains were gathered by women with rolled up skirts, early in the morning. According to the concept of the time, witches collected morning dew at dawn on St. George's Day in order to cast a spell.

Initdially, float-grass was not a plant connected with Hungarians. It is not present among the archaeological seed finds in Hungary. The habit of eating it spread only in the modern times. It is probable that this habit was borrowed from Slavic nations: " It is known in the cuisine of the Polish, Czech and other kind of Slavic nations, but in the case of pure Hungarians, I have not yet experienced the habit of eating it instead of common millet and rice mush. However, I have seen it grow freely at the water dams and bridges" – RAPAICS (1934) quotes Veszelszki. According to NAGYVÁTI (1791) it is not eaten and not grown anywhere in Hungary, "although, we would have suitable places for it." In spite of this, there are data that it was part of peasant dues in 1603 in Szeged. The same situation was dominant in Transylvania. In 1707, four cubic cords are mentioned in Nagymihály, Zemplén county. MÁTYUS (1787) however recognizes it: "it is propagated by sowing the seeds in May: it was seen that the grains would become more beautiful and plentiful this way." At the Grassalkovich estate in Gödöllő it was included among the crops grown. BERZEVICZKY (1802) regards it as a major crop in Hungary. It was also included in the list of goods in 1879.

According to the descriptions, the gruel was sweet ("manna mush") (RAPAICS 1934). Although no bread was baked from it because the dough did not rise or hardly rose, yet "cooked in milk or water and a little fat and salt added, it becomes an excellent dish, being such a stodge that you would be fed with a little" – says FORGÓ (1817). According to János NAGYVÁTI (1791), the pulp made of it is even tastier when wine is poured on. Sometimes it was mixed with wheat flour used for making bread. It was held that it is an excellent filling for sausages. Béla GUNDA (1975) and Iván BALASSA (1991) collected a number of modes of preparation: shepherds along the river Berettyó made it like millet mush, at Nagysárrét, Bihar, it was first boiled to become thick and then put into milk or eaten with sour-cream, while the pulp made of it in Bodrogköz was baked in a tray. The archaic mode of preparation was seen with fishers, shepherds and marshmen living beside waters, who fried or roasted the pulp on stones smeared with fat heated on fire.

It was an excellent fodder as well. Benefits, according to NAGYVÁTI (1791), included the excellent straw that it provided after harvesting, as many as three times a season. Hogs and horses liked it as well.

In Germany, it was called "swingel" (swine mush). Poultry was fed on it. Allegedly, healing properties were also attributed to it. The pressed juice of sweet grass was used to heal pneumonia, mitigate fever and was put on abscesses or tumours (MÁTYUS 1787).

The history of the sweet grass reflects the process of becoming a cultivar. The gathering of the wild crop found in nature did not meet demands any more. As a first step in deliberate production, it was not only gathered, but the habitat visited frequently and protected to some extent. Later on, it was sown. Sooner or later this might have led to a cultivar. It was on good terms with millet, but new gruel plants such as the rice and maize (corn) abruptly put an end to the process. Another potential cereal crop less. Only folk tradition and pages of obscure books preserve knowledge of sweet grass.

8.3.5. Other cereals

It seems that humans in ancient times did not detest weeds as much as people do today. Occasionally they were even deliberately eaten. Even the otherwise toxic seeds of korn-rade (*Agrostemma githago* L.) were baked into the bread in small amounts and liquor was distilled from them (RAPAICS 1934). They were also given to horses.

In prehistoric times, certain kind of weed species, just like those of cereals, were put under selection pressure ("secondary cultivars"), but with the passing of the culture dealing with them, their development ceased at an incompletely domesticated. Losing their cultural traits, they have remained as weeds up to the present day.

Rye brome (*Bromus secalinus* L.), black-bindweed (*Fallopia convolvulus* (L.) A. Löve), cornflower (*Centaurea cyanus* L.), nipplewort (*Lapsana communis* L.), catchweed, cleavers (*Galium aparine* L.), false cleavers (*G. spurium* L.) and potentially spring wild-oat (*Avena fatua* L.) were all considered to be widespread cereal weeds in Neolithic Europe. These species constituted the earliest weed community of Europe, called "*Bromo-Lapsanetum praehistoricum*" (KNÖRZER 1971).

In addition to ca. 65 cm^3 of carbonised, stored naked barley (*Hordeum vulgare* var. *nudum*), grains found in a burnt house at the Late Bronze Age site of Gording Hede, Denmark, there were also ca. 30 cm^3 of carbonised seeds of weeds (HELBAEK 1951). Other Iron Age sites in Denmark provide similar instances (HELBAEK 1954). The fact that these widely grown weed seeds were stored for human consumption is supported by evidence of stomach content analysis of Late Bronze Age and Roman marshland corpses: Tollund, Grauballe (HELBAEK 1950, 1959b, 1959c), and Borremore (BRANDT 1950).

The stomach content analysis of the Iron Age marshland body uncovered in Tollund, Denmark, revealed that the man ate a grist gruel made of unroasted, coarsely chopped seeds: the composition included barley (*Hordeum vulgare* L.), flax (*Linum usitatissimum* L.), gold-of-pleasure (*Camelina sativa* (L.) Cr.), buckwheat family (*Polygonum* spec.), buckwheat (*Fagopyrum* spec.), oats (*Avena* spec.), common millet (*Panicum miliaceum* L.), fan-hen or white goosefoot (*Chenopodium album* L.), bearbind (*Convolvulus arvensis* L.), hemp-nettle (*Galeopsis* spec.) and ball mustard (*Neslea paniculata* (L.) Desf.) (HELBAEK 1963).

The gruel found in the Grauballe body consisted of spelt, rye, barley, oats, flax and weed seeds. The Tollund man lacked bread wheat and rye, while the Borremore find consisted entirely of weeds. Common weeds species identified are as follows: sheep's sorrel (*Rumex acetosella* L.), redshank (*Polygonum persicaria* L.), black-bindweed (*Fallopia convolvulus* (L.) A. Löve), fan-hen or white goosefoot (*Chenopodium album* L.), spergula (*Spergula arvensis* L.). Also found were: soft-grass (*Holcus lanatus* L.), soft brome (*Bromus mollis* L.), gold-of-pleasure (*Camelina alyssum* (Mill.) Thell.), field pansy (*Viola arvensis* Murr.), plaintain (*Plantago major* L.), ribwort-plantain (*P. lanceolata* L.), curltop lady's-thumb (*Polygonum lapathifolium* L.)

Echinochloa (*Echinochloa crus-galli* (L.) P. B.) grains found in the Low Rhine region were intended for human consumption, maybe even grown (?) (KNÖRZER 1991). The proportion of rye brome (*Bromus secalinus* L.) grains found in certain Hallstatt sites is unusually high, sometimes as much as 1% of the cereals. In addition, they were mixed among spelt (Kamieniec: TOMCZYNSKA – WASYLIKOWA 1988) and emmer (Smuszewo: KLICHOWSKA 1977). They might have been eaten together with these wheat varieties.

Rye brome propagates quickly, especially in acidic soils. It can be found mostly in grain producing areas. It is one of the oldest weeds and its original form is not yet known. Rye brome acquired the traits of solid ear spindles and large seeds as a weed of common bread wheat, but this was not enough to turn it into a cultivar.

It should be noted that a similar phenomenon in Germany was observed in the case of the Bölcske-Vörösgyír Middle Bronze Age tell where einkorn, emmer and other cereal grains included 1% of field and rye brome grains that were naked, i.e., cleaned (K BERZSÉNYI – GYULAI 1998). Probably they were eaten together with the cereals here as well.

No ordinary cereal grain was found during the 1992 excavation of Zalaszentbalázs-Szőlőhegy meadow's Late Neolithic – Early Copper Age (Late Lengyel culture – Balaton-Lasinja culture) settlement (excavation by Eszter Bánffy). However, the obviously cleaned stock of common bulrush (*Schoenoplectus lacustris* (L.) Palla), incremented with earth and soil, is more informative. Obviously, it was used to replace cereals, which were difficult to grow in the adverse climate of the Copper Age, and processed into mush. The achene of bulrush is rich in nutrients. Probably, it served as an emergency food in times of need, a plant that still occurs in streamlets and marshes as dominant stock and in high rush-beds as a common plant. A number of bulrush achenes were found in the botanical material of the excavation of the Csepel, Hollandi Road settlement of the bell-beaker population in the Early Bronze Age (excavation by Rózsa Kalicz-Schreiber 1973). It can be assumed that achenes were used as a replacement for cereals.

It is also likely that the prehistoric cultures living in the territory of Hungary made an attempt to domesticate not only certain glumaceae, but also legumes not grown previously and other kind of plants, soon making dishes of them. In Tószeg- Laposhalom in 1928, at excavations of the final phase of the Nagyrévi culture, carbonised seeds were found on the bottom of a dish, resembling a mush. These were identified in 1949 by Zoltán Zsák as weeds: narrow-leaved vetch and summer vetch (*Vicia* cf. *angustifolia*) (in: HARTYÁNYI – NOVÁKI – PATAY 1967–68).

8.4. On the nutritional value of ancient grains

The current level of analytic chemistry allows us to compare cereals of ancient times with those of today not only in morphological terms but also in terms of nutritional value. One way to learn this is to seek out those "ancient" species and varieties that still occur nearly unchanged for centuries or even millennia. Another way of doing so is to analyse carpological finds available for archaeobotanical research. Unfortunately, the "resurrection" of these fossils seems to be hopeless, although there are some reassuring results reported in this field as well (see Chapter 1.6.).

The germinability of seeds and bearings decreases over time exponentially. The reason for this is that wax accumulates in the wall of the testa, enzymes are inactivated, while fats and oils in the endosperm oxidize as time passes, i.e., they become rancid. ("Sensational reports" on germination of "mummy wheats" and other, several-thousand-year-old seeds are unscientific and false. Eventually, they always prove to be a hoax.)

Under special conditions, however, cereals are able to preserve their germinability if not for centuries but for a relatively long time. According to well-documented and verified data, barley and oat grains found in the foundation stone of the City Theatre of Nürnberg (1832) were still able to germinate in 1955 when the theatre was demolished and they are still in production to maintain the variety (AUFHAMMER – FISCHBECK 1964). As a result of yield measurements made in culture vessels, and lately on experimental parcels (such as nutrient supply, grain content of ears, hectolitre mass, mass per thousand grains, yield, and so on) these old cultivars have become comparable to intensive high yield barley and oat varieties of today.

Of the findings of the 15-year-long set of experiments and determinations aimed at evaluation of nutritional values, barley varieties of Nürnberg fell behind recent cultivars such as Wisa or Una in terms of pure yields by 40–50%, but their raw protein levels exceeded those in modern varieties by 25–30%. Further comparative production experiments were made using the barley of Nürnberg in the Soil and Crop Production Research Institute, Bavaria, between 1979 and 1982, in order to establish yields, size of grains, grain/straw ratio, malt quality and resistance to diseases. According to personal communication by M. Baumer, results of raw protein measurements concurred with the former findings.

The oldest germinating wheat grains known are somewhat younger than this, dating to 1877. Different kinds of cereals (barley, bread wheat, rye) were found in airtight containers in 1967 when the seed collection of the Crop Production and Evaluation Institute at the University of Vienna (Institut für Pflanzenbau und Pflanzenzüchtung an der Hochschule für Bodenkultur) was re-catalogued. Common bread wheat (*Triticum aestivum*) grains were tiny and red (var. erytrospermum) (RUCKENBAUER 1971). According to their label, they were of the "Székesfehérvár" (Stuhlweissenburger) variety (most probably of Bánát origin). When experimenting with germination at low temperatures, some of the grains, which have been in deep rest for 90 years, could be revived. It was confirmed in sowing tests that they are all autumn wheats. They were propagated and kept in variety maintenance ever since. Comparison of nutritional values with contemporary varieties showed that the old cereals contained less protein and essential amino acid. However, their baking value was better than that of modern varieties. The Institute for Agrobotany at Tápiószele also received some of the precious material and, by the end of the vegetation period in 2002, it turned out that both naked and hulled forms were found among the grains and also the height of the stalks varied (personal communication by László Holly and Gábor Vörösváry). At the same time, they have a great deal of similarity with the comparatively small ears of the pure country wheat of 1937, to be found in the seed collection of the Museum of Agriculture.

According to personal communication by A. C. Zeven (Wageningen), they made successful tests in Austria using local bread wheat varieties originating in 1900 and 1930. B. Bartha (Basel) carried out production tests in 1988–89 using a medieval local rye variety found in France still in production. In Switzerland (Zürich-Rekkenholz, Changins), a number of cereal cultivars collected from the cantons are still held in variety maintenance production. Local varieties are subjected to the most modern analytical tests. According to the polyacrylamid-electrophoresis (PAGE) tests, certain local varieties (land races) of autumn wheat express significant differences even in comparison to each other. Results are used for the evaluation of the varieties as well (SCHMID 1989).

KÖRBER-GROCHNE (1989) carried out comparative experiments in Stuttgart-Hohenheim and the Schwäbisch-Alb with einkorn, emmer, spelt, durum wheat, common or bread wheat, two rowed naked barley, rye and oat species. Nutritional values of the grains (such as mass per thousand grains, raw protein, carbohydrate, raw fat, ash, raw fibre, dry matter) proved to be suitable for qualitative assessment of these individual species. According to examinations, more ancient hulled grains (einkorn, emmer, spelt) have an average raw protein content ca. 50% higher than that of common wheat, the most widespread wheat variety, but at the same time the raw protein content of common wheat does not show any significant difference when compared to barley, rye or oats. This is especially typical in the case of einkorn and spelt, where protein accounts for a fifth of all dry matter in the grains. Raw fat content is approximately 30% higher in hulled wheats. With oats, this value is extremely high. As to the amount of carbohydrates, naked barley and rye are best. Similarly, there is a difference in the amino acid composition of hulled and naked grain wheats as well.

The mineral element content of ancient cereals, especially of ancient einkorn, exceeds that of today's varieties; in certain cases, such as iron and zinc content, by orders of magnitude (Bocz 2001). This parameter relates to the increasing amount of mass per thousand grains and, parallel to this, to the reduction of skin layer thickness, where minerals tend to accumulate. There is a correlation between microelement composition and the set of amino acids. Higher microelement ratios increase amino acid masses. All this means that, in terms of pure yields, ancient cereals fall behind contemporary ones, but their nutritional value, resistance and flavour are of higher quality.

It is important to note that the carbohydrate and raw protein content of grains depend on the nitrogen content of the soil (nitrogen fertilisers increase nitrogen content), on the climate (warm, dry weather increases the values), and also on the shape of the grains (protein content of flat grains is higher than that of rounded grains) (Brouwer 1972). These relationships were confirmed by the experiments of Körber-Grohne.

However, protein analysis provides a potential for further classification. Antigen features and the so-called reserve proteins in seeds and fruits are specific for a species and genus, are therefore suitable for identification purposes (Konarev, Gavriljuk – Gubareva 1987). They can be analysed using high-pressure liquid chromatography (HPLC) (Wrigley *et al.* 1987).

In Gatersleben, Germany, protein analysis was carried out using all the land races of all bread wheat species and their species and genus hybrids between 1970 and 1974. Some ten thousand taxons were grouped on the basis of their raw protein and lysine content (Lehmann *et al.* 1978).

Gel electrophoresis opened up news ways in detecting iso-enzymes (izosims). This analytical procedure may provide new findings in terms of origin and genetic relationships (Pozsár 1971).

Along with the analysis of nutritional values in recent seeds and fruits described above, sub-fossil materials, albeit in a very limited amount, were also subjected to analytical examinations. According to tests made by Brocq-Rousseu – Gain (1907, 1911), non-carbonised wheat can be active (i.e. viable) for 200 years, but no peroxidase activity was measured in any older einkorn and common wheat. Based on the analysis of Täckholm – Täckholm (1941), grain from ancient Egypt had starch granules in their endosperm that retained iodic reactivity, but no protein could be detected. Barton-Wright – Booth – Pringle (1944) found riboflavin and nicotic acid in carbonised barley grains in the tomb of Tuthankhamon (1350 BC), preserved most probably due to anaerobic conditions. Helbaek (1959c) detected protein and albumin from seeds of curltop lady's-thumb (*Polygonum lapathifolium*) achenes and other kinds of glumaceae, accounting for a large part of the stomach content of the marshland corpse dated to the 3th–5th centuries AD, found in Grauballe (Denmark). Leek (1973) reports that he found protein in each of the seven samples taken from Egyptian breads dated between 3000–1400 BC.

Zeven – Doekes – Kislev (1975) tested fourteen einkorn, emmer, spelt and common bread wheat finds from various ages using electrophoresis in order to clarify how long protein can be detected in grains. One sample contained carbonised emmer grains: Ermelo (Netherlands) (510 ± 65 years). Of the wheat samples from various archaeological periods the only finds that were not carbonised were emmer from Saqqara (Egypt) (300 BC) and Massada (Israel) (100 BC to 73 AD). The rest of the samples came from grains preserved in seed collections set up in the 19th century in the Netherlands.s The first three samples mentioned did not contain enough protein, but the electrophoretic pattern of modern wheat varieties was characteristic of gliadin, albumin and globulin.

As a result of ten-year-long nutrient supply experiments, BOCZ (1998) arrived at the conclusion that protein content is usually increased due to an increase of nitrogen, phosphorus and potassium fertiliser supply with the only exception of species belonging to the vexillate (Papilionaceae) family. However, increased amounts of artificial fertilisers have an impact on the macro and micronutrient content of the plants as well. While minerals usually increase, essential microelement uptake might drop to a third. Diminishing microelement content may have an impact on the human and animal body. In order to determine the genetically encoded amount of microelements of each species as a basis for comparison, ancient wheat cultivars, einkorn and extensive ancient common bread wheat were obtained, found deep in the Island mountain range in Transylvania, and in the land of the Mokány (Móc). Bocz compared the macro and microelement ratios found in einkorn and ancient common bread wheat, to be traced back to the 18th century steppe wheat, with those in contemporary bread wheat (BOCZ 1988). He found that as plant improvement progresses, nitrogen/phosphorus, nitrogen/potassium and nitrogen/magnesium ratios drop in the grains while nitrogen/calcium, nitrogen/manganese, nitrogen/zinc and nitrogen/copper ratios increase.

Comparative assessment of the nutritional value in ancient common bread wheat, taken as reference, and in modern cultivars shows that zinc content in recent wheat varieties is 350% lower (!), while amino acids are only half or third of those in ancient wheats.

It is also worth considering why the thickness of the testa in today's wheats has been reduced during the improvement process. (The testa of einkorn is the thickest.) As ploidisation takes places, this is also reduced. In the case of einkorn, this is 14.1% when expressed in total grain mass, while contemporary common bread wheat cultivars have only 5.2-8.4%.

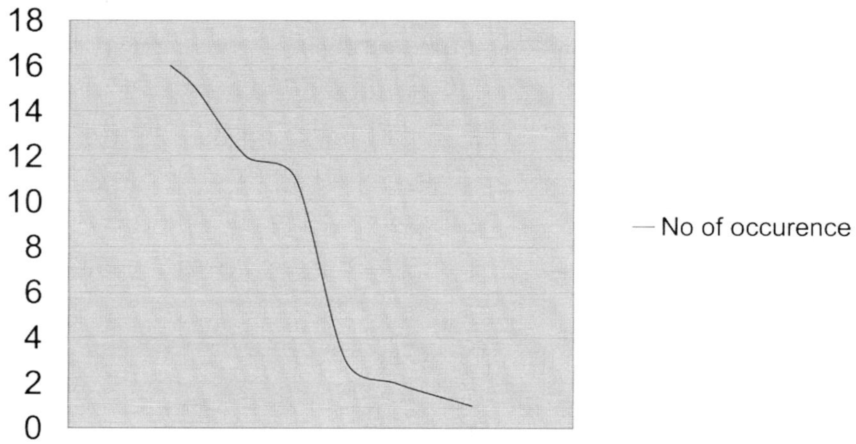

Gene erosion of einkorn around Kalotaszeg, based on PÉNTEK – SZABÓ 1981.

Minerals accumulate in the testa of the grains which come as bran when the grain is milled, just like toxins of eventually occurring pathogens. A thinner testa might represent the risk that persisting pesticide residues make their way into the endosperm. Consumption of bread made of this kind of cereal – whether branless or whole grain bread – might pose serious health risks. Namely, by eating bread baked from flour without bran, we get less than the necessary requirement of the essential microelements, while bread made of whole grain flour supplies us with too many pesticide residues, or, in the case of an unfortunate year, toxins of bacteria or microscopic pathogen fungi, which have toxic effect in minor amounts. When ancient land races are grown organically, all these complications can be avoided.

Recent investigations confirm what was stated above. Nitrogen and macroelement levels of an einkorn population coming from an ancient habitat correlate with those of current common wheats, but the microelement levels were significantly higher: sulphur, copper, iron + 15-148%, zinc + 291% (BOCZ 2001). While nutrient supply has an effect on the macroelement content of grains, and this – as opposed to earlier views – is not higher than 30%, the microelement composition does not change. At the same time, a correlation could be shown between microelement composition and the set of amino acids. Higher microelement levels in einkorn increased the amount of amino acids by 150%. Another, even more interesting finding is that with increased ploidisation, microelement levels are reduced compared to that in ancient einkorn. This might have something to do with increased yields.

According to measurements made by BÁLINT *et al.* (2000), the mineral content of hulled wheats is different, but diminishing with higher degrees of ploidisation. This is related to the mass of thousand grains and, parallel to this, with the thinning of the testa, where minerals accumulate. The highest mineral contents (copper, zinc, iron, calcium, magnesium) were measured in diploid einkorn, which has the lowest mass in one thousand grains but the thickest testa.

KISS *et al.* (2001), after examining minerals in wheat species coming from various geographic regions and having different degrees of ploidity concluded again that the potassium, calcium, zinc and magnesium contents of wheats were reduced significantly with the increase of ploidity. However, in the case of species or subspecies with identical ploidity, the habitat has significant impact on the mineral contents and composition of grains.

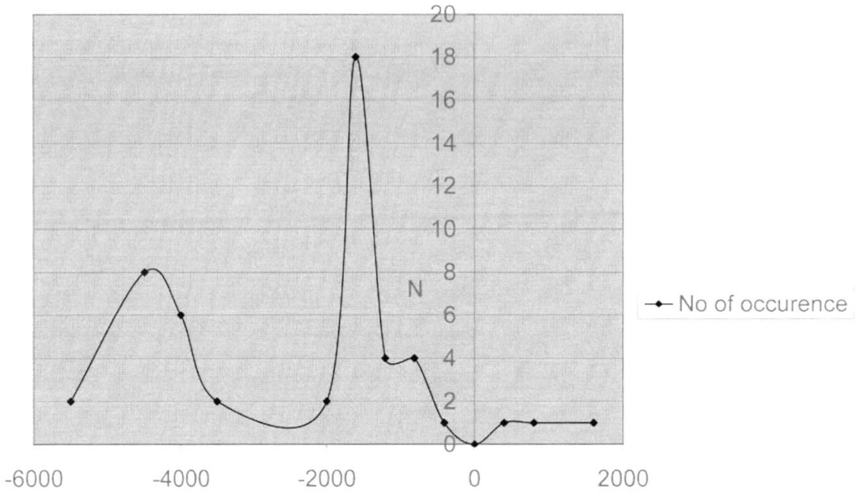

Gene erosion of einkorn in the Carpathian Basin during the historical ages, based on PÉNTEK – SZABÓ 1981.

8.5. Ethnobotanical research of einkorn and an attempt to save it

Hulled wheats (einkorn, emmer, spelt) could only be replaced by common bread wheat over many millennia. Expert authors mention that einkorn can still be found in the Balkans, the Pyrenees and Anatolia. As a result of ethnobototanical research, data are also known from Transylvania. However, einkorn production there is practically non-existent today. According to archaeobotanical research findings and written data, einkorn was in production in Northern Hungary from the prehistory to the end of the 19th century.

KRAUSS (1943) indicated the presence of einkorn in several locations in the region inhabited by the Saxons of Transylvania. GUNDA (1966b) also mentioned that einkorn production was drastically reduced in Transylvania. ZSEMLYEI (1974) (in: PÉNTEK – SZABÓ 1981) describes it in the region of the Small Szamos. The comprehensive work of PÉNTEK – SZABÓ (ibid.) evaluated the situation of einkorn in Transylvania. Gene erosion started during World War I and was accelerated during the collectivisation period between 1955 and 1960. By the end of the seventies, this ancient cultivar occurred very rarely in purebred stocks. Ethnobotanists found it used increasingly in a mixed crop with common wheat, vetch and rye. Einkorn was part of oat and vetch fodder in Kiskapus.

Eight thousand years of production of this famous wheat variety have come to an end. In most places, einkorn is no longer produced and has become a weed again. In Pányi and Egerbegy it occurs only as a weed (PÉNTEK – SZABÓ 1981). In certain parts of Transylvania it is still found sporadically (BOCZ 1998). It is found mainly at higher elevation in thinner soils and forest edges prone to destruction by game (personal communication by István Már and Kata Frendl). It is fed to animals and the straw is used for making hats. More recently however, interest was raised in relation to a possible improvement process and production of ancient diploid wheats such as einkorn and emmer all over Europe.

Composition of an einkorn test in Kiskapus (Transylvania) Sakamoto (1980) in: PÉNTEK – SZABÓ 1981.

Species	Spiculums, No of the grains	Weight (g)	%
Triticum monococcum	2652	112,3	51,67
Triticum aestivum	1699	42,1	19,37
Avena sativa	2608	57,8	26,58
Hordeum vulgare	5	0,1	0,04
Lolium temulentum	35	0,3	0,13
Avena fatua	4	0,1	0,04
Other wild Grasses	6	0,02	0,009
Wild Legumes	148	4,4	2,02
Other wild Species	25	0,2	0,09

Due to its role played in the cultural history of humanity, einkorn deserves more detailed discussion. Einkorn was mainly given to animals, to hogs and sheep in the Székely regions. It was eaten less frequently. It is very nutritious inasmuch as, among wheat cultivars, it is einkorn that has the highest raw protein content expressed in percent of dry matter (PÉNTEK – SZABÓ 1981). It has excellent parameters also in terms of lysine and other essential amino acids (MAQBOOL – ARAIN – SIDDIQUI 1997).

GUNDA (1966b) demonstrated that an ancient cake, pancake and bread were baked from it very early in several places in Transylvania. Bread made from einkorn was easy to digest. "The bread is like a rye-bread... According to people from Szék, pigs feeding on einkorn will mast well but do not become fatty. Cattle are given only the groats and meal" (PÉNTEK – SZABÓ 1981).

Einkorn straw has no value as bedding material for livestock and, lately, it has not been used for feeding animals. At best, it was used as an emergency food. However, straw hat makers preferred its thin yet strong, yellowish straw. Typical straw hats in Kalotaszeg were made of einkorn.

An einkorn production project was started in October 1995 at the biohistorical site of Szarvasgede (Northern Hungary, county Nógrád) (GYULAI 1999c). With the production, we would like to revitalize einkorn culture, which was formerly widespread in Hungary. Einkorn is the bread crop of ancient cultures, and it might also be the ideal cereal for "reform cuisine" because of its valuable amino acid content, microelement composition and low carbohydrate levels. Food or fodder made of it is easy to digest and it has low ballast content. Beneficial characteristics expressed in production are hardiness, resistance against weeds, stiff stalks, high level of resistance against pathogens and pests, all of which allow organic growing. We baked bread of it several times. The bread made of whole grain flour obtained from husked einkorn grain grown in our project was brownish in colour and compact, its flavour resembled that of the "nut bread" so fashionable in Switzerland.

KOVÁCS – BÁLINT – SUTKA (2001) hold the view that it is high time to deal with the exploitation of the valuable features and genetic variability hidden in einkorn, as the available einkorn cultivars preserved the natural gene pool of the species. According to their preliminary results, there are a number of einkorn populations that would be suitable for production based on their agronomic and qualitative characteristics. For this purpose, they identified 50 different parameters in 30 different einkorn batches and compared them. Great variance was demonstrated. The only thing they had in common was a high level of resistance against mildew, leaf-blight and wheat rust. Resistance against viruses varies and needs further testing.

Very promising results were obtained from producing einkorn gene bank batches in small parcels at the Agrobotany Institute of Tápiószele (HORVÁTH – HOLLY – SIMON 2001). Examination of forty different batches was made by sowing them into sandy and meadow soils both in Autumn and Spring. The highest yields (4423 kg/ha) were found with an einkorn population in meadow soil sown in Autumn. At the same place, other wheat species and their wild relatives are used for production experiments.

The largest problem in the processing of einkorn is husking. (We have forgotten what was a resolved problem in prehistory.) Of the hulled wheats, einkorn has the lowest grain/glumella ratio, because of the ancient nature of the grain. In addition, husks are attached to the grains even more strongly than in the case of the other hulled wheats. According to personal communication by László Kozma, the worst yield was given by einkorn when husking (chaffing) ratios were measured in a laboratory (66%). Emmer has given a better value and the best ratio was reached with spelt. As opposed to spelt, large-scale mechanised chaffing of einkorn is still a problem to be resolved.

8.6. Modern examinations of ancient cereals

Macro and microelement as well as amino acid determination processes used habitually in food and fodder chemical analysis were applied to determine nutritional value parameters in carbonised cereal grains coming from Roman excavations around Lake Balaton (Keszthely-Fenékpuszta, 5th century), Migration Period (Fonyód-Bélatelep, 8th century) and late medieval surveys (Pogányszentpéter, 16th century) (examinations by János Csapó in: GYULAI 1995c).

The tests were not inhibited by carbonisation, which, in spite of the appearance, was not complete. The process might have been long and occurred sparingly, otherwise organic molecules, which are sensitive to heat, could not have been preserved even in traces.

Samples from different periods made it possible to examine and compare different parameters of nutritional value. As this was so far a unique investigation made on macroscopic remains from various archaeological periods, checking the results and findings needs further systematic investigations. Lacking a suitable set of data, we can only make limited observations. As these tests provide information concerning crop production and eating habits of past ages, which are not to be obtained any other way, it would be desirable that morphological and metric analyses of botanical finds be regularly supplemented by such kind of experiments.

Macroelements evalutated in analytical tests carried out under the direction of János Csapó suggested large molecules of organic matter. A portion of these were emitted into the environment when they decayed, but they could also be washed out easily. Thus it can be understood easily why they can only be found in minor amounts in grains of subfossil cereals. Their presence, for the reasons mentioned above, does not have any significance in terms of information. Microelements are much more stable. According to our studies, the microelement contents of ancient cereals exceed current cereal values considerably (30% on average) (cf. TARJÁN – LINDNER 1981). However, no extreme values were measured, which would have indicated toxicity.

The amino acid tests provided unexpected results. The fact that free amino acids were found gives rise to the conclusion that the grains were not burnt, for no amino acids could be detected from carbon. Therefore, it is very probable that in each of the cases, carbonisation would have taken place slowly under anaerobic conditions, but certainly sparingly, under fire.

Assuming that the grains were not soaked in any substance with high nitrogen content, and were not preserved deliberately and that groundwater could not deliver artificial fertilisers to them, it is very likely that in these grains, some 85–90% of all amino acids are degraded in 400 years and within 1700–2000 years about 93–95% of them disappear.

When proteins disintegrate, free amino acids and ammonia are created. It seems that ammonia could not leave the material and it might have been bound in the dominating carbohydrate macromolecules. This is supported by the extremely high raw protein content. (Ammonia – being a nitrogen compound – is assessed as raw protein.) Since individual amino acids have various decay speeds, their proportional ratio changes over time (CSAPÓ – TÓTH-PÓSFAI – CSAPÓ-KISS 1986).

Evaluating the amount of amino acid/100 g sample, it can be stated that the oldest samples contained the least amount of amino acid. The levels increased as samples become younger. It seems that the amino acid content of cereals was reduced over time and assuming that the samples taken from different locations survived the centuries under similar conditions. This fact could be used for helping the age determination of cereals preserved in assumedly similar conditions and coming from different periods.

Results are shown for common or bread wheat on the graph. As quite a number of curves can be fitted to three points, graphic extrapolation is easier to do when amino acid composition of the unknown sample is projected into the horizontal axis where the age can be read directly. Naturally, verification of the relationship found here cannot be done with the ten samples tested here. It requires the analysis of an additional 50–100 samples to verify the results. However, we do not have the means for the time being to do so.

Comparing the amount of aminoacid/100 g protein with the amino acid composition of recent bread wheat and rye revealed that the microelement content of ancient cereals significantly exceeded (30% in general) that of contemporary cereals (cf. SOUCI – FACHMANN – KRAUT 1987). According to the analysis made by János Csapó (in: GYULAI 1995c) protein in modern common wheat contains considerably more (28–32%) glutamic acid and 8–11% more prolin than those tested by us. As opposed to this, proteins of recent wheats contain less threonin (2.7–3.5%), cystine (1.3–1.8%), methionin (1.5–1.8%) and lysine (2.8–3.2%).

It has to be noted, however, that these are only conditional values with respect to cereals from past centuries, because we assumed that individual amino acids decayed over time with the same speed. Knowing, however, the behaviour of different amino acids, it is very likely that this process takes place in very different ways under different conditions. At any rate, it is interesting to see that cystine, methionin and threonine, all disintegrating very rapidly under various environmental impacts, were still to be found in detectable amounts in the samples. Thus, assessment of the original amino acid composition seems to be extremely difficult, if not impossible. In any case there is a need for testing further sets of samples statistically.

The current technology of plant biotechnology and genetics makes it possible to regenerate plant individuals from only one cell (DUDITS – HESZKY 2000). If only one cell of the seed tested remained intact and untouched over the centuries, plant regeneration is technically possible (GYULAI *et al.* 2001). Plants can be developed from cells of surviving callus, provided the DNA content of the nucleus is not impaired. For this purpose, common millet, muskmelon, cantaloupe, watermelon, peach, cherry and sour cherry samples were taken first from the well dated to the beginning of the 15th century, found under the building of the former Military Headquarters, Teleki Palace in Buda Castle, (excavation by Zoltán Bencze, Dóra B. Nyékhelyi, András Végh 1998–99), as well as from the well dated also to the beginning of the 15th century in Budapest I., Kapucinusok Street (excavation by András Végh 2000). The seeds were incubated using the same procedure as with recent plants, on culture medium supplemented with growth promoting hormones. Germinating tissues were found only in common millet, muskmelon, cantaloupe and watermelon cells, unfortunately however the "resurrected" cells were soon killed by endogen infections (mycoplasms, phytoplasms).

DNA was isolated based on the methods developed by GYULAI *et al.* (2000) for molecular genetic studies only from material suitable for the extraction of the genetic material (common millet, muskmelon, cantaloupe, watermelon). Comparative genetic assessment of the material was tested using the PCR-method (WILLIAMS *et al.* 1990). The genetic tests carried out confirmed that medieval common millet, muskmelon, cantaloupe and watermelon seeds theoretically might contain intact surviving cells. Even though the plant regeneration experiment was not successful this time, the plants still contained a large amount of extractable DNA. The PCR method confirmed that the DNA extracted is of plant origin and does not come from the decomposing bacteria and fungi. PCR reactions verified easy to reproduce DNA, suitable for genome analysis and cultivar comparison. It seems to be an important result that DNA extracted from the nearly 700–year-old seeds has a pattern different from those in the current common millet, muskmelon and watermelon varieties. Further molecular genetic analyses might reveal the genetic relationships between today's and 15th-century plants (common millet, muskmelon, cantaloupe and watermelon), their heritage and genealogy as well as the origin of today's varieties. Results in the longer run could be used for genetic improvement aiming at resistance.

Fig. 448. Pot fragment with burnt food remains from the Copper Age site on the Zalaszentbalázs-Szőlőhegy meadow. Photograph by the author.

Fig. 449. Pot fragment with burnt food remains from the Albertfalva site of the Bell-Beaker Csepel-group. Photograph by Anna Endrődi.

Fig. 450. Carbonised crab apple (*Malus silvestris*) fruits from the Albertfalva site of the Bell-Beaker Csepel-group. Photograph by Anna Endrődi.

Fig. 452. Bread remains from the floor level in a house in the Túrkeve-Terehalom Middle Bronze Age tell. Front view. Photograph by Tibor Kádas.

Fig. 454. Grain mush remains from the Bölcske-Vörösgyír Middle Bronze Age tell settlement. Photograph by Tibor Kádas.

Fig. 451. Bread remains from the floor level in a house in the Túrkeve-Terehalom Middle Bronze Age tell. Rear view. Photograph by Tibor Kádas.

Fig. 453. Millet grains (*Panicum miliaceum*) cooked for mush from theLate Bronze Age waste pit of Balatonmagyaród-Hídvégpuszta. Photograph by the author.

Fig. 455. Millet mush remains from the Late Bronze Age settlement of Balatonmagyaród-Hídvégpuszta. Photograph by Tibor Kádas.

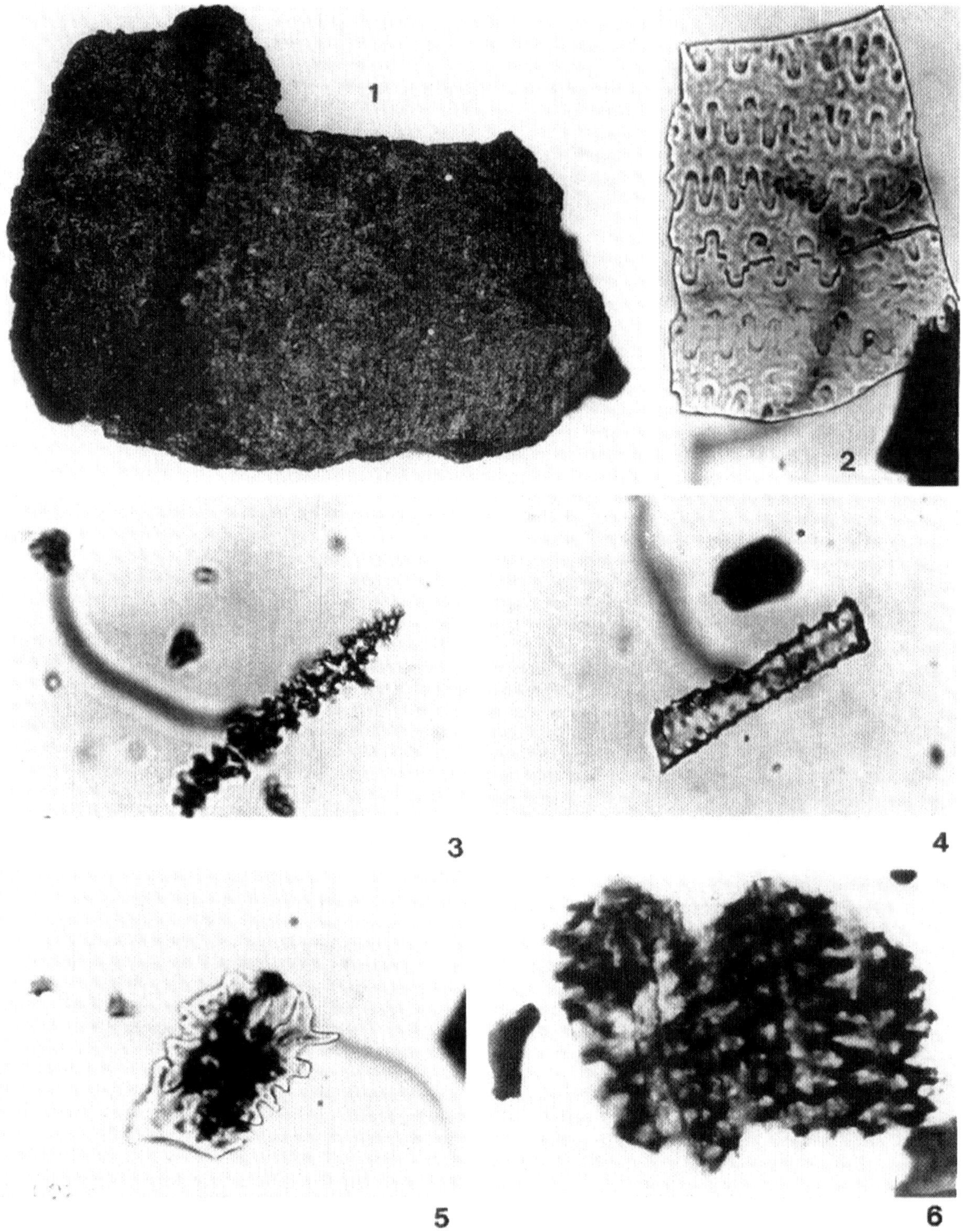

Fig. 456. Carbonised remains of a strawberry cake from the waste pit of theLate Bronze Age settlement of Balatonmagyaród-Hídvégpuszta (Photograph by the author).

1: Remains of strawberry cake from Balatonmagyaród; 2: Epidermis of *Triticum sp.* of strawberry cake; 3: Phytolith of *Triticum sp.* of strawberry cake; 4: Phytolith of *Panicum miliaceum* of strawberry cake; 5: Phytolith of *Panicum miliaceum* of strawberry cake; 6: Phytolith of *Panicum miliaceum* of strawberry cake.

The zoological finds of the Balatonmagyaród (Hungary) from the Bronze Age (analysed by I. Takács)

Shells of mushels or snails (Unio/ Anodonta sp.) fragment of 1 mm^2:	1 piece	1 individual
Insect 1-1.5 mm long wing cover, yellowish red, with black spot in the middle:	1 piece	1 individual
Frog (Anura sp.)		
vertebra:	1 piece	1 individual
scapula:	1 piece	
tibia:	1 piece	
metapodium:	1 piece	
Fishes (Pisces)		
bleak (Abramis sp.) III. vertebra:	1 piece	
caudal vertebrae:	5 pieces	2 individual
pike (Esox lucius) dentale fr., dextra:	1 piece	1 individual
overgrown tooth:	1 piece	
calcinated vertebra:	1 piece	
pisces sp. (two species) small vertebra:	1 piece	2 individual
very small vertebra:	1 pieces	
other bones indet.		
rib fragments:	11 pieces	
other fishbone:	2 pieces	
Birds (Aves) three species		
small bird (sparrow size) long bone fragment of adult individual:	1 piece	1 individual
toe bone of bigger individual and its joint surface - medium size:	2 pieces	1 individual
bird (pigeon size) calcinated long bone and caudal vertebra:	2 pieces	1 individual
Mammalia (four species)		
left side humerus of mouse sized individ.	1 piece	1 individual
caudal vertebra:	1 piece	
long bone of bigger (rat size) juvenile individual (femur?):	1 piece	1 individual
metapodium of adult individual, bigger than mouse:	1 piece	1 individual
rib fragment of medium size mammal (cat size):	1 piece	1 individual
unidentified bone fragments:	18 pieces	
Unidentified mixed fragments:	50 pieces	
14 taxa altogether:	111 pieces	16 individuals

Fig. 457. The zoological finds in Balatonmagyaród-Hídvégpuszta. Analyzed by István Takács.

The botanical finds of the Balatonmagyaród (Hungary) from the Bronze Age

Avena fatua L. (oat):	1 piece of seeds (cum caryopse corticata)
Hordeum vulgare L. (barley):	1 piece of seeds (cum caryopse corticata)
Lathyrus sativus L. (grass pea):	37 pieces of seeds
Lens culinaris MEDIK. (lentil):	4 pieces of seeds
Panicum miliaceum L. (millet):	660 pieces of seeds (caryopsys nuda) 66 pieces gruelfragment
Papilionaceae:	2614 pieces of seeds
Pisum sativum L. (pea):	678 pieces of seeds
Vicia angustifolia GRUFBG.:	1 piece of seeds
Vicia ervilia (L.)WILLD. (bitter vetch):	261 pieces of seeds
Vicia hirsuta L.(GRAY) (haire tare):	1 piece of seeds

Fig. 458. The archaeobotanical remains in Balatonmagyaród-Hídvégpuszta.

Results of the analysis of the two samples of the Balatonmagyaród (Hungary) from the Bronze Age

1.a. Treatment: boiled in cc. chloralhydrate, embedded in naphrax.

Found: 1 piece of charred animal hair, 1 piece of longish wool thread, phytolits and tissues of different cereals and grinding stone remains.

1.b. Treatment: treated with potassiumchlorate and 65 % nitric acid, embedded in naphrax.

Found: phytolits of cereals, aristae of wheat and/or strawberries, plant tissues, remains of emmer.

This means that for the production of the cake, wheat flour was complemented by essential amount of emmer flour as well.

2. Boiled for longer time in cc. chloralhydrate, subsequently placed in cc. hidrogen peroxide + cc. aamonium hydroxide, embedded in naphrax.

Found: plant tissues, wheat epidermis, fishbones, animal hair, fragments of pine-needles, freshwater diatomes and relatively large amount of grease- or wax globules.

Fig. 459. Results of the analysis of the two samples of the Balatonmagyaród (Hungary) from the Bronze Age.

Element mg/kg	Obtained %	Per 100% dry matter
Ca	78330	79200
P	4670	4720
Mg	7170	7250
K	2630	2660
Na	323	326
Mn	488	493
Cu	13	13,1
Zn	33	33,4
Fe	2430	2460
Se	96,3	97,4
% in dry matter	98,9	

Fig. 460. Elemental composition of the Celtic fish-soup remains found in Keszthely-Fenékpuszta. Aanalysis by János Csapó.

Fatty acid	Fatty acid %	C atoms per double binding
Mirisztin acid	0,9	14:00
Mirisztoil acid	0,31	14:01
Palmitin acid	4,16	16:00
Palmitoil acid	0,51	16:01
Stearin acid	22,94	18:01
Oil acid	58	18:01
Linol acid	13,18	18:02

Fig. 461. Comparative fatty acid analysis of the Celtic food remains found in Keszthely-Fenékpuszta. Analysis by János Csapó.

Fatty acid %	carpmeat	pigfat	sunflower oil	Fenékpuszta sample
Mirisztin acid	1,77	3,62	0	0,9
Mirisztoil acid	0,14	0	0	0,31
Palmitin acid	21,34	24,13	6,13	4,16
Palmitoil acid	8,67	3,2	0,08	0,51
Stearin acid	5,31	12,6	4,8	22,94
Oil acid	39,64	38,77	19,24	58
Linol acid	8,33	11,6	68,15	13,18

Fig. 462. Comperative study of fatty acids of Celtic fishsoup remains from Keszthely-Fenékpuszta. Analysis by János Csapó.

Fig. 463. Fragments of a sacrificial mush or bread from the Early Iron Age tumulus at the Fehérvárcsurgó-Eresztvény forest. Photograph by the author.

Amino acid	g AA/100g sample	g AA/100 gprotein
Aszparagin acid	0,012	8,9
Treonin	0,004	2,7
Szerin	0,004	3,2
Glutamin acid	0,012	9,1
Prolin	0,004	3,2
Glicin	0,021	15,9
Alanin	0,008	6,0
Cisztin	0,003	1,9
Valin	0,005	3,5
Methionin	0,001	0,9
Izoleucin	0,001	1,0
Leucin	0,004	2,8
Tirozin	0,005	3,4
Fenilalanin	0,008	6,2
Lizin	0,003	2,4
Hisztidin	0,002	1,2
Arginin	0,004	3,1
Troptofán		
Ammónia	0,032	24,6
Summed	0,133	100,0
N % x 6,25	0,4	
Dry matter	98,9	

Fig. 464. Amino acid composition of the Celtic fish-soup remains found in Keszthely-Fenékpuszta. Analysis by János Csapó.

Fig. 465. Cooked mush one-course dish (cereal gruel) non-leavened fragment from the Kiskundorozsma-Nagyszék Sarmatian settlement exploration. Sample 1, enlargement. Photograph by Endre Tóth.

Fig. 466. Cooked mush one-course dish (in flour baked millet (*Panicum miliaceum*) using fat and blood) gruel lumps from the Kiskundorozsma-Nagyszék Sarmatian settlement exploration. Sample 2 front- and rearview. Photograph by Endre Tóth.

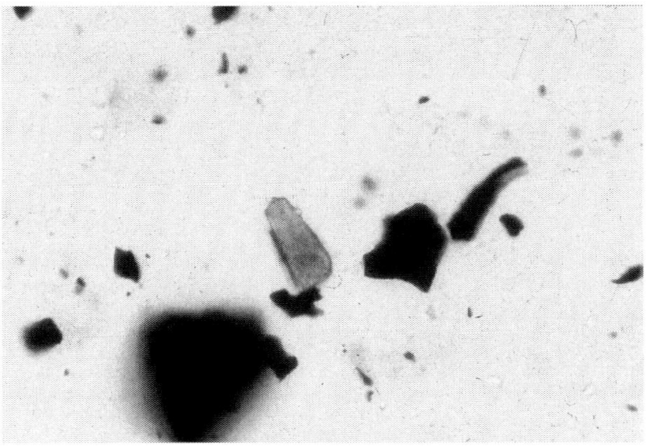

Fig. 467. Flour particles from the aleuron layer of cereals (200x enlargement), from the examination of the Kiskundorozsma-Nagyszék Sarmatian settlement food remains. Photograph by the author.

Fig. 468. Cooked mush one-course dish (cereal gruel) non-leavened fragment sample 3 from the Kiskundorozsma-Nagyszék Sarmatian settlement exploration. Photograph by Endre Tóth.

Fig. 469. Cooking pot containing food remains, seen in situ at the Borsod motte exploration, from the Conquest Period. Photograph by Mária Wolf.

Fig. 470. Cooking pot containing food remains, seen in situ at the Borsod motte exploration, from the Conquest Period. Enlargement. Photograph by Mária Wolf.

	wheat	millet	honey	cattleblood	Dorozsma1	Dorozsma2
Na	110	60	100	1420	7330	8980
K	4450	5000	360	48,5	883	1130
Ca	480	700	1400	100	23320	24320
Mg	1300		30	24,5	16660	14630
Fe	50	77	10	90000	500	324
P	4200	2700	200	60	891	1460
Cu	5,3	8,9	0,6	0,9	20,2	15,3
Zn	21,9	33	0,9	1,15	53	49
Mn	35,6	15	0,3	0,02	26,1	21,3
Ash		3,10%	0,2 g/100 g		13,20%	12,70%

Fig. 471. Comparative analysis of minerals on the mush remains
of the Kiskundorozsma-Nagyszék Sarmatian site. Analysis by János Csapó.

	wheat	millet	cattleblood	fishmeal	carp	tuna	meatmeal	beef	pig	horse-flesh	hazelnut	walnut	Dorozsma1	Dorozsma2
Asp	5,4	8,8	10,3	9,6	9,7	10	9	8,8	8,9	11,6	8,1	8,6	6,5	4,7
Thr	3,2	4,2	5,1	4,5	4,8	4,7	4,3	4	5,1	4,7	2,6	2,9	1,7	3,4
Ser	5	5,3	5,7	4,4	3,2	3	4	3,8	4	4,3	4,4	5	1	3,4
Glu	32,7	20,3	9,2	14,3	15	17,1	15,7	14,4	14,5	17	22,6	19,8	6,8	6
Pro	10,9	6,7	4,5	4,5	2,8	2,3	6,6	5,4	4,6	6,7	6,2	5,7	2,7	1,3
Gly	4,4	4,1	4,3	7,6	11,1	6,5	4,9	7,1	6,1	4,4	3,6	3,6	2,4	2,7
Ala	4	8,7	8	7	6,9	5,2	5,5	6,4	6,3	5,5	4,2	3,4	5,8	6,7
Cys	2,8	2,6	1	1,3	1,7	0,7	0,7	1,4	1,3	1	1,2	1,3	0,7	0,7
Val	4,8	6	8,6	5,4	3,5	3,3	5	5,7	5	3,6	4,8	3,8	2,4	2
Met	1,6	2,7	1,5	2,9	3,2	3,5	3,5	2,3	2,5	2,4	0,9	0,8	0,7	0,7
Ile	3,6	4,5	1,1	4,6	3,5	3,4	5	5,1	4,9	3,8	3	2,9	1,7	2,7
Leu	7,3	10,4	12,5	7,7	12,6	15,3	8	8,6	7,5	7,4	8,2	7,5	3,1	3,4
Tyr	3,3	3,6	3,6	3,3	2,8	3,7	3,6	3,2	3	3,6	2,6	2,6	1	1,3
Phe	4,9	5,3	7,5	4,1	3,4	2,9	4	4	4,1	3,8	5,7	5,2	2	2
Lys	3,1	3,7	9,1	8,3	8,2	9,9	8,4	8,4	7,8	8,3	3,3	3,5	5,1	6
His	2,5	2,7	5,7	2,7	2,3	2,1	3,1	2,9	3,2	3,9	2,8	3,3	1,7	2,7
Trip	1,2	2,1	1,3										0	0
Arg	5	5,8	4	6,6	3,5	4	6,5	6,6	6,4	6	14,4	18,1	3,8	4,7

Fig. 472. Comparative analysis of amino acids on the mush remains
of the Kiskundorozsma-Nagyszék Sarmatian site. Analysis by János Csapó.

No. carbon atom	Fatty acid	Fatty acid, %				
double band		carpmeat	pigfat	sunflower	Dorozsma1	Dorozsma2
10:00	Caprinacid	0	0,21	0		
12:00	Laurinacid	0	0,25	0		
14:00	Miristinacid	1,77	3,62	0		
14:01	Miristoilacid	0,14	0	0		
15:00	Pentadekanacid	0,45	0,13	0		
16:00	Palmitinacid	21,34	24,13	6,13	43,55	34,6
16:01	Palmitoilacid	8,67	3,2	0,08		
17:00	Margarinacid	0,29	0,73	0,05		
18:00	Stearinacid	5,31	12,6	4,8	22,03	16,98
18:01	Oilacid	39,64	38,77	19,24	34,42	34,27
18:02	Linolacid	8,33	11,6	68,15		14,15
20:01	Eikosenacid	0	1,28	0		
18:03	Linolenacid	5,59	0,89	0,37		
20:02	Eikosa-dienacid	0,79	0	0		
20:03	Eikosa-trienacid	1,08	0,26	0,77		
20:04	Arachidonacid	1,95	0,54	0		
20:05	Eikosa-pentaenacid	1,92	0,23	0		
22:00	Behenacid	0	0,96	0		
22:04	Dokoza-tetraenacid	0,17	0	0		
22:05	Dokoza-pentaenacid	0,83	0,35	0		
22:06	Dokoza-hexaenacid	1,76	0,25	0		

Fig. 473. Comparative analysis of fatty acids on the mush remains of the Kiskundorozsma-Nagyszék Sarmatian site. Analysis by János Csapó.

Fig. 474. Cooking pots found in the oven of the 8th wooden house at the Borsod motte exploration, from the Conquest Period. Photograph by Mária Wolf.

Fig. 475. Remains of the one-course dish (meat-grual) found in 1998 at the Edelény-Borsod motte settlement from the Conquest Period. Photograph by Endre Tóth.

Fig. 476. Millet (*Panicum miliaceum*) baked in cereal-gruel (fine milling product) found in 1992 at the Edelény-Borsod motte settlement from the Conquest Period. Enlargement. Photograph by the author.

Fig. 477. Millet (*Panicum miliaceum*) baked in cereal-gruel (fine milling product) found in 1992 at the Edelény-Borsod motte settlement from the Conquest Period. Photograph by Endre Tóth.

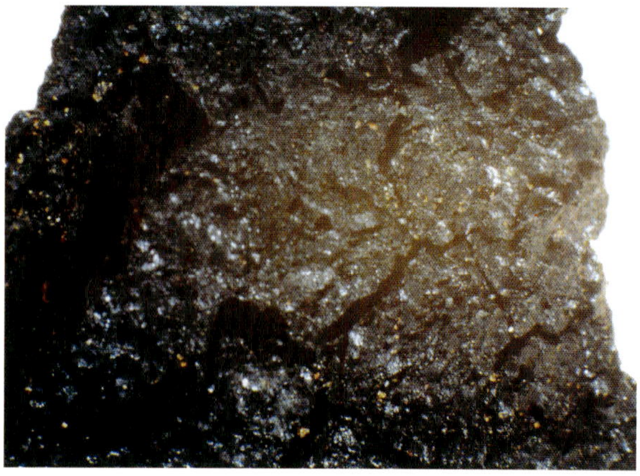

Fig. 478. Remains of the one-course dish (meat-grual) found in 1998 at the Edelény-Borsod motte settlement from the Conquest Period. Enlargement. Photograph by Endre Tóth.

Fig. 479. Wild berries and seeds of lictarium (thin jam) found in 1998 at the Edelény-Borsod motte settlement from the Conquest Period. Photograph Anna Endrődi.

Fig. 480. Remains of a dish: in flour baked millet (*Panicum miliaceum*), foxtail millet (*Setaria italica*), field pea (*Pisum sativum* subsp. *arvense*) using fat by cooking found in 1992 at the Edelény-Borsod motte settlement from the Conquest Period, including millet, foxtail millet and pea embedded in grain meal. Photograph by the author.

Fig. 481. Lictarium (thin jam) made from sloe (*Prunus spinosa*), wild pear (*Pyrus achras*), crab apple (*Malus silvestris*), field rose (*Rosa silvestris*) and cornelian cherry (*Cornus mas*) found in 1998 at the Edelény-Borsod motte settlement from the Conquest Period. Photograph by Endre Tóth.

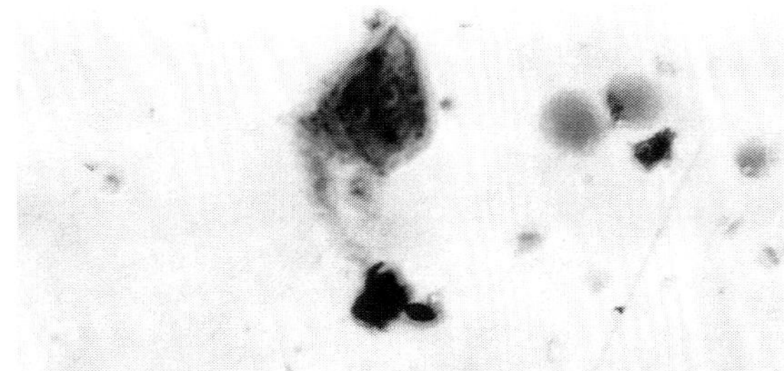

Fig. 482. Nematoda cocon from the dish remain samples (in flour baked millet (*Panicum miliaceum*), foxtail millet (*Setaria italica*), field pea (*Pisum sativum* subsp. *arvense*) used fat by cooking) found at the 1992 excavation of the Edelény-Borsod motte, Hungarian Conquest Period. Magnification: 400x. Photo by the Author.

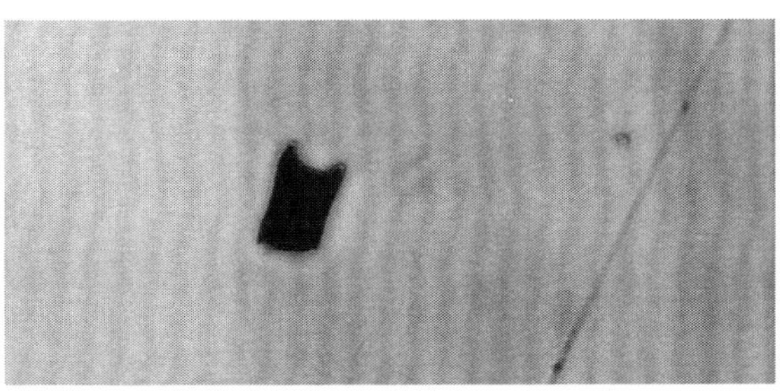

Fig. 483. Phytolith of millet (*Panicum miliaceum*) from lamb-gruel samples found at the 1992 excavation of the Edelény-Borsod motte, Hungarian Conquest Period. Magnification: 400x. Photo by the Author.

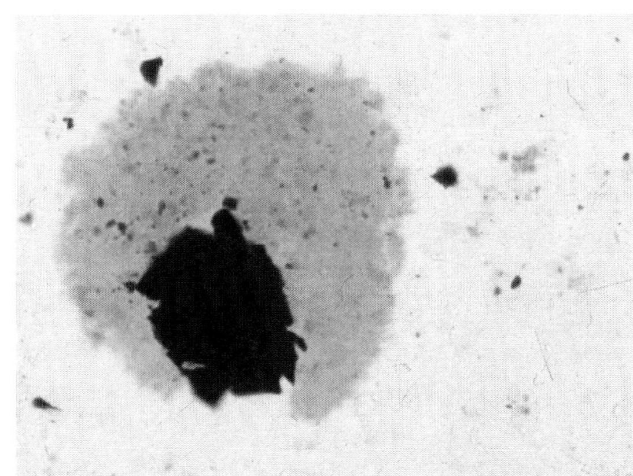

Fig. 484. Exposed flour particles from the cereals aleuron layer from the meat-grual (goulash sop mad from dough, vegetables, onion/barlic and lamb) samples found at the 1998 excavation of the Edelény-Borsod motte, Hungarian Conquest Period. Magnification: 200x. Photo by the Author.

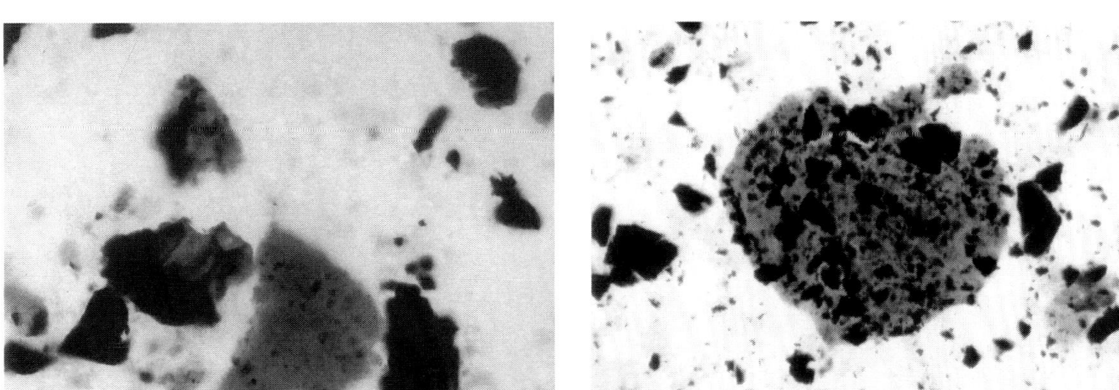

Fig. 485. Exposed flour particles from the cereals aleuron layer from lamb-gruel samples found at the 1992 excavation of the Edelény-Borsod motte, Hungarian Conquest Period. Magnification: 400x. Photo by the Author.

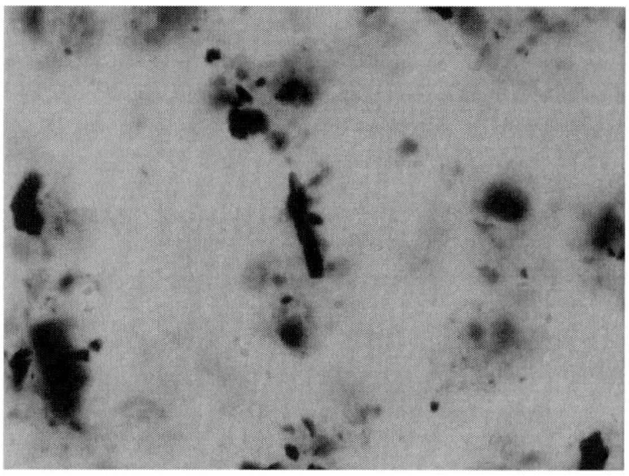

Fig. 486. Rye (*Secale cereale*) phytolith from the meat-gruel (goulash soup made from dough, vegetables, onion/barlic and lamb) samples found at the 1998 excavation of the Edelény-Borsod motte, Hungarian Conquest Period. Magnification: 400x. Photo by the Author.

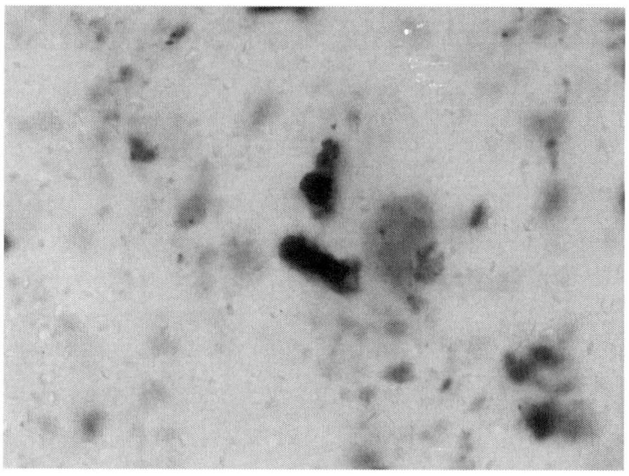

Fig. 487. Trapezoid phytolith of common bread wheat (*Triticum aestivum* subsp. *vulgare)* from the meat-gruel (goulash soup made from dough, vegetables, onion/barlic and lamb) samples found at the 1998 excavation of the Edelény-Borsod motte, Hungarian Conquest Period. Magnification: 400x. Photo by the Author.

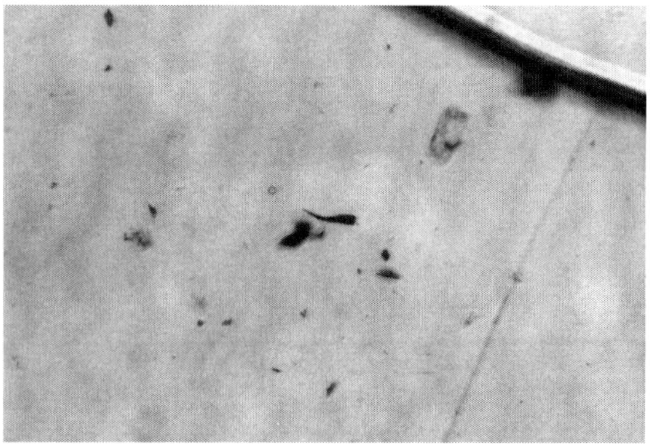

Fig. 488. Planthair from the meat-gruel (goulash soup made from dough, vegetables, onion/barlic and lamb) samples found at the 1998 excavation of the Edelény-Borsod motte, Hungarian Conquest Period. Magnification: 400x. Photo by the Author.

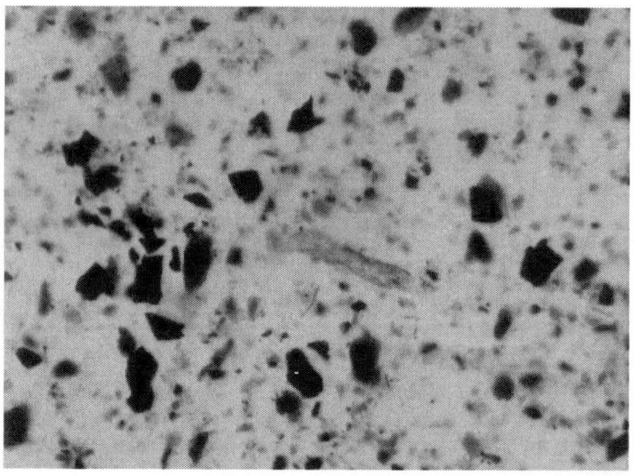

Fig. 489. Fragments of vascular bundles originating from the incrassated storing tap-root of a root-vegetable (garden parsnip or carrot) from the meat-gruel (goulash soup made from dough, vegetables, onion/barlic and lamb) samples found at the 1998 excavation of the Edelény-Borsod motte, Hungarian Conquest Period. Magnification: 400x. Photo by the Author.

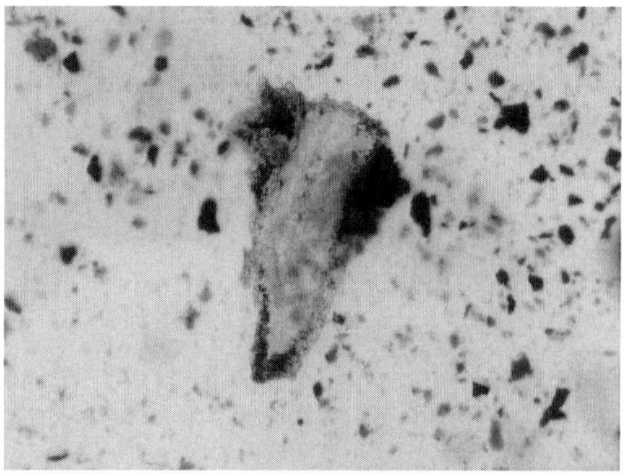

Fig. 490. By starting hypoderm cells, a piece of the upper epidermis in a onion/garlic leave from the meat-gruel (goulash soup made from dough, vegetables, onion/barlic and lamb) samples found at the 1998 excavation of the Edelény-Borsod motte, Hungarian Conquest Period. Magnification: 400x. Photo by the Author.

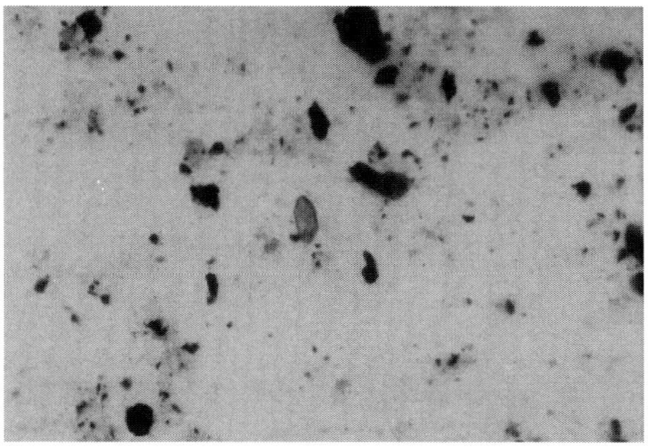

Fig. 491. Cocons of human whipworm (*Trichuris trichiura* or *Trichocephalus trichiuris*) from the meat-gruel (goulash soup made from dough, vegetables, onion/barlic and lamb) samples found at the 1998 excavation of the Edelény-Borsod motte, Hungarian Conquest Period. Magnification: 400x. Photo by the Author.

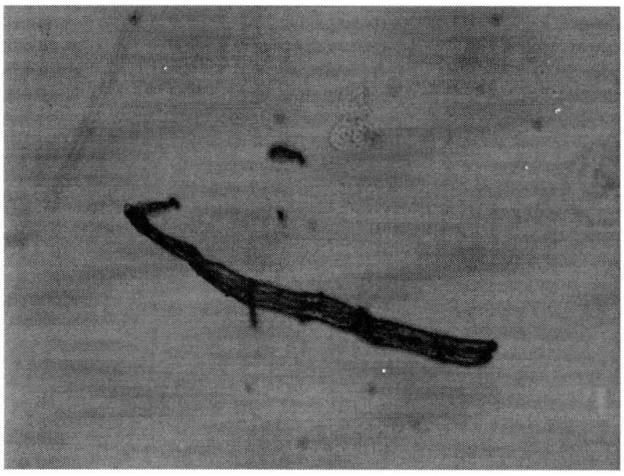

Fig. 492. Roothair from the meat-gruel (goulash soup made from dough, vegetables, onion/barlic and lamb) samples found at the 1998 excavation of the Edelény-Borsod motte, Hungarian Conquest Period. Magnification: 400x. Photo by the Author.

Fig. 493. Tricalpellat pollen from the lictarium remains (jam) 5th samples found at the 1998 excavation of the Edelény-Borsod motte, Hungarian Conquest Period. Magnification: 400x. Photo by the Author.

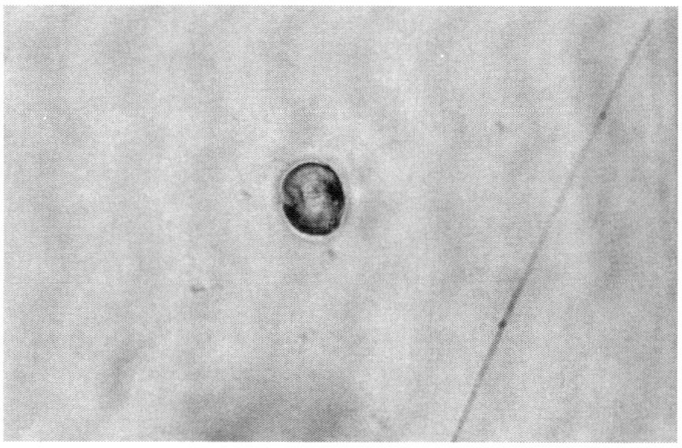

	wheat	millet	honey	cattleblood	sample 1.	sample 2.	sample 3.
Na	110	60	100	1420	75	64	67
K	4450291	5000	360	48,5	357	289	386
Ca	480	700	1400	100	44270	36190	39630
Mg	1300		30	24,5	1540	1710	1560
Fe	50	77	10	90000	750	646	1051
P	4200	2700	200	60	357	230	468
Cu	5,3	8,9	0,6	0,9	12,3	11,9	9,9
Zn	21,9	33	0,9	1,15	111	104	92
Mn	35,6	15	0,3	0,02	63,9	83,1	85,4
Ash		3,10%	0,2 g/100 g		15,50%	12,60%	19,10%

Fig. 494. Comparative analysis of minerals on the one-course dish remains found at the 1998 excavation of the Edelény-Borsod motte, Hungarian Conquest Period. Analysis by János Csapó.

	wheat	millet	cattleblood	fishmeal	carp	tuna	meatmeal	beef	pig	horse-flesh	hazelnut	walnut	sample 1.	sample 2.	sample 3.
Asp	5,4	8,8	10,3	9,6	9,7	10	9	8,8	8,9	11,6	8,1	8,6	5,1	6,5	6,2
Thr	3,2	4,2	5,1	4,5	4,8	4,7	4,3	4	5,1	4,7	2,6	2,9	5,4	5,9	3,3
Ser	5	5,3	5,7	4,4	3,2	3	4	3,8	4	4,3	4,4	5	3,4	8	3,6
Glu	32,7	20,3	9,2	14,3	15	17,1	15,7	14,4	14,5	17	22,6	19,8	15,6	16,2	9,8
Pro	10,9	6,7	4,5	4,5	2,8	2,3	6,6	5,4	4,6	6,7	6,2	5,7	1,4	3,1	2,7
Gly	4,4	4,1	4,3	7,6	11,1	6,5	4,9	7,1	6,1	4,4	3,6	3,6	7,5	6,8	5,3
Ala	4	8,7	8	7	6,9	5,2	5,5	6,4	6,3	5,5	4,2	3,4	5,9	7,1	5
Cys	2,8	2,6	1	1,3	1,7	0,7	0,7	1,4	1,3	1	1,2	1,3	0,9	1,2	1,2
Val	4,8	6	8,6	5,4	3,5	3,3	5	5,7	5	3,6	4,8	3,8	7,6	5,6	3,3
Met	1,6	2,7	1,5	2,9	3,2	3,5	3,5	2,3	2,5	2,4	0,9	0,8	0,6	1,1	1,5
Ile	3,6	4,5	1,1	4,6	3,5	3,4	5	5,1	4,9	3,8	3	2,9	4,4	3,3	2,1
Leu	7,3	10,4	12,5	7,7	12,6	15,3	8	8,6	7,5	7,4	8,2	7,5	11,2	8,5	5,6
Tyr	3,3	3,6	3,6	3,3	2,8	3,7	3,6	3,2	3	3,6	2,6	2,6	3,3	3,5	3
Phe	4,9	5,3	7,5	4,1	3,4	2,9	4	4	4,1	3,8	5,7	5,2	6	4,9	3
Lys	3,1	3,7	9,1	8,3	8,2	9,9	8,4	8,4	7,8	8,3	3,3	3,5	5,2	4,3	4,7
His	2,5	2,7	5,7	2,7	2,3	2,1	3,1	2,9	3,2	3,9	2,8	3,3	1,7	2	1,5
Trip	1,2	2,1	1,3										0	0	0
Arg	5	5,8	4	6,6	3,5	4	6,5	6,6	6,4	6	14,4	18,1	4,3	3,5	1,8

Fig. 495. Comparative analysis of amino acids on the one-course dish remains found at the 1998 excavation of the Edelény-Borsod motte, Hungarian Conquest Period. Analysis by János Csapó.

No. carbon atom	Fatty acid	Fatty acid, %					
double band		carpmeat	pigfat	sunflower oil	sample 1.	sample 2.	sample 3.
10:00	Caprinacid	0	0,21	0			
12:00	Laurinacid	0	0,25	0			
14:00	Miristinacid	1,77	3,62	0			
14:01	Miristoilacid	0,14	0	0			
15:00	Pentadekanacid	0,45	0,13	0			
16:00	Palmitinacid	21,34	24,13	6,13	50,99	48,27	36,01
16:01	Palmitoilacid	8,67	3,2	0,08			
17:00	Margarinacid	0,29	0,73	0,05			
18:00	Stearinacid	5,31	12,6	4,8	15,87	16,31	17,31
18:01	Oilacid	39,64	38,77	19,24	33,14	35,42	34,67
18:02	Linolacid	8,33	11,6	68,15			12,01
20:01	Eikosenacid	0	1,28	0			
18:03	Linolenacid	5,59	0,89	0,37			
20:02	Eikosa-dienacid	0,79	0	0			
20:03	Eikosa-trienacid	1,08	0,26	0,77			
20:04	Arachidonacid	1,95	0,54	0			
20:05	Eikosa-pentaenacid	1,92	0,23	0			
22:00	Behenacid	0	0,96	0			
22:04	Dokoza-tetraenacid	0,17	0	0			
22:05	Dokoza-pentaenacid	0,83	0,35	0			
22:06	Dokoza-hexaenacid	1,76	0,25	0			

Fig. 496. Comparative analysis of fatty acids on the one-course dish remains found at the 1998 excavation of the Edelény-Borsod motte, Hungarian Conquest Period. Analysis by János Csapó.

9. The beginnings of plant-based food

We have only assumptions about the beginnings of using cereal crops as food. Ethnological, ethnobotanical and religious historical data are of assistance in assessing this. The habit of eating cereal grains as a food might have started by roasting grains found in torn-off ears. Roasted grain is more tasty, proteins (gluten) are precipitated, and starch becomes easier to digest. There are a number of passages in Biblical Hebrew texts about this: when hiding, David was given roasted grain (Samuel or the Book of Kings I:27–29), and in Leviticus III. 23:14, roasting and grinding of grains in green ears is described in relation to grain sacrifice. The custom of eating roasted cereal grains, a tradition sustained since prehistoric times survived until the most recent times. TAGÁN (1941) found that people roast green ear grains in Galimdsa, the southern parts of the Taurus mountain range, which are then consumed in a caramelised state. In northern Syria, ears of green einkorn, emmer, or even common and durum wheats are still regularly torn off and roasted in piles (WILLIAMS – EL-HARAMEIN 1985). Caramelised grains, falling or rubbed out easily from the ears this way, constitute "frikeh" a favourite article of goods. When fresh it is cooked like rice, as a side-dish for meat. But it is also used as a filling of vine leaves, egg-plants, or squash. Dried frikeh can be preserved for a long time.

The tradition of roasted grain prevails in the Carpathian Basin as well: the "bodag" of the Palóc is wheat grains taken out from ears roasted on fire, squashed into pulp, knead into dough, torn into pieces and finally fried on amber (VAJKAI 1943). The "turf cake" of the Hungarian folk tales preserves the same tradition.

Beside eating cereals in grain form, milling them to have meal, flour and bran became increasingly significant. All wheats and ryes are excellent for this purpose, but barley and millet also serve as supplements or complements, although the latter were well-known and much beloved gruel cereals alone as well (LELLEY 1980).

We only have very few bits of information about the taste of gruel and bread made of ancient cereal grist. Barley was mainly grown for feeding animals and not humans. Excellent middling can be obtained from awny barley after husking and milling. Barley bread mentioned in the Bible (John 6:6–8) was most probably made of naked grain barley. White, salted groats of barley are frequently encountered in Greek and Roman mythology. This was used as a holy substance to throw at the forehead of the beast to be sacrificed (Homer: Odyssey III, Ilias I. Vergilius: Aeneis II, Horace: Sermonum lib. alter II.). The expression "barley mush" in Hungarian folk tales is also rooted in the distant past.

Pastry and bread made of einkorn are conveniently sweet with good flavor. GUNDA (1966a) stated that the bread of einkorn flour is difficult to bake and takes long to dry but lasts long. Emmer flour is of lower quality (BALÁS – HENSCH 1888). Groats was cooked in soups at the end of the nineteenth century (CSERHÁTI 1906).

The lowest quality hulled wheat is spelt (due to its lower protein content, which is still about one and a half that of common wheat!). It has never been a major crop in the Carpathian Basin and was never grown in Hungarian prehistory as a separate crop.

When vegetable foods lose their water content during cooking, they will burn onto the bottom or the sides of the pot or dish. Their surface will build up a characteristically multiangular structure full of cracks, typical for this kind of food. This structure represents a trait for identification.

Housewives know all too well that einbrennsuppe made of fine groats or flour can be cooked easily until it becomes thick as a result of water vapour. As soon as the bottom of the pot reaches the temperature of 100°C, the part of the food directly in touch with the bottom will burn. The burnt part acts as a thermal insulator, in other words the substance above it keeps on boiling, as long as it thickens and sticks to the bottom without being burnt. Further drying shrinks the surface which becomes multiangular due to the cracks, which is called "typically gruel-shaped." The inside becomes porous and filled with bubbles. Such structure is present in prehistoric gruel remains as well. Obviously, these burnt, dried food remains, which were unfit for human consumption, were removed and poured into waste pits, often together with the pot.

Macroscopic observations are decisive in case of gruel and bread examinations (WÄHREN 1988). The form and surface structure of the remains refer to the type of the food, the drilling cores from the inside refer to the internal structure (quality of the flour used), while the size and distribution of the pores indicate leavening.

9.1. The significance of gruel-type food

The view that Neolithic and Bronze Age people had no higher culture of cooking still prevails. This view allows for cooked gruel at best, eventually for cake-bread. According to MAURIZIO (1927) and STOKAR (1951), the most ancient plant-based dish of humans was mush. Gruel is made of grains, previously ground or squashed. As opposed to bread, these were subjected to no fermentation. The baking of dough after lactic acid fermentation, in other words the "invention" of leavened bread took only place at the end of the Neolithic. To do this, the creation of an oven providing steady and permanent temperature about 250°C was necessary. However, gruel retained its important role. The significance of this kind of food was sustained up to the end of the Middle Ages, and, what is more, it was even enhanced. It was replaced by gruel plants, e.g. maize and potato, of the New World, and the dishes made from them.

Cereal grains have been used for making dishes since ancient times. Barley was detected in a pot unearthed in Lebanon and dated to 7500 BC. In Portal, beside Lake Neuenburg (4900 BC) there is a site that yielded a pot with 2-mm-thick dough remains made from very finely ground flour and showing the traces of lactic acid fermentation (WÄHREN 1985b).

WÄHREN (1985b) examined some one hundred Neolithic pot fragments with food remains dated between 3830 and 3080 BC. These contained several overlapping layers. Nearly 90% of them had einbrennsuppe remains. The question why einbrennsuppe and groats gruel were so widespread in the prehistory and whether there was any kind of development cannot be answered at this time. We only have assumptions. Beside customs, obviously practical purposes played a role. First, such food must have been made in porous ceramic pots, which stopped further leaking. Namely, these unglazed clay pots and dishes were not watertight. Sealing of the pores was made most effectively by einbrennsuppe made from fine flour and thus pots were not washed up. Gruel was only cooked afterwards. Sometimes one can find several food layers in these prehistoric pots. The reason for this can only be guessed at: maybe to prevent further leaking, maybe to achieve more even heat distribution? Examinations showed that einbrennsuppe was eaten just as often as was mush.

A large part of Neolithic food remains found on the walls of the pots found in the lake dwellings at different Neolithic excavation sites (Burgäschisee-Süd, Yverdon, Hornstaad) were identified in the Archaeological Laboratory, Tubingen as einbrennsuppe and gruel remains (SCHLICHTHERLE 1983). The fineness of milled grist used as base material was different. In a few cases, gruel was made of unmilled grains pressed together. Gruel was not necessarily made from cereals in each case. For instance, in Yverdon, a pot contained crushed flax seeds (SCHLICHTHERLE 1983).

WÄHREN (1985b) analysed fried gruel remains in Neolithic lake dwellings and tried to reconstruct the way they were prepared. The most ancient type was made by wetting the cereal grains, throwing flour in them, squeezing them together and putting them on a hot fireplace. This way a five-cm-long piece of gruel could be prepared (Twann, early Cortaillod culture, 3830–3730 BC). In the second stage of development, gruel was made in a similar way but in order to have it better baked, it was covered with hot ash (Twann, upper Cortaillod culture, 3600–3500 BC). The third way of baking gruel used coarsely milled cereal that was wetted and mixed with flour. The dough was poured onto a heated large stone, sometimes covered with ash (Twann, upper Cortaillod culture, 3600–3500 BC). Dough became dry and hard during the baking process and it was usually eaten fresh. It could also be stored for a while when dried, as "dry gruel concentrate" and re-used for instance by dunking into a soup.

Cooked gruel consisted also of grains and seeds rich in carbohydrates. The raw material for gruel was grain of hulled crops: einkorn (*Triticum monococcum*), emmer (*T. turgidum* subsp. *dicoccum*), barley (*Hordeum vulgare*), or naked common bread wheat (*T. aestivum* subsp *vulgare*). Typically it was made with some kind of liquid (water or milk) and was not fermented. It was used in a liquid state or boiled until thick, almost solid. The flavouring of gruel was achieved by adding legumes and vegetables, apples and other fruits, meat and fish to the cereals.

Such a cooked gruel enriched with ingredients definitively constitutes a one-course dish. We have analysed prehistoric food remains burnt onto the inner wall of pots. In 1989, the relatively thick lower part (ca 0.5 cm) of the remains found on the inner wall of fragments unearthed from the Neolithic (Schnurkeramik) settlement in Zürich Mozartstrasse were identified as groats gruel soup, while the upper layer proved to be "a prehistoric one-course dish." The one-course dish was made from mush, cooked with fish and parsnip. As proof, melanin granules originating from fish skin and garden parsnip vascular bundle fragments were identified.

The best term for pieces found in Twann (3700–3600 BC) is dry gruel cake. This type can be considered an interim stage in the long period between liquid gruel and bread.

Gruel cake must have been a widespread food in prehistory. Carbonised remains found at Lake Boden in the Hornstaad early Neolithic site and Cognac Neolithic settlement were both identified in 1990–1991 as wheat groats gruel (forebear of bread).

The oldest leavened bread comes from beside Bern, Twann (Cortaillod-culture). According to radiocarbon dating, it is from 3560–3530 BC (WÄHREN 1985a). At the same place, cake and gruel remains were also found. This bread was made by adding cereal flour, water, salt and leavening. The sugar content of the baked bread must have been high and it was apparently elastic. The small loaf (its length today is 7 cm) was slightly bent, its mass in the current carbonised state is 25.2 g and it was baked in an oven. The original diameter is assumed to be 17 cm, mass 250 g. Dough raised as a result of lactic acid fermentation. Leavening was inferred from X-ray photographs made of the loaf and from analysis of the samples obtained by boring. Microscopic examinations showed that the dough was kneaded from finely ground wheat flour.

Beside gruel and bread, confectionery was also made as early as the Neolithic. Such a cake baked on birch bark was found from the age of the Upper Cortaillod culture (3600–3500 BC) (WÄHREN 1985b). These cakes might have been eaten enriched with raspberry, strawberry or blackberry.

Gruel foods were at least as popular in archaeological periods in the Carpathian Basin as well. Hungarian food remains shown in this book demonstrate how favored this type of food was from the Neolithic up to the Middle Ages (*Table 22*).

10. Beverage remains

Beverage remains almost never appear from archaeological explorations. The reason for this is that drinks are made of substances that are very sensitive to environmental conditions and perish easily, such as water, ethanol and other organic matter. However, there is a possibility for them to be preserved under special conditions. As an additional problem, it emerges that recognition of these remains is extremely difficult and analysis requires extensive experience. Certainly, beverage investigations based on chemical analysis are inseparable from archaeobotanical and macroscopic and microscopic food remains examination. In fact, drink remains come every time with plant macro remains and sometime food remains. In spite of the difficulties outlined above, some drink remains can still be described.

GRÜSS (1930, 1935) found phosphate traces (which could have come from milk as well) and honey on internal wall remains of different pots. Prehistoric beer and wine remains were found in the Zagros Mountains, Godin Tepe (Iran) site (3500 BC) (WILFORD 1992). Tartaric acid tests demonstrated wine remains from a pot found in Northern Iran Hajji Firuz Tepe Neolithic site (5400–5000 BC) (MCGOVERN 1999). According to HELCK (1971), one can reckon with the consumption of beer in ancient Egypt since 2900 BC. He does not exclude the possibility of the beer coming from even more ancient times in the Near East, appearing there together with crop production. GAÁL (1988) mentioned several Egyptian written and iconographic sources in relation to beer.

The most ancient human beverage is *beer*. One can assume that it has been known since grain crops were grown. The consumption of beer can only be assumed in Hungarian prehistory because no convincing evidence is available as yet. Therefore, it would be worthwhile to examine seed and fruit finds from the aspect of finding germinated grains, a sign of beer production (KÜSTER 1989). As a matter of fact, malt can be obtained not only from barley, but from all wheat species, common millet, and other kinds of grain as well. It is also possible that people got liquid suitable for fermentation by soaking bread like the Egyptians. Barley has the highest sugar content of the grains, and, beside barley, wheat species are also able to provide malt, although the higher protein content of the latter does not improve quality. Therefore, beer fermented using barley is the most likely assumption.

Such ancient beer must have had low alcohol content, and must have been a turbid, tartish drink, similar to "kvas", provided it was not sweetened with some kind of syrup.

It is assumed that marmalade and juice were made, maybe wine fermented from wild fruits and from honey as well since earliest times. Provided drinks made of honey or fermented from honey were made (such as *mead* in the Middle Ages) were consumed, it can be detected by pollen analysis.

In 1994, at the Kompolt Late Copper Age site in an oven a large number of cornel-berries (*Cornus mas*) and European wild grape (Vitis vinifera subsp. silvestris) were found in carbonised state, indirect evidence for fermenting alcoholic beverages (GYULAI 1997a). While cornel berries are suitable for making both marmalade and liquor, wild grapes provide only wine. According to ethnobotanical research in the Caucasus, wine made from wild grapes is very tasty (personal communication by András Terpó). Cornel berries can be processed into excellent wine by adding honey and water. In our view, both species were used for this purpose. Therefore, the statements discarding such finds as the large scale "fermenting" bowl of the Nagyrévi culture, which could provide indirect evidence for wine making, need reconsideration (NOVÁKI 1975).

A well-known opinion is that the Celts loved alcoholic drinks. Those who could not afford to procure assumedly expensive import wines must have been content with local beverages, beer or honey wine (mead).

We have somewhat more information on Roman beverage remains. Germinated barley grains, to be associated with brewing are known from the Roman Period from Central Europe and England (KREUZ

1992–93). Germinated grains were also found among carbonised grain remains during the excavation of Lussonium (today: Dunakömlőd) Late Roman fortress. As a local, albeit Romanised, population was found here, it is very likely that drinking beer was common, as opposed to drinking wine according to Roman practice.

From this period, beverage remains are also available. Remains found at the Intercisa site (today Dunaújváros) in a sealed bottle were identified as oil mixed with water (personal communication by Füzes). Probably, resin and oil had been used as flavouring agents of wine. We are also aware of a wine find from Szekszárd thanks to its hermetically sealed state (SÁGI – FÜZES 1966). In 1990–1991, we analysed some wine remains found in Romania in an ancient Greek amphora, and the examination of a Hungarian Roman sealed bottle is still under way. Moreover, analytical tests showed that one of the 5th-century bowls coming from the Seeraj (Sindh) site in Pakistan, excavated by Paolo Biaggi, once contained wine (GYULAI – KÁLLAY 1998–99).

According to written sources from the Middle Ages, the flavouring and preservation of beer was enhanced by hops (*Humulus lupulus*) in Western and Central Europe, and a variety of wax berry, *Myrica gele,* in Northern Europe. Their archaeobotanical occurrence can be associated therefore with brewing (BEHRE 1984). Remains of common hop occur earliest at the Viking sites of Haithabu (BEHRE 1983) and Elisenhof (BEHRE 1976). Beside brewing, they might have been used as an herb (relaxant) (VARRÓ 1991). The bitter substances of hop called amara narcotise is irrespective of the alcohol content. Resin like lupulin, on the other hand, extracted from the pistillate inflorescence of common hop has a sleep-inducing effect. Lately, common hop achenes were found in the medieval wells of Buda Castle, which can be associated with brewing.

Below, an account follows of the complex archaeobotanical and food chemistry tests that were used to reconstruct an Iron Age burial-feast three thousand years ago (*Figs. 463–464*). Between 1983 and 1987, nine Early Iron Age (Hallstatt C, 700–600 BC) tumuli were explored under the leadership of Éva F. Petres, Tibor Kovács and Béla Jungbert in the Fehérvárcsurgó-Eresztvény forest. Of these, grave No I., found to be intact, is possibly an elite burial of a prince. A pot, i.e. an urn, in this mound contained burnt human bones. This burial also yielded several fragments of other pots that had collapsed under the weight of the funerary-fire. In some of the pots and among the remains of the fire, plant residues, barley grains, cornelian cherry stones, and carbonised food remains were found (K. BERZSÉNYI – GYULAI in print). Macroscopic and analytical investigations (macro- and microelements, amino acids and fatty acids) confirmed that the carbonised remains, hardly reaching the size of a barley grain, were indeed of plant origin. They were fragments of a gruel or bread made from cereal middling. This find is unique as we have very few organic remains from burnt graves in the Carpathian Basin to help us better understand death rituals involving food and drink offerings.

During processing, very thin, gum-like small ochre coloured organic matter residues were observed on the sidewalls of two pots. Analytical tests indicated wine remains (GYULAI – KÁLLAY 2000) (*Fig. 498*). We found all the non-volatile ingredients of wine: tartaric acid and its salts, polyphenolic compounds, metal complexes, and all the metals and proteins characteristic of wine (*Fig. 497*). In other words, the "sacrificial" wine poured into the pot standing at the side of the balefire did not burn, but slowly concentrated at a relatively low temperature, during which all the volatile compounds, such as the products of alcoholic fermentation (alcohols, aromatic substances and glycerol) and other constituents (such as maleic acid) disappeared, while non-volatile parts remained (GYULAI – KÁLLAY in print).

The beverage find in the Early Iron Age princely grave is the earliest example of wine remains from Central Europe. In addition, it comes from the time when European wild grape (*Vitis vinifera* subsp. *vinifera*) still occurred in the Carpathian Basin. The literally "royal drink" was probably imported from the Mediterranean, together with other objects, as is suggested by the archaeological connections of the site.

In 2001, yet another occurrence of wine remains was successfully identified, this time from the Roman Period. In 2001, the archaeologist Gábor Lassányi found a jug in a burnt Roman grave No 1–2 in Aquincum (Kunigunda St.) that had a thin cemented layer on its sides and bottom (*Fig. 499–500*). The jug, slightly damaged due to soil pressure, was once sealed. Analytical tests carried out by Miklós Kállay showed a composition indicating wine dry matter content, in particular tartaric acid, which is typical only for grape and wine (GYULAI 2002c) (*Fig. 497*). Polyphenols and proteins, along with metal ions (sodium, potassium, calcium and magnesium) present support this view. A higher than usual calcium content might have come from the wall of the jug.

Although it did not produce direct evidence of a beverage, a small Avar storage pot found in Budakalász in 1987 must have been used for storing drinks (*Fig. 501*). It was found by archaeologists, Adrienn Pásztor and Tivadar Vida, in a grave in an Avar cemetery. As a result of comparative morphological examinations, the material of this *in situ* conserved pot, made of organic matter, was identified as animal skin (GYULAI 2001b) (*Fig. 502*). Avars, like any other nomadic people, used leather containers systematically for storing and transporting, eventually even for serving drinks (TOMKA 1997).

Determination of tartaric acid	
Sample	Fehérvárcsurgó tartaric acid (g/g)
watery	0,50
acidic	0,47
alkaline	0,57

Determination of tartaric acid	
Sample	Budapest, Kunigunda u. tartaric acid (g/g)
1.	0,03
2.	0,02
3.	0,03

Measurement of metal component	
Metal	Fehérvárcsurgó (mg/g)
Ca	17,00
K	0,20
Mg	1,20
Na	0,90
Cu	0,01
Fe	2,70
Al	47,71
Si	18,94
Sr	0,04
Pb	–

Measurement of metal component			
Metal	Kunigunda u. 1. (mg/g)	Kunigunda u. 2. (mg/g)	Kunigunda u. 3. (mg/g)
Ca	10,50	13,30	9,30
K	0,22	0,24	0,20
Mg	1,81	1,56	1,44
Na	0,20	0,29	0,22
Cu	0,007	–	–
Fe	1,33	2,13	0,85
Al	0,85	6,27	1,21
Si	0,90	1,99	1,55
Sr	0,73	0,39	0,32
Pb	–	–	–
P	0,17	0,91	0,24

Fig. 497. Analytical results of the drink remains found in Fehérvárcsurgó,
Hallstatt settlement and in the 2nd-century AD Roman cubiculum, Budapest, Kunigunda Str.
Analysis by Miklós Kállay

Fig. 498. Wine remains from a bowl of the Early Iron Age princely cubiculum at the Fehérvárcsurgó-Eresztvény erdő. Photograph by Miklós Kállay.

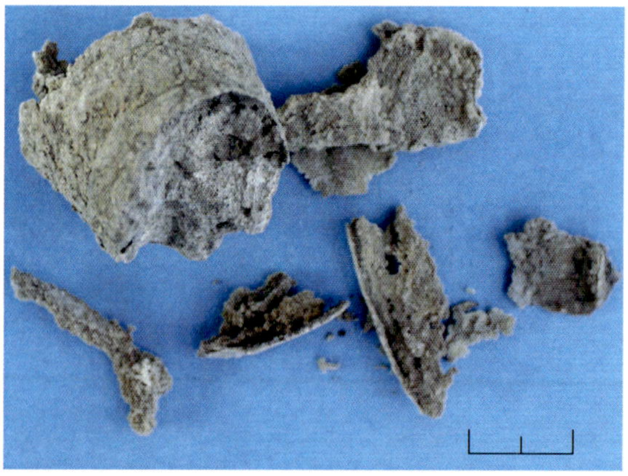

Fig. 499. Wine remains of the jar found in the 2nd-century AD Roman cubiculum, Budapest, Kunigunda Str. Photograph by Tünde Kovács.

Fig. 500. Sample of wine remains mixed with soil from the neck of a jar found in a 2nd-century AD Roman cubiculum, Budapest, Kunigunda Str. Photograph by Tünde Kovács.

Fig. 501. Leather (drinking?) bowl from a grave in the Avar graveyard in Budakalász. Photograph by the author.

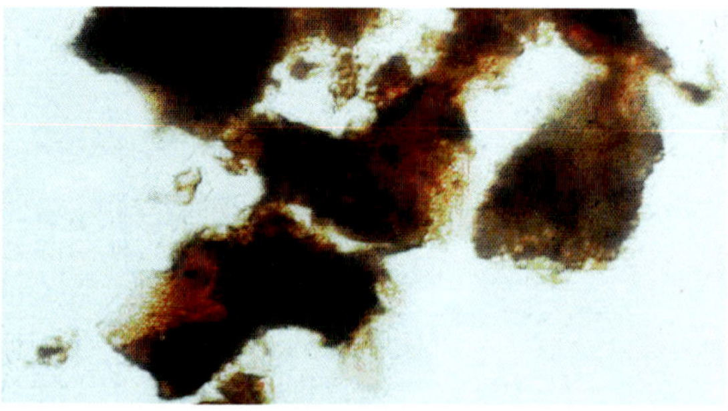

Fig. 502. Microscopic image of the leather bowl from a grave in the Avar graveyard in Budakalász Photograph by the author.

11. Food remains from the archaeological periods of the Carpathian Basin

11.1. Neolithic food remains

As a result of the interdisciplinary development of archaeological excavations in recent years, food remains were analysed on both Hungarian and foreign find materials. Systematic processing resulted recently in a readable chronicle of gastronomy of the past eight thousand years, to be deciphered from the ever-growing body of evidence about food remains.

One of the earliest known food remains in Hungary comes from the Tiszapolgár/Csőszhalom Middle Neolithic site. A significant number of soil samples were collected and later floated between 1992–1995 from the excavation site led by Pál Raczky. Among the seeds and fruits, carbonised gruel and bread were also found.

In 1996, in the Middle Neolithic and Bronze Age structures of Regéc (excavation by John Chapman) samples revealed the presence of barley grain-sized cooked gruel fragments. This was a porous piece, a fragment of a larger organic increment. The breaking surface was full of pores and cracks. Its coarse porous structure is more a feature of cooked mush. A different texture is shown by another carbonised, similar sized example, which has a shiny surface due to fatty acids that precipitated and crystallised as a result of the impact heat at the time. This might well be a remnant of a cooked one-course dish.

We also have very remarkably solid food fragments from the Linearbankeramik culture in Transdanubia. The flotated samples of the Mosonszentmiklós-Pálmajor excavation, led by András Figler, brought to light in 1993–1994 two fragments that have different porosity than is customary with mush. Their surface is shiny as a result of the fatty acids that precipitated and crystallised during baking. Therefore, they might well be the remains of a plant-based one-course dish.

11.2. Copper Age food remains

Eszter Bánffy discovered a few traces of food remains on ceramic fragments coming from the Zalaszentbalázs-Szőlőhegy meadow Late Neolithic – Early Bronze Age (Late Lengyeli culture – Balaton-Lasinja culture) settlement excavation in 1992 (*Fig. 448*). (Ceramic tiles and fragments found at sites should be scrutinised before washing and acidification. Otherwise the food remains are destroyed by the restorer.) Carbonised spots on the internal wall of the fragments were created by cooking and they were the remnants of the last dish prepared in the pot. A bit was scraped off from the thin layer, digested with an oxidising agent, and finally a microscopic preparation was made of it with the help of a highly refractive embedding agent. In the process of light microscopic examinations, smaller and larger pieces of yellowish brown matter were found, which originated from a carbohydrate-rich cereal grist (flour, groats). As no phytoliths were found, it was concluded that glumellae were carefully removed from the grains to be milled. Food was produced from wheat grist, since we found longitudinal and cross-cells originating from the grain testa of bread wheat (Triticum spec.) and cells from the aleuron. Thus, the food residue was identified as a groats-gruel, one-course dish made from hulled bread wheat (probably emmer) (Gyulai 1995a).

Also from the Late Copper Age come the carbonised mush/bread fragments found in the soil sample taken from the Boleráz-culture layer of Győr-Szabadrétdomb. The tiny, carbonised mush/bread fragments or rather crumbs can be seen as proof of the use of cereals in the Copper Age. They were selected under a stereomicroscope after the flotation of soil samples taken from fireplaces of the Baden culture in the area of Csepel Water Works (excavation by Anna Endrődi 1999). The fragments are full of smaller or larger holes that were created by kneading to create dough or by mixing to create gruel of the milled cereals with different fineness (flour, middling). The surface of the porous carbonised pieces is shiny with fatty acids exuded during baking. In all probability, they were made of hulled grain grist because the botanical survey of the settlement revealed mainly hulled grains (barley, einkorn, emmer) in a carbonised state. Naked bread wheat (club wheat) was found only sporadically.

The carbonised food residue fragment, to be characterised as mush, which was found in one of the structures explored by Zoltán Bencze at the route of M0 motorway at Rákospalota, is also dated to the Middle Copper Age.

The Middle Copper Age (Furchenstich culture) Albertfalva site also revealed food remains in the fill material of a large pit (excavation by Anna Endrődi 2001). Tiny carbonised fragments reveal the structure of cooked mush. No grains were found. It can be concluded from this that even if crop production declined in the Middle Copper Age due to impaired climate, or was even abandoned, cooked food and mainly gruel was still eaten.

11.3. Bronze Age food remains

Food remains from Budapest XXIth district, Hollandi Road No. 33/b (excavation by Rózsa Kalicz-Schreiber 1973) lead us to the Early Bronze Age. One of the pits was sampled and the sample set aside for several decades. Analysis of the sample, affiliated with the Bell-Beaker culture, revealed carbonised mush/bread fragments.

There were several pottery enclosures in the Kiskundorozsma Early Bronze Age cemetery graves (excavation by Gábor Lőrinczy 1999). Flotation of the fill material in the pots provided carbonised pieces of food remains. These evenly carbonised cereal gruel pieces must have been made of relatively fine grist. The surface is not interrupted by cracks or holes or other signs of leavening, but are evenly porous. Most probably, they belonged to the burial rite.

In 2001, in addition to cereal grains, we found a dozen very tiny fragments of gruel made of grain grist in the Albertfalva prehistoric settlement exploration, originating from a soil sample of an Early Bronze Age workshop (*Figs. 449–450*).

Middle Bronze Age earthworks are rich not only in remains of crops, but provide a number of food remnants as well. Cereal storage remains suggest that they were kept for human consumption and eaten as dishes. Supporting evidence is seen in carbonised mush/bread fragments found during processing of the soil samples coming from Százhalombatta-Földvár (excavation by Ildikó Poroszlai 1995).

Examinations of a ceramic pot found in the Bölcske-Vörösgyír Middle Bronze Age tell settlement (excavation by Gyula Nováki 1965–67) have not been completed, yet the results obtained so far indicate food remains (*Fig. 454*).

Flotation of samples taken from the floor level of a burnt house in Túrkeve-Terehalom Ottomány culture (1600 BC) (excavation by Marietta Csányi and Judit Tárnoki 1990) revealed remains of carbonised bread and a lesser amount of mush. Porosity, in other words the size and position of the pores, indicates the structure of leavened bread (*Figs. 451–452*). A high level of eating culture is inferred from these remains of the oldest bread known in Hungary.

Flotation of soil samples taken from pits in the Ludas-Varjú lane Late Bronze Age Kyjatice culture settlement resulted in different kinds of food fragments intermingled with other organic residues (excavation by László Domboróczky, Csilla Ács, Károly Tankó and Simon Gall 2001). Millet gruel fragments consisting of a few pieces of grains, groats gruel and some kind of leavened dough remains, rich in holes (maybe bread) occur here. The earth fill of one of the urn graves contained remains which were very similar to cooked mush.

The few fragments, probably gruel or bread, found at the excavation of the Gór-Kápolnadomb Late Bronze Age site, led by Gábor Ilon in 1989–1993 allow a glimpse into the kitchen secrets of prehistoric housewives. Albeit only a small piece, carbonised Late Bronze Age mush/bread fragments found near Győr, at the Börcs-Paphomlok site (excavation by András Figler 1994–96) are very significant.

Doubtlessly, the most important among prehistoric food remains is the find coming from one of the waste pits of the Balatonmagyaród-Hídvégpuszta Late Bronze Age tumulus culture site (1200 BC) (GYULAI 1996c). At a depth of about 2 metres, on the bottom of the pit, carbonised organic remains were found at the excavation led by László Horváth (*Figs. 453, 455*). In the soil sample taken, seeds of all the Bronze Age vegetables occurred: pea (*Pisum sativum*), bitter vetch (*Vicia ervilia*), and bitter vetch (*Lathyrus sativus*) (*Fig. 458*). Beside this, millet gruel remains and gruel crumbs consisting of a few naked, i.e. husked, millet grains were found. The archaeozoologist István Takács identified a number of bones bearing traces of frying (intact and chopped up bones of bream, pike, smaller birds, and larger mammals) (*Fig. 457*). The surfaces of the bird bones were browned by frying. At the same location relatively many porous organic fragments were also found, filled with holes. Archaeobotanical tests, macroscopic and microscopic analysis by Max Währen and Benno Richter (1989), instrumental analytical investigations carried out by János Csapó (macro- and microelements, amino acid and fatty acid analysis 1989–90) provided the conclusive evidence that this is the residue of a strawberry cake, made using bakery grade bread wheat and millet flour with the addition of swine fat and eggs (*Figs. 456, 459*). Phytolith tests were also made to determine that the dough of the cake was two-thirds common millet flour and one third hulled bread wheat flour. The sponge cake was filled with strawberries. Namely, seeds and fragments of seeds of wild strawberry (*Fragaria vesca*) were found on the surface of the carbonised remains. The dough was made using extremely fine flour obtained by repeated milling. Tiny fragments of stone with freshly broken, sharp edges seen under a stereomicroscope might have been derived from a manual grinding stone. We infer that flour was separated from bran when it was sieved after grinding. The pore size of the sieve must have been smaller than 1 mm. Flour prepared and used for making the dough was of identical quality as bakery grade flour today. The Bronze Age strawberry cake of Balatonmagyaród is an important piece of culture historical heritage. This oldest confectionery product of Europe allows a glimpse into the everyday life of prehistoric people.

11.4. Iron Age food remains

There is only very sporadic information available concerning the mode of preparation of crops in the Iron Age. As was shown above, the number of crops grown was extended during the Iron Age, therefore it can be assumed that their preparation was also very diverse. Common millet was a well-known gruel cereal. It is easy and quick to prepare; when boiled in water or milk, it becomes thick. One can eat it cold. It can be sweetened with honey and flavoured by wild berries or fruits. Certainly, groats gruel was also eaten. Coarsely ground bread wheat and millet grains provide highly nutritious groats mush. Such a groats mush, boiled in milk, flavoured with honey and wild berries must have been a very tasty dish. Further food remains, waiting for analysis are known from Hallstatt Age sites around Bratislava, were stuck to the inner wall of ceramic pots (personal communication by Etelka Studeníková).

Of the crops grown by humans, only bread wheat and rye can be used to bake bread. As both kinds appear in prehistory, it can be assumed that bread was also baked customarily. The remains of leavened bread can be identified in the Neolithic. In the Ipweger marsh in Northern Germany, a kind of "cake" was found, made of millet and barley, shaped more like a pancake than bread, with a diameter of 15.4 cm and a height of 4 cm. It can be dated to 713 BC (BEHRE 1992b; KREUZ 1992–93). Although the flour of barley (hulled and naked barley) is not really good for bread making, it was still used for making cakes and loafs in its pure form or mixed with wheat during the Middle Ages (HJELMQVIST 1990). According to the description of PASTERNAK (1991), bread was made of two-rowed hulled barley in Jordan in the following way: "Flour is kneaded for fifteen minutes with a little water and salt to get dough. A fire is set under an approximately 3 mm thick plate. The dough is kneaded flat and fried on the plate for about ten minutes. The result is a heavy, salty bread, where glumallae can be distinguished. It is eaten warm and keeps hunger away for a long time."

PASTERNAK (1991) mentions that two-rowed barley in South Jordan was mainly used for animal fodder. However, in bad years, when bread wheat perished, cattle was slaughtered and barley, otherwise grown for fodder, was used to make flat loafs of bread and mush. One cannot exclude the possibility that the Iron Age "cake" could be an emergency food rather than dessert. Since dough-generating wheats were already known in the Iron Age, there was no need to use barley for human consumption.

The protein content of the vexillate plants (common lentil, pea, horsebean) grown in the Iron Age is high. They can be used for making soups, stews, or served as a side dish.

A potential way of using legumes is raised by KREUZ (1992–93) when he assumes the possibility of southern connections. Ancient legumes like horsebean, common lentil, pea, vetchling or bitter vetch have been often eaten in the Mediterranean since the earliest times. A famous local dish around Trentino is "minestra di orzo", a one-course dish made of an ancient pea that has carrots, mushrooms and maybe a piece of smoked bacon or cured meat as ingredients.

Einbrennsuppe might have been known as well, since freshly fired ceramic pots were tightened by cooking einbrennsuppe in them as early as in the Neolithic.

It is very difficult to say whether fresh vegetables and salads were consumed in the Iron Age or the entire prehistory for that matter, since of these plants only leaves, tubers, flowers and stalks were used, which can be preserved only under special conditions such as in cloacae or wells. No such finds are known from this age. However, it is probable that species used included fan-hen or white goosefoot (*Chenopodium album*), livid amaranth (*Amaranthus lividus*), narrow-fruited cornsalad (*Valerianella dentata*), pigeon salad (*V. locusta*), orach (*Atriplex* spec.), dock (*Rumex* spec.), nettle (*Urtica* spec.). Salads can be made from the leaves of a number of other species, such as wild turnip (*Brassica campestris*), black mustard (*B. nigra*), dandelion (*Taraxacum officinale*), ramsons, bear's-garlic (*Allium ursinum*) and so on. These species are known from the Neolithic and are registered as potential vegetable plants.

Root plants might have been used as well, as the south German and Swiss Neolithic and Bronze Age lake dwellings and north German Late Bronze Age botanical finds include garden parsnip (*Pastinaca sativa*), wild celery (*Apium graveolens*) and carrot (*Daucus carota*). Daucus carota achenes found at a site near Cologne (ca. 100 BC) were seeds or fruits of wild carrot or improved garden carrot (KNÖRZER 1987). Further vegetables included fruited cornsalad (*Valerianella rimosa*) and naven (*Brassica* cf. *campestris*).

Crop production and nutrition of the Celts can be traced back through the Hallstatt Age to the urn grave culture. Consumption of meat and other kinds of animal products was varied by oil and kitchen-stuff as well as fruits.

Beside animal fat, some vegetable oils were also used for cooking. A widespread oil plant was gold-of-pleasure (*Camelina sativa*). It is also likely that flax, known as a fibre plant, was also used as an oil resource. Finds suggest that opium poppy production was reduced at this time, as compared to the Neolithic.

The oily seeds of gold-of-pleasure, just like Oriental sesame seeds today, might have been used for decoration of cakes and bread in prehistory.

The main sweetener was wild honey. Food was flavoured by salt. Spices must have been wild plants. Caraway (*Carum carvi*) is identified in the Neolithic.

Wild fruits were also eaten in large quantities; such as, blackberry (*Rubus fruticosus* and *R. caesius*), raspberry (*R. idaeus*), wild strawberry (*Fragaria vesca*). Hazelnut (*Corylus avellana*), sloe (*Prunus spinosa*), crab-apple (*Malus silvestris*), sometimes cornel-berry (*Cornus mas*) and rose-hips (*Rosa* sp.) were widely gathered (KNÖRZER 1987). Wild fruits were eaten fresh and stored as vitamin and trace element supplies. Crab apple and wild pear can be preserved cut in half and dried. Put in storage pits among grains, their taste and freshness is preserved for a long time. Such dried crab apple pieces were found at the bottom of a pot in 2002 in Albertfalva, when a further section of the Bell-Beaker Csepel-Group settlement was explored (excavation by Anna Endrődi 2002).

Marmalade and juice was made of wild fruits and berries. Alcoholic fermentation of wild fruits yielded tasty fruit wines. The making of honey wine carries us back far into the distant past.

It is certain that baked fruits of European chestnut (*Castanea sativa*), a species native in the sheltered, mild climate areas of Central European forests (such as the Tessin), was also put on the table. Harsher landscapes yielded oak (*Quercus* sp.) acorns. Based on ethnographic analogies it can be stated that the inhabitants of forests, such as herdsmen supplemented their diet with roasted acorns. Otherwise, it was a well-known emergency food. European beech (*Fagus silvatica*) acorns are not found in this age, only in Haithabu, from the Early Medieval Period (BEHRE 1983).

Östergötland, Sweden (2035 ± 100 BC) revealed – beside grains – stored juniper (*Juniperus* sp.) seeds and Norway spruce (*Picea* sp.) seeds and needles, maybe for the purpose of making wine or drinks (EKROTH 1986). It is also possible that they were destined for milling.

A few Iron Age food remains from Hungary are known. It has science historical significance that food remains were identified at the very beginning of Hungarian archaeobotanical research. Jenő Nyári started excavations in 1876–77 in the "Bone house" and "Corridor" parts of the Aggtelek-Baradla cave (NYÁRY 1881). Botanical finds here were mostly from the Iron Age and less so from the Bronze Age. Imre Deininger found a "fist sized" piece of bread, carbonised and covered with gold-of-pleasure (*Camelina sativa*) seeds (in: HARTYÁNYI – NOVÁKI – PATAY 1967–68).

A 60–cm-wide black ash layer was found when a pit dated to the Celtic Period was unearthed at the Hungarian-Soviet excavation at Keszthely-Fenékpuszta in 1980, led by István Erdélyi. The archaeozoologist István Takács collected a large amount of fishbone remains and half a litre of black ash. Analysis of the animal bones was completed by archaeozoologist László Bartosiewicz. The sample taken from the black ash layer of the pit contained a number of fish bones. These were kitchen waste, yet they allow for certain conclusions with regard to the eating habits of those people who lived here, and the diverse, rich ichtiological fauna of the Celtic Period around Lake Balaton. Bones included fatty carps as well as better quality fish, such as pike.

The grayish-black ash layer was sampled for analysis. Stereo-microscopically, a number of fish bones, scales and fragments were seen. Microscopic observations on samples digested with oxidizing agents and embedded in highly refractive agents revealed orange-fluorescent melanophores (pigment cells) and the melanin particles which are the colouring agents of fish skin. In other words, the fish-bones were covered with meat and skin when they were put into the pit.

The elemental composition as well as the amino acid and fatty acid content of the sample were analysed by János Csapó (*Fig. 462*). The fact that free amino acids were detectable indicates that grains were not burnt, since one could not detect amino acids and other compounds containing nitrogen such as raw protein, the latter in high concentrations, from carbon. It can be assumed that a very slow and anaerobic carbonisation took place. Fatty acid tests gave decisive results. Oleic acid, stearine and linoleic acid occurred in highest levels among the detected fatty acids. These results were compared with fatty acid composition of carp meat, swine fat and common sunflower oil, all tested earlier on. The fatty acid composition of the sample, especially the high ratio of oleic acid, identified it as carp meat. However, high stearine and linoleic acid ratios allow for the presence of swine fat. It can be concluded that swine fat was used for preparing the fish dish (GYULAI 1998c).

Archaeozoological, macroscopic and microscopic observations and instrumental analytical investigations (macro and micro element, amino and fatty acid analysis, 1989–90) resulted in the conclusion that the fish bones found in the blackish-grey ash layer were the remains of a fish dish (fish supper). Certainly, the food, including the fish fragments, was put in the pit either all at once or in several parts. Maybe they were burned during cooking or burnt with other objects some other way, and the unpalatable, burnt remains were discarded into the waste pit.

Fragmented food remains found in the soil samples of a house explored in 1993 and 1994 by András Figler at Mosonszentmiklós-Pálmajor is also from the Celtic Period. Macroscopic observations suggest that a tiny carbonised fragment was bread or mush.

Food remains were also found along with plant remains in Budapest-Nagytétény at the Celtic site excavations under the Campona Shopping Center (excavation by Gábor Szilas 2001). Some one hundred minor fragments were pieces of only one larger item. In terms of their function, they could have been remains of mush, bread or cake.

11.5. Roman Period food remains

Based on written and iconographic sources, a high level of food and drink culture can be assumed in Pannonia in the Roman Period. Food remains provide evidence that cereals were milled, bread baked or gruel cooked. They were uncovered from graves without exemption: Bécsi Road No. 44 (excavation by Patrice Bertin 1997), Corvin Square No. 1 (excavation by Patrice Bertin 1997), Szentendrei Road No. 101 (excavation by Paula Zsidi 1997), therefore it can be stated for certain that these finds were put into the graves at the same time the dead were buried (as a burnt sacrifice, cf. burial rituals of the Hittites). The carbonised remains in fact are only chips of larger food remains. They bear the typical porous structure of bread, created by lactic acid fermentation. The surface is made shiny by exuded fatty acids.

11.6. Migration Period food remains

Csanád Bálint excavated a burnt Gepid house at an Eperjes-Csikóstábla site in 1976. Soil samples taken from the floor level were floated producing burnt and gruel-like pieces.

So far archaeologists have found an intact bread only once in the Carpathian Basin. This carbonised but entire bread, found in Devín, Slovakia, near Bratislava, among cereal remains in a demolished fireplace, is from the Early Migration Period. At the same place, several bread pieces were found (PIETA 1988; PIETA – PLACHÁ 1989). These might have belonged to the Danube Germans, maybe Quads, settled along the limes of the time in the 5th century. Hajnalová says that leavened bread was mainly made of common wheat and had a round shape (diameter: 23 cm, height: 5.5 cm) (HAJNALOVÁ 1989b). The bread was made mainly of common bread wheat boxings, but rye, barley, common millet meal was also added as ingredients. The bread shows the features of duff fermented using leaven.

We have a number of food remains from the Late Migration Period, right before the Hungarian conquest. Tiny, carbonised gruel fragments were found in soil samples taken from the Slavic settlement dated to the 9th–10th centuries in Panyola-Ásottfok, excavated by István Erdélyi in 1996 (GYULAI 1996e). As all these come from samples taken at the floor level of houses and barley grains were also found at the same place, but no other cereals, the conclusion seems apparent that they are remains of barley mush. According to ethnological analogies, excellent gruel can be made from barley using either salt or honey.

In 1994, Ágnes Ritoók and Béla Miklós Szőke explored the site of a parking lot in Zalavár-Vársziget. According to recovered material, the 9th-century Carolingian centre was inhabited by a mixed population, mainly Avar common people. Most of the seeds come from the fill of a well, which was probably a storage pit lined with wooden planks. Gruel and bread fragments found here provide an important contribution to the history of gastronomy (Gyulai 1998b).

11.7. Food remains from steppe cultures

The archaeological heritage of eastern steppe cultures has always been the closest to Hungarians from among all the peoples living in the Carpathian Basin. Their costumes, lifestyle, warfare and even their cooking culture show similar features and it is not difficult to recognise in them the traditions of the Eastern European steppe. The migration of peoples into the Carpathian Basin can be traced back to climatic and social reasons. Settlements of the Scythians coming from the grassland and forested steppe in the Late Bronze Age, the Sarmatians, contemporaries of the Romans, and later on, the waves of the Huns, Avars and finally the Hungarians during the Migration Period was all restricted to the sandy lands of the Carpathian Basin (Bodrogköz, Nyírség, Duna-Tisza-köze, Mezőföld, Kisalföld). All this can be explained by persistent preference for the customary natural environment and lifestyle. However, the landscape structure of the Carpathian Basin at the time did not permit maintenance of nomadic lifestyles. If they wanted to survive, the new peoples were forced to settle down and live on livestock husbandry mixed with cropping. However, costumes, warfare and diet were preserved for a long time.

The quickly moving nomadic or semi-nomadic peoples originating from the steppe were basically livestock raisers albeit they produced some crops as well. This dichotomy left an imprint on the diet as well. However, the level of cropping methods and the kinds of crops show marked differences compared to non-steppe cultures, which apparently had an impact on gastronomic aspects as well.

The view denying any higher level of culinary culture among steppe nomadic peoples is still maintained. As opposed to the faulty interpretation of "tender meat under the saddle," we present a different inventory of what steppe peoples living in the Carpathian Basin ate over the centuries. Gastronomic knowledge of two thousand years can be analysed here by ages and cultures. All the dishes are credible, based on scientific evidence.

Beside gruel-based dishes originating from the Scythian, Sarmatian and Avar Periods, remains of one-course dishes from the age of the conquering Hungarians also have a significant place in the inventory. After so much guesswork, we finally have a credible and accurate image based on direct evidence of the food eaten by the ancient Hungarians that dramatically changes the history of gastronomy. At the same time, they provide a great deal of information and lessons for contemporary people who are preoccupied with the formulation of correct nutritional parameters and the search for a better and healthier lifestyle.

The Scythians, speaking an Iranian language and arriving here as the first of a succession of peoples, still lived on the steppes at the Black Sea in the 7th century BC. Their territory extended to the lower sections of the river Danube in the 2nd century BC. Hungarian historians and chronicles called them the *Szittya*, and considered them as the ancestors of Hungarians.

The only Scythian set of botanical finds is known from Rákoskeresztúr-Újmajor (excavation by Anna Endrődi 1996). The key crops were six-rowed barley and common millet. Barley and common millet dominate, which is a sign of quickly moving pastoral peoples. The site also yielded carbonised cereal gruel fragments.

People in the Migration Period were regarded by Romans as "Barbarians" if they had a set of values, ideology and economic model different from theirs. Consequently, their crops, livestock, food and eating habits were different from those of the Romanised populations. The Romans fought for a long time – and, as it is known, in vain – against the dominance of the Barbarian cultural model. They despised and rallied at the excessive eating, much drinking and mainly meat-eating peoples. Julius Capitolius scorns at Maximinus (Thrax), progeny of Barbarian parents observing: "He drank a whole amphora of wine (ca. 20 l) a day, and was able to devour forty pounds of meat, or even sixty", and "never tasted any vegetables", ... "did not eat anything else but the meat of boar, ducks, geese and all kinds of venison". Yet they had to admit that "meat was the food for the strong" (in: MONTANARI 1996).

The Sarmatians were also tribes of Iranian origin. They set up a large tribal empire between the 7th and 3rd centuries BC extending from the northern parts of the Black Sea, and including the Hungarian Plains as well. The Sarmatians settled on the Great Hungarian Plain in the 1st century BC and became tillers by the 3rd–4th centuries AD. They only preserved their nomadic traditions in their costumes, dishes, eating habits.

The food remains unearthed at the excavation of Csaba Szalontai and Katalin Tóth in 1998 and 1999 at Kiskundorozsma-Nagyszék can also be connected to the Sarmatians who lived here in the 3rd–4th centuries AD. After flotation of soil samples taken from pits of different purposes, 1200 pieces (!) of carbonised food remains were found (*Figs. 465, 468*). Notably many carbonised food remains were came to light from two of the pits. The remains were almost identical. The common feature of the carbonised, separated (i.e. not stuck to the wall of the pot) remains was that they all were made of some kind of cereal grist. The smaller or larger pieces must have been burnt or otherwise damaged parts of the same cooking procedure, therefore they amply provided the amount required for testing.

Macroscopic investigations revealed the presence of apparently cooked, husked millet grains (the germ having fallen out) both on the surface and in the inner realm of the fragments, which were embedded into some kind of fine, groat-like grist with an average particle size of 0.1 millimetre. No traces of gas bubbles (holes, pores) indicating the leavening of dough, i.e. the occurrence of lactic acid fermentation, were found. Consequently, the dough was not fermented but cooked. Samples taken from these were carefully digested with oxidising agents and the samples such digested were embedded. Preparations completed this way were scrutinised under a light microscope.

The microscopic observations revealed relatively many flour particles originating from the cereals' aleuron layer (*Fig. 467*). A number of different cereals: barley, common bread wheat, emmer phytoliths and epidermis remains were found. This suggested that the basic material of the groats, which contained the millet grains, was made of mixed cereal grist. It is very probable that they were not set up on the basis of some recipe before milling, much rather the cereal stock itself was inhomogeneous. This finding confirms the assumption that the beginnings of Hungarian crop production were characterised by mixed cropping. It is noted here that during the archaeobotanical survey of the site in 1999, grains of all the crops mentioned here were identified in the soil samples taken from the pits.

The results above and the instrumental analytical chemical tests made by János Csapó finally helped us to determine the origin of the food remains (*Figs. 471–472*). The potassium and nitrogen content of the samples is relatively low, while the calcium level was high. High iron levels can be associated with the use of animal blood. However, the strikingly high magnesium level seems to confirm the cereal origin of the food. The zinc content was proportional with the values measured in hulled grains, mainly in emmer. High sodium might indicate common salt (NaCl). (This was the first occasion that readily leached common salt could be detected in a food remnant.)

Values of the two samples tested do not show any significant differences when compared to each other. On the other hand, they are very similar to the average element composition of cereals.

Some of the amino acids were just beyond the limit of detectability, meaning that the protein content of the sample was destroyed almost entirely (*Fig. 473*). When the amino acid composition of the samples tested is compared with that of recent crops and other kinds of food, they resemble most in their proportions to millet and (cattle) blood (*Fig. 466*). It is noted here that the only piece of information on the gastronomy of the Sarmatians comes from Pliny the Second (*Nat. hist.* 18, 100), maintaining that their favourite food was millet gruel mixed with horse milk or blood.

As fatty acids are a lot more stable than amino acids, and were present more amply in the sample tested in any case, they could be compared to the fatty acid composition of recent materials. High values of lanolin and oleic acids suggest plant origin because this is not typical for foodstuff of animal origin. A relatively low stearine acid content also supports this view. However, the relatively high palmitic acid ratio contradicts this. A resolution would have been easier if arachidonic acid had been detected, because it occurs only in products of animal origin, but it was not present at all. The size of the samples was not sufficient to make cholesterol tests. Even if some cholesterol had been present in the material, it decayed by now. As the oleic and linoleic acid content of the samples was high, which was supposed to have oxidised or decayed under such conditions, it can be assumed that even if part of the food was of animal origin, it could not have been a major part in these well preserved remains. Fatty acid content of the samples was similar to that of sheep tallow, but it was even more like lard.

With the results of the versatile analysis of food remains found at the Kiskundorozsma-Nagyszék site, a reconstruction can also be attempted. The mixture of husked millet grains and cereal groats was not fermented, but cooked or baked, and supplemented with other ingredients.

We managed to clarify through analytical tests that the basic material of the gruel was groats of mixed crop milling (emmer, common bread wheat, barley) or dough kneaded which was first roasted on fat (most probably lard) (see the flour particles from the grains aleuron layer). As a result of macroscopic, microscopic and analytical tests, the process of food preparation can be outlined as follows: husked millet grains were scattered onto fine cereal middling and the well mixed (kneaded?) dough was combined with animal fat, meat(?), blood(?) and/or salt. After the addition of water, it was cooked over a fire.

This cooked, eventually baked and salted mush, enriched with animal blood (horse, cattle), consisting of millet grains mixed into cereal groats must have been a typical food for the Sarmatians living here.

The Huns were a nomadic people originating from Inner Asia, transformed into an alliance of tribes for warfare during the 3rd century BC. Their regular attacks triggered the construction of the Great Wall in China, for defence. In the 4th century, the expansion of the Hun empire reached to the Roman Empire and they even conquered Pannonia. In 433, Attila became King, who led campaigns to Gaul and Italy. Although Attila was defeated in the battle of Catalaunum (451), Rome became a taxpayer to the Huns. After the death of Attila (453) the Hun alliance collapsed.

We know very little of the eating culture of the Huns. Finds associated with them are sparse at any rate, and contain no food remains. However, it is known from the description of Priscos Rhetor, the ambassador to the court of Attila in 448 on assignment from Constantine the Porphyrogenite, that the food of the Huns consisted of millet, barley "kamon" and honey "medos", and they also drank wine. Attila ate only meat and drank wine. A great deal of discussion emerged about the function of the large crowned Hun bowls like the one found in Törtel with a volume of more than 100 litres. It is possible that it was used to burn sacrifices for warriors fallen elsewhere, but it is also possible that drinks were mixed in it. Csaba Nyers, master chef for instance, uses it to cook millet mush.

The Avars were horseback nomads of Turkish-Tatar origin who arrived in the territory of contemporary Hungary from the area of the Lower Don, first in 568 AD, then again after 670. The "griffin-sarmentose" population, arriving during the second Avar conquest, was different from those people arriving in the 6th century both in terms of anthropology and decorative art. Gyula László, who developed the theory of the "double conquest," rightly calls them "proto-Hungarians".

The Avars populated the Great Hungarian Plain and the eastern and southern belt of Transdanubia. By the 7th and 8th centuries, the Avars lived in pit-houses and carried on a sedentary tiller lifestyle. Finally, it was Charles the Great who broke the power of the more than two-century old Avar empire. The Avars, having been defeated by the Franks, survived until the Hungarian conquest.

According to archaeological findings, the general utility tools of the Avars and the Slavs living in Transdanubia became very similar by the end of the Migration Period (9th century). A single agricultural population emerged from the set of finds.

In the Late Migration Period, one can find a number of different crops (common bread wheat, barley, rye, oats, common millet), wild forest fruits (hawthorn, bramble, wild strawberry, cornel berry) and improved fruits (cherries, sour cherry, plums, peach, apricot, pear, walnut, grapes) but very few kitchen-stuff items (only pea and common lentil). Maybe the steppe traditions con in continued to dominate their diet? The carbonised late Migration Period garlic (*Allium sativum*) segments found in Zalavár are likely connected with food preparation.

More recently, we have food remains connected with the Avars. We found food remains when soil samples from the Szegvár-Oromdűlő Early Avar cemetery graves and ceramics were floated (excavation by Gábor Lőrinczy 1990). Food remains of a population having nomadic crop production traditions and retaining a similarly nomadic diet but carrying settled crop production practices were found here, not in a customary place, i.e. site excavation, but in graves, where they must have been placed for cultic purposes. These very many and tiny, carbonised mush/bread fragments are seen as important evidence of the use of cereals. Evidently, it must have been a favourite food as it was found in a dozen graves. All this is the hallmark of a common widespread eating habit and a parallel ritual of the dead. These tiny fragments, rather crumbs, are very similar to each other. The surfaces of the fragments are interwoven with smaller and larger holes, which were created during the kneading of finely ground cereals (flour, groats) to make dough or to mix mush. No bubbles or blisters were observed. Their carbonised state suggests that they got into the pot and the graves in a burnt form. The only exception was one of the graves where ten of the hundred remains counted contained only browning, and no carbonisation. (Maybe this is the result of slow burning). Although the number of fragmented pieces of gruel or cake bread reached one thousand, their size and mass was negligible. The only exception was represented by remnants in one of the grave ceramics that were markedly larger than the others. In the case of one of the larger pieces, the structure was also well distinguishable: we found 2 millet grains baked in cereal groats grist.

It is not known whether it was of any significance, yet it is noted that no food remains, merely seeds were found in the soil samples taken from the graves of the Late Avar Period graveyard, Pitvaros-water reservoir (excavation by Gábor Lőrinczy 1993), not very far from this area. We can only guess at the reason for this. Maybe the ritual of the dead changed. No more carbonised food remains got into the pots from the burial-feast, or if they did, they were degraded in the soil because they were not carbonised.

11.8. Eating habits of the conquering Hungarians

The Hungarians, consisting of the alliance of seven Hungarian and one Kabar tribes, settled in the Carpathian Basin at the end of the 9th century. They were basically a semi-nomadic population, dealing with migrating large livestock husbandry and limited tilling. At any rate, the conquering Hungarians arrived into the Carpathian Basin with the knowledge of crop production. Beside the linguistic evidence, archaeological finds also support this assumption.

The source of meat dishes of the conquering Hungarians were cattle, sheep, goat, pig, chicken, goose and duck while plants used included common millet, barley, common bread wheat, rye, pea, common lentil, onion, garlic, certain roots (such as carrot, garden parsnip), spices, forest fruits, berries, and grapevine. We expect the use of mushrooms and other edible plants found in the wild, but these left no traces. Cooking and baking were known. The heritage of conquering Hungarians includes round bottom clay pots, bowls and mugs coming from grave enclosures. Foods included gruel and meat mush. The most common way of preserving food was sundrying, cooked and salted beef or mutton, then grinding it into powder. It was practical, used little space and lasted long. They took it with them when they started off for longer journies. When poured into water, the meat powder provided a quickly raising damper.

The legend of "tender meat under the saddle" comes from Marcellinus (330–390 AD) and did not even refer to the Hungarians: "The Huns ... eat the raw meat of all kinds of animals, which is put on the back of their horses, under their thighs and is thus softened to become warm". János Schiltberger (1394) wrote of the Kirgiz that they keep thin slices of meat under their saddle, eat the slices, but first they salt them. However, Bergman, who went to visit the Kalmük along the river Volga, mentioned another way of use: chafing on the back of horses was healed by putting raw meat soaked in salty water on it.

We have a number of archaeobotanical remains from this age, which indicate the kind of dishes of plant origin eaten by the conquering Hungarians. In the grave of a chief unearthed in Zemplén in the 1940s, carbonised common millet was found. Though the seed finds identified at the 1993 excavation site of Miklós Takács at the 10th-century settlement in Lébény-Billedomb were mainly six-rowed barley and millet, yet common bread wheat and rye grains, both requiring higher level of crop production skills, were also there, together with pea. Gruel remains were also found.

Most important Finno-Ugric words in relation to food preparation in the conquest period include köles (common millet), ed (cereal crop), kenyér (bread) (mush) and they refer to the consumption of these kinds of food. Some of the loanwords arriving into the language in Levedia from the Turks relate to milk processing: tej (milk), túró (cottage cheese), sajt (cheese), író (whey), köpü (churn) (butter) tarhó (yoghurt). (Yoghurt itself is a late arrival, coming form the Ottoman Turks.) The old Hungarian term for the inoculum of tarhó (which contains lactic acid bacteria) was tarhómag (i.e. yoghurt seed). The steppe people, and hence the conquerors did not know cheese making based on the technology of curdling milk in a sweet state, therefore their kind of cheese was more like dried curd, like the one made by the Kazakhs today.

The easiest way to process milk is to make cottage cheese. Yoghurt is more difficult and kumis, fermented mare milk, is most complicated. *Kumis* is made by letting raw milk stand after butter is churned out. It could not be consumed raw due to its strongly laxative effect. The process of lactic acid fermentation was regulated by stirring. Consumption of this white, tartish sour-sweet liquid, with low alcohol content of 1–3%, by the Scythians was mentioned as early as the 5th century BC by Herodotus. An 11th-century Russian chronicle mentions it as the drink of the Kumans of Polevc ("Turkish-Tatar tribe on the Eastern European steppe"). It was also described by Rubruk, a Western monk who visited the Mongols in the 13th century. Although the originally Turkish word "kumis" does not occur in the written documents of ancient Hungarian, one cannot exclude that it was known to our ancestors, maybe by another name.

Beer drinking cannot be excluded, either, as steppe people usually drank beer. You can make beer from almost all the cereal grains. Either by malting germinating barley or wheat first and mashed, or ground and baked as gruel or bread first and mashed later. Cumans (1519) mashed beer (boza) from millet first cooked as mush. Boza beer was a sour, yellowish white drink with low alcohol content. Conquering Hungarians must have known it just like méd, the honey beer or honey wine fermented from wild honey. However, the issue of wine making and drinking is a much more complicated question in the case of the conquering Hungarians. Hungarians arrived into the Carpathian Basin with a whole body of knowledge on fructiculture and viticulture. Here they might have found grapevine cultivars of Roman origin and Pannonian viticulture was carried on from father to son. A large part of the words in Hungarian relating to viticulture, fructiculture and wine making are of Bulgarian and Turkish origin. They got into Hungarian during the co-operation period with the Kazar Kaganate. Vine grapes was encountered by Hungarians during their stay at their former residence, Levedia, and most probably it was cultivated in Etelköz as well. Possibly, Anonymous refers to this in Gesta Hungarorum when he describes what a great festivity (áldomás) was made by the chiefs when they sacrificed a fat horse on the mountain of Tarcal, according to their heathen habits ("magnum aldamas fecerunt").

However, more than any written documents, the many archaeological finds tell a tale about the beginnings of Hungarian drinking culture; such as, remains of jugs or the drinking set of the treasure of Nagyszentmiklós.

11.9. Food remains of the conquering Hungarians

The heritage of steppe people can be clearly seen in the eating habits of the conquering Hungarians. Carbonised food remains preserved a message from the distant past. The revival of such a diet based on these traditions could help us to live more healthily.

More than one book was written on the imaginary cuisine of the conquering Hungarians, based on ethnographical and archaeological analogies. However, it was not very long ago that the first food remains associated with the conquering Hungarians were found at the 10th-century site of Lébény-Billedomb (excavation led by Miklós Takács). These were carbonised fragments of mush/bread (GYULAI 1998e). Similar remains were found from the Early Árpádian Period site of Gyomaendrőd (excavation by Dénes B. Jankovich 1987–90) and from the Árpádian Period site found at the exploration of the M0 motorway in Rákospalota-Újmajor (excavation by Zoltán Bencze 1995–96). These gruel-like foods prepared from common millet imply the survival of nomadic eating habits.

When exploring Edelény-Földvár in 1998, Mária Wolf found clay pots placed on their sides near the oven of a burnt house from the 10th century (*Figs. 474*). On the side of one of the pots burnt food remains were discovered (*Figs. 469–470*). The three samples received for analysis came from different parts of the pot. They had homogenous structures mixed evenly (*Figs. 475, 478*). Neither grains, nor other kind of seeds or grists were distinguishable in them. They were made of evenly mixed, fine middling-like cereal meals. The mixture thus prepared was not fermented but cooked right away. No traces of leavening of the dough, in other words lactic acid fermentation, were found.

Microscopic examinations confirmed the observations. Indeed, a large number of flour particles were found in the aleuron layer of the grains (*Fig. 484*). It was also possible to determine the material of the flour or rather the grist particles: they included common bread wheat (*Trititcum aestivum* subsp. *vulgare*) epidermis and phytolith fragments, rye (*Secale cereale*) phytoliths (*Figs. 486–487*). Not only archaeobotanical finds, but also written sources confirm the notion that the mixed cropping of common bread wheat and rye (called abajdoc, meslin, Triticum mixtum) was common as early as the Árpádian Period and continued through the Middle Ages until the modern times.

Fragments of vascular bundles originating from the incrassated storing taproot of a root-vegetable (garden parsnip or carrot) were also observed. Consequently these kinds of roots were also cooked in the pot (*Figs. 488–489, 492*). By starting hypodermic cells, a piece of the upper epidermis in an onion/garlic leave was also identified (*Fig. 490*). Cocoons of a certain nematoda: *trichocephalus* (*Trichuris trichiura*) were also found. This worm lives in the caecum and colon of people and swine (*Fig. 491*). Probably it did not get into the food when it was cooked but later on, when it was buried.

Analytic studies were conducted by János Csapó (*Fig. 494*). The potassium and nitrogen content of the samples was relatively low, as opposed to calcium, which was found to be extremely high. The same holds true for magnesium and manganese. The zinc and iron content was also high. The element composition of all the three samples was similar and only in the case of iron was a single deviating value found. High calcium and iron contents refer to another kind of foods present in the pot, maybe bone meat beside the grist of grains. High magnesium levels seem to confirm that the origin of the remains was partly grain. The zinc content is more or less proportional to that found in other grain crops such as wheat and rye.

Provided common salt was used (NaCl), which is very likely, groundwater must have leached it and it must have been removed through the movement of the groundwater table. N o common salt was detected.

Some of the amino acids were just at the edge of detectability (*Fig. 495*). The amino acid composition of the samples – disregarding the extremely high ammonium content – proved to be very similar to that of contemporary cereals (bread wheat). When the amino acid composition of the samples evaluated is compared to that of recent cereals and other kinds of foodstuff, the conclusion is that they are the closest to cereals (common bread wheat) and meat, horse meat being the most likely candidate due to a high glutamic acid value.

The level of unsaturated fatty acid (oleic acid and linoleic acid) was high in the samples indicating that even if some of these well-preserved food remains were of animal origin, it must not have been a dominant amount (*Fig. 496*). Although the fatty acid composition of the samples taken from the food remains found in the Edelény-motte is similar to that of swine lard, the similarity is even greater when compared with mutton tallow.

Investigations led to the conclusion that a one-course dish found in a pot was made of cereal groats of common bread wheat and rye, which was first roasted in fat (most probably in mutton tallow) (see the flour particles from the cereal aleuron layer). Onions and/or garlic (based on epidermal sections of onion leaf) and roots (parsnip or carrot vascular bundle fragments) were added. The mixture was then thinned, apparently with water. Microelement ratios and the amino acid composition suggest that the one-course dish had meat in it as well.

In 2001 three more carbonised food remnants were examined, all coming from the excavation carried out in the area of the Edelény-Borsod motte. All three specimens were taken from an in situ exposed pot. Two of them came from 1992 excavations and one from 1998. The first of the 1992 remains consisted of many carbonised fragments with no contamination as it was homogenous (*Fig. 476*). The surface of the fragment was shiny due to the heat impact (*Fig. 477*). On the larger pieces, one could distinguish a compacted external skin, which followed the shape of the pot, and a porous internal layer. The internal parts of the fragments showed a typical gruel surface filled with holes and bubbles (*Fig. 480*). Embedded in fine groats, husked millet grains were also observed. A few field pea seeds were also found, burned into the mush.

The two remains from the 1992 exploration were obviously similar. They consisted of seeds stuck together as in a gruel and embedded in fine groats. The burnt fragments had a characteristic surface created during cooking. They contained evenly mixed grains embedded in some fine kind of grist. One of them contained exclusively chaffed grains of common millet (*Panicum miliaceum*), while the other had more foxtail millet (*Setaria italica*) (also husked), and less common millet. The latter also contained a few field pea (*Pisum sativum* subsp. *arvense*) seeds.

The mixture thus prepared was not fermented but cooked. No traces of leavening of the dough, in other words lactic acid fermentation (i.e. bubbles) were observed.

The third remain from the year 1998 was different from the rest at the first sight. Seeds and carbonised forest fruits fell out at the slightest mechanical impact from the smaller and larger incremented pieces, which were also mixed with daub (*Fig. 479*). Most of the remains originated from sloe (*Prunus spinosa*). Beside the stones, mummified sloe, blackthorn fruits were also found in great abundance. We also found wild pear (*Pyrus* spec.) seeds, crab apple (*Malus silvestris*) fruit mummy, rose (*Rosa* spec.) mummified hips, cornelian cherry or dogwood (*Cornus mas*) and carbonised shells as well (*Fig. 481*). The fragments also contained a stone of danewort (*Sambucus ebulus*). However, its reddish colour indicated it did not come from the food but rather from the daub.

For the purposes of microscopic examinations, a tiny bit of the food remains was removed and digested using oxidative agents until an orange colour, indicating the end stage of the process, appeared. Following this, it was washed, decanted, centrifuged and dried. Finally, microsections were made of the preparations using high light refraction index value agents for embedding. (Some of them were coloured.) Under light microscope, using 200–400x enlargement and phase contrast inset, many exposed flour particles were found in the 1992 samples that originated from the cereals aleuron layer (*Fig. 485*). The flour or rather grist was identified through the presence of common millet phytoliths (*Fig. 483*).

Digestion and microscopic examination of the 1998 sample did not reveal any cereal phytoliths. On the contrary, many microscopic botanical remains (pulp, hair and so on) came from the fruits described above (*Fig. 482*). Pollen grains were also found (*Fig. 493*). As the presence of fruits refer to autumn, when the pollen production of plants is reduced, it can be assumed that the pot might also contain honey, an essential sweetener for making sour fruits more palatable. Thus, tricarpellate type pollen, the one found in the sample is associated with a vexillar plant, Dutch clover (*Trifolium repens*), which is found frequently in honey made in Hungary.

This remnant is different from any other tested so far. It seems that forest fruits (mainly sloe or blackthorn, but also crab apple, wild pear, rose, cornelian cherry as well) were used to make lictarium, a favourite delicacy in the Middle Ages, somewhat thinner than marmalade today; or peszmeg, a kind of fruit juice preserved by cooking, sweetened by honey, which required left constant stirring. Cooking was still underway – not all the fruits had flesh and stone had been separated – when the operation had to be abandoned suddenly for some unknown reason. The event must have happened around October as indicated by the assortment of the fruit species present that can be found together only in autumn. Lippay (1664) describes all the fruits found here noting that excellent lictarium can be made of them with the use of honey, spices and wine.

Analytical chemical examinations of the two 1992 food remains were also made by János Csapó. Samples taken from remains No. 1 and 2 have relatively low potassium and nitrogen content, which can be attributed to leaching. The same can be said of calcium, magnesium and manganese. The zinc level was average but iron was high. The first sample may have contained ingredients of animal origin as well.

A number of the amino acids were just at the detectability level. Sample No. 1 contained more amino acids than sample No. 2. Here, almost 10% of all nitrogen was present in the form of amino acids. This means that proteins did not disintegrate completely. The composition of sample 1 expressed in terms of grams of amino acid/100 g protein was most similar to that of common millet. The amino acid composition of sample No. 2 did not resemble any of the basic materials it was compared with.

Fatty acid analysis indicated palmitic acid, stearic acid, oleic acid, linoleic acid and arachic acid in sample No. 1. Lipid tests of sample No. 2 led to similar results, with the exception of arachic acid. Unsaturated fatty acids in sample No. 2 (oleic and linoleic acids) were near 50%. More than 12% linoleic acid suggests vegetable origin because it is not typical in animals. High oleic acid content, around 33%, also suggests vegetables. Since arachic acid is found only in animals, its absence also supports vegetable origin. When the fatty acid content of sample No. 2 is compared with that of some recent foodstuffs, it is not similar to any of them. Sample No. 1 has typical arachic acid levels. Cholesterol was not detected possibly decaying completely over time. As sample No. 1 also contained high levels of palmitic acid and oil, it is certain that animal fat was present in the food. When the fatty acid composition of a few fats is compared with the sample values, they seem to be similar to mutton.

Though the two 1992 samples contain two different foods, they are remains of the same one-course dish. A common feature is that fine cereal flour/grist was mixed with millet (in the second case, setaria). Findings of sample 1 proved that the gruel was enriched with meat. Based on the fatty acid composition, this must have been mutton.

11.10. Medieval food remains

Gruel continued to be a most favourite food in the Middle Ages. Such gruel remains were found during the archaeobotanical processing of the Muhi community, a village that perished in the Late Middle Ages (excavation by József Laszlovszky and Tamás Pusztai 1995).

Similarly, food remains uncovered from the medieval settlement explored in 1998 near Baj Öregkovács-hegy in the forest, excavated by Sándor Petényi, were also dated to the Middle Ages. Two pot fragments contained unusually large millet grains, incremented due to burning into the wall of the pot. (Large grains always refer to improved sowing seeds.) No glumella or chaff was observed on them. Consequently, before cooking they must have been carefully cleaned. Burnt grains were stuck together in smaller or larger clumps, obviously solidified in this way during cooking. The millet gruel found here characterised Hungarian food culture in the Middle Ages. Millet gruel used to be a traditional staple food for Hungarians, which can be traced back to the times before the conquest.

It is known from a number of sources that in the life of Hungarians, just as in the case of any other European people, the consumption of gruel made of husked millet played a decisive role. Millet gruel was a common dish, irrespective of social rank and privileges, and it continued to be so up to the 18th–19th centuries until the spread of maize ("Turkish wheat") coming from the New World provided a new gruel plant.

Some of the Baj pot fragments showed blackish burnt spots originating from baking or cooking. Food remains on the bottom fragment and the wall fragment were created when the contents of the pot "incremented" onto the warmer parts on the bottom, the sides and the edges, which were in direct contact with the fire. Using a stereomicroscope one can clearly see that the surface of the remains is covered with an alveolate mesh of cracks. These are the remains of one-course dishes that were produced when fluid-containing food was cooked. No unevenness or protrusions can be seen on the surface, they show a homogenous, evenly smooth structure. No bones or other remains were found in them.

Without exception, the food remains had only one layer; the pot was used only once, or this is the remains of the last dish prepared. The fatty acid analysis conducted from the food samples burned on the two pot walls suggests the presence of lard, while amino acid tests indicate the use of a meat-dish.

It is noted here that in food remains uncovered in Hungary from various archaeological periods, the kind of fat usually found is lard, provided there is enough substance available for the analysis. This means that, in contrast to the information obtained from the analysis of finds in Western Europe, cooking was done in the Carpathian Basin using fat rather than vegetable oil. Botanical finds contain with very few oily seeds. Along with tradition, the reason might have been climatic as well.

In 1998–99, during the exploration works of the Teleki Palace and the building of the former Military Headquarters in Buda Castle, both mentioned earlier on with regard to medieval plants (excavation by Zoltán Bencze, Dóra B. Nyékhelyi, András Végh), cisterns and wells were found from the 15th century, which contained the carbonised bits and pieces of mush/bread/cake. These and many other pot fragments with burnt food remains may get us closer to a better understanding of the medieval gastronomic culture, once they are analysed.

Finally, let us pay tribute to the gastronomic knowledge of the prehistoric housewife by presenting the recipe for the already described Bronze Age strawberry cake, providing evidence that people living here possessed high-level culinary skills several thousand years ago. The reconstruction of the food remnants originating from the Tumulus culture site at Balatonmagyaród-Hídvégpuszta in 1200 BC also offers a noteworthy example for linking archaeological and natural historical (scientific) examinations and for the exploitation of the potential provided by an interdisciplinary approach. In spite of all the attempts, it seems to be an insurmountable endeavour to revive all the circumstances of making the fine cake and hence its long past flavours, which was once prepared using hulled wheat and common millet flour fine ground on milling stones, wild bird eggs, lard and fat of an ancient swine species and baked in pots put into clay ovens. Yet the operations included in the recipe below may be close to reality.

Recipe of a Bronze Age strawberry cake:

Ingredients: 5 eggs, 100 g honey, 800 g fine cereal flour, 200 g common millet flour, a bit of salt, 250 g wild forest strawberry.
Time needed for preparation: 40 minutes
Time needed for baking: 10–15 minutes in pre-heated oven at 200°C
Time needed for cooling: 1–2 hours

Flour and salt are mixed and poured into a sieve. Egg white is whipped to make hard cream. Meanwhile, the egg-yolk is mixed with the full ration of honey and poured onto the cream with care. The full ration of flour is sieved on top while stirring. The mass created thus is poured into a baking-dish smeared with fat and baked in the pre-heated oven at 200°C. Meanwhile, strawberry is boiled at low fire in its own juice with a little bit of water. The cake is filled with the purée made of the strawberries.

12. References

AMMERMAN, A. J., BUTLER, J. J., DIAMOND, G., MANOZZI, P., PALS, J. P., SEVINK, J., SMIT, A. – VOORRIPS, A. 1978

Report on the excavations at Monte Leoni. A Bronze Age settlement in the Parma Valley. Hellinium 18.

ANDERSON, P. C. 1998

History of Harvesting and Threshing Techniques for Cereals in the Prehistoric Near East. The Origins of Agriculture and Crop Domestication. *Proceedings of the Harlan Symposium. 10–14 May 1997, Aleppo, Syria.* 45–159. ICARDA, IPRGRI, FAO.

ANDRASOVSZKY, J. 1925

Szőlő (The grapewine). In: Jávorka, S. (ed.), *Magyar flóra.* 1307.

ANDRÁSFALVY, B. 1957

A vörösbor Magyarországon [Der Rotwein in Ungarn]. *Néprajzi Értesítő* 39, 46–69.

ANDRÁSFALVY, B. 1961

Pekmez. Adatok törökkori szőlőkultúránk ismeretéhez. In: Babics, A. (ed.), *Kisebb tanulmányok.* 87–95. MTA Dunántúli Tudományos Intézet, Pécs.

ANDRÁSFALVY, B. 1973

A Sárköz és a környező Duna-menti területek ősi ártéri gazdálkodása és vízhasználatai a szabályozás előtt. Vízügyi Történeti Füzetek 6, Budapest.

ANDREWARTHA, H. G. – BIRCH, L. C. 1954

The Distribution and Abundance of Animals. The University of Chicago Press, Chicago, Illinois.

ANONYMUS

Anonymus – Gesta Hungarorum. Ford.: Pais Dezső. Budapest 1977.

AUFHAMMER, G. – FISCHBECK, G. 1964

Ergebnisse von Gefäss- und Feldversuchen mit dem Nachbau keimfähiger Gersten- und Haferkörner aus dem Grundstein des 1832 errichteten Nürnberger Stadttheater. *Zeitschrift für Pflanzenzüchtung* 51, 354–378.

AUGUSTIN, B., JÁVORKA, S., GIOVANNINI, R. – ROM, P. 1948

A magyar gyógynövények (*The Hungarian herbs*). Budapest.

ÁRENDÁS, V. 1982

A magyarországi archeobotanikai adatok összehasonlító értékelése [The Comparative study of archaeobotanical data from Hungary]. *Agrártörténeti Szemle* 1982/1–2, 1–52.

BAJI, B. 2000

Új, potenciális kultúrnövények az Agrobotanikai Intézet Génbank gyűjteményében [New potential cultureplants in the genebank collection of the Institute of Agrobotany]. In: Gyulai, F. (ed.), *Az Agrobiodiverzitás megőrzése és hasznosítása.* 140–144. Tápiószele.

BAKAY, K. 1978

A magyar államalapítás. Gondolat, Budapest.

BAKAY, K., KALICZ, N. – SÁGI, K. 1966

Magyarország régészeti topográfiája I. [*The Hungarian Archaeological Topography I.*]. *Veszprém megye régészeti topográfiája. A keszthelyi és a tapolcai járás.* Budapest.

BAKAY, K., KALICZ, N. – SÁGI, K. 1970
Magyarország régészeti topográfiája III. [*The Hungarian Archaeological Topography III.*]. *Veszprém megye régészeti topográfiája III. A devecseri és a sümegi járás.* Budapest.

BAKELS, C. C. 1979
Linearbandkeramische Früchte und Samen aus den Niederlanden. *Archaeo-Physika* 8, 1–10.

BALASSA, I. 1963–64
Kezdetleges gabonatisztító eljárások (Primitive Getreidereinigungsverfahren). *A Magyar Mezőgazgasági Múzeum Közleményei* 1964, 41–60.

BALASSA, I. 1965–66
Az eke szántóföldre történő kiszállításának módjai. A Magyar Mezőgazgasági Múzeum Közleményei 1966.

BALASSA, I. 1973
Az eke és a szántás története Magyarországon (The history of plough and ploughing in Hungary). Akadémiai Kiadó, Budapest.

BALASSA, I. 1985
A néprajztudomány és a magyar őstörténet. In: Szombathy, V. (ed.), *Az őshazától a Kárpátokig.* 72–108. Budapest.

BALASSA, I. 1991
A harmatkása. In: Ujváry, Z. (ed.), *Történelem, régészet, néprajz.* 147–162. Ethnica, Debrecen.

BALÁS, Á. 1876
Általános és különleges mezőgazdasági növénytermelés alapvonalai. Budapest.

BALÁS, Á. – HENSCH, Á. 1888
Általános és különleges növénytermelés I–II. Magyar-Óvár.

BALL, B., BROTHERSON, T., JACK, D. – GARDNER, S. J. 1996
Identifying Phytoliths Produced by the Inflorescence Bracts of Three Species of Wheat (Triticum monococcum L., T. dicoccum Schrank., and T. aestivum L.) Using Computer-Assisted Image and Statistical Analyses. *Journal of Archaeological Science* 23, 619–632.

BALOGH, J. 1966
A művészet Mátyás király udvarában I–II. Budapest.

BANNER, J. 1942
Das Tisza-, Maros-, Körös-Gebiet bis zur Entwicklung der Bronzezeit. Szeged–Leipzig.

BARABÁS, Z. 1987 (ed.)
A búzatermesztés kézikönyve. Budapest. Mezőgazdasági Kiadó.

BARKER, G. – GAMBLE, C. 1985
Beyond Domestication in Prehistoric Europe. Investigations in Subsistence Archaeology and Social Complexity. Academic Press, London etc.

BARTON-WRIGHT, E. C., BOOTH, R. G. – PRINGLE, W. J. S. 1944
Analysis of barley from King Tutankhamen's tomb. *Nature* 153, 288.

BARTOSIEWICZ, L. 1995
Archaeozoological studies from the Hahót basin, SW Hungary. In: Szőke, B. M. (ed.), *Archaeology and Settlement History in the Hahót Basin, South-West Hungary.* 307–368. Antaeus 22. Budapest.

BARTOSIEWICZ, L. 1999
Őskori állattartás és környezet a Kárpát-medencében. In: Füleky Gy. (ed.), *A táj változásai a Kárpát-medencében.* 195–200. GATE, Gödöllő.

BARULINA, H. 1930
Lentils of the U. S. S. R. and other countries. *Bulletin of Applied Botany and Plant Breeding* 40, 1–319.Leningrad.

BÁLINT, A. 1939
A mezőkovácsházi középkori település emlékei. *Dolgozatok* 15, 146–164.

BÁLINT, A. 1960–62
A középkori Nyársapát lakóházai. *Móra Ferenc Múzeum Évkönyve* 1960–62, 39–115. Szeged.

BÁLINT, CS. 1980
Természetföldrajzi tényezők a honfoglaló magyarok megtelepedésében. *Ethnographia* 91, 35–51.

BÁLINT, A., KOVÁCS, G., ERDEI, L. – SUTKA, J. 2000
A szemtermés ásványi összetételében megfigyelhető különbségek eltérő ploidszintű Triticum L. és Aegilops L. fajoknál. In: Gyulai, F. (ed.), *Az agrobiodiverzitás megőrzése és hasznosítása. Szimpózium Jánossy Andor emlékére. Budapest, május 4–6.* 296–300.

BÁNFFY, E., T. BÍRÓ, K. – VADAY, A. 1997
Újkőkori és rézkori telepnyomok Kompolt 15. számú lelőhelyen. *Agria* 33, 19–58.

BÁNYAI, L. 1971
Kölesfajták agrobotanikai vizsgálata. *Agrobotanika* 11, 39–60. Tápiószele.

BEHRE, K.-E. 1970
Die Entwicklungsgeschichte der natürlichen Vegetation im Gebiet der unteren Ems und ihre Abhängigkeit von den Bewegungen des Meeresspiegels. *Probleme der Küstenforschung im südlichen Nordseegebiet* 8, 13–47.

BEHRE, K.-E. 1976
Die Pflanzenreste aus der frühgeschichtlichen Wurt Elisenhof. Mit einem Beitrag von Fritz Koppe. Elisenhof: die Ergebnisse der Ausgrabung der frühgeschichtlichen Marschensiedlung beim Elisenhof in Eiderstedt 1957/58 und 1961/64. Bd. 2. Studien zur Küstenarchaeologie Schleswig-Holsteins. Serie A. Bern [etc.], Lang.

BEHRE, K.-E. 1980
Zur mittelalterlichen Plaggenwirtschaft in Nordwestdeutschland und angrenzenden Gebieten nach botanischen Untersuchungen. In: Beck, H., Denecke, D. - Jahnkuhn, H. (eds), *Untersuchungen zur eisenzeitlichen und frühmittelalterlichen Flur in Mitteleuropa und ihrer Nutzung II.* 30–44. Abhandlungen der Akademie der Wissenschaften in Göttingen 116.

BEHRE, K.-E. 1983
Ernährung und Umwelt der wikingerzeitlichen Siedlung Haithabu. Die Ergebnisse der Untersuchungen der Pflanzenreste. Die Ausgrabungen in Haithabu 8. Wachholtz, Neumünster.

BEHRE, K.-E. 1984
Zur Geschichte der Bierwürzen nach Fruchtfunden und schriftlichen Quellen. 6th Symposium Palaeoethnobotany (Groningen) 1983. In: Zeist W. van – Casparie, W. A. (ed.), *Plants and Ancient Man.* 115–122. Boston.

BEHRE, K.-E. 1988
Die Umwelt prähistorischer und mittelalterlicher Siedlungen. Rekonstruktion aus botanischen Untersuchungen an archäologischem Material. Siedlungsforschung. *Archäologie–Geschichte–Geographie* 6, 57–80.

BEHRE, K.-E. 1992a
Zum Brotfund aus dem Ipweger Moor, Ldkr. Wesermarsch. *Berichte zur Denkmalpflege Niedersachsen* 1, 9.

BEHRE, K.-E. 1992b
The history of rye cultivation in Europe. *Vegetation History and Archaeobotany.* 1/3, 1992, 141–156.

BEIJERINCK, W. 1947
Zadenatlas der nederlendische Flora. Wageningen.

BELEA, A. – FEJÉR, O. 1977
Újabb adatok a búza (Triticum L.) származásához. *Növénytermelés* 26, 213–221.

BELEA, A. 1986
Faj és nemzetségkeresztezések a növényvilágban. Budapest.

BELEA, A., KŐSZEGI, B. – KRAMARIKNÉ, K. J. 1991
A Triticum L. Nemzetség evolúciójának néhány kérdése. *Növénytermelés* 40/1, 79–87.

BEN ZE'EV, N. – ZOHARY, D. 1973
Species relationships in the genus Pisum L. *Israel Journal of Botany* 22, 73–91

BENCZE, Z. 1999
Beszámoló Rákospalota-Újmajor területén 1995–1997 között folytatott régészeti feltárások középkori eredményeiről (Bericht über die mittelalterlichen Ergebnisse der archäologischen Ausgrabungen durchgeführt 1995–1997). In: Bencze, Z., Gyulai, F., Sabján, T. – Takács, M. (eds), *Egy Árpád-kori veremház feltárása és rekonstrukciója. (Ausgrabung und Rekonstruktion eines Grubenhauses aus der Árpádenzeit.)* 9–72. Monumenta Historica Budapestinensia 10.

BERENOVÁ, M. 1986
Die Archäologie über die Pflanzenproduktion bei den Hunnen, Awaren und Protobulgaren. *Památky archeologické* 77, 81–103.

BERENOVÁ, M. 1991
Archäologie und Botanik aus der Sicht der Archäologen (Anbau von Kulturpflanzen). Palaeoethnobotany and Archaeology. *International Work-Group for Palaeoethnobotany 8th Symposium Nitra-Nové Vozokany 1989.* 37–50. Nitra.

BERTSCH, K. 1932
Die Pflanzenreste der Pfahlbauten von Sipplingen und Langenrain im Bodensee. *Badische Fundberichte* 2, 305–320.

BERTSCH, K. 1941
Der Obstbau im vor- und frühgeschichtlichen Deutschland. *Germanenerbe* 6, 103–113.

BERTSCH, K. 1942
Lehrbuch der Pollenanalyse. Stuttgart.

BERTSCH, K. – BERTSCH, F. 1949
Geschichte unserer Kulturpflanzen. Stuttgart.

BERZEVICZKY, G. 1802
Ungarns Industrie und Commerz. Weimar.

BERZSÉNYI, B. = K. BERZSÉNYI, B.

K. BERZSÉNYI, B. 2000
A települési vegetáció keletkezése a régészeti-növénytani adatok alapján. In: Füleky Gy. (ed.), *A táj változásai a Kárpát-medencében a történelmi események hatására.* 26–30. Budapest–Gödöllő.

K. BERZSÉNYI, B. – GYULAI, F. 1998
The Archaeobotanical Analysis of the Middle Bronze Age Settlement at Bölcske-Vörösgyír. *31st International Symposium on Archaeometry. Budapest, 27 April–1 May, Lectures, Abstracts.*

K. BERZSÉNYI, B. – GYULAI F. (in print)
Fehérvárcsurgó-Eresztvényi erdő kora vaskori halomsírjainak növénymaradványai. *Alba Regia,* Székesfehérvár.

K. BERZSÉNYI, B. – DÁLNOKI, O. 2005
Plant Cultivation and Crop Processing at the Formative LBK Settlement of Szentgyörgyvölgy-Pityerdomb in Transdanuba (6th Millennium BC). *Antaeus* 28, 261–270.

BOCZ, E. 1988
Ősi búzafajok minőségi vizsgálata. Előadás a keszthelyi Agrártudományi Egyetemen, Keszthely 1988.03.28.

BOCZ, E. 1998
Búza: ősi fajtával - teljes minőséget. *Magyar Mezőgazdaság* 53/14, 12–13.

BOCZ, E. 2001
A búza evolúciójának kihatása a búza nemzetségek termőképességének növekedésére, illetve a mikroelemek tömegének csökkenésére. *7. Növénynemesítési Tudományos Napok, Budapest. Összefoglalók,* 48.

BODNÁR, M. 1999
Tájvédelmi Körzet négy folyó ölelésében – a Kesznyéteni Tájvédelmi Körzet. In: Füleky, Gy. (ed.), *A táj változásai a Kárpát-medencében.* 25–32. Gödöllő.

BOELICKE, U., VON BRANDT, D., LÜNING, J., STEHLI, P. – ZIMMERMANN, A. 1988
Der bandkeramische Siedlungsplatz Langweiler 8. Gemeinde Aldenhoven, Kreis Düren. 2 Bände und 1 Mappe. Habelt, Bonn.

BOGAARD, A., BENDING, J. – JONES, G. 2004
Early Neolithic crop husbandry in the Great Hungarian Plain: Evidence from the Körös site of Ecsegfalva *23, Co. Békés.* In: Buxo, R. – Piqué, R. (ed.), *13th Symposium of the International Work Group for Palaeoethnobotany, Girona 16th–22th May 2004. Abstracts,* 89. Girona.

BOGDÁN, I., PAPP, ZS. – SZABÓ, M. 1963
Kétszázéves gabonaszemek az Országos Levéltárban. *Agrártörténeti Szemle* 5, 50–63.

BOLLINGER, T. – JACOMET, S. 1981
Resultate der Samen- und Holzanalysen aus den Cortaillodschichten (ohne verkohlte Kulturpflanzenreste). *Die neolithischen Ufersiedlungen von Twann* 14, 35–68.

BOROJEVIC, K. 1987
Analiza ugljenisanog semenja sa pokapiteta Svetinja. *Starinar* 38, 65–72.

BORBÁS, V. 1900
A Balaton tavának és partmellékének növényföldrajza és edényes növényzete. Budapest.

BÓNA, I. 1971
Honfoglalás kori magyar sír Dunaújvárosban. *Archaeologiai Értesítő* 98, 170–175.

BÖKÖNYI, S. 1959
Die frühalluviale Wirbeltierfauna Ungarns. *Acta Archaeologia Academiae Scientiarum Hungaricae* 11, 39–102.

BÖKÖNYI, S. 1968
Az állattartás történeti fejlődése Közép és Kelet-Európában. *Agrártörténeti Szemle* 10, 277–335.

BÖKÖNYI, S. 1974
History of Domestic Mammals in Central and Eastern Europe. Budapest.

BRADLEY, R. 1978
Prehistoric field systems in Britain and northwest Europe: a review of some recent work. *World Archaeology* 9, 265–280.

BRANDT, I. 1950
Planterester i et moselig fra Borremosre. *Aarbøger for nordisk Oldkyndighed Historie* 1950, 342–349.

BRECHER, GY. 1960
A magismeret atlasza. Budapest.

BROCQ-ROUSSEU – GAIN, E. 1907
Sur l'existence d'une peroxydiastase dans les graines sèches. *Compte rendu hébdomadaire des séances de l'Académie des sciences* 145, 1297–1298. Paris.

BROCQ-ROUSSEU – GAIN, E. 1911
La durée des peroxydiastase des graines. Annales du Service des Antiquiteés de l'Egypte. *Le Caire* 11, 40–43.

BRONGERS, J. A. 1976
Air photography and Celtic field research in the Netherlands. Nederlandse Oudheden 6.

BROUWER, W. 1972
Handbuch des Speziellen Pflanzenbaues I. (Weizen, Roggen, Gerste, Hafer, Mais). Berlin.

BRÓZIK, S. 1962
Csonthéjasgyümölcsűek. Őszibarack. Termesztett gyümölcsfajtáink 2. Budapest.

BURGA, C. A. 1988
Swiss vegetation history during the last 18000 years. *New Phytologist* 110, 581–602.

BURIÁN, B. 1983
A vetőmagtermesztés fejlődése és magyarországi története. Budapest.

BUURMAN, J. 1988
Economy and environment in Bronze Age West-Friesland, Noord Holland (from wetland to wetland). In: Murphy, P. – French, Ch. (eds), *The Exploitation of Wetlands*. 267–292. British Archaeological Reports 186., Oxford

DE CANDOLLE, A. 1894
Termesztett növényeink eredete. Budapest.

CÂRCIUMARU, M. 1983a
Contribuţii la cunoaşterea agriculturii geto-dacilor pe baza unor determinări de seminţe carbonizate şi analize sporopolinice din aşezarea de la Cârlomănesti (judetul Buzău). In: Ialomiţa, I. (ed.), *Materiale de istorie agrara a Romaniei.* 237–242. Slobozia.

CÂRCIUMARU, M. 1983b
Consideraţii palaeoethnobotanice şi contribuţii la agricultură geto-dacilor. *Thraco-Dacica* 4 (1–2), 126–134.

CHAMBERS, F. M. – JONES, M. K. 1984
Antiquity of rye in Britain. *Antiquity* 58, 219–224.

CHEBEN, I. – HAJNALOVÁ, É. 1997
Neolithische und äneolithische Öfen in der Slowakei aus der Sicht des Archäologen und Archäobotanikers. *Archaeologia Austriaca* 81, 41–52. Wien,

CHILDE, V. G. 1936
Man Makes Himself. London.

CLARK, G. 1949
A vadembertől a civilizációig. Budapest.

COLES, J. M. 1988
A Wetland Perspektive. In: Purdy, B. A. (ed.), *Wet Site Archaeology.* 1–14. Caldwell, New Jersey.

COSTANTINI, L. 1977
Le piante. In: Tucci, G. (ed.), *La città bruciata del deserto salado.* 159–171. Venice-Mestre.

COSTANTINI, L. 1981a
The beginning of agriculture in the Kachi Plain: The evidence of Mehrgarh. In: Allchin, B. (ed.), *South Asian Archaeology 1981. Proc. 6th Int. Conf. Assoc. South Asian Archaeologists in Western Europe.* 29–33. Cambridge University Press.

COSTANTINI, L. 1981b
Semi e carboni del mesolitico e neolitico della Grotta dell'Uzzo, Trapani. *Quaternaria* 23, 233–247.

COSTANTINI, L. – COSTANTINI-BIASINI, L. 1985
Agriculture in Baluchistan between the 7th and the 3rd millennium BC. *Newsletter of Baluchistan Studies* 2, 16–30. Instituto Universitario Orientale, Naples.

CULIKOVA, V. 1981
Rostlinné makroz bytky ze stredovekého Mostu. *Archeologické rozhledy* 33, 649–675.

CZUCZOR, G. – FOGARASI, J. 1868
A magyar nyelv szótára IV. Budapest.

CSALLÁNY, D. 1961
Archäologische Denkmäler der Gepiden im Mitteldonaubecken (454–568 u.Z.). *Archaeologia Hungarica* 38, Budapest.

CSALLÁNY, G. 1914
Szentes őskori képe. In: Sima, L. (ed.), *Szentes város története I.* 325. Szentes.

CSAPÓ, J., TÓTH-PÓSFAI, I. – CSAPÓ-KISS, ZS. 1986
Optimization of Hydrolysis at Determination of Amino acid Content in Food and Feed Products. *Acta Alimentaria* 15 (1), 3–21.

CSERHÁTI, S. 1906
A gabonafélék termesztése. Az Erdélyi Gazdasági Egylet Könyvkiadó Vállalatának L.-ik füzete. Kolozsvár.

CSOMA, ZS. 1994–95
Szőlészeti, borászati hagyományok, a megújulás és a közösség kötelékében. Kapcsolatok, hatások, konfliktusok Dunántúl és Európa között a 17. század végétől a 20. század elejéig. Centrál-Európa Alapítványi Könyvek 3. Debrecen–Budapest.

CSOMA, ZS. 1996
A középkori magyarországi szőlőhegyek arculatának kialakulása és a szőlő-bortermelés kettős gyökere. In: Füleky, Gy. (ed.), *A táj változásai a Honfoglalás óta a Kárpát medencében. Tudományos konferencia. Gödöllő 1996 június 24–26.* 127–160.

CSŐRE, P. 1980
A magyar erdőgazdálkodás története (középkor). Budapest.

DANERT, S., FUKAREK, F., HANELT, P., HELM, J., KRUSE, J., LEHMANN, CH. O. – SCHULZE-MOTEL, J. 1980–1981
Urania Növényvilág I-II. (Urania Pflanzenreich I-II). Budapest.

DARBY, W. J., GHALIOUNGUI, P. – GRIVETTI, L. 1977
The gift of Osiris, 2 vols. Academic Press, London–New York–San Francisco.

DARWIN, CH. 1859
On the Origin of Species. Murray, London.

DAVIS, P. H. 1970
Pisum L. In: Davis, P. H. (ed.), *Flora of Turkey.* Vol. 3. 370–373. Edinburgh University Press.

DAVYDOVA, A. V. 1968
The Ivolga Gorodishche. (A Monument of the Shiung-nu Culture in the Trans-Baikal Region). *Acta Archaeologica Hungarica Academiae Scientiarum Hungaricae* 20, 209–245

DAX, M., ÉRI, I., MITHAY, S., PALÁGYI, SZ. – TORMA, I. 1972
Magyarország Régészeti Topográfiája IV. Veszprém megye régészeti topográfiája. A pápai és a zirci járás. Budapest.

DÁLNOKI, O. 1998
Következtetések Pannonia provincia szőlőművelésére az Aquincum-Kaszásdűlőről előkerült szőlőmaradványok alapján, különös tekintettel Aquincum territoriumára. Szakdolgozat. Eötvös Lóránd Tudományegyetem Régészettudományi Szak Budapest. Kézirat.

DÁLNOKI, O. 2000
Kelta telepek környezeti képéről a Budapest-Corvin tér keltáinak növénytermesztése alapján. In: Füleky, Gy. (ed.), *A táj változásai a Kárpát-medencében a történelmi események hatására.* 42–50. Kulturális Örökség Igazgatósága Budapest – Szent István Egyetem, Gödöllő.

DÁLNOKI, O. – JACOMET, St. 2002
Some aspects of Late Iron Age agriculture based on the first results of on archaeobotanical investigation at Corvin ter, Budapest, Hungary. *Vegetation History and Archaeobotany* 11, 9–16.

DEININGER, I. 1881
Deiniger Imre jelentése. In: Nyáry, J. (ed.), *Az Aggteleki barlang, mint őskori temető.* 55–64. Budapest.

DEININGER, I. 1890 (ed.)
Tanulmány a búza fölött. Tan-Intézeti Kiadványok 26. Nagykanizsa.

DEININGER, I. 1892
Adatok kultúrnövényeink történetéhez. A Lengyel-i őskori telep növénymaradványai (Data to the history of cultivated plants. Plant remains from the Lengyel prehistoric settlement). *Keszthelyi Magyar Királyi Gazdasági Tanintézet Évkönyve* 1891, 1–31. Nagykanizsa.

DEININGER, M. E. 1877
II. Les graines de ceréals sont. In: Nyáry, E. (ed.), *Les hommes de L'Age de la Pierre dans la Caverne D'Aggtelek, Comté de Gömör.* 630–632. Congres International d'Préhistoriques Comte-Rendu de la Hitieme Session a Budapest I.

DERNSCHWAM, H. 1984
Erdély, Besztercebánya. Törökországi útinapló. 497–501. Budapest.

DIENES, I. 1972
A honfoglaló magyarok. Corvina, Budapest.

DOMBAY, J. 1960
Die Siedlung und das Gräberfeld in Zengővárkony. Archaeologia Hungarica 33.

DOMBORÓCZKY, L. 1997
Füzesabony-Gubakút. Újkőkori falu a Kr. e. VI. évezredből. (Neolithic village from the 6th Millennium BC.) In: Raczky, P., Kovács, T. – Anders, A. (eds), *Utak a múltba. (Paths in to the Past.) Az M3-as autópálya régészeti leletmentései. (Rescue Excavations on the M3 Motorway.)* 19–27. Magyar Nemzeti Múzeum, ELTE Régészettudományi Intézet. Budapest.

DOMBORÓCZKY, L. 2005
A Körös-kultúra északi elterjedési határának problematikája a Tiszaszőlős-Domaháza-pusztán végzett ásatás eredményeinek alapján. *Archeometriai Műhely* 2005/2, 5–15.

DUDITS, D. – HESZKY, L. 2000
Növénybiotechnológia. Agroinform, Budapest.

EHRENDORFER, F. 1973
Liste der Gefässpflanzen Mitteleuropas. Stuttgart.

EHRICH, R. W. 1992
Chronologies in Old World Archaeology. London.

EKROTH, H. 1986
Förkolnat botaniskt materal från en skärvstenshög i Östergötland. In: Larsson, T. (ed.), *Skärvstenshög och skärvstensvall.* 24–29. Södermanlands Museum, rapports, Nyköping.

ELLENBERG, H. 1979
Zeigerwerte der Gefässpflanzen Mitteleuropas. *Scripta Geobotanica* 9.

ENDRŐDI, A. 1998
Késő rézkori leletek Óbudáról (Late Copper Age Finds from Óbuda). *Budapest Régiségei* 32, 105–123.

ENDRŐDI, A. 2002
Present stage of Bell Beakers research in Hungary (characteristics of settlement structure). In: Czebreszuk, J. – Szmyt, M. (eds), *The Northeast Frontier of Bell Beakers. Symposium, 26–29 May 2002, Poznań.* Abstracts, 20.

ENDRŐDI, A. – GYULAI, F. 1998–2000
Hearths and Other Finds of the Late Copper Age Baden Culture at Budapest-Csepel Island (Gynaecomorphic vessels, Archaeobotanical Remains). *Archaeologiai Értesítő* 125, 9–44.

ENDRŐDI, A. – GYULAI, F. 1999
Soroksár-Várhegy a fortified bronze age settlement in the outskirts of Budapest. Plant cultivation of middle bronze age fortified settlements. *Communicationes Archaeologicae Hungaricae* 1999, 6–34.

ENDRŐDI, A. – VÖRÖS, I. (in print)
The beliefs of the Baden culture as reflected by finds from Budapest. In: Kovács, T. (ed.), *Studien zur Metallindustrie und Glaubenswelt der Kupferzeit Mitteleuropas. Festschrift für Pál Patay zum 85. Geburtstag.* Budapest.

ENTZ, G. – SEBESTYÉN, O. 1942
A Balaton élete. Budapest.

ERDÉLYI, I. 1982
Nomád népek – letelepült népek kapcsolatának és viszonyának problémája a Kárpát-medencében és Kelet-Európában. Veszprém Megyei Múzeumok Közleményei 16.

P. ERMÉNYI, M. 1975–77
Forrástanulmány a régészeti korokból származó csonthéjas gyümölcsleletekről Közép-Európában (Quellenstudie über die Steinfruchtfunde in Mitteleuropa aus archäologischen Zeiten). *Magyar Mezőgazdasági Múzeum Közleményei,* 135–165.

ERROUX, J. 1981
Étude des graines de sites préhistoriques des Causses, La Poujade, St. Etienne de Gourgas, Pompignan. *Paléobiologie continentale* 12, 273–278.

EVANS, G. M. 1976
Rye. In: Simmonds, N. W. (ed.), *Evolution of crop plants.* 108–111. London.

ÉGETŐ, M. 1993
Az alföldi paraszti szőlőművelés. Budapest.

FACSAR, G. 1970
Összehasonlító morfológiai vizsgálatok kerti szőlőfajták magjain I. (Vergleichende morfologische Untersuchungen der Samen von Gartenrebensorten I.). *Botanikai Közlemények* 57/3.

FACSAR, G. 1973
Agricultural-Botonical Analysis of the Medieval grape seeds from the Buda Castle Hill, Budapest. Mitteilungen des Archaeologischen Institut der Ungarischen Akademie der Wissenschaften 4.

FACSAR, G. 1987
A Sopron-Krautacker vaskori telep ásatásának karpológiai vizsgálata. *Praenorica Folia Historico-naturalis* 2, 35–38.

FACSAR, G. – JEREM, E. 1985
Zum urgeschichtlichen Weinbau in Mitteleuropa. Rebkernfunde von Vitis vinifera L. aus der urnenfelder-, hallstatt- und latènezeitlichen Siedlung Sopron-Krautacker. *Wissenschaftliche Arbeiten aus dem Burgenland* 71, 121–144.

FARKAS, S. 1890
Csongrádi bronzkori leletről. *Archaeologiai Értesítő* 10, 356.

FEINDT, F. S. M. 1989
Organische Reste an einem Bronzeschwert aus Flintbek. Archäobotanik. *Dissertationes Botanicae* 133, 81–88.

FEKETE, G. 1981
A növényi populációk. In: Hortobágyi, T. – Simon, T. (eds), *Növényföldrajz, társulástan és ökológia.* 169–191. Budapest.

FEKETE, K. 1999
Az alakor rekultivációja. Kísérlet a legősibb búza újratermesztésére Magyarországon. Szakdolgozat. Gödöllői Agrártudományi Egyetem Mezőgazdasági Főiskolai Kar, Mezőtúr, Kézirat.

FERENCZY, E. 1958
A magyar föld népeinek története a honfoglalás koráig. Budapest.

FETTICH, N. 1964
A jutasi avarkori temető revíziója. *Veszprém Megyei Múzeumok Közleményei* 2, 79–107.

FEYÉR, P. 1981
A szőlő és bortermelés Magyarországon 1848-ig. Budapest.

FIGLER, A., BARTOSIEWICZ, L., FÜLEKY, GY. – HERTELENDI, E. 1997
Copper Age Settlement and the Danube Eater System: a case study from North Western Hungary. In: Chapman, J. – Dolukhanov, P. (eds), *Landscapes in Flux Central and Eastern Europe in Antiquity.* 209–230. Colloquia Pontica 3, Oxbow Books.

FIRBAS, P. 1949
Spät- und nacheiszeitliche Waldgeschichte Mitteleuropas nördlich der Alpen. Jena.

FORGÓ, GY. 1817
Rendkívül való szükség' idején, a' közönségesen szokásban lévő Gabona fajokon kívül, miből készíthetni még Kenyeret Hazánkban, 's mit találhatni még, a' mivel ollyankor táplálhassa magát a' Szegénység? *Tudományos Gyűjtemény* 10, 41–57.

FOWLER, P. J. – EVANS, J. G. 1967
Plough marks, lynchets and early fields. *Antiquity* 41, 289–301.

FÖLDES, L. 1983
"Telkek" és költözködő falvak a honfoglaló és Árpád-kori magyarság gazdálkodásában. In: Tőkei, F. (ed.), *Nomád társadalmak és államalakulatok.* 327–349. Akadémiai Kiadó, Budapest.

FRÁTER, E., BIRÓ, M. – KÓSA, G. 2000
A tájhasználat változása a Békés megyei Fáspusztán. In: Füleky, Gy. (ed.), *A táj változásai a Kárpát-medencében a történelmi események hatására.* 266–268. Budapest–Gödöllő.

FRISNYÁK, S. 1990
Magyarország történeti földrajza. Budapest.

FURGER, A. R. – HARMANN, F. 1983
Vor 5000 Jahren... So lebten unsere Vorfahren in der Jungsteinzeit. Haupt, Bern-Stuttgart.

FÜZES, M. 1963
A vörsi longobard temető növényleletei. *Veszprém Megyei Múzeumok Közleményei* 1, 311–340.

FÜZES, M. 1964
A jutasi és ősküi avarkori temetők növényleletei. *Veszprém Megyei Múzeumok Közleményei* 2, 437–458.

FÜZES, M. 1971
Régészeti növénytani megjegyzések Moór E.: A bor és a szőlő c. cikkéhez. *A Veszprém Megyei Múzeumok* Közleményei 10, 115–126.

FÜZES, M. 1972
Előzetes jelentés az 1967. évi pogányszentpéteri kolostor-ásatás XVI. század eleji gabonaleletéről (Präliminarer Bericht über die Getreidefunde der Klosterausgrabung in Pogányszentpéter im Jahre 1967). *A nagykanizsai Thury György Múzeum Jublieumi Évkönyve* 285–290. Nagykanizsa.

FÜZES, M. 1977
A növénytermesztés kezdete hazánkban, különös tekintettel a fenékpusztai növényleletekre (The beginnings of plant cultivation in Hungary with special regard to the plant finds from Fenékpuszta). Egyetemi doktori értekezés. Agrártudományi Egyetem Keszthely. Kézirat. (Doctoral dissertation, Keszthely. Manuscript.)

FÜZES, M. 1978
Egy római katonai expedició növényi bizonyítékai. *Élet és Tudomány* 25, 787–790.

FÜZES, M. 1987
Majackoje vára (Szovjetúnió) feltárásának archaeobotanikai tanulságai. Előadás a Zala megyei közgyűjtemények szakmai napján 1987.10.15. Kézirat.

FÜZES, M. 1990
A földmívelés kezdeti szakaszának (neolitikum és rézkor) növényleletei Magyarországon (Archaeobotanikai vázlat) [Plant remains from the early phase of plant cultivation (Neolithic and Copper Age) in Hungary. An archaeobotanical outline]. *Tapolcai Városi Múzeum Közleményei* 1, 139–238.

FÜZES, M. 1991
A Dunántúl korai növénytermesztése és növényleletei. A Starčevo kultúra és a "Tapolcai csoport" (Die frühe Pflanzenzucht und die Pflanzenfunde in Transdanubien. – Die Starčevo-Kultur und die "Tapolca-Gruppe"). *Bibliotheca Musei Tapolcensis* 1, 267–362.

FÜZES, M. – SÁGI, K. 1968
A balatoni szőlőkultúra pannon gyökerei. *Filológiai Közlemények* 14, 347–363.

FŰKÖH, L. 1990
A magyarországi holocén Mollusca-fauna fejlődéstörténete az elmúlt tízezer év során. Kandidátusi értekezés. 1990. Kézirat.

GAÁL, E. 1988
Sör az ókori Egyiptomban és Mezopotámiában (*Beer in Ancient Egypt and Mesopotamia*). Budapest.

GAÁL, L. 1978
A Magyar növénytermesztés múltja. Akadémiai Kiadó, Budapest.

GASSNER, G. 1989
Mikroskopische Untersuchung pflanzlicher Lebensmittel. Gustav Fischer Verlag, Stuttgart.

GERÓCS, F., KOVÁCS, V. – TORMA, A. 1995
További adatok Magyarország fosszilis mag- és termésleleteihez Gór kora vaskori, Dunakömlőd-Lussonium római kori és Szécsény 16–17. sz.-i települések régészeti feltárása alapján. Tudományos Kiákköri Dolgozat. Janus Pannonius Tudományegyetem Bölcsészettudományi Kar Pécs. Kézirat.

GHIMESSY, L. 1984
A tájpotenciál. Táj, víz, ember, energia. Budapest.

GLASER, L. 1937
A Kelet-Dunántúl a honfoglalás és a vezérek korában. Fejér vármegye kialakulása. Különnyomat: A magyar városok és vármegyék 22. Fejér Vármegye. Cegléd.

GOLDMANN, K. 1982
Märkischer Weizen für Byzanz – zum Reisebericht Ibrahim Ibn Jakubs von 965/66. *Ausgrabungen in Berlin* 6, 197–210.

GOLDMANN, K. 1987
Überlegungen zur Demographie der Urnenfelderkulturen. In: Czopek, S. (ed.), *Die Urnenfelderkulturen Mitteleuropas. Symposium Libnice, 21–25.10.1985. Praha*, 53–57.

GOMBOCZ, Z. 1960
Honfoglalás előtti bolgár-török jövevényszavaink. Nyelvtudományi Értekezések 24, Budapest.

GOWLETT, J. A. J. 1993
Ascent to Civilization. McGraw-Hill, New York etc.

GÖBEL, W. – KNÖRZER, K. H. 1973
Naturwissenschaftliche Untersuchungen an einer späthallstattzeitlichen Fundstelle bei Langweiler, Kr. Düren. *Bonner Jahrbuch* 173, 301–315.

GREGUSS, P. 1937
A szőregi bronzkori kaláris fonalának és a kiszombori avar sír szövetmaradványának anyaga (The materials of the Szőreg Bronze Age string and the textile remains from the Kiszombor Avar grave). *Botanikai Közlemények* 34, 60.

GREGUSS, P. 1969
Magyarországi kövült fák. [*Tertiary angiosperm woods in Hungary.*] (Translated by S. Országh). Akadémiai Kiadó, Budapest.

GRIEG, J. 1981
The investigation of a Medieval barrel latrine from Worcester. *Journal of Archaeological Science* 8, 265–282.

GRÜSS, J. 1930
Inhaltreste aus der vor- und frühgeschichtlichen Zeit. *Der Naturforscher* 9, 156.

GRÜSS, J. 1935
Über Honigreste aus prähistorischer Zeit. *Forschung und Fortschritte* 11, 260–261.

GRYNAEUS, T. 1996
Ősök, táltosok, szentek. In: Pócs, É. – Voigt, V. (eds), *Tanulmányok a honfoglaláskor és Árpád-kor folklórjából.* 121–137. MTA Néprajzi Kutatóintézet

GUNDA, B. 1961
Altertümliche Mahlsteine in den Karpaten. *Acta Ethnographica* 10, 3–27.

GUNDA, B. 1966a
Ethnographia Carpathica. Budapest.

GUNDA, B. 1966b
A sütőkövek és ősi kenyérfélék. In: Gunda, B.: *Ethnographica Carpathica.* 296–332. Budapest.

GUNDA, B. 1975
Ősi gyűjtögető tevékenység a mocsárvilágban. In: Bencsik J. (ed.), *Hansági Múzeum Évkönyve 2.* 169–178. Hajdúböszörmény.

GUNST, P. – LÖKÖS, L. 1982
A mezőgazdaság története. Mezőgazdasági Kiadó, Budapest.

GYÖRFFY, GY. 1963, 1987
Az Árpád-kori Magyarország történeti földrajza I, III. Budapest.

GYÖRFFY, GY. 1983
A magyar állam félnomád előzményei. In: Tőkei, F. (ed.), *Nomád társadalmak és államalakulatok.* 365–390. Akadémiai Kiadó, Budapest.

GYÖRFFY, GY. – ZÓLYOMI, B. 1994
A Kárpát-medence és Etelköz képe egy évezred előtt. In: Kovács, L. (ed.), *Honfoglalás és régészet.* 13–37. Budapest.

GYÖRFFY, GY. – ZÓLYOMI, B. 1996
A Kárpát-medence és Etelköz képe egy évezred előtt. *Magyar Tudomány* 103, 899–918.

GYÖRFFY, I. 1977
István király és műve. Gondolat, Budapest.

GYULAI, F. 1986
A növénytermesztés emléke Fonyód-Bélatelep Árpád-kori kori településéről (Pflanzenbaureste aus der Árpáden-Zeit im Siedlungsgebiet des Plattensees). *Veszprém Megyei Múzeumok Közleményei* 18, 135–158.

GYULAI, F. 1987
A honfoglaló magyarság gabonatermesztése I. (Grain Production by the Early Hungarians I.). *Zalai Gyűjtemény* 26, 65–74.

GYULAI, F. 1988
Obst- und Weinfunde aus der Árpádenzeit (10. Jahrhundert) in Fonyód-Bélatelep am Balaton. In: Hopf, M. (ed.), *Der prähistorische Mensch und seine Umwelt. Festschrift für Udelgard Körber-Grohne.* 395–402. Stuttgart.

GYULAI, F. 1989
A honfoglaló magyarság gabonatermesztése II. (Grain Production by the Early Hungarians II.). *Zalai Gyűjtemény* 28, 5–17.

GYULAI, F. 1993
Environment and Agriculture in Bronze Age Hungary. Archaeolingua, Series Minor 4. Budapest.

GYULAI, F. 1994
A Kárpát-medence haszonnövényei a 9–10. században. In: Györffy Gy. (ed.), *Honfoglalás és régészet.* 247–258. Balassi Kiadó, Budapest.

GYULAI, F. 1995a
The plant and food remains from the Copper Age settlement at Zalaszentbalázs-Szőlőhegyi mező. In: Szőke, B. M. (ed.), *Archeological and Settlement History in the Hahót Basin, South-West Hungary.* 145–157. Antaeus 22.

GYULAI, F. 1995b
Natural environment and climate. Plant exploitation and agriculture. In: Bartosiewicz, L. (ed.), *Animals in the Urban Landscape in the Wake of the Middle Ages. A case study from Vác, Hungary.* 3–4, 93–110. British Archaeological Reports 609, Oxford.

GYULAI, F. 1995c
The analytical study of carbonized grain remains from the lake Balaton region (Hungary). *Annali Botanica* 53, 251–260. Roma.

GYULAI, F. 1996a
Using image analysis in the evaluation of plant imprints found on sherds from the Neolithic site of Bicske. In: Makkay, J., Starnini, E. – Tulok, M. (eds): *Excavations at Bicske-Galagonyás (Par. III). The Notenkopf and Sopot-Bicske Cultural Phases.* 258–263. Svevo, Trieste.

GYULAI, F. 1996b
Előzetes jelentés Százhalombatta középső bronzkori tell 1991. évi ásatás növényleleteinek vizsgálatáról. In: Poroszlai, I. (ed.), *Ásatások Százhalombattán.* 16–24. Százhalombatta.

GYULAI, F. 1996c
Balatonmagyaród-Hídvégpuszta késő bronzkori település növényleletei és élelmiszermaradványai (Die Planzenfunde und Lebensmittelreste aus der spätbronzezeitlichen Siedlung von Balatonmagyaród-Hídvégpuszta). *Zalai Múzeumok* 6, 169–195.

GYULAI, F. 1996d
Umwelt und Pflanzenbau in Transdanubien während der Urnenfelder-, Hallstatt- und Latènekultur. In: Jerem, E. – Lippert, A. (eds), *Die Osthallstattkultur. Akten des Internationalen Symposiums, Sopron, 10–14. Mai 1994.* 127–136. Archaeolingua, Budapest.

GYULAI, F. 1996e
Előzetes jelentés Panyola-Ásottfok késő népvándorláskori telep 1996. évi ásatásából származó növényleletek viszgálatáról. In: Erdélyi, I. (ed.), *Panyola. Településtörténeti kutatási eredmények 1991–1996.* 53–57. Student, Budapest.

GYULAI, F. 1997a
Kompolt-Kistéri tanya 15. számú lelőhely növénymaradványai, mint a hazai italkészítés legkorábbi (rézkori) közvetett bizonyítékai (Proof of Copper Age drink production based on the botanical remains found at site number 15, Kompolt-Kistér farm). *Agria* 33, 59–76.

GYULAI, F. 1997b
A honfoglaló magyarság ételeinek régészeti-növénytani forrásai. "Nyereg alatt puhítjuk...? Vendéglátási és ételkészítési szokások a honfoglaló magyaroknál és a rokon kultúrájú lovas népeknél." Kereskedelmi, Vendéglátó és Idegenforgalmi Főiskola Tudományos Közlemények II. *Ómagyar kultúra* 10, 113–134.

GYULAI, F. 1997c
Szenült szemtermések analitikai vizsgálata (Analytic examination of carbonified crops.) *Zalai Múzeum* 8, 176–190.

GYULAI, F. 1998a
Archaeobotanika, élelmiszermaradványok vizsgálata (Archaeobotany and Foodremains Analysis.) In: Ilon, G. (ed.), *Régésztechnikus kézikönyve I.* Panniculus. Ser. B.3, 225–238. Szombathely.

GYULAI, F. 1998b

A Kis-Balaton térségének archaeobotanikai kutatási eredményei (Archaeobotanical Investigation in the region Kis-Balaton.) In: Pomogyi, P. – Dömötörfy, Zs. (eds), *A Kis-Balaton térségének magasabbrendű növényzetével kapcsolatos kutatási eredmények.* 17–24. Magyar Hidrológia Társaság és a NYUVIZIG konferenciája, Keszthely 1998. március 24.

GYULAI, F. 1998c

The study of organic remains from the Celtic period site of Keszthely-Fenékpuszta. In: Anreiter, P., Bartosiewicz, L., Jerem, E. – Meid, W. (eds), *Man and the Animal World. Studies in Archaeozoology, Archaeology, Anthropology and Palaeolinguistics in memoriam Sándor Bökönyi.* 275–283. Budapest.

GYULAI, F. 1998d

Budapest, Bécsi út 38–42., a Badeni kultúra lelőhelyének növénytani anyaga. In: Endrődi A. (ed.), Késő rézkori leletek Óbudáról (Late Copper Age Finds from Óbuda). *Budapest Régiségei* 32, 105–123.

GYULAI, F. 1998e

Archaeobotanical Sources in Investigating the Diet of Conquering Hungarians. In: Laszlovszky J. (ed.), *Tender Meat under the Saddle. Costoms of Eating, Drinking and Hospitality among Conquering Hungarians and Nomadic Peoples.* 120–156. Krems.

GYULAI, F. 1999a

A Rákospalota-Újmajor 1. lelőhelyről származó növényleletek archaeobotanikai feldolgozása (Analysis of the Botanical Remains Collected from the 1996 Excavations at the Motorway M0 site No. 1 Rákospalota-Újmajor.) In: Bencze, Z., Gyulai, F., Sabján, T. – Takács, M. (eds), *Egy Árpád-kori veremház feltárása és rekonstrukciója.* (*Ausgrabung und Rekonstruktion eines Grubenhauses aus der Árpádenzeit.*) 73–91. Monumenta Historica Budapestinensia 10.

GYULAI, F. 1999b

Az agrobiodiverzitás változása a Kárpát-medencében (*The changing of the Agrobiodiversity in the Carpathian Basin*). Fenntartható Fejlődés Bizottsága, Budapest.

GYULAI, F. 1999c

The einkorn as living part of an ancient landscape. In: Jerem, E. – Poroszlai, I. (eds), *Archaeology of the Bronze and Iron Age. Environmental Archaeology, Experimental Archaeology, Archaeological Parks. Proceedings of the International Archaeological Conference, Százhalombatta, 3–7 October, 1996.* 297–302. Archaeolingua, Budapest.

GYULAI, F. 2000a

A gabonafélék diverzitásának változása a Kárpát-medencében. Archaeobotanikai áttekintés (The Changing of the Cereals-Diversity in the Carpathian Basin. Archaeobotanical Review.) In: Gyulai F. (ed.), *Az agrobiodiverzitás megőrzése és hasznosítása. Szimpózium Jánossy Andor emlékére. Budapest, május 4–6.* 278–287.

GYULAI, F. 2000b

Klíma, táj, kultúrák: összefüggések és különbözőségek a Kárpát-medencében. In: Füleky Gy. (ed.), *A táj változásai a Kárpát-medencében a történelmi események hatására.* 9–13. Budapest–Gödöllő.

GYULAI, F. 2001a

Archaeobotanika. Jószöveg Kézikönyvek. Budapest.

GYULAI, F. 2001b

Avar kori szerves eredetű edény anyagának vizsgálata. In: Kiss, M. – Lengvári, I. (eds), *„Együtt a Kárpát-medencében". A népvándorláskor fiatal kutatóinak VII. összejövetele. Pécs 1996. szeptember 27–28.* 69–72.

GYULAI, F. 2002a

Archaeobotanical remains and environment of Bell Beaker Csepel-Group. In: Czebreszuk, J. – Szmyt, M. (eds), *The Northeast Frontier of Bell Beakers. Symposium, 26–29 May 2002, Poznań.* Abstracts.

GYULAI, F. 2002b

Die Pflanzenreste des in Dunakeszi–Székesdűlő freigelegten spätbronzezeitlichen Röhrenbrunnens. *Budapest Régiségei* 36, 301–307.

GYULAI, F. 2002c

A Kárpát-medencei szőlő- és borkultúra régészeti növénytani emlékei. In: Benyák, Z. – Benyák, F. (eds), *Borok és korok.* 101–113. Hermész kör, Budapest.

GYULAI, F. 2003

Die Pflanzenreste eines spätbronzezeitlichen Röhrenbrunnens. In: Endrődi, A. – Kalicz, N. (eds), In memoriam Rózsa Kalicz-Schreiber (1929-2001). *Budapest Régiségei* 36, 305–312.

GYULAI, F. 2004

A fajtahasználat legkorábbi bizonyítékai a Kárpát-medencében? *Növénytermelés (Crop Production)* 53/4, 305–401.

GYULAI, F. (in print)

Túrkeve-Terehalom középső bronzkori tell település leégett konyhájának magleletei és ételmaradványai. In: Bóta, L. – Haász, A. (eds): *Kovász. Tanulmánykötet Dr. Csizmadia László 70. születésnapjára.* Budapesti Gazdasági Főiskola, Budapest.

GYULAI, F., HERTELENDI, E. – SZABÓ, I. 1992

Plant remains from the early medieval lakeshore settlement Fonyód-Bélatelep (Lake Balaton, Hungary) with especial emphasis on the history of fruit cultivation in Pannonia. *Vegetation History and Archaeobotany* 1, 177–184.

GYULAI, F. – KÁLLAY, M. 1998–99

A drink-remain from Seeraj (Sindh, Pakistan). In: Shaikh, N. – Biagi, P. (eds), *Ancient Sindh Annual Journal of Research* 5. 29–34. Shah Abdul Latif University Khairpur (Sindh).

GYULAI, F. – KÁLLAY, M. 2000

The remains of Hallstatt Period Beverages from Western Hungary. Analyzing the earliest remains of wine from Central Europe. In: *32nd International Symposium Archaeometry.* May 15–19, 2000. Mexico City. 44–45. British Archaeological Reports.

GYULAI, F. – KÁLLAY, M. 2002

Italmaradványok a Kárpát-medencében. In: Benyák, Z. – Benyák, F. (eds), *Borok és korok.* 113–114. Hermész kör, Budapest.

GYULAI, F – KÁLLAY, M. (in print)

Közép-Európa legkorábbi bormaradványának archaeometriai vizsgálata. Hallstatt-kori italmaradvány Nyugat-Magyarországról. In: Bóta, L. – Haász, A. (eds): *Kovász. Tanulmánykötet Dr. Csizmadia László 70. születésnapjára.* Budapesti Gazdasági Főiskola, Budapest.

GYULAI, F. – TORMA, A. 1993
Az Urnasíros kultúra góri településének növényleletei. Nyugat-Dunántúl bronzkora. *Pápai Múzeumi Értesítő (Acta Musei Papensis)* 6, 277–286.

GYULAI, F. – TORMA, A. 1999
Die Pflanzenfunde einer Siedlung der Urnenfelderkultur in Gór. *Savaria* 24, 359–367.

GYULAI, G., GÉMESNÉ, J. A, SÁGI, ZS., HESZKY, L., VENCZEL, G., PINTÉR, P., KRISTÓF, Z., TÖRJEK, O., HESZKY, L., BOTTKA, S., KISS, J. – ZATYKÓ, L. 2000
Doubled haploid development and PCR-analysis of F1 hybrid derived DH-R_2 paprika (*Capsicum annuum L.*) Lines. *Journal of Plant Physiology* 156, 168–174.

GYULAI, G., MAGDA, A., KISS, J., GYULAI, F., HOLLY, L. – HESZKY, L. 2001
DNS izolálás és PCR-amplifikáció 700 éves növénymagvakból. *7. Növénynemesítési Tudományos Napok, Budapest.* Összefoglalók, 89.

GYULAI, G., TÓTH, Z., SZABÓ, Z., GYULAI, F., LÁGLER, R., KOCSIS, L., – HESZKY, L. 2009
Domestication Events of Grape (Vitis vinifera) from Antiquity and Middle Ages in Hungary from Growers' Viewpoint. *Hungarian Agricultural Research* 2009/3–4, 8–12.

HAJDÚ, P., KRISTÓ, GY. – RÓNA TAS, A. 1976
Bevezetés a magyar őstörténet kutatásainak forrásaiba I. Tankönyvkiadó, Budapest.

HAJNALOVÁ, E. 1979a
Paleobotanické doklady pol'nohospod rstva a pouzívania drevín v oblasti Liptova v dobe halstatskej, laténskej a rímskej. *Slovenska Archeologia* 27, 437–474.

HAJNALOVÁ, E. 1979b
Einige Erkenntnisse über die Züchtung und Ausnützung von Pflanzen in der Vergangenheit auf dem Gebiet Bulgariens und Rumäniens aufgrund archäobotanischer Funde In: *Rapports, Co-Rapports, Communications Tchécoslovaques pour le IVe Congres de l'Association Internationale d'Étude du Sud-Est Européen.* 61–70. Académie Tchécoslovaque des Sciences. Prague.

HAJNALOVÁ, E. 1989a
Katalóg zvyŝkov semien a plodov v archeologických náleyoch na Slovensku. *Acta Interdisciplinaria Archaeologica VI. Nitra,* 3–12.

HAJNALOVÁ, E. 1989b
Evidence of a carbonized loaf of bread and cereals from Bratislava-Devin. *Slovenska Archeologia* 37, 89–104.

HAJNALOVÁ, E. 2001
Ovocie a ovocinárstvo v archeobotanických nálezoch na Slovensku. *Acta Interdisciplinaria Archaeologica.* 68, 46–47.

HAJNALOVÁ, E., FURMÁNEK, V. – MARKOVÁ, K. 1999
Eicheln aus dem Objekt 21/86 aus der Siedlung in Včelince. In: Bátora, J. – Peška, J. (eds), *Aktuelle Probleme de Erforschung der Frühbronzezeit in Böhmen und Mähren und in der Slowakei.* 231–239. Nitra.

HAMPEL, J. 1897
A régibb középkor (IV–X. század) emlékei Magyarhonban 2. Budapest.

HANSEN, J. M. 1978
The earliest seed remains from Greece: Palaeolithic through Neolithic at Franchthi Cave. *Berichte der Deutschen Botanischen Gesellschaft* 91, 39–46.

HARASZTI, S. – PETHŐ, T. 1963
Útikalandok a régi Magyarországon. Budapest.

HARDING, A. F. 1987
Social and Economic Factors in the Origin and Development of the Urnfield Cultures. *Symposium Liblice 21.–25.10.1985. Die Urnenfelderkulturen Mittel-Europas.* 37–41. Praha

HARLAN, J. R. – ZOHARY, D. 1966
Distribution of wild wheats and barley. *Science* 153, 1074–1080.

HARMATTA, J. 1952
A hun birodalom felbomlása. *A Magyar Tudományos Akadémia II. Osztály Közleménye* 2, 147–192.

P. HARTYÁNYI, B. 1976
Későrómai kori növényi maradványok a II-III. sz.-i aquincumi legiostábor retenturájában. *Budapest Régiségei* 24/1, 125–135.

P. HARTYÁNYI, B. 1975–77
Középkori budai lakóház mellékgödrében talált növényi maradványok (In der Nebengrube eines mittelalterlichen Wohnhauses in Buda vorgefundene Pflanzenreste). *Magyar Mezőgazdasági Múzeum Közleményei,* 15–51.

P. HARTYÁNYI, B. 1982
Tiszaalpár-Várdomb bronzkori lakótelepéről származó mag- és termésleletek (Die von der bronzezeitlichen Wohnsiedlung Tiszaalpár-Várdomb stammenden Korn- und Fruchtfunde). *Cumania* 7, 133–286.

P. HARTYÁNYI, B. 1981–83
Kora Árpád-korból származó búza a Hont-i ispánsági várból (Aus der frühen Árpádenzeit stammender Weizen der Honter Gespannschaftsburg). *Magyar Mezőgazdasági Múzeum Közleményei,* 95–113.

P. HARTYÁNYI, B. 1988–89
Növényleletek a battonya-parázstanyai neolitikus lakótelepen (Pflanzenfunde der neolithischen Wohnsiedlung bei Battonya – Parázs-Gehöft). *Magyar Mezőgazdasági Múzeum Közleményei,* 39–67.

P. HARTYÁNYI, B. – SZ. MÁTHÉ, M. 1980
A Berettyóújfalu-Szilhalom-i későneolitikus lakótelep 1976-ban feltárt növényleletei (An die Oberfläche 1976 gebrachte pflanzliche Überreste einer Wohnsiedlung in Szilhalom bei Berettyóújfalu aus der späteren Zeit des Neolithikums). *Magyar Mezőgazdasági Múzeum Közleményei,* 125–145.

P. HARTYÁNYI, B. – NOVÁKI, GY. 1971
Gabonalenyomatok a Körös-csoport edényein (Getreideabdrücke auf den Gefässen der Körös-Gruppe). *Móra Ferenc Múzeum Évkönyve* 2, 5–8.

P. HARTYÁNYI, B. – NOVÁKI, GY. 1973–74
Növényi mag- és termésleletek Magyarországon az újkőkortól a XVIII. sz.-ig II. (Samen- und Fruchtfunde in Ungarn von der Jungsteinzeit bis zum XVIII. Jahrhundert II.). *Magyar Mezőgazdasági Múzeum Közleményei,* 23–73.

P. HARTYÁNYI, B. – NOVÁKI, GY. 1975
Samen- und Fruchtfunde in Ungarn von der Jungsteinzeit bis zum 18. Jahrhundert. *Agrártörténeti Szemle* 17, 1–22. Supplementum.

P. HARTYÁNYI, B., NOVÁKI, GY. – PATAY, Á. 1967–68
Növényi mag- és terméleletek Magyarországon az újkőkortól a XVIII. sz.-ig I. (Samen- und Fruchtfunde in Ungarn von der Jungsteinzeit bis zum XVIII. Jahrhundert I.). *Magyar Mezőgazdasági Múzeum Közleményei*, 5–85.

P. HARTYÁNYI, B. – PATAY, Á. 1970
A dunaföldvári öregtoronynál előkerült régészeti növények vizsgálata. *Szekszárdi Balogh Ádám Múzeum Évkönyve*, 209–222.

HEER, O. 1865
Die Pflanzen der Pfahlbauten. Neujahrsblatt der Naturforsch. *Gesellschaft Zürich (88) auf das Jahr* 1866.

HEGYI, K. 1976
Egy világbirodalom végvidékén. Budapest.

HEHN, V. 1877
Kulturpflanzen und Haustiere in ihrem Übergang aus Asien nach Griechenland und Italien sowie das übrige Europa. Berlin.

HELBAEK, H. 1950
Tollund Mandens sidste Maaltid. Et botanisk bidrag til belysning af oldtidens kost. Aarbøger for nordisk Oldkynd. *Historie* 1950, 311–341.

HELBAEK, H. 1951
Urkudtsfro som Naeringsmiddel i forromersk Jernalder. (Seeds of weeds as food in the fore-Roman Iron Age.) *Kuml* 1951, 65–74.

HELBAEK, H. 1954
Prehistoric food plants and weeds in Denmark. A survey of archeological research 1923–1954. *Danmarks Geologiske Undersogelse, ser* IV, 3(6), 250–261.

HELBAEK, H. 1959a
Domestication of food plants in the Old World. *Science* 130, 365–372.

HELBAEK, H. 1959b
Grauballemandens sidste Måltid. (The last meal of Grauballe man.) *Kuml* 1959, 83–116.

HELBAEK, H. 1959c
The last meal of Grauballe man: an anylysis of food remains in the stomach. *Kuml* 1959, 111–116.

HELBAEK, H. 1963
Palaeoethnobotany. In: Brothwell, D. – Higgs, E. (eds), *Science in Archaeology*. 177–185. London.

HELBAEK, H. 1964a
Early Hassunan vegetable at Es-Sawwan near Samarra. *Sumer* 20, 45–48.

HELBAEK, H. 1964b
First impressions of the Çatal Hüyük plant husbandry. *Anatolian Studies* 14, 121–123.

HELBAEK, H. 1966
Pre-pottery Neolithic farming at Beidha. Palestine Exploration. *Quarterly* 98, 61–66.

HELBAEK, H. 1969
Plant collecting, dry-farming and irrigation agriculture in prehistoric Deh Luran. In: Hole, F., Flannery, K. V. – Neely, J. A. (eds), *Prehistory and human ecology of the Deh Luran Plain.* 383–426. Memoirs of the Museum of Anthropology 1, University of Michigen, Ann Arbor.

HELBAEK, H. 1970
The plant husbandry of Haçilar. In: Mellaart, J. (ed.), *Excavations at Haçilar.* 189–244. Edinburgh University Press. Edinburgh.

HELCK, W. 1971
Das Bier im Alten Ägypten. Berlin.

HELM, M. H. 1982
Feld-, Wald- und Wiesenkochbuch. Zürich.

HENNIG, E. 1966
Beobachtungen zum Mahlvorgang an ur- und frühgeschichtlichen Getreidemühlen. *Ethnographisch-Archäologische Zeitschrift* 7, 71–87.

HERTELENDI, E., SÜMEGI, P. – SZÖÖR, GY. 1991
Geochronologic and Paleoclimatic Characterzation of Quaternary Sediments in the Great Hungarian Plain. Budapest.

HERTELENDI, E., SVINGOR, É., RACZKY, P., HORVÁTH, F., FUTÓ, I., BARTOSIEWICZ, L. – MOLNÁR, M. 1997
Radiocarbon chronology of the Neolithic and time span of tell settlements in Eastern Hungary based on calibrated radiocarbon dates. In: Költő, L. – Bartosiewicz, L. (eds), *Archaeometrical Research in Hungary II.* 61–70. Hungarian National Museum, Budapest.

HESZKY, L., BÓDIS, L. – KISS, E. 1999
A kulturflóra biodiverzitása Magyarországon. (Plant biodiversity of agroecosystems in Hungary.). *Növénytermelés* 48, 435–443.

HILLMAN, G. C. 1975
The plant remains from Tell Abu Hureyra. A preliminary report. *Proceedings of the Prehistoric Society* 41, 70–73.

HILLMAN, G. C. 1978
On the origin of domestic rye – Secale cereale: the finds from aceramic Can Hasan III inTurkey. *Anatolian Studies* 28, 157–174.

HILLMAN, G. C. – DAVIES, M. S. 1990
Measured Domestication Rates in Wild Wheats and Barley Under Primitive Cultivation, and Their Archaeological Implications. *World Prehistory* 4/2, 157–222.

HJELMQVIST, H. 1955
Die Älteste Geschichte der Kulturpflanzen in Schweden. *Opera Botanica* 1/3, 1–186.

HJELMQVIST, H. 1984
Botanische Analyse einiger Brote. In: Arwidsson, G. (ed.), *Birka II:I. Systematische Analyse der Gräberfunde.* 263–288. Almqvist and Wiksell International, Stockholm.

HJELMQVIST, H. 1990
Über die Zusammensetzung einiger prähistorischer Brote. Fornvännen. *Tidskrift för Svensk Antikvarist Forskning* 85, 9–21.

HO, PING-TI 1977
The indigenous origins of Chinese agriculture. In: Reed, C. A. (ed.), *Origens of agriculture.* 413–483. Mouton.

HOLL, I. 1966
Mittelalterliche Funde aus einem Brunnen von Buda. Studia Archaeologica 4. Budapest.

HOLLY, L. 2000
A kultúrnövény diverzitás megőrzésének lehetőségei. In: Gyulai, F. (ed.), *Az agrobiodiverzitás megőrzése és hasznosítása.* 39–43. Tápiószele.

HOPF, M. 1978
Frühe Kulturpflanzen in Südosteuropa. *Berichte der Deutschen Botanischen Gesellschaft* 91/1, 31–38.

HOPF, M. 1983
Jericho plant remains. In: Kenyon, K. M. – Holland, T. A. (eds), *Excavations at Jericho 5.* 576–621. British School of Archaeology in Jerusalem, London.

HOPF, M. 1985
Bronzezeitliche Sämereien aus Ouroux-Marnay, Dép. Saône-et-Loire. *Jahrbuch des Römisch-Germanischen Zentralmuseums* 32, 255–264.

HOPF, M. 1989
Pflanzliche Reste aus Smolenice Molmír, Kr. Trnava. *Acta Interdisciplinaria Archaeologica* 6, 193–216.

HOPF, M. 1991
South and Southwest Europe. In: Zeist, W. van, Wasylikowa, K. – Behre, K. E. (eds), *Progress in Old World Palaeoethnobotany.* 241–279. Rotterdam.

HOPF, M. – BLANKENHORN, B. 1984
Vor- und frühgeschichtliche Kulturpflanzen in Süddeutschland. Jahrbuch der Bayerischen Bodendenkmalpflege 22.

HORTOBÁGYI, T. – SIMON, T. 1981 (ed.)
Növényföldrajz, társulástan és ökológia. Budapest.

HORVÁTH, E. 1975
Későglaciális és óholocén növénymaradványok Szombathely környékéről. Előadás a Botanikai Szakosztály 1014. Szakülésén 1975. 05. 06.

HORVÁTH, L., HOLLY, L. – SIMON, A. 2001
Triticum ssp. génbanki populációk összehasonlító vizsgálati eredményei 1999–2000-ben. *7. Növénynemesítési Tudományos Napok, Budapest.* Összefoglalók, 93.

HÖNEISEN, M. 1990
Die Ausbreitung frühester bäuerlicher Kultur in Europa. In: Schweizerisches Landesmuseum (ed.), *Die ersten Bauern. Pfahlbaufunde Europas 2.* 14–26. Zürich.

HUNYADI, K. 1988
Szántóföldi gyomnövények és biológiájuk. Budapest.

ILON, G. – SABJÁN, T. 1989
XV. sz.-i cserépkályhák Külsővatról (Kachelöfen aus dem XV. Jahrhundert aus Külsővat). *Pápai Múzeumi Értesítő 2. Acta Musei Papensis.* 77–140.

INTERNATIONAL CODE OF BOTANICAL NOMENCLATURE (Tokyo Code 1993).
Regnum Vegetabile 131, Koeltz Scientific Books, Königstein.

JACOMET, S. 1983
Agrargeschichte im Überblick: von den Anfängen bis zur Dreifelderwirtschaft. *Technik und Wirtschaft in ur- und frühgeschichtlicher Zeit* 22/23, 27–33.

JACOMET, S., BROMBACHER, CH. – DICK, M. 1989
Archäobotanik am Zürichsee. Ackerbau, Sammelwirtschaft und Umwelt von neolitischen und bronzezeitlichen Seeufersiedlungen im Raum Zürich. Berichte der Zürcher Denkmalpflege 7.

JACOMET, S. – KREUZ, A. 1999
Archäobotanik. Ulmer, Stuttgart.

JACQUAT, C., PAWLIK, B. – SCHOCH, W. 1982
Die mittelalterlichen Pflanzenfunde. Der Münsterhof in Zürich. *Schweizer Beiträge zur Kulturgeschichte und Archäologie des Mittelalters* 10, 267–278.

JANUSHEVICH, Z. V. 1975
Fossil Remains of Cultivated Plants in the South West Soviet Union. *Folia Quaternaria* 46, 23–30.

JANUSHEVICH, Z. V. 1976
Kulturnije rastenija jugu-sapada SSSR po paleobotanicheskim issledovanijam (Cultivated plants in south western USSR according to paleobotanical investigations). Kishinev, USSR.

JANUSHEVICH, Z. V. 1978
Prehistoric food plants in the south-west of the Soviet Union. *Berichte der Deutschen Botanischen Gesellschaft* 91, 59–66.

JANUSHEVICH, Z. V. 1984
The specific composition of wheat finds from ancient agricultural centres in the USSR. In: Zeist, W. van – Casparie, W. A. (eds), *Plants and ancient man.* 267–276. Rotterdam–Boston.

JARRIGE, J.-F. – MEADOW, R. H. 1980
The antecedents of civilization in the Indus Valley. *Scientific American* 243/2, 102–110.

JÄGER, K.-D. 1965
Verkohlte Samen aus einem bronzezeitlichen Grabgefäss von Tornow, Kr. Calau. Ein Beitrag zur Anbaugeschichte der Ackerbohne (Vicia faba L.) in Mitteleuropa. *Ausgrabungen und Funde* 10, 131–138.

JEREM, E., FACSAR, G., KORDOS, L., KROLOPP, E, – VÖRÖS, I. 1984
A Sopron-Krautackeren feltárt vaskori telep régészeti és környezetrekonstrukciós vizsgálata I. *Archaeologiai Értesítő* 111, 141–170.

JEREM, E., FACSAR, G., KORDOS, L., KROLOPP, E, – VÖRÖS, I. 1985
A Sopron-Krautackeren feltárt vaskori telep régészeti és környezetrekonstrukciós vizsgálata II. *Archaeologiai Értesítő* 112, 3–24.

JEREM, E. – POROSZLAI, I. 1999 (eds)
Archaeology of the Bronze and Iron Age – Environmental Archaeology, Experimental Archaeology, Archaeological Parks. Proceedings of the International Archaeological Conference, Százhalombatta, 3–7 October, 1996. Archaeolingua, Budapest.

JESZENSZKY, Á. – KÁRPÁTI, I. 1963
A füge. Magyarország kultúrflórája 18.

KALICZ, N. 1968
Die Frühbronzezeit in Nordost-Ungarn. Archaeologia Hungarica N.S. 45.

KALICZ, N. 1991
Beiträge zur Kenntnis der Kupferzeit im ungarischen Transdanubien. In: Lichardus, J. (ed.), *Die Kupferzeit als historische Epoche.* 347–387. Bonn.

KALICZ, N. 1993
The Early Phases of the Neolithic in Western Hungary (Transdanubja). *Poročilo oraziskovanju paleolita, neolita i euneolita v Sloveniji* XXI, 85–135. Ljubljana,

KALICZ-SCHREIBER, R. 1974
Die Probleme der Glockenbecherkultur in Ungarn. In: Lanting, J. N. – Waas, J. D. van der (eds), *Glockenbecher Symposion. Oberried 1974.* 183–216. Fibula-Van Dishoeck, Bussum–Haarlen.

KALICZ-SCHREIBER, R. – KALICZ, N. 1998
Die Somogyvár-Vinkovci-Kultur und die Glockenbecher in Ungarn. In: Dobiat, D. – Leidorf, K. (eds), *Tradition und Innovation. Festschrift für Christian Strahms.* 325–347. Studia Honoraria 3. Marie Leidorf, Rahden/Wesf.

KAPÁS, S. 1997
Növényfajták és növénynemesítők. Budapest.

KARG, S. 1989
Verkohlte Pflanzenreste aus der urnenfelderzeitliche Siedlung. *Fundberichte aus Baden-Württemberg* 14, 247–249.

KARG, S. – JACOMET, S. 1991
Pflanzliche Makroreste als Informationsquelle zur Ernährungsgeschichten des Mittelalters in der Schweiz und Süddeutschland. In: Tauber, J. (ed.), *Methoden und Perspektiven der Archäologie des Mittelalters.* 121–143. Archäologie und Museum 20, Liestal.

KASSAI, K. 1991
Recent palaeoethnobotanical research in Hungary. In: Renfrew, J. (ed.), *New light on early farming. Recent Developments in Palaeoethnobotany.* 277–279. Edinburgh.

KATONA, J. – DÖMÖTÖR, J. 1963
Magyar borok - borvidékek. Budapest.

KAUS, K. 1987
Weinbau im Burgenland vor 2700 Jahren! Pannonische Weinblätter. Post der Burgenländisch-Pannonischen Weinritterschaft 7.

KÁLDY-NAGY, GY. 1970
Magyarországi török adóösszeírások. Budapest.

KÁLDY-NAGY, GY. 1985
A budai szandzsák 1546–1590. évi összeírásai. Demográfiai és gazdaságtörténeti adatok Pest megye múltjából 6. Budapest.

KÁRPÁTI, Z. – BÁNYAI L. 1980
A pohánka és a tatárka. Magyarország kultúrflórája VII/10.

KÁRPÁTI, Z., GÖRGÉNYI, L.-NÉ – TERPÓ, A. 1968
Kertészeti növénytan I–II. Budapest.

KEIMER, L. 1924
Die Gartenpflanzen im alten Ägypten. (1967 reprint). Vol. 1, Hamburg/Berlin

KEREKES, L., HORN, J., KOCHNÉ KLOPPAR, E. – RAPAICS, R. 1937
Pomológia. Őszibarack, alma. Budapest.

KERTÉSZ, R. – SÜMEGI, P. 2001
Theories, critiques and a model: Why did the expansion of the Körös-Starčevo culture stop in the centre of the Carpathian Basin? In: Kertész, R. – Makkay, J. (eds), *From the Mesolithic to the Neolithic. Proceedings of the International Archaeological Conference held in the Damjanich Museum of Szolnok, September 22–27, 1996.* 225–246. Budapest, Archaeolingua.

KIMBER, G. – FELDMAN, M. 1987
Wild Wheat. Special Report 353. College of Agriculture, University of Missouri-Columbia.

KIRILLY, ZS.-NÉ 1968
A jobbágyság gabonatermelésének vizsgálata a maghozam szempontjából. *Agrártörténeti Szemle* 10, 70–78.

KISLEV, M. E. 1979–80
Triticum parvicoccum sp. oov., the Oldest Naked Wheat. *Israel Journal of Botany* 28, 95–107.

KISLEV, M. E. 1989
Pre-Domesticated Cereals in the Pre-Pottery Neolithic A Period. In: Hershkovitz (ed.), *People and Culture in Change.* 147–151. British Archaeological Reports 508.

KISS, A. 1964
A Kárpát-medence kora középkori szőlőművelésének kérdéséhez. *Agrártörténeti Szemle* 6, 143–148.

KISS, A. S., BELEA, A., VÖRÖSVÁRY, G. – STEFANOVITS-BÁNYAI, É. 2001
Eltérő ploiditású és földrajzi tájakról származó búzafajok ásványielem vizsgálata. 7. *Növénynemesítési Tudományos Napok, Budapest.* Összefoglalók, 101.

KISS, T. – BÓDIS, K. 2000
Hosszabb ideig antropogén hatás alatt álló félig kötött homokformák kiválasztása morfometriai mérések alapján. In: Füleky, Gy. (ed.), *A táj változásai a Kárpát-medencében a történelmi események hatására.* 205–210. Budapest–Gödöllő.

KLICHOWSKA, M. 1964
Jakie owoce i nasiona roslin znaleziono w Krakówie na Wawelu w czasie prac archeologicznych w latach 1950–1954. *Sprawozdania Archeologiczne* 16, 431–432.

KLICHOWSKA, M. 1977
Rostliny uprawne i dziko rosnace z grodzinska ludnosci kultury luzyckiej w Smuszewie, woj. pilskie. *Fontes Archaeologici Posnanienses* 28, 36–44.

KLICHOWSKA, M. 1984
Struktury uprawne w epoce brazu i we wczesnej epoce zelaza na ziemiach polskich w swietle badan archeobotanicznych. *Archeologia Polski* 29, 69–108.

KNIEZSA, I. 1938
Magyarország népei a XI. században. In: Serédy, J. (ed.), *Szent István emlékkönyv II.* 365–472. Budapest.

KNIEZSA, I. 1955
A magyar nyelv szláv jövevényszavai. Franklin Társulat. Budapest.

KNÖRZER, K. H. 1971
Urgeschichtliche Unkräuter im Rheinland. Ein Beitrag zur Einstehungsgeschichte der Segetalgesellschaften. Vegetatio 23.

KNÖRZER, K. H. 1972
Subfossile Pflanzenreste aus der bandkeramischen Siedlung Langweiler 3 und 6, Kreis Jülich, und ein urnenfelderzeitlicher Getreidefunde innerhalb dieser Siedlung. *Bonner Jahrbuch* 172, 395–403.

KNÖRZER, K. H. 1983
Mittelalterliche Pflanzenfunde unter dem Alten Markt in Duisburg. In: Milz, J. – Krause, G. (eds), *Duisburg im Mittelalter. Begleitschrift zur Austellung, 4. Sept.–27. Nov. 1983*, 78–87.

KNÖRZER, K. H. 1984
Veränderungen der Unkrautvegetation auf rheinischen Bauernhöfen seit der Römerzeit. *Bonner Jahrbücher* 184, 479–503.

KNÖRZER, K. H. 1985
Zur Auswertung von Wildpflanzenfunden aus dem Neolithikum des Rheinlandes. *Archäologische Informationen* 8, 8–17.

KNÖRZER, K. H. 1987
Geschichte der synanthropen Vegetation von Köln. *Kölner Jahrbuch für Vor- und Frühgeschichte* 20, 271–388.

KNÖRZER, K. H. 1989
Das Auftreten thermophiler Pflanzen im Niederrheingebiet während des Postglazials. *Archäobotanik. Dissertationes Botanicae* 133, 1–10.

KNÖRZER, K. H. 1991
Deutschland nördlich der Donau. In: Zeist, W. van, Wasylikowa, K. – Behre, K.-E. (eds), *Progress in Old World Palaeoethnobotany.* 189–206. Rotterdam.

KOLLAUTZ, A. – MIYAKAWA, H. 1970
Geschichte und Kultur eines völkerwanderungszeitlichen Nomadenvolkes. Die Jou-Jan der Mongolei und die Awaren in Mitteleuropa. I. Teil. Geschichte. II. Teil. Die Kultur. Aus Forschung und Kunst. Hrsg. v. Geschichtsverein für Kärnten, 10. Klagenfurt.

KOMÁROMI, J. 1955
A miskolci Sötétkapu melletti ásatás jobbágyházai. *Miskolci Múzeum Közleményei* 1, 18–21.

KOMLÓDI, M. = JÁRAINÉ KOMLÓDI, M.

JÁRAINÉ KOMLÓDI, M. 1966
Adatok az Alföld negyedkori klíma- és vegetációtörténetéhez I. (Quaternary climatic changes and vegetational history of the Great Hungarian Plain I.). *Botanikai Közlemények* 53, 191–201.

KONAREV, V. G., GAVRILJUK, I. P. – GUBAREVA, N. K. 1987
Storage proteins in the identification of species, cultivars and lines. *Seed Science and Technology* 15, 675–678.

KOPPÁNY, T. – SÁGI, K. 1967
A kereki Fehérkő vár története. Somogyi Múzeum Füzetei 9.

KORDOS, L. 1977
Holocén klímaváltozások kimutatása Magyarországon a "pocok hőmérő" segítségével. *Földrajzi Közlemények* 25, 222–229.

KORDOS, L. 1982
Történeti éghajlattan. In: Hajdú, P., Kristó, Gy. – Róna-Tas, A. (eds), *Bevezetés a magyar őstörténet kutatásainak forrásaiba 4.* 98–129. Budapest.

KOREK, J. 1943
A Szentes-kajáni avar temető. *Dolgozatok* 19, 1–88.

KORNAS, J. 1972
Rozmieszczenie i ekologia rozsiewania sie chwastow w zespolach polnych w Gorcach. (Distribution and dispersal ecology of weeds in segetal plant communities in the Gorce Mts. - Polish Western Carpathians.) *Acta Agrobotanica* 25, 1.

KORNAS, J. 1968
A geographical-historical classification of synanthropic plants. Materialy Zakladu Fitosochologie Stosowanej, UW, Warszawa-Bialowieza 25, 33–41.

KOVALOVSZKI, J. 1957
Régészeti adatok Szentes környékének településtörténetéhez. *Régészeti Füzetek* 5, 40.

KOVÁCS, G. – BÁLINT, F. A. 2002
Őskori gabonakísérletek a százhalombattai régészeti parkban. Előadás a IV. Tud. Tájtörténeti Konferencián, Gödöllő 2002.07.03.

KOVÁCS. G., BÁLINT. F. A. – SUTKA. J. 2001
Különböző alakor (Triticum monococcum L.) génbanki tételek genetikai variabilitásának jellemzése és agronómiai tulajdonságaik vizsgálata többváltozós statisztikai módszerekkel. *7. Növénynemesítési Tudományos Napok, Budapest.* Összefoglalók, 49.

KOVÁCS, L. – VESZPRÉMY, L. 1995 (eds)
A honfoglaláskor írott forrásai. Budapest.

KOVÁCS, V. 1994
Pannonia mezőgazdasága az archaeobotanikai eredmények tükrében. Szakdolgozat. Janus Pannonius Tudományegyetem Természettudományi Kar, Pécs. Kézirat.

KÖLTŐ, L., MÜLLER, R., SZENTPÉTERI, J. – SZŐKE, M. B. 1996
Kelet és Nyugat határán. In: Költő, L. – Vándor, L. (eds), *Évezredek üzenete a láp világából. Régészeti kutatások a Kis Balaton területén (1979–1992).* 104–143. A Somogy Megyei Múzeumok Igazgatósága – Rippl-Rónai Múzeum kiadványa, Kaposvár–Zalaegerszeg.

KÖRBER-GROHNE, U. 1983
Über die Notwendigkeit einer Registrierung und Dokumentation wilder und primitive Fruchtbäume, zu deren Erhaltung und zur Gewinnung von Vergleichsmaterial für paläo-ethnobotanische Funde. 6th Symposium Palaeo-ethnobotany (Groningen) 1983. *Plants and Ancient Man. Studies in Palaeoethnobotany* 18, 237–241.

KÖRBER-GROHNE, U. 1987
Nutzpflanzen in Deutschland. Kultugeschichte und Biologie. Stuttgart.

KÖRBER-GROHNE, U. 1989
Nährstoffinhalte und andere Stoffe in Körnern von Emmer, Einkorn und weiteren Getreidearten, ermittelt in kontrollierten Feldversuchen. *Archäobotanik. Dissertationes Botanicae* 133, 41–50.

KRASILOV, V. A. 1975
Paleoecology of Terrestrial Plants. Basic Principles and Techniques. New York, Toronto, Jerusalem.

KRAUSS, F. 1943
Nösnerländische Pflanzennamen. Ein Beitrag zum Wortschatz der Siebenbürger Sachsen. Beszterce-Bistritz.

KREUZ, A. 1992–93
Frühlatènezeitliche Pflanzenfunde aus Hessen als Spiegel landwirtschaftlicher Gegebenheiten des 5.–4. Jh. v. Chr. *Berichte der Kommission für Archäologische Landesforschung in Hessen* 2, 147–170.

KROLL, H. 1979
Kulturpflanzen aus Dimini. *Archaeo-Physika* 8, 173–189.

KROLL, H. 1980
Mittelalterlich/frühneuzeitliches Steinobst aus Lübeck. *Lübecker Schriften zur Archäologie und Kulturpflanzengeschichte* 3, 167–173.

KROLL, H. 1982
Kulturpflanzen aus Tiryns. Archäolog. Anzeiger, 467–485.

KROLL, H. 1983
Kastanas. Ausgrabungen in einem Siedlungshügel der Bronze- und Eisenzeit Makedoniens 1975–1979. Die Pflanzenfunde. Prähistorische Archäologie in Südosteuropa 2, Berlin.

KROLL, H. 1987
Vor- und frühgeschichtlicher Ackerbau in Archsum auf Sylt. Eine botanische Grossrestanalyse. *Römisch-Germanische Forschungen* 44, 51–158.

KROLL, H. 1990a
Melde von Feudvar, Vojvodina. *Praehistorische Zeitschrift* 65, 46–48.

KROLL, H. 1990b
Saflor von Feudvar, Vojvodina. *Archaeologische Korrespondenzblatt* 20, 41–46.

KROLL, H. 1991
Südosteuropa. In: Zeist, W. van, Wasylikowa, K. – Behre, K.-E. (eds), *Progress in Old World Palaeoethnobotany.* 161–178. Rotterdam.

KUCKUCK, H. 1959
On the findings on the arising of Triticum spelta L. in Iran and on the arising Triticum aestivum types through crossing of different Spelta types. 9–10. Wheat Information Service, Kyoto.

KUCKUCK, H. 1962
Vavilov's Genzentrentheorie in heutiger Sicht. 21–28. EUCARPIA. Paris.

KUCKUCK, H. 1964
Experimentelle Untersuchungen zur Entstehung der Kulturweizen. *Zeitschrift für Pflanzenzüchtung* 51, 97–140.

KUNKEL, G. 1984
Plants for Human Consumption. An Annotated Checklist of the Edible Phanerogams and Ferns. Königstein.

KÜHN, F. 1977
Botanicky rozbor obilí z halstatské doby z Rajhradu. *Přehled Výskomú* 1977, 43–46.

KÜHN, F. 1979
Rozbory nalezú polních plodín. *Přehled Výskomú* 1979, 75–79. Brno.

KÜSTER, H. 1985a
Herkunft und Ausbreitungsgeschichte einiger Secalietea- Arten. *Tüxenia* 5, 589–598.

KÜSTER, H. 1985b
Neolithische Pflanzenreste aus Hochdorf, Gemeinde Eberdingen (Kreis Ludwigsburg). *Forschungen und Berichte zur Vor- und Frühgeschichte in Baden-Württemberg* 19 (Hochdorf I), 13–83. Stuttgart.

KÜSTER, H. 1986
Sammelfrüchte des Neolithicums. *Abhandlungen* 48, 433–440. Münster.

KÜSTER, H. 1988a
Urnenfelderzeitliche Pflanzenreste aus Burkheim, Gemeinde Vogtsburg, Kreis Breisgau-Hochschwarzwald (Baden-Württemberg). In: Küster, H. (ed.), *Der prähistorische Mensch und seine Umwelt. Festschrift für Udelgard Körber-Grohne zum 65. Geburtstag.* 261–268. Stuttgart.

KÜSTER, H. 1988b
Pflanzenreste der Späthallstatt-/Frühlatènezeit aus Niedererlbach (Niederbayern). *Bayerische Vorgeschichtsblätter* 53, 77–82.

KÜSTER, H. 1989
Die Urgeschichte der Gerste. *Jahrbuch der Gesellschaft für die Geschichte und Bibliographie des Brauwesens E.V.* 198–211.

KÜSTER, H. 1991
Mitteleuropa südlich der Donau, einschliesslich Alpenraum. In: Zeist, W. van, Wasylikowa, K. – Behre, K.-E. (eds), *Progress in Old World Palaeoethnobotany*. 179–188. Rotterdam.

LA BAUME, W. 1961
Frühgeschichte der europäischen Kulturpflanzen. *Giessener Abhandlung zur Agrar- und Wirtschsftsforschung des europäischen Ostens* 16, 1–56.

LACKÓ, D. – RHÉ, GY. 1912
Balácza. Veszprém.

LACZA, J. = SZÚJKÓNÉ LACZA, J.

SZÚJKÓNÉ LACZA, J. 1991
Reconstruction of the history of the postglacial flora and vegetation in the region between Danube and Tisza and Tiszántúl. Kijev.

LADIZINSKY, G. 1979
The origin of Lentil and its wild genepool. *Euphytica* 28, 179–187.

LADIZINSKY, G., BRAUN, D., GOSCHEN, D. – MUEHLBAUER, F. J. 1984
The biological species of the genus Lens L. *Botanical Gazette* 145, 253–261.

LAEGDSMAND, K. 1968
The Ard-Head from Mors. Tools and Tillage.

LAKÓ, GY. 1967–78
A magyar szókészlet finnugor elemei I. 1967; II. 1971; III. 1978. Budapest.

LANDOLT, E. 1977
Ökologische Zeigerwerte zur Schweizer Flora. Geobotanisches Institut der ETH, Zürich.

LÁNG, G. 1976
Szántóföldi növénytermesztés. Budapest.

LÁSZLÓ, GY. 1974
Vértesszőlőstől Pusztaszerig. Budapest.

LEEK, F. F. 1973
Further studies in ancient Egyptian bread. *Journal of Egyptian Archaeology* 59, 199–204.

LEHMANN, C. O., RUDOLPH, A., HAMMER, K., MEISTER, A., MÜNTZ, K. – SCHOLZ, F. 1978
Eiweissuntersuchungen am Getreide- und Leguminosen-Sortiment Gatersleben. Teil 1: Gehalt an Rohprotein und Lysin von Weizen sowie Weizen-Art- und -Gattungsbastarden. *Kulturpflanze* 26, 133–161.

LEHOCZKY, T. 1883
A szádelővölgyi sziklavölgy természetrajzi és régészeti tekintetben. *Archaeolo*giai Értesítő 3, 140.

LEHOCZKY, T. 1898
Őskori őrlőkövekről. *Archaeologiai Értesítő* 18, 90.

LELLEY, J. 1980
Ember és kenyere. Budapest.

LELLEY, J. – MÁNDY, GY. 1963
A búza. Triticum aestivum L. Magyarország Kultúrflórája 8. Budapest.

LELLEY, J. – RAJHÁTHY, T. 1955
A búza nemesítése. Budapest.

LENZ, H. O. 1859
Botanik der alten Griechen und Römer. Gotha.

LEONARD, N. J. 1974
Die ersten Ackerbauer. Time-Life International (Nederland).

LIGETI, L. 1986
A magyar nyelv török kapcsolatai a honfoglalás előtt és az Árpád-korban. Akadémiai Kiadó, Budapest.

LINDAU, G. 1917
Növényi maradványok a tószegi Laposhalom őskori leleteiben. Archaeologiai Értesítő 37.

LIPPAY, J. 1664
Posoni kert. Poson.

LIPS, J. 1962
A dolgok eredete. Gondolat Kiadó, Budapest.

LISITSINA, G. N. 1978
Main types of ancient farming on the Caucasus, on the basis of a palaeo-ethnobotanical research. *Berichte der Deutschen Botanischen Gesellschaft* 91, 47–57.

LISITSINA, G. N. 1984
The Caucasus – A centre of ancient farming in Eurasia. In: Zeist, W. van – Casparie, W. A. (eds), *Plants and ancient man.* 285–292. Rotterdam.

LISITSINA, G. N. – PRISEPENKO, L. V. 1977
Paleo-etnobotanicseszkie nahodki Kavkaza i Blizsnevo Vosztoka. Moszkva.

LOCHTE, TH. 1951
Grundriss der Entwicklung des menschlichen Haares. Beiträge zur Haut-, Haar- und Fellkunde 5, Frankfurt.

LOCHTE, TH. 1954
Tafeln zur Haarkunde. Beiträge zur Haut-, Haar- und Fellkunde 6, Leipzig.

LÜNING, J. 2000
Steinzeitliche Bauern in Deutschland. Die Landwirtschaft im Neolithikum. Universitätsforschungen zur Prähistorischen Archäologie 58. R. Habelt GmbH. Bonn.

LÜNING, J. – STEHLI, P. 1989
Die Bandkeramik in Mitteleuropa: von der Natur- zur Kulturlandschaft. In: *Siedlungen der Steinzeit. Spektrum der Wissenschaft.* 110–120. Heidelberg.

MACKEY, J. 1966
Species relationships in Triticum. In: Mackey, J. (ed.), *Proceedings of the 2nd International Wheat Genetics Symposium, Lund*, Sweden, 1963. Hereditas, Suppl. 2, 237–276.

MACKEY, J. 1975
The boundaries and subdivision of the genus Triticum. In: *Proceedings of the 12th International Botanical Congress*. Leningrad.

MADARASSY, A. = V. MADARASSY, A.

V. MADARASSY, A. 2000
A természetkímélő mezőgazdálkodás szerepe az agrobiodiverzitásban. In: Gyulai, F. (ed.), *Az agrobiodiverzitás megőrzése és hasznosítása.* 25–27. Tápiószele.

MAGYAR ÉRTELMEZŐ SZÓTÁR
Magyar Értelmező Szótár VI. Budapest 1962.

MAGYAR NÉPRAJZI LEXIKON 1977
Ortutay Gy. (ed.), *Magyar Néprajzi Lexikon I–V.* Budapest.

MAIER, U. 1983
Nahrungspflanzen der späten Mittelalters aus Heidelberg und Ladenburg nach Bodenfunden aus einer Fäkaliengrube und einem Brunnen des 15./16. Jahrhunderts. *Forschungen und Berichte der Archäologie des Mittelalters in Baden-Württemberg* 8, 139–183.

MAKKAI, L. 1968
Az idő sodrában. Budapest.

MAKKAI, L. 1980
Az őshazából az újhazába. In: Hanák, P. (ed.), *Hogyan éltek elődeink?* 7–21. Budapest.

MAKKAY, J. 2001
Textile impressions and related finds of the Early Neolithic Körös culture in Hungary. Published by the author. Budapest.

MANSFELD, R. 1986
Verzeichnis landwirtschaftlicher und gärtnerischer Kulturpflanzen (ohne Zierpflanzen) I–IV. Berlin.

MAQBOOL, A., ARAIN, MA. – SIDDIQUI, K. A. 1997
Screening of Aegilops, Triticum and Hordeum species for grain weight, protein and lysine content. *Wheat Information Service* 85, 7–13.

MASSZON, V. M. 1978
Egy kőkori település Közép-Ázsiában, Dzsejtun. Budapest.

MATOLCSI, J. 1967–68
Avarkori háziállatok maradványai Gyenesdiáson (Überreste von Haustieren aus dem Zeitalter der Awaren in Gyenesdiás). *Magyar Mezőgazdasági Múzeum Közleményei* 1968, 85–122.

MAURIZIO, A. 1916a
Botanisch-chemisches zur Getreidenahrung der Pfahlbauer. *Anzeiher für Schweizerische Altertumskunde* 18, 183–185.

MAURIZIO, A. 1916b
Die Getreide-Nahrung im Wandel der Zeiten. Zürich.

MAURIZIO, A. 1927
Die Geschichte unserer Pflanzennahrung von den Urzeiten bis zur Gegenwart. Berlin.

MÁNDY, GY. 1972
Hogyan jöttek létre kultúrnövényeink? Budapest.

MÁNDY, I. – KISS, GY. 1971
A lencse (The lentils). *Magyarország kultúrflórája* 33/15, 76–84.

MÁTHÉ, G. 1999
Az alakor újrahonosítása. Kísérlet az agrobiodiverzitás megőrzésére. Szakdolgozat. Kossuth Lajos Tud. Egy. Természetvéd. ökológus posztgrad. szak, Debrecen. Kézirat.

MÁTYUS, I. 1787
Ó és Uj Diatetica. Pozsony.

MCGOVERN, E. P. 1999
Expedition Magazina 39, no. 1.

MEDZRIHRADSZKY, ZS. 1996
Some palynological data for the phytogeographical evaluation of Gyoma 133. In: Bökönyi, S. (ed.), *Cultural and Landscape Changes in South-East Hungary II. Prehistoric Roman Barbarian and Late Avar Settlement at Gyoma 133 (Békés County Microregion).* 163–166. Budapest.

MEHLHORN, H. – PIEKARSKI, G. 1989
Grundriss der Parasitenkunde. Stuttgart.

MERKLIN, R. L. 1969
A Method for Paleosynecological Investigation of Geosynclinal Strata. In: *Biostratigrafiya, fauna i flora kainizoya severo- zapadnio chasti Tikhookeanskogo podviznogo poyasa.* 140–149. Moskau.

MESSIKOMMER, H. 1913
Die Pfahlbauten von Robenhausen. Zürich.

MEURERS-BALKE, J. 1985
Experimente zum Anbau und zur Verarbeitung prähistorischer Getreidearten. *Archäologische Informationen* 8, 8–17.

MÉRI, I. 1964
Árpád-kori népi építkezésünk feltárt emlékei Orosháza határában. *Régészeti Füzetek Ser. II.* 12. 40.

MIHÁLCHNÉ FARAGÓ, M. 1983
Palynológiai vizsgálatok a Balaton fenékmintáin. *Magyar Állami Földtani Intézet Évi jelentése az 1981. évről.* 439–448. Budapest.

MOHÁCSY, M. – MALIGA, P. 1959
Cseresznye és meggytermesztés. Budapest 1959.

MOLNÁR, E. 1949
A magyar társadalom története az Árpád-kortól Mohácsig. Budapest.

MOLNÁR, E. 1954
A magyar nép őstörténete. Budapest.

MOLNÁR, É. 1997
Egy jobbágyfalu gazdasági életének vizsgálata a török korban Szarvasgede példáján és faluképe a 20. század végén. Szakdolgozat. Budapesti Közgazdaságtudományi Egyetem Vezetőképző Intézete Budapest. Kézirat.

MOLNÁR, N. 1961
A magyar gazdasági növénynevek történetéből. Agrártudományi Egyetem, Gödöllő.

MOLNÁR, ZS. 1999
Ősi és másodlagos (szikes) puszták a Tiszántúlon. In: Füleky, Gy. (ed.), *A táj változásai a Kárpát-medencében.* 231–233. Gödöllő.

MONTANARI, M. 1996
Éhség és bőség (La fame e l'abbondanza). Budapest.

MOÓR, E. 1943
A magyar őstörténet fő problémái. Szegedi Múzeumi Kiadványok I/12.

MÓRA, F. 1908
Szegedvidéki leletekről. *Archaeologiai Értesítő* 28, 362.

MÜLLER, K. 1997
A százhalombattai régészeti park környezetrekonstrukciós terve. Kézirat.

MÜLLER, R. 1982
A mezőgazdasági vaseszközök fejlődése Magyarországon a késővaskortól a törökkor végéig I–II. (Die Entwicklung der eisernen Agrargeräte in Ungarn von der Späteisenzeit bis zum Ende der Türkenherrschaft I–II.). Zalai Gyűjtemény 19, Zalaegerszeg.

MÜLLER, R., GROPENGIEßER, E., JENSEN, I. – PFAFF, P. 1989
Sieben Jahrtausende am Balaton. Von der Ur- und Frühgeschichte bis zum Ende der Türkenkriege. Reiss Museum, Mannheim.

MÜLLER-WILLE, M. 1965
Eisenzeitliche Fluren in den westländischen Nordseegebiet. Münster.

NAGY, B. 2000

Hordalékkúp térszínének változása az emberi tevékenység következtében a Sajó-Hernád hordalékkúp példáján. In: Füleky, Gy. (ed.), *A táj változásai a Kárpát-medencében a történelmi események hatására.* 35–38. Budapest–Gödöllő.

NAGYNÉ BODOR , E. 1988

A Balaton pannóniai és holocén képződményeinek palynológiai vizsgálata. *Magyar Állami Földtani Intézet Évi jelentése az 1986. évről.* 535–568. Budapest.

NAGYVÁTI, J. 1791

A szorgalmatos mezei-gazda I–II. Pest.

NETOLITZKY, F. 1926

Anatomie der Angiospermen-Samen. Handbuch der Pflanzenanatomie 10, Berlin.

NETOLITZKY, F. 1927

Arbeitsmethoden zur mikroskopischen Untersuchung verkohlter Körper. *Mikrokosmos* 20, 178–183.

NETOLITZKY, F. 1931

Unser Wissen von den alten Kulturpflanzen Mitteleuropas. *Römisch-Germanisches Kommission* 20, 14–76.

NEUWEILER, E. 1905

Die prähistorischen Pflanzenreste Mitteleuropas mit besonderer Berücksichtigung der schweizerischen Funde. *Vierteljahrsschritt der Naturforschenden Gesellschaft Zürich* 50, 23–132.

NEUWEILER, E. 1931

Die Pflanzenreste aus dem spätbronzezeitlichen Pfahlbau "Sumpf" bei Zug. *Vierteljahrsschritt der Naturforschenden Gesellschaft Zürich* 76, 116–121.

NÉMETH, P. 1973

Újabb eredmények a honfoglaló magyarság települési rendjének kutatásában. *Szabolcs-Szatmári Szemle* 18/4, 67–69.

NOVÁKI, GY. 1965–66

Koravaskori sarlólelet Kőszegről (Früheisenzeitliche Sichelfunde aus Kőszeg). *Magyar Mezőgazdasági Múzeum Közleményei*, 67–72.

NOVÁKI, GY. 1969

Änderungen der Weizenarten in Ungarn von der Bronzezeit bis zum Mittelalter. *Móra Ferenc Múzeum Évkönyve* 2, 39–45.

NOVÁKI, GY. 1975

Die Geschichte des Pflanzenbaus in Ungarn von der Neusteinzeit bis zum Beginn des Mittelalters. *Agrártörténeti Szemle* 17, Suppl., 22–88.

NOVÁKI GY. 1975–77

Régi szántóföldek nyomai a Börzsönyben (Spuren von einstigen Ackerfeldern im Börzsöny-Gebirge). *Magyar Mezőgazdasági Múzeum Közleményei*, 53–79.

NOVÁKI, GY. 1981–83

Régészeti és palaeoethnobotanikai adatok a "gabonásvermek" kérdéséhez. (Archäologische und palaeoethnobotanische Angaben zur Frage der "Getreidegruben"). *Magyar Mezőgazdasági Múzeum Közleményei*, 57–94.

NOVÁKI GY. 1984–85

Szántóföldek maradványai a XIV-XVI. századból a Sümeg-Sarvalyi erdőben (Überreste der Ackerfelder des 14.–16. Jahrhunderts aus dem Wald Sümeg-Sarvaly). *Magyar Mezőgazdasági Múzeum Közleményei*, 19–32.

NYÁRY, J. 1881

Az Aggteleki barlang mint őskori temető. Budapest.

NYÚJTÓ, F. – SURÁNYI, D. 1981

Kajszibarack. Budapest.

OPRAVIL, E. 1966

Rostleny ze Stredovékych objektú v Plzni Solni ullice. *Casopis Národní Musea oddil prirod.* 135, 84–88.

OPRAVIL, E. 1972

Rostliny z velkomoravského hradiste v Mikulčicich. (Les végétaux de bourgwall de la Grande-Moravie a Mikulčice.) Studie Archeologického Ustavu Československe Akademie VĚD I/2.

OPRAVIL, E. 1974

Z historije pohanky. *Vlastivedny sbornik okresu Novy Jicin* 14, 51–56.

OPRAVIL, E. 1977

Vinnì réva na uzmeni Ceskoslovenska u praceku. (Die Weinrebe auf dem Gebiete der CSSR in der Vorzeit). *Archeologické rozhledy* 29, 361–365. Praha.

OPRAVIL, E. 1984

Rostlinné zbytky ze dvora kupeckého domu v Olomouci (13–17. stol.). *Archeologické rozhledy* 36, 194–202.

OPRAVIL, E. 1985

Rostlinné zbytky z odpadní jímky v Tábore c. p. 6. *Archeologické rozhledy* 37, 186–194.

ŐRÖSI, P. Z. 1968

Méhek között. Budapest.

PAAP, N. 1984

Botanische Analysen in Lübeck. Eine Zwischenbilanz. *Lübecker Schriften zur Archäologie und Kulturgeschichte* 8, 49–55.

PARÁDI, N. 1959

Technikai vizsgálatok népvándorláskori és Árpád-kori edényeken. Régészeti Füzetek 12.

PASTERNAK, R. 1991

Getreide: Lebensmittel. Abfall oder Futtermittel? Analysen von heutigem Dreschgut aus Jordanien als Interpretationshilfen zur Bewertung archäologischer Getreide-Funde. *Archäologisches Korresponsenblatt* 21, 151–158.

PATAY, Á. – SZ. PÓCZY, K. 1964

Gyümölcsmaradványok az aquincumi múmiasírból (Obstrelikte im Mumiengrab von Aquincum). *Magyar Mezőgazdasági Múzeum Közleményei* 1964, 135–147.

PATAY, R. – FEKETE, Z. 2006

Régészeti kutatások másfélmillió négyzetméteren. *Pest Megyei Múzeumi Füzetek* 7, 16–18.

PÁKAY, ZS. – SÁGI, K. 1971
A szőlőművelés hatása a Balatonkörnyék népeinek életére és településére. *A Veszprém Megyei Múzeumok Közleményei* 10, 95–114.

PÁLMÁNY, B. 1984
Szécsény legrégebbi lakosságnévsora az 1551. évi bordézsmajegyzék alapján. Szécsényi Honismereti Híradó.

PÁLÓCZI HORVÁTH, A. 1999
A visegrádi palota kertje mint környezettörténeti forrás. In: Füleky, Gy. (ed.), *A táj változásai a Kárpát-medencében.* 96–100. Gödöllő.

PÁLÓCZI HORVÁTH, A. – TORMA, A. 1999
Environmental archaeological research at Visegrád in the medieval garden of the Royal Palace. In: Jerem, E. – Poroszlai, I. (eds), *Archaeology of the Bronze and Iron Age. Experimental Archaeology, Environmental Archaeology, Archaeological Parks. Proceedings of the International Archaeological Conference Százhalombatta, 3–7 October 1996.* 343–350. Archaeolingua, Budapest.

PÁRDUCZ, M. 1971
Einige Probleme der Sarmatenforschung des Karpatenbeckens. In: *Actes du VIIIe Congres International des Sciences Préhistoriques et Protohistoriques.* Tome I, 267–274. Beograd.

PÁRDUCZ, M. – KOREK, J. 1958
Császárkori telep Ózdon. *Archaeologiai Értesítő* 85, 18–36.

PEARSALL, D. M. 1989
Paleoethnobotany. A Handbook of Procedures. San Diego.

PERCIVAL, J. 1921
The wheat plants. Duckworth Co. London.

PETIT, D. 1983
Saint-Pierre-Lentin 1977–1978. Rapport préliminaire. *Revue Archéologique du Loiret* 9, 33–34.

PÉNTEK, J. – SZABÓ, T. A. 1981
Az alakor (Triticum monococcum L.) Erdélyben. *Ethnographia* 92, 259–277.

PÉNZES, A. 1950
Kökény-szilva (Prunus) tanulmányok. *Agrártudományi Egyetem Kert- és Szőlőgazdaságtudományi Kar Évkönyve* 153–173.

PIENING, U. 1983
Verkohlte Pflanzenreste der Frühlatènezeit von Lauffen am Necker, Kreis Heilbronn. *Fundberichte aus Baden-Württemberg* 8, 47–54.

PIENING, U. 1988a
Kultur- und Wildpflanzenreste aus Gruben der Urnenfelder- und Frühlatènezeit von Stuttgart-Mühlhausen. In: Küster, H. (ed.), *Der prähistorische Mensch und seine Umwelt. Festschrift für Udelgard Körber-Grohne zum 65. Geburtstag.* 269–280. Stuttgart.

PIENING, U. 1988b
Neolitische und hallstattzeitliche Pflanzenreste aus Freiberg-Geisingen (Kreis Ludwigsburg). In: Küster, H. (ed.), *Der prähistorische Mensch und seine Umwelt. Festschrift für Udelgard Körber-Grohne zum 65. Geburtstag.* 213–228. Stuttgart.

PIETA, K. 1988
Die Slowakei im 5. Jahrhunder. Germanen, Hunnen und Awaren. In: Bott, G. – Meier-Arendt, W. (eds), *Schätze der Völkerwanderungszeit.* 385–417. Germanisches Nationalmuseum, Nürnberg.

PIETA, K. – PLACHA, V. 1989
Getreide- und Brotfunde aus der Völkerwanderungszeit in Devin. *Slovenska Archeologia* 37, 69–88.

PIPERNO, D. R. 1987
Phytolith Analysis. Academic Press, Inc. London.

PLETNJOWA, S. A. 1978
Die Chassaren. Mittelalterliches Reich an Donau und Wolga. Leipzig.

POULSEN, J. 1983
Landwirtschaft und Bevölkerungsverhältnisse in der dänischen Bronzezeit. *Zeitschrift für Archäologie* 17, 145–158.

POZSÁR, B. I. 1971
Modern Trends in the Isozymes Research, with Relation to Archeobotany. *Acta Museorum Agriculturae Pragae* 6 (1–2), 15–16.

PÓR, J. – FALUBA, Z. 1982
Cseresznye és meggy. Budapest.

PRÉCSÉNYI, I. 1980 (ed.)
Ökológiai stabilitás. Acta Invenum Debreceniensis, Studium 10/1.

PRICE, D. T. – FEINMANN, M. G. 1993
Images of the Past. Mountain View, California, Mayfield.

PRISZTER, SZ. 1986
Növényneveink (Plant names). Budapest.

PRISZTER, SZ. 1990
Húsos som (Cornus mas L.). Magyarország kultúrflórája IV/2. Budapest.

PRÓNAI, I. 1999
Különleges és hétköznapi ehető növények. Különös tekintettek a Magyarországon vadon előforduló növényekre. Szeged.

PULSZKY, F. 1897
Magyarország archaeológiája I. [*The archaeology of Hungary I.*]. Budapest.

QUITTA, H. – KOHL, G. 1969
Neue Radiocarbondaten zum Neolithikum und zur frühen Bronzezeit Südosteuropas und der Sowjetunion. *Zeitschrift zur Archäologie* 3, 223–255. Berlin.

RADEMACHER, B. 1968
Gedanken zur Fortentwicklung der Unkrautforschung und Unkrautbekämpfung. *Zur Pflanzenkrankheit und Pflanzenschutz, Sonderheft* 4, 11–22.

RAPAICS, R. 1934
A kenyér és táplálékot szolgáltató növényeink története. Népszerű természettudományi könyvtár 16. Budapest.

RAPAICS, R. 1940
A magyar gyümölcs. Budapest.

RAPAICS, R. 1943
Termesztett növényeink eredete. Kincsestár 89, Budapest.

RÁCZ, L. 1993
Éghajlati változások a középkori és kora újkori Európában. In: R. Várkonyi, Á. – Kósa, L. (eds), *Európa híres kertje.* 67–83. Orpheusz, Budapest.

RÁCZ, L. 1997
A Kárpát-medence történeti ökológiája a középkor és az újkor idején. In: Füleky, Gy. (ed.), *A táj változásai a Honfoglalás óta a Kárpát-medencében.* 18–25. Gödöllő.

RÁCZ, G., RÁCZ-KOTILLA, E. – LAZA, A. 1984
Gyógyszerismeret. Budapest.

RÁPÓTI, J. – ROMVÁRY, V. 1983
Gyógyító növények (Medicinal plants). Budapest.

RENFREW, J. M. 1966
A report on recent finds of carbonized cereals grains and seeds from prehistoric Thessaly. *Thessalika* 5, 21–36.

RENFREW, J. M. 1973
Palaeoethnobotany. The prehistoric food plants of the Near East and Europe. London.

RENFREW, C. – BAHN, P. 1999
Régészet. Elmélet, módszer, gyakorlat. Budapest.

REYNOLDS, P. J. 1991
Butser Ancient Farm. Guide Book. Leicestershire.

RICHTER, B. 1987
Mikroskopische Untersuchungen an Speiseresten. In: Suter, P. J. (ed.), *Zürich Kleiner Hafen. Tauchgrabungen 1981–1984.* 180–184. Zürcher Denkmalpfege, Monografien 3.

RICHTER, B. 1988
Mikroskopische Analyse vorgeschichtlicher Nahrungsreste. *Mikrokosmos* 77/4, 112–116.

RINDOS, D. 1984
The Origins of Agriculture. An Evolutionary Perspective. Academic Press, London.

ROGERIUS 1981
Rogerius mester Siralmas éneke. In: Katona, T. (ed.), *A tatárjárás emlékezete.* 111–152. Budapest.

ROTHMALER, W. 1953
Probleme der Kulturpflanzengeschichte. *Beiträge zur Frühgeschichte der Landwirtschaft* 1, 83–93.

ROTHMALER, W. – NATHO, I. 1957
Bandkeramische Kulturpflanzenreste aus Thüringen und Sachsen. *Beiträge zur Frühgeschichte der Landwirtschaft* 3, 73–98.

ROTTLÄNDER, R. C. A. 1983a
Einführung in die naturwissenschaftlichen Methoden in der Archäologie. Tübingen.

ROTTLÄNDER, R. C. A. 1983b
Chemische Analyse prähistorischer Gefässinhalte. In: Schuh, F. (ed.), *Enzyklopädie Naturwissenschaft und Technik* 1983, 72–80.

ROTTLÄNDER, R. C. A. – SCHLICHTHERLE, H. 1980
Gefässinhalte. Eine kurz kommentierte Bibliographie. Naturwissenschaftliche Beiträge zur Archäologie. *Archaeo-Physika* 7, 61–70.

ROVNER, I. – GYULAI, F. 1999
Morphometric Atlas for Seed Identification and Analysis in Archaeobotany. National Science Foundation (USA) – Hungarian Academy of Sciences. Annual Report. National Science Foundation.

ROVNER, I. – GYULAI, F. 2007
Computer-Assisted Morphometry: A New Method for Assessing and Distinguishing Morphological Variation in Wild and Domestic Seed Populations. *Journal of Economic Botany* 61/2, 1–19.

ROVNER, I. – RUSS, J. 1992
Darwin and design in phytolith systematics: Morphometric methods for mitigating redundancy. In: Mulholland, S. – G. Rapp, Jr. (eds), *Phytolith Systematics: Emerging Issues.* 253–276. Plenum Press, New York and London.

ROWLEY-CONWY, P. 1987
The interpretation of ard marks. *Antiquity* 61, 263–266.

RÖSCH, M. 1991
Archäobotanik und Pflanzensoziologie. Auswertungsmöglichkeiten subfossiler Floren (Beispiel Hagnau-Burg, Urnenfelderkultur, Hornstaad-Hörnle IA, Jungneolithikum). *Palaeoethnobotany and Archaeology. International Work-Group for Palaeoethnobotany 8th Symposium Nitra-Nové Vozokany 1989.* 29–36. Nitra.

RUCKENBAUER, P. VON 1971
Keimfähiger Winterweizen aus dem Jahre 1877. *Beobachtungen und Versuche die Bodenkultur* 22, 372–386.

SABJÁN, T. 1999
A veremház rekonstrukciója (Die Rekonstruktion des Grubenhauses). In: Bencze, Z., Gyulai, F., Sabján, T. – Takács, M. (eds), *Egy Árpád-kori veremház feltárása és rekonstrukciója. (Ausgrabung und Rekonstruktion eines Grubenhauses aus der Árpádenzeit.).* 131–176. Monumenta Historica Budapestinensia 10.

SAKAMOTO, S. 1980
Variation of Cultivated Plants and their Wild Relatives in Rumania. Preliminary Report of Field Survey on the Agrico-pastoral Peoples in South-Western Eurasia. In: Tani, Y. (ed.), *Research Institute for the Humanistic Studies.* 87–94. Kyoto University.

SÁGI, K. 1972
Magyar néphagyományok a második világháború katonáinak tudatában. *Veszprém Megyei Múzeumok Közleményei* 11, 447–456.

SÁGI, K. – FÜZES, M. 1966
A régészeti-növénytan alapelemei és néhány módszertani kérdése (The fundamentals of archaeobotany and some methodological problems). Múzeumi Módszertani Füzetek 5, Budapest.

SÁGI, K. – FÜZES, M. 1967
Régészeti és archaeobotanikai adatok a pannoniai kontinuitás kérdéséhez. *Agrártörténeti Szemle* 9, 79–97.

SCHERMANN, SZ. 1966
Magismeret I-II. (Seed identification I-II.). Akadémiai Kiadó, Budapest.

SCHIEMANN, E. 1932
Entstehung der Kulturpflanzen. Handbuch der Vererbungswissenschaften 3.

SCHIEMANN, E. 1953
Vitis im Neolithicum der Mark Brandenburg. *Der Züchter* 23, 318–327.

SCHLICHTHERLE, H. 1983
Mikroskopische Untersuchungen an neolithischen Gefässinhalten aus Hornstaad, Yverdon und Burgäschisee-Süd. In: Müller-Beck, H. – Rottländer, R. (eds), *Naturwissenschaftliche Untersuchungen zur Ermittlung Prähistorischer Nahrungsmittel. Ein Symposionsbericht.* 39–61. Archaeologica Venatoria.

SCHMID, J. E. 1989
Einsatz biotechnischer Methoden bei der Erforschung alter Getreidelandsorten. *Bulletin der Eidgenössischen Technischen Hochschule Zürich* 219, 19–21.

SCHNEIDER, M. 1991
Charred Plant Remains from Late Bronze Age Stillfried (Austria). In: Hajnalová, E. (ed.), *Palaeoethnobotany and Archaeology. International Work-Group for Palaeoethnobotany 8th Symposium Nitra-Nové Vozokany 1989.* 295–298. Nitra.

SCHOCH, W., PAWLIK, B. – SCHWEINGRUBER, F. H. 1988
Botanische Makroreste. Ein Atlas zur Bestimmung häufig gefundener und ökologisch wichtiger Pflanzensamen. Bern und Stuttgart.

SCHULTZE-MOTEL, J. 1972
Literatur über archäologische Kulturpflanzenreste (1969). *Kulturpflanze* 19, 265–282.

SCHULTZE-MOTEL, J. (ed.) 1986
Rudolf Masfeld's Verzeichnis landwirtschaftlicher und gärtnerischer Kulturpflanzen (ohne Zierpflanzen) 1. Berlin.

SCHULTZE-MOTEL, J. – GALL, W. 1967
Prähistorische Kulturpflanzen aus Thüringen. Alt-Thüringen 9.

SCHWANITZ, F. 1973
A kultúrnövények keletkezése; az egész növényvilág evolúciós modellje. Budapest.

SCHWARZBACH, M. 1950
Das Klima der Vorzeit. Stuttgart.

SCHWEINFURT, G. 1884
Über Pflanzenreste aus altägyptischen Gräbern. *Berichte der Deutschen Botanischen Gesellschaft* 2, 351–371.

SHACKLEY, M. 1981
Environmental Archaeology. London.

SKOFLEK, I. 1982
Das mittelalterliche Dorf Sarvaly. Fontes Archaeologici Hungariae. Budapest.

SKOFLEK, I. 1985
Mag- és terméslеletek Sümeg-Sarvalyról a XVI. századból (Samen und Fruchtfunde vom 16. Jahrhundert aus dem Wald von Sümeg-Sarvaly). *Magyar Mezőgazdasági Múzeum Közleményei* 1984/85, 33–44.

SKOFLEK I. – ÁRENDÁS V. 1969
Levél- és terméslenyomatok császárkori égett agyagdarabokon. *Botanikai Közlemények* 56, 57–63.

SKOFLEK I., – ÁRENDÁS, V. 1971
Botanische Untersuchung der aus den kaiserzeitlichen Siedlungen stammenden Lehmbewurfe (Methode und Ergebnisse). *Mitteilungen des Archäologischen Instituts der Ungarischen Akademie der Wissenschaften* 2, 119–129.

SKOFLEK, I. – HORTOBÁGYI, I. 1973
Medieval seed and fruit finds from the Castle Hill of Buda. *Mitteilungen des Archäologischen Instituts der Ungarischen Akademie der Wissenschaften* 4, 135–156.

SOMOGYI, S. 1984
A Kárpát-medence természeti viszonyainak változásai a honfoglalás előtt (Environmental changes in the Carpathian Basin before the Hungarian Conquest). In: Székely, Gy. (ed.), *Magyarország története I/1. Előzmények és magyar történet 1242-ig (The history of Hungary I/1. Antecedents and Hungarian history until 1242).* 25–69. Budapest.

SOÓ, R. 1964–1985
A magyar flóra és vegetáció rendszertani-növényföldrajzi kézikönyve. I. 1964; II. 1966; III. 1968; IV. 1970; V. 1973; VI. 1980; VII. 1985. Budapest.

SOÓ, R. – JÁVORKA, S. 1951
A magyar növényvilág kézikönyve I-II. Budapest.

SOÓ, R. – KÁRPÁTI, Z. 1968
Növényhatározó I-II. (Plant identification I-II.). Budapest.

SOUCI, S. W., FACHMANN, W. – KRAUT, H. 1987
Die Zusammensetzung der Lebensmittel-Nährwerte. Tabellen 1986/87. Stuttgart.

SÓS, Á. = CS. SÓS, Á.

CS. SÓS, Á. 1973
Die slawische Bevölkerung Westungarns im 9. Jahrhundert. München

CS. SÓS, Á. 1985
Zalavár a középkorban (Zalavár in the Middle Age). Zalaegerszeg.

STEENSBERG, A. 1976
Virgil's wheel-ard and the two mouldstrokers. In: O'Danachair, C. (ed.),: *Folk & Farm. Essays in Honour of A. T. Lucas.* Royal Society of Antiquaries of Ireland. 265–272. Dublin.

STOKAR, VON W. 1951
Die Urgeschichte des Hausbrotes. Leipzig.

STRAKA, H. 1975
Pollen- und Sporenkunde. Stuttgart.

STUMMER, A. 1911
Zur Urgeschichte der Rebe und des Weinbaues. *Mitteilungen der Anthropologischen Gesellschaft in Wien* 41, 283–296.

SURÁNYI, B. 2002
Tájfajták a Kárpát-medencében (XVIII. sz.–1950). (Agrártörténeti vázlat a népi növénynemesítésről). *Agrártörténeti Szemle* 2002, 321–406.

SURÁNYI, D. 1985
Kerti növények regénye. Budapest.

SURÁNYI, D. 1985–86
A magyar gyümölcs évszázadai 1–21. Kertészet-Szőlészet.

SÜMEGI, P. 2000
A középkori Kárpát-medence éghajlati és környezeti viszonyai. In: Bende, L. – Lőrinczy, G. (eds), *A középkori magyar agrárium.* 9–25. Ópusztaszer.

SÜMEGI, P., HERTELENDI, E., MAGYARI, E. – MOLNÁR, M. 1998
Evolution of the environment in the Carpathian Basin during the last 30.000 BP years and its effects on the ancient habitats of the different cultures. In: Bartosiewicz, L. – Költő, L. (eds), *Archaeometrical Research in Hungary II.* 183–193. Budapest–Kaposvár.

SÜMEGI, P., KERTÉSZ, R. – RUDNER, E. 2003
Palaeoenvironmetal History of Hungary. In: Visy, Zs. (ed.), *Hungarian Archeology at the Turn of Millenium.* 51–56. Budapest.

SZABÓ, I. 1963
A prédium I. Agrártörténeti Szemle 5.

SZABÓ, I. 1966
A falurendszer kialakulása Magyarországon X–XV. század. Akadémiai Kiadó, Budapest.

SZABÓ, I. 1969
A középkori magyar falu. Akadémiai Kiadó, Budapest.

SZABÓ, L. 1982
A zab. Magyarország kultúrflórája 9/1, Budapest.

SZABÓ, L. GY. 1987
A búza alaktana és fejlődése. In: Barabás, Z. (ed.), *A búzatermesztés kézikönyve.* 45–87. Budapest.

SZABÓ, T. A. – PÉNTEK, J. 1996
Ezerjófű. Etnobotanikai útmutató. Budapest.

SZAMOTA, I. 1891
Régi utazások Magyarországon és a Balkán-félszigeten. Budapest.

SZAMOTA, I. – ZOLNAI, GY. 1902–1906
Magyar oklevél-szótár. Budapest.

SZIKSZAI FABRICZIUS, B. 1590
Nomenclatura. In: Melich, J. (ed.), *Szikszai Fabriczius Balázs latin-magyar szójegyzéke 1906.*

SZKAZKIN, SZ. D. 1979
A parasztság a középkori Nyugat-Európában. Gondolat, Budapest.

SZŐCS, Z. 1995
Az amaránt. (The amaranth.) Középeurópai Egyetem (CEU), Budapest.

SZŐKE, B. 1962
A honfoglaló és kora Árpád-kori magyarság régészeti emlékei. Budapest.

SZŐKE, B. M. 1980
Zur awarenzeitlichen Siedlungsgeschichte des Körös-Gebietes in Südost-Ungarn. *Acta Archaeologica Academiae Scientiarum Hungaricae* 32, 181–203.

SZUTÓRISZ, F. 1905
A növényvilág és az ember. Budapest.

TAGÁN, G. 1941
Néprajzi megfigyelések a kisázsiai Taurusz hegység déli részén. *Ethnographia* 52, 230–245.

TAKÁCS, M. 1999
Lakóház-rekonstrukciók az Árpád-kori telepkutatásban (Tudománytörténeti áttekintés). (Wohnhaus-Rekonstruktionen in der árpádenzeitlichen Siedlungsforschung – wissenschaftlicher Überblick.) In: Bencze, Z., Gyulai, F., Sabján, T. – Takács, M. (eds), *Egy Árpád-kori veremház feltárása és rekonstrukciója. (Ausgrabung und Rekonstruktion eines Grubenhauses aus der Árpádenzeit.)* 93–131. Monumenta Historica Budapestinensia 10.

TAKÁCS, S. 1917
Rajzok a török világból I–III. Budapest.

TARDY, J. 1994 (ed.)
Természetvédelem. Budapest.

TARJÁN R. – LINDNER K. 1981
Tápanyagtáblázat. Medicina, Budapest.

TASKIN, V. S. 1968
Materialy po istorii Sjunnu (po kitajskim istocnikam). Moskva.

TÄCKHOLM, V. – DRAR, M. 1954
Flora of Egypt Vol. III, Allium in ancient Egypt. 93–106. Cairo University Press, Cairo.

TÄCKHOLM, V. – TÄCKHOLM, G. 1941
Flora of Egypt. 1. Cairo.

TEMPÍR, Z. 1961
Archeologické nálezy obilnin na uzemi Ceskoslovenska. *Vĕdecké Práce Československy Zemĕdĕlské Muzeum.* 1, 157–200. Praha.

TEMPÍR, Z. 1964
Beiträge zur ältesten Geschichte des Pflanzenbaus in Ungarn. *Acta Archaeologica Academiae Scientiarum Hungaricae* 16, 70–73.

TEMPÍR, Z. 1966
Výsledky paleobotanického studia pĕstováni zemĕdelských rostlin na území ČSSR. (Results of paleoethnobotanic studies on the cultivation of agricultural plants in the CSSR). *Vĕdecké Práce Československy Zemĕdĕlské Muzeum* 4, 27–144. Praha.

TEMPÍR, Z. 1968
Archeologické nálezy zemedelskych rostlin a plevelu v Cechách a na Morave. *Vĕdecké Práce Československy Zemĕdĕlské Muzeum* 8, 15–88. Praha.

TEMPÍR, Z. 1973
Nálezy pravĕkých a středovĕkých zbytkú pĕstovaných a užitkových rostlin a plevelú na nĕkterých lokalitách na Moravĕ. *Vĕdecké Práce Československy Zemĕdĕlské Muzeum* 13, 19–47. Praha.

TEMPÍR, Z. 1979
Kulturpflanzen im Neolithikum und Äneolithikum auf dem Gebiet von Böhmen und Mähren. *Archaeo-Physika* 8, 302–308.

TERPÓ, A. 1976
The carpological examination of wildgrowing vine species of Hungary I. *Acta Botanica Academiae Scientiarum Hungaricae* 22, 209–247.

TROELS-SMITH, J. 1944
Fund of Vitis silvestris pollen in Danmark. Dansk Geologisk Forening, 10, 4.

TUZSON, J. 1913
Adatok a délorosz puszták összehasonlító flórájához. *Botanikai Közlemények* 12, 181–202.

TUZSON, J. 1915
A magyar Alföld növényföldrajzi tagolódása (Die pflanzengeographische Gliederung der ungarischen Tiefebene). Mathematikai Természettudományi Értesítő 33.

UBRIZSY-SAVOIA, A. – GYULAI, F. 1993
Elementi mediterranei nel giardino imperiale ungherese del XV secolo. *Atti del III Convegno (primo internationale) "Paesaggi e Giardini del Mediterraneo" 1 La conoscenza. Pompei, 4–5–6 giugno 1993.* 259–270.

UDVARDY, L. 2000
Archaikus gabonagyomjaink mint dísznövények. In: Gyulai, F. (ed.), *Az agrobiodiverzitás megőrzése és hasznosítása.* 415–418. Tápiószele.

ÚJVÁROSI, M. 1957
Gyomnövények, gyomírtás. Budapest.

ÚJVÁROSI, M. 1973
Gyomnövények, gyomírtás I–II. Budapest.

ÚJVÁROSI, M., FEKETE, A., REISINGER, P. – PFEIFER, D. 1979
A gyomfelvételezések feldolgozásának számítógépes módszere. Mezőgazdasági és Élelmezésügyi Minisztérium Növényvédelmi és Agrokémiai Központ, Budapest.

UNGER, F. 1851
Über die in Salzberge zu Hallstatt in Salzkammergute vorkommenden Pflanzentrümmer. *Sitzungsberichten der mathematisch-naturwissenschaftlichen Klasse der Kaiserlichen Akademie der Wissenschaften zu Wien* 7, 148–156.

UNGER, F. 1862
Botanische Streifzüge auf dem Gebiete der Culturgeschichte. Inhalt eines alten ägyptischen Ziegels an organischen Körpern. *Sitzungsberichten der mathematisch-naturwissenschaftlichen Klasse der Kaiserlichen Akademie der Wissenschaften zu Wien* 45, 75–88.

VAJKAI, A. 1943
A magyar nép életmódja. In: Bartucz, L. (ed.), *A magyar nép.* 136–200. Budapest.

VALKÓ, E. 1971
Botanische Angaben aus der Bronzezeit im Gebiet Ungarns. *IIIième Congres International des Museés d'Agriculture, Budapest. Résumeés des communications présentées.* 203–206.

VALKÓ, E. – STIEBER, J. 1969
The first evidence of prehistoric vine growing in Hungary. *Acta Agronomica* 18, 432–435.

VARRÓ, A. B. 1991
Gyógynövények gyógyhatásai. Marksped, Kaposvár.

VASS, E. 1983
Vác a török korban. In: Sápi, V. (ed.), *Vác története I.* 77–120. Studia Comitatensia 13, Szentendre.

VAVILOV, N. I. 1928
Geographische Genzentren unserer Kulturpflanzen. In: Verhandlungen des V. Internationalen Kongresses für Veterbungswissenschaft Berlin 1927. *Zeitschrift für induktive Abstammungs-und Vererbungslehre, Shupplementband* 1, 342–369.

VAVILOV, N. I. 1934
Le problème de l'origine des plantes cultivées. *Annales Institut National de la Recherche Agronomique* 36, 239–240.

VAVILOV, N. I. 1950
The origin, variation immunity and breeding, of cultivated plants. Chronica Botanica, Waltham, Mass, USA, 13.

VÁCZY, P. 1958
A korai magyar történet néhány kérdéséről. *Századok* 92, 265–345.

VÁCZY, P. 1974
A frank és avar nép. *Századok* 108/1–2, 1041–1052.

VÁLYI, K. 2002
Ember és környezet Szerbmonostor fénykorában. Egy 13. századi kút archaeobotanikai és archaeozoológiai leletei. In: Novák, L. F. (ed.), *Gazdálkodás az Alföldön.* 43–52. Arany János Múzeum Közleményei 9.

VÁNDOR, L. 1986 (ed.)
Régészeti kutatások a Kis-Balaton térségében. Zalaegerszeg.

VÁRKONYI, Á. = R. VÁRKONYI, Á.

R. VÁRKONYI, Á. 1992
Történeti ökológia és a művelődés históriája. *Magyar Tudomány* 11, 1296–1309.

VEREBES, A. 2000
Spätbronzezeitliche Speisereste aus Isztimér-Csőszpuszta. *Alba Regia* 29, 61–64.

VILLARET- VON ROCHOW, E. 1958
Die Pflanzenreste der bronzezeitlichen Pfahlbauten von Valeggio am Mincio. Bericht über das Geobot. Forschungsinst. *Rübel in Zürich für das Jahr* 1957, 96–144.

VILLARET- VON ROCHOW, E. 1971
Samenanalysen aus der spätneolitischen Grabung Auvernier (Neuenburger See). *IIième Congres International des Museés d'Agriculture, Budapest. Résumeés des communications présentées.* 206–208.

VILSTEREN, V. T. VAN 1984
Botanisch onderzoek bij de opgraving. In: Het Kasteel Voorst. Macht en val een Overijsselse brucht circa 1280–1362 naar aanleiding van een opgraving. Vereeniging tot Beoefening van Overijsselsch Regt en Geschiedenis. *Werken* 36, 133–143.

VÖRÖS, I. 1980
Zoological and palaeoeconomical investigation on the archaeozoological material of the early Neolithic Körös culture. *Folia Archaeologica* 31, 35–62.

VÖRÖS, I. 1982/83
Gyöngyöshalász – Encspuszta késő rézkori település állatcsontleletei. *Agria* 19, 35–61.

VÖRÖS, I. 1987
A Tiszalúc-sarkadi rézkori település állatcsontleletei. (Előzetes jelentés). *Folia Archaeologica* 38, 121–127.

VÖRÖS, I. 2000
Adatok az Árpád-kori állattartás történetéhez. In: Bende, L. – Lőrinczy, G. (eds), *A középkori magyar agrárium.* 71–120. Ópusztaszer.

WASYLIKOWA, K. 1973
Cereals from Early Medieval fortifield settlement in Lubomia, district Wodzslaw Slaski, Southern. Poland. *Polia Quarternaria* 43, 55–61.

WASYLIKOWA, K., CARCIUMARU, M., HAJNALOVÁ, E., P. HARTYÁNYI B., PASHKEVICH, G. A. – YANUSHEVICH, Z. V. 1991
East-Central Europe. In: Behre, K. E. (ed.), *Progress in Old World Palaeoethnobotany.* 207–239. A. A. Balkema, Rotterdam–Brookfield.

WÄHREN, M. 1984
Brote und Getreidebrei von Twann aus dem 4. Jahrtausend vor Christus. *Archäologie der Schweiz* 7/1, 2–6.

WÄHREN, M. 1985a
Unser täglich Brot in der Geschichte un dim Volksgebrauch. Verlag des Schweizerischen Bäcker- und Konditorenmeisterverbandes, Bern.

WÄHREN, M. 1985b
Die Entwicklungsstationen vom Korn zum Brot im 5. und 4. Jahrtausend. *Getreide, Mehl und Brot* 39, 373–379. Bochum.

WÄHREN, M. 1987
Das Brot in der Bronzezeit und älteren Vorrömischen Eisenzeit nördlich der Alpen unter besonderer Berücksichtigung von Brotfunden aus Kreisgrabenfriedhöfen des Münsterlandes. AFWL *Ausgrabungen und Funde in Westfalen-Lippe (Münster)* 5, 23–71.

WÄHREN, M. 1988
Jungsteinzeitliche Speisereste aus dem Kanton Freiburg. *Archéologie Fribourgeoise*, 85–95.

WÄHREN, M. 1989a
Brot und Gebäck von der Jungsteinzeit bis zur Römerzeit. *Helvetia Archaeologica* 20, 82–116.

WÄHREN, M. 1989b
Identifizierung von gesäuertem Brot in Knochenasche-Kristallen einer urnenfelderzeitlichen Bestattung in Bellenberg, Ldkr. Neu-Ulm. *Kataloge der Prähistorischen Staatsammlung München* 23, 59–65.

WELLMANN, I., MÁNDY, GY. – MESCH J. 1963
Száznegyven esztendős búzakalász-lelet. *Agrártörténeti Szemle* 1963/4, 1–43.

WENKE, J. R. 1999
Patterns in Prehistory. New York, Oxford.

WERNECK, H. L. 1949
Ur- und frühgeschichtliche Kultur- und Nutzpflanzen in den Ostalpen und am Ostrande des Böhmerwaldes. Wels.

WERNECK, H. L. 1954
Kulturpflanzen aus Lauriacum Lorch bei Enns. *Forschungen in Lauriacum* 2, 85–96.

WERNECK, H. L. 1955
Der Obstweihefund im Vorraum des Mithraeums zu Linz Donau, Oberösterreich. *Naturkundliches Jahrbuch der Stadt Linz* 41–54.

WERNECK, H. L. 1961
Ur- und frühgeschichtliche sowie mittelalterliche Kulturpflanzen und Hölzer aus den Ostalpen und dem südlichen Böhmerwald (Nachtrag 1949–1960). *Archaeologia Austriaca* 30, 68–117.

WERNECK, H. L. 1969
Pflanzenreste aus der Stadt auf dem Magdalenensberg bei Klagenfurt in Kärnten. Kärtner Museumsschr. 48.

WIETHOLD, J. 1995
Reis, Pfeffer und Paradieskorn: Planzenreste des 16. und 17. Jahrhunderts aus der Kloake der Patrizierfamilie von Dassel aus Lüneburg. *Archäologie und Bauforschung in Lüneburg* 1, 129–166.

WILFORD, J. N. 1992
Bibulous Days at Sumerian Base. Intern. Herald Tribune, November 6.

WILLCOX, G. 1991
Carbonised plant remains from Shortughai, Afghanistan. In: Renfrew, J. (ed.), *New light on early farming. Recent Developments in Palaeoethnobotany.* 277–279. Edinburgh.

WILLERDING, U. 1969
Ursprung und Entwicklung der Kulturpflanzen in vor- und frühgeschichtlicher Zeit. *Deutsche Agrargeschichte* 1, 188–233.

WILLERDING, U. 1970
Vor- und frühgeschichtliche Kulturpflanzenfunde in Mitteleuropa. *Neue Ausgrabungen und Forschungen in Niedersachsen* 5, 287–375.

WILLERDING, U. 1978
Paläo-ethnobotanische Befunde an mittelalterlichen Pflanzenresten an Süd-Niedersachsen, Nord-Hessen und dem östlichen Westfalen. *Deutsche Botanische Gesellschaft* 91, 129–160.

WILLERDING, U. 1979
Palaeo-ethnobotanische Untersuchungen über die Entwicklung von Pflanzengesellschaften. In: Wilmanns, O. – Tüxen, R. (eds), *Werden und Vergehen von Pflanzengesellschaften. Berichte der Internationalen Symposien der internationalen Vereinigung für Vegetationskunde.* 61–109. Vaduz.

WILLERDING, U. 1980
Anbaufrüchte der Eisenzeit und des frühen Mittelalters, ihre Anbauformen, Standortverhältnisse und Erntemethoden. In: Beck, H., Denecke, D. – Jahnkuhn, H. (eds), *Untersuchungen zur eisenzeitlichen und frühmittelalterlichen Flur in Mitteleuropa und ihrer Nutzung II.* 126–196. Göttingen.

WILLERDING, U. 1983a
Zum ältesten Ackerbau in Niedersachsen. Wegner, G.: Frühe Bauernkultur in Niedersachsen. *Archäologische Mitteilungen aus Nordwestdeutschland, Staatliches Museum für Naturkunde und Vorgeschichte Oldenburg* 1, 179–219.

WILLERDING, U. 1983b
Paläo-ethnobotanische Befunde und schriftliche sowie ikonographische Zeugnisse in Zentraleuropa. Plants and Ancient Man. Rotterdam. *Studies in palaeoethnobotany* 5, 75–88.

WILLERDING, U. 1984
Funde mittelalterlicher Pflanzenreste aus der Altstadt von Göttingen. In: Schütte, S. (ed.), *Das neue Bild des alten Göttingen. 5 Jahre Stadtarchäologie.* 57–62.

WILLERDING, U. 1986a
Landwirtschaftliche Produktionstrukturen im Mittelalter. In: Hermann, B. (ed.), *Mensch und Umwelt im Mittelalter.* 245–255. Stuttgart.

WILLERDING, U. 1986b
Paläo-ethnobotanische Befunde zum Mittelalter in Höxter/Weser. *Neue Ausgrabungen und Forschungen in Niedersachsen* 17, 319–346.

WILLERDING, U. 1986c
Zur Geschichte der Unkräuter Mitteleuropas. Neumünster.

WILLERDING, U. 1987
Die Palaeo-Ethnobotanik und ihre Entwicklung in deutschsprachigen Raum. *Berichte der Deutschen Botanischen Gesellschaft* 100, 81–105.

WILLERDING, U. 1988
Zur Entwicklung von Ackerunkrautgesellschaften im Zeitraum vom Neolithikum bis in die Neuzeit. In: Küster, H. (ed.), *Der prähistorische Mensch und seine Umwelt. Festschrift Udelgard Körber-Grohne.* 31–41. Stuttgart.

WILLIAMS, P. C. – EL-HARAMEIN, F. J. 1985
Frekeh making in Syria – a small but significant local industry. *Rachis* 4, 25–27.

WILLIAMS, J. G. K., KUBELIK, A. R., RAFALSKI, K. J. – TINGEY, S. V. 1990
DNA polymorphisms amplified by arbitrary primers are useful as genetic markers. *Nucleic Acids Research* 18, 6531–6535.

WILLIS, K. J. 1997
The impact of early agriculture upon the Hungarian landscape. In: Chapman, J. – Dolukhanov, P. (eds), *Landscapes in Flux. Central and Eastern Europe in Antiquity.* 193–207. Oxford: Colloquia Pontica 3.

WILSIE, A. C. 1969
A termesztett növények alkalmazkodása és elterjedése a Földön. Budapest.

WITTMACK, L. 1890
Samen aus den Ruinen von Hissarlik. *Zeitschrift für Ethnologie* 22, 614–620.

WITTMACK, L. 1903
Die in Pompeji gefundenen pflanzlichen Reste. *Gaertenflora* 2, 114–150.

WOSINSKY, M. 1889
A kölesdi őstelep (Tolna megyében). *Archaeologiai Értesítő* 9, 35.

WOSINSKY, M. 1896
Tolnavármegye az őskortól a honfoglalásig I. Budapest.

WRIGLEY, C. W., BATEY, I. L., CAMPBELL, W. P. – SKERRITT, J. H. 1987
Complementing traditional methods of identifying cereal varieties with novel procedures. *Seed Science and Technology* 15, 679–688.

YAKAR, J. 1991
Prehistoric Anatolia. Tel Aviv.

ZALOTAY, E. 1932
Csongrád vármegye őskori települése. *Dolgozatok* 8, 76.

ZEIST, W. VAN 1970a
The Oriental Institute excavations at Mureybit, Syria: Preliminary report on the 1965 campaign. Part III. The Paleobotany. *Journal of Near Eastern Studies* 29, 167–176.

ZEIST, W. VAN 1970b
Prehistoric and early-historic Food Plants in the Netherlands. *Palaeohistoria* 14, 14–173.

ZEIST, W. VAN 1971
Plant cultivation in the Near East: prelude and beginning. *Acta Museorum Agriculturae* 6, 1–9.

ZEIST, W. VAN 1972
Palaeobotanical results in the 1970 season at Çayönü, Turkey. *Helinium* 12, 3–19.

ZEIST, W. VAN 1975
Preliminary report on the botany of Gomolava. *Journal of Archaeological Science* 2, 315–325.

ZEIST, W. VAN 1983
Fruit in foundation deposits of two temples. *Journal of Archaeological Science* 10, 351–354.

ZEIST, W. VAN – BAKKER-HEERES, J. A.H. 1979
Some economic and ecological aspects of the plant husbandry of Tell Aswad. *Paléorient* 5, 161–169.

ZEIST, W. VAN – BAKKER-HEERES, J. A. H. 1986
Archaeobotanical studies in the Levant. 3. Late Palaeolithic Mureybit. *Palaeohistoria* 26, 171–199.

ZEIST, W. VAN – BOTTEMA, S. 1966
Palaeobotanical investigations at Ramad. *Annales Archéologiques Syriennes* 16, 179–180.

ZEIST, W. VAN – CASPARIE, W. A. 1968
Wild einkorn wheat and barley from Tell Mureybit in Northern Syria. *Acta Botanica Neerlandica* 17, 44–53.

ZEIST, W. VAN – NEEF, R. 1983
Plantenresten uit vroeg-historisch Leeuwarden. In: van Zeist, W., Neef, R., Brinkhuizen, D. C. – Jager, S. (eds), *Planten-visen vogelresten uit vroeghistorisch Leeuwarden.* Uitgegeven door de Commissie Archeologisch Stadskernonderzoek Leeuwarden 1983, 6–18.

ZEVEN, A. C., DOEKES, G. J. – KISLEV, M. 1975
Protein in Old Grains of Triticum sp. *Journal of Archaeological Science* 2, 209–213.

ZHUKOVSKY, P. M. 1964
Kulturnie rastenia i ich sorodiczy. Leningrad.

ZICHY, I. 1923
A magyarság őstörténete és műveltsége a honfoglalásig. *A Magyar Nyelvtudomány Kézikönyve* I/5. Budapest.

ZIMMERMANN, W. H. 1976
Die eisenzeitliche Ackerflur von Flögeln-Haselhörn, Kr. Wesermünde. *Probleme der Küstenforschung* 11, 79–90.

ZIMMERMANN, W. H., KÖRBER-GROHNE, U. – KROLL, H. 1984
Kulturverhältnisse. Archäologische und naturwissenschaftliche Untersuchungen an Siedlungen im deutschen Küstengebiet I. Ländliche Siedlungen. Weinheim, 246–274.

ZOHARY, D. – HOPF, M. 1988
Domestication of Plants in the Old World. Clarendon, Oxford.

ZÓLYOMI, B. 1952
Magyarország növénytakarójának fejlődéstörténete az utolsó jégkortól (Die Entwicklungsgeschichte der Vegetation Ungarns seit dem letzten Interglazial). *MTA Biológiai Tudományok Osztályának Közlem*ényei 14, 491–543.

ZÓLYOMI, B. 1953
Die Entwicklungsgeschichte der Vegetation Ungarns seit dem letzten Interglazial. *Acta Botanica Academiae Scientiarum Hungaricae* 4, 367–430.

ZÓLYOMI, B. 1971
6000jährige Geschichte der Agricultur in der Umgebung des Balaton-Sees aufgrund von pollenstatistischen Untersuchungen der Seesedimenten. IIIième Congrès International des Musées d'Agriculture. *Magyar Mezőgazdasági Múzeum Közleményei* 1971. 194–195. Budapest

ZÓLYOMI, B. 1980
Landwirtschaftliche Kultur und Wandlung der Vegetation im Holozän am Balaton. *Phytocoenologia* 7, 121–126.

ZÓLYOMI, B. – PRÉCSÉNYI, I. 1985
Pollenstatistische Analyse der Teichablagerungen des mittelalterlichen Klosters bei Pilisszentkereszt. *Acta Archaeologica Academie Scientiarum Hungaricae* 37, 153–158.

ZSEMLYEI, J. 1974
Kis-Szamos vidéke (Kolozs megye) magyar tájszólásának kölcsönszavai II. Kolozsvár. Doktori Értekezés. Kézirat.

13. List of Figures

Fig. 47. Emmer (*Triticum turgidum* subsp. *dicoccum*) naked grains from Pári-Altäcker dülő, Linearbandceramic culture, Inventory of the Hungarian Agricultural Museum, Budapest.

Fig 48. Emmer (*Triticum turgidum* subsp. *dicoccum*) naked grains from Pári-Altäcker dülő, Linearbandceramic culture, Inventory of the Hungarian Agricultural Museum, Budapest.

Fig. 49. Emmer (*Triticum turgidum* subsp. *dicoccum*) naked grains from Pári-Altäcker dülő, Linearbandceramic culture, Inventory of the Hungarian Agricultural Museum, Budapest.

Fig. 50. Wheat (*Triticum* spec.) spicelets (*furca bicornis*) from Pári-Altäcker dülő, Linearbandceramic culture, Inventory of the Hungarian Agricultural Museum, Budapest.

Fig. 51. a. Einkorn (*Triticum monococcum*), b. Emmer (*Triticum turgidum* subsp. *dicoccum*) naked grains from Dévaványa-Réhelyi gát, Tisza culture, Inventory of the Hungarian Agricultural Museum, Budapest.

Fig. 52. Barley (*Hordeum vulgare*) grains found in a storage vessel from Dévavanya-Réhelyi gát, Tisza culture, Inventory of the Hungarian Agricultural Museum, Budapest.

Fig. 53. Naked barley (*Hordeum vulgare* var. *nudum*) from Dévavanya-Réhelyi gát, Tisza culture, Inventory of the Hungarian Agricultural Museum, Budapest.

Fig. 54. Imprint of einkorn (*Triticum monococcum*) grain from Szenyér-Mesztegnyő. After FÜZES 1990.

Fig. 55. Bicske-Galagonyás. Digitized images of seed imprints found on potsherds (left) and the positive image obtained by computer (right).

Fig. 56. Common millet (*Panicum miliaceum*) naked grains from the Linearbandceramic culture settlement in Zánka. After FÜZES 1990.

Fig. 57. Club wheat (*Triticum aestivum* subsp. *compactum*) grains from the Linearbandceramic culture settlement in Zánka. After FÜZES 1990.

Fig. 58. Korn-rade (*Agrostemma githago*) seed from the Linearbandceramic culture settlement in Zánka. After FÜZES 1990.

Fig. 59. Hazelnut (*Corylus avellana*) shell from the Linearbandceramic culture settlement in Zánka. After FÜZES 1990.

Fig. 60. Einkorn (*Triticum monococcum*) grains from the Linearbandceramic culture settlement in Zánka. After FÜZES 1990.

Fig. 61. Field pea (*Pisum sativum* var. *arvense*) seeds from the Late Neolithic settlement of Berettyóújfalu-Szilhalom. After P. HARTYÁNYI – SZ. MÁTHÉ 1980.

Fig. 62. Einkorn (*Triticum monococcum*) grains from the Late Neolithic settlement of Berettyóújfalu-Szilhalom. After P. HARTYÁNYI – SZ. MÁTHÉ 1980.

Fig. 63. Crab apple (*Malus silvestris*) charcoal fruits from the Late Neolithic settlement of Berettyóújfalu-Szilhalom. After P. HARTYÁNYI – SZ. MÁTHÉ 1980.

Fig. 64. Common bread wheat (*Triticum* cf. *aestivum* subsp. *vulgare*) grains from the Late Neolithic settlement of Berettyóújfalu-Szilhalom. After P. HARTYÁNYI – SZ. MÁTHÉ 1980.

Fig. 65. Common bulrush (*Scheonoplectus lacustris*) from a pit of the Lengyel-Balaton-Lasinja culture in Zalaszentbalázs-Szőlőhegyi mező. Photograph by the author.

Fig. 66. Charlock (*Sinapis arvensis*) seeds from Öcsöd-Kendereshalom, Bodrogkeresztur culture settlement, Inventory of the Hungarian Agricultural Museum, Budapest.

Fig. 67. Cornelian cherry or dogwood (*Cornus mas*) stones from the Kompolt-Kistéri hamlet Late Bronze Age pit-oven. Photograph by the author.

Fig. 68. Woodland European grape (*Vitis vinifera* subsp. *silvestris*) seeds from the Kompolt-Kistéri hamlet Late Bronze Age pit-oven. Photograph by the author.

Fig. 69. Hazelnut (*Corylus avellana*) shell fragments from the Kompolt-Kistéri hamlet Late Bronze Age pit-oven. Photograph by the author.

Fig. 70. Oak (*Quercus* spec.) whole acorns from the Csepel Waterworks Late Copper Age Baden culture pit fireplace. Photograph by Ferenc Hegyi.

Fig. 71. Einkorn (*Triticum monococcum*) grains from the Csepel Waterworks Late Copper Age Baden culture pit fireplace. Photograph by Ferenc Hegyi.

Fig. 72. Oak (*Quercus* spec.) half acorns from the Csepel Waterworks Late Copper Age Baden culture pit fireplace. Photograph by Ferenc Hegyi.

Fig. 73. Oak (*Quercus* spec.) acorns fragments from the Csepel Waterworks late Copper Age Baden culture pit fireplace. Photograph by Ferenc Hegyi.

Fig. 74. Beech (*Fagus* spec.) acorns from the Csepel Waterworks Late Copper Age Baden culture pit fireplace. Photograph by Ferenc Hegyi.

Fig. 75. Single-seed hawthorn (*Crataegus monogyna*) stones from the Csepel Waterworks Late Copper Age Baden culture pit fireplace. Photograph by Ferenc Hegyi.

Fig. 76. Plant finds from the Bronze Age.

Fig. 77. Settlements of the Bell Beaker culture.

Fig. 78. Distribution of useful plant species in Bell Beaker settlements.

Fig. 79. Ecological distribution of species in Bell Beaker settlements.

Fig. 80. Ecological distribution of the plant remains in Bell Beaker settlements.

Fig. 81. Distribution of plant remains from Bell Beaker settlements.

Fig. 82. Common bread wheat (*Triticum aestivum* subsp. *vulgare*) grains from Ároktő-Dongóhalom, Hatvan-Füzesabony culture, Inventory of the Hungarian Agricultural Museum, Budapest.

Fig. 83. Creeping thistle (*Cirsium arvense*) acheniums from Ároktő-Dongóhalom, Hatvan-Füzesabony culture, Inventory of the Hungarian Agricultural Museum, Budapest.

Fig. 84. Field bindweed (*Convolvulus arvensis*) seeds from Ároktő-Dongóhalom, Hatvan-Füzesabony culture, Inventory of the Hungarian Agricultural Museum, Budapest.

Fig. 85. Hedge buckwheat (*Fallopia dumetorum*) nutlets from Ároktő-Dongóhalom, Hatvan-Füzesabony culture, Inventory of the Hungarian Agricultural Museum, Budapest.

Fig. 86. Cultivated plants of Bronze Age tell settlements.

Fig. 87. Distribution of wheat species of Bronze Age sites (updated map after Nováki 1969).

Fig. 88. Danewort (*Sambucus ebulus*) seed from Ároktő-Dongóhalom, Hatvan-Füzesabony culture, Inventory of the Hungarian Agricultural Museum, Budapest.

Fig. 89. Vetch (*Vicia* spec.) seed from Ároktő-Dongóhalom, Hatvan-Füzesabony culture, Inventory of the Hungarian Agricultural Museum, Budapest.

Fig. 90. Six rowed barley (*Hordeum vulgare* subsp. *hexastichum*) hulled grains from Alpár-Várdomb, Vatya culture, Inventory of the Hungarian Agricultural Museum, Budapest.

Fig. 91. Barley (*Hordeum vulgare*) ear spindles from Alpár-Várdomb, Vatya culture, Inventory of the Hungarian Agricultural Museum, Budapest.

Fig. 92. Barley (*Hordeum vulgare*) ear spindles fragments from Alpár-Várdomb, Vatya culture, Inventory of the Hungarian Agricultural Museum, Budapest.

Fig. 93. Four-rowed barley (*Hordeum tetrastichum*) hulled grains from Alpár-Várdomb, Vatya culture, Inventory of the Hungarian Agricultural Museum, Budapest.

Fig. 94. Many rowed barley (*Hordeum polystichum*) hulled grains from Alpár-Várdomb, Vatya culture, Inventory of the Hungarian Agricultural Museum, Budapest.

Fig. 95. Many rowed barley (*Hordeum polystichum*) hulled grains from Alpár-Várdomb, Vatya culture, Inventory of the Hungarian Agricultural Museum, Budapest.

Fig. 96. Naked barley (*Hordeum vulgare* var. *nudum*) grains from Alpár-Várdomb, Vatya culture, Inventory of the Hungarian Agricultural Museum, Budapest.

Fig. 97. Barley (*Hordeum vulgare*) spiculum fragments from Alpár-Várdomb, Vatya culture, Inventory of the Hungarian Agricultural Museum, Budapest.

Fig. 98. Einkorn (*Triticum monococcum*) naked grains from Alpár-Várdomb, Vatya culture, Inventory of the Hungarian Agricultural Museum, Budapest.

Fig. 99. Einkorn (*Triticum monococcum*) naked grains from Alpár-Várdomb, Vatya culture, Inventory of the Hungarian Agricultural Museum, Budapest.

Fig. 100. Emmer (*Triticum turgidum* subsp. *dicoccum*) naked grains from Alpár-Várdomb, Vatya culture, Inventory of the Hungarian Agricultural Museum, Budapest.

Fig. 101. Emmer (*Triticum turgidum* subsp. *dicoccum*) naked grains from Alpár-Várdomb, Vatya culture, Inventory of the Hungarian Agricultural Museum, Budapest.

Fig. 102. Emmer (*Triticum turgidum* subsp. *dicoccum*) naked grains from Alpár-Várdomb, Vatya culture, Inventory of the Hungarian Agricultural Museum, Budapest.

Fig. 103. Emmer (T*riticum turgidum* subsp. *dicoccum*) longshaped grains from Alpár-Várdomb, Vatya culture, Inventory of the Hungarian Agricultural Museum, Budapest.

Fig. 104. Emmer (*Triticum turgidum* subsp. *dicoccum*) short and thick-set grains from Alpár-Várdomb, Vatya culture, Inventory of the Hungarian Agricultural Museum, Budapest.

Fig. 105. Common bread wheat (*Triticum aestivum* subsp. *vulgare*) grains from Alpár-Várdomb, Vatya culture, Inventory of the Hungarian Agricultural Museum, Budapest.

Fig. 106. Club wheat (*Triticum aestivum* subsp. *compactum*) earspindles from Alpár-Várdomb, Vatya culture, Inventory of the Hungarian Agricultural Museum, Budapest.

Fig. 107. Club wheat (*Triticum aestivum* subsp. *compactum*) earspindles from Alpár-Várdomb, Vatya culture, Inventory of the Hungarian Agricultural Museum, Budapest.

Fig. 108. Grass pea wine (*Lathyrus sativus*) seeds from Alpár-Várdomb, Vatya culture, Inventory of the Hungarian Agricultural Museum, Budapest.

Fig. 109. Small seed lentil (*Lens culinaris* subsp. *microsperma*) from Alpár-Várdomb, Vatya culture, Inventory of the Hungarian Agricultural Museum, Budapest.

Fig. 110. Field pea (*Pisum sativum* var. *arvense*) seeds from Alpár-Várdomb, Vatya culture, Inventory of the Hungarian Agricultural Museum, Budapest.

Fig. 111. Flax (*Linum usitatissimum*) seeds from Alpár-Várdomb, Vatya culture, Inventory of the Hungarian Agricultural Museum, Budapest.

Fig. 112. Wild strawberry (*Fragaria vesca*) seeds from Alpár-Várdomb, Vatya culture, Inventory of the Hungarian Agricultural Museum, Budapest.

Fig. 113. Crab apple (*Malus silvestris*) seeds from Alpár-Várdomb, Vatya culture, Inventory of the Hungarian Agricultural Museum, Budapest.

Fig. 114. Brome grass (*Bromus arvensis*) naked grains from Alpár-Várdomb, Vatya culture, Inventory of the Hungarian Agricultural Museum, Budapest.

Fig. 115. White goosefoot or fan-hen (*Chenopodium album*) seeds from Alpár-Várdomb, Vatya culture, Inventory of the Hungarian Agricultural Museum, Budapest.

Fig. 116. Field gromwell (*Lithospermum arvense*) nutlets from Alpár-Várdomb, Vatya culture, Inventory of the Hungarian Agricultural Museum, Budapest.

Fig. 117. Blackberry or bramble (*Rubus fruticosus*) seeds from Alpár-Várdomb, Vatya culture, Inventory of the Hungarian Agricultural Museum, Budapest.

Fig. 118. Raspberry (*Rubus idaeus*) from Alpár-Várdomb, Vatya culture, Inventory of the Hungarian Agricultural Museum, Budapest.

Fig. 119. Common elder (*Sambucus nigra*) seed from Alpár-Várdomb, Vatya culture, Inventory of the Hungarian Agricultural Museum, Budapest.

Fig. 120. Woody nightshade (*Solanum dulcamara*) from Alpár-Várdomb, Vatya culture, Inventory of the Hungarian Agricultural Museum, Budapest.

Fig. 121. Einkorn (*Triticum monococcum*) naked grains from Balatonboglár, Middle Bronze Age, Inventory of the Hungarian Agricultural Museum, Budapest.

Fig. 122. Emmer (*Triticum turgidum* subsp. *dicoccum*) naked grains from Balatonboglár, Middle Bronze Age, Inventory of the Hungarian Agricultural Museum, Budapest.

Fig. 123. Barley (*Hordeum vulgare*) grain from a house in Baracs, Nagyrév culture, Inventory of the Hungarian Agricultural Museum, Budapest.

Fig. 124. Clover (*Trifolium* spec.) seed from a house in Baracs, Nagyrév culture, Inventory of the Hungarian Agricultural Museum, Budapest.

Fig. 125. Korn-rade (*Agrostemma githago*) seeds from a house in Baracs, Nagyrév culture, Inventory of the Hungarian Agricultural Museum, Budapest.

Fig. 126. Barren brome (*Bromus sterilis*) hulled grains from a house of Baracs, Nagyrév culture, Inventory of the Hungarian Agricultural Museum, Budapest.

Fig. 127. Hare's-ear mustard (*Conringia orientalis*) seed from Baracs-Bottyán sánc, Nagyrév culture, Inventory of the Hungarian Agricultural Museum, Budapest.

Fig. 128. False cleavers (*Galium spurium*) achenes from Baracs-Bottyán sánc, Nagyrév culture, Inventory of the Hungarian Agricultural Museum, Budapest.

Fig. 129. Redshank (*Polygonum persicaria*) nutlets from Baracs-Bottyán sánc, Nagyrév culture, Inventory of the Hungarian Agricultural Museum, Budapest.

Fig. 130. Einkorn (*Triticum monococcum*) naked grains from the Bronze Age settlement Dunaújváros (Dunapentele), Inventory of the Hungarian Agricultural Museum, Budapest.

Fig. 131. Einkorn (*Triticum monococcum*) naked grains from the Bronze Age settlement Dunaújváros (Dunapentele), Inventory of the Hungarian Agricultural Museum, Budapest.

Fig. 132. Einkorn (*Triticum monococcum*) naked grains from the Bronze Age settlement Felsődobsza, Inventory of the Hungarian Agricultural Museum, Budapest.

Fig. 133. Einkorn (*Triticum monococcum*) naked grain from the Bronze Age settlement Felsődobsza, Inventory of the Hungarian Agricultural Museum, Budapest.

Fig. 134. Einkorn (*Triticum monococcum*) naked grain from the Bronze Age settlement Felsődobsza, Inventory of the Hungarian Agricultural Museum, Budapest.

Fig. 135. Einkorn (*Triticum monococcum*) naked grains from the Middle Bronze Age settlement Jászdózsa-Kápolnahalom, Inventory of the Hungarian Agricultural Museum, Budapest.

Fig. 136. Einkorn (*Triticum monococcum*) naked grains from the Middle Bronze Age settlement Jászdózsa-Kápolnahalom, Inventory of the Hungarian Agricultural Museum, Budapest.

Fig. 137. Emmer (*Triticum turgidum* subsp. *dicoccum*) naked grains from the Middle Bronze Age settlement Jászdózsa-Kápolnahalom, Inventory of the Hungarian Agricultural Museum, Budapest.

Fig. 138. Einkorn (*Triticum monococcum*) naked grains from the Bronze Age settlement Mende-Leányvár, Inventory of the Hungarian Agricultural Museum, Budapest.

Fig. 139. Einkorn (*Triticum monococcum*) spikelets from the Middle Bronze Age tell settlement Százhalombatta-Téglagyár, Vatya culture, Inventory of the Hungarian Agricultural Museum, Budapest.

Fig. 140. Emmer (*Triticum turgidum* subsp. *dicoccum*) naked grains from the Bronze Age settlement Mende-Leányvár, Inventory of the Hungarian Agricultural Museum, Budapest.

Fig. 141. Einkorn (*Triticum monococcum*) naked grain from the Middle Bronze Age tell settlement Százhalombatta-Téglagyár, Vatya culture, Inventory of the Hungarian Agricultural Museum, Budapest.

Fig. 142. Emmer (*Triticum turgidum* subsp. *dicoccum*) naked grains from the Middle Bronze Age tell settlement Százhalombatta-Téglagyár, Vatya culture, Inventory of the Hungarian Agricultural Museum, Budapest.

Fig. 143. Brome grass (*Bromus arvensis*) naked grains from the Middle Bronze Age tell settlement Százhalombatta-Téglagyár, Vatya culture, Inventory of the Hungarian Agricultural Museum, Budapest.

Fig. 144. Einkorn (*Triticum monococcum*) naked grains from the Bölcske-Vörösgyír Middle Bronze Age tell settlement. Photograph by Tibor Kádas.

Fig. 145. Einkorn (*Triticum monococcum*) naked grains from the Bölcske-Vörösgyír Middle Bronze Age tell settlement. Photograph by Tibor Kádas.

Fig. 146. Emmer (*Triticum turgidum* subsp. *dicoccum*) naked grains from the Bölcske-Vörösgyír Middle Bronze Age tell settlement. Photograph by Tibor Kádas.

Fig. 147. Emmer (*Triticum turgidum* subsp. *dicoccum*) naked grains from the Bölcske-Vörösgyír Middle Bronze Age tell settlement. Photograph by Tibor Kádas.

Fig. 148. Club wheat (*Triticum aestivum* subsp. *compactum*) grain from the Bölcske-Vörösgyír Middle Bronze Age tell settlement. Photograph by Tibor Kádas.

Fig. 149. Barley (*Hordeum vulgare*) grains from the Bölcske-Vörösgyír Middle Bronze Age tell settlement (Photograph by Tibor Kádas)

Fig. 150. Common millet (*Panicum miliaceum*) naked grains from the Bölcske-Vörösgyír Middle Bronze Age tell settlement. Photograph by Tibor Kádas.

Fig. 151. Field bindweed (*Convulvulus arvensis*) seeds from the Bölcske-Vörösgyír Middle Bronze Age tell settlement. Photograph by Tibor Kádas.

Fig. 152. Bitter vetch (*Vicia ervilia*) seeds from the Bölcske-Vörösgyír middle Bronze Age tell settlement. Photograph by Tibor Kádas.

Fig. 153. Sloe or blackthorn (*Prunus spinosa*) stones from a house in the Bölcske-Vörösgyír middle Bronze Age tell. Photograph by the author.

Fig. 154. Horsebean (*Vicia faba*) seeds from the Bölcske-Vörösgyír Middle Bronze Age tell settlement. Photograph by Tibor Kádas.

Fig. 155. Danewort (*Sambucus ebulus*) seeds from a house in the Bölcske-Vörösgyír Middle Bronze Age tell. Photograph by the author.

Fig. 156. Common or red elder (*Sambucus nigra* or *racemosa*) seeds from the Bölcske-Vörösgyír Middle Bronze Age tell settlement. Photograph by Tibor Kádas.

Fig. 157. Rye brome (*Bromus secalinus*) grains from a house in the Bölcske-Vörösgyír Middle Bronze Age tell. Photograph by the author.

Fig. 158. Emmer (*Triticum turgidum* subsp. *dicoccum*) naked grains from a house in the Túrkeve-Terehalom Middle Bronze Age tell. Photograph by Tibor Kádas.

Fig. 159. Common bread wheat (*Triticum aestivum* subsp. *vulgare*) grains from a house in the Túrkeve-Terehalom Middle Bronze Age tell. Photograph by Tibor Kádas.

Fig. 160. Einkorn (*Triticum monococcum*) spikelets from a house in the Túrkeve-Terehalom Middle Bronze Age tell. Photograph by Tibor Kádas.

Fig. 161 Emmer (*Triticum turgidum* subsp. *dicoccum*) spicelet (furca bicornis) from a house in the Túrkeve-Terehalom Middle Bronze Age tell. Photograph by Tibor Kádas.

Fig. 162. Barley (*Hordeum vulgare*) grains from a house in the Túrkeve-Terehalom Middle Bronze Age tell. Photograph by Tibor Kádas.

Fig. 163. Safflower (*Carthamus tinctorius*) acheniums from a house in the Túrkeve-Terehalom Middle Bronze Age tell. Photograph by Tibor Kádas.

Fig. 164. Knotgrass (*Polygonum aviculare*) nutlets from a house in the Túrkeve-Terehalom Middle Bronze Age tell. Photograph by Tibor Kádas.

Fig. 165. Danewort (*Sambucus ebulus*) seed fragment from a house in the Túrkeve-Terehalom Middle Bronze Age tell. Photograph by Tibor Kádas.

Fig. 166. Small seed lentil (*Lens culinaris* subsp. *microsperma*) seeds from a house in the Túrkeve-Terehalom Middle Bronze Age tell. Photograph by Tibor Kádas.

Fig. 167. Pea (*Pisum sativum*) seeds from a house in the Túrkeve-Terehalom Middle Bronze Age tell. Photograph by Tibor Kádas.

Fig. 168. Bitter vetch (*Vicia ervilia*) seeds from a house in the Túrkeve-Terehalom Middle Bronze Age tell. Photograph by Tibor Kádas.

Fig. 169. Narrow-leaved vetch (*Vicia angustifolia*) seed from a house in the Túrkeve-Terehalom middle Bronze Age tell. Photograph by Tibor Kádas.

Fig. 170. Rye brome (*Bromus secalinus*) grain fragment from a house in the Túrkeve-Terehalom Middle Bronze Age tell. Photograph by Tibor Kádas.

Fig. 171. White mustard (*Sinapis alba*) seed from the Túrkeve-Terehalom Middle Bronze Age tell settlement. Photograph by Tibor Kádas.

Fig. 172. Spring wild-oat (*Avena fatua*) naked grains from the Túrkeve-Terehalom Middle Bronze Age tell settlement. Photograph by Tibor Kádas.

Fig. 173. Common orache (*Atriplex patula*) from a house in the Túrkeve-Terehalom Middle Bronze Age tell. Photograph by Tibor Kádas.

Fig. 174. White goosefoot or fan-hen (*Chenopodium album*) seeds from a house in the Túrkeve-Terehalom Middle Bronze Age tell. Photograph by Tibor Kádas.

Fig. 175. False cleavers (*Galium spurium*) achenes from a house in the Túrkeve-Terehalom Middle Bronze Age tell. Photograph by Tibor Kádas.

Fig. 176. Brome (*Bromus* spec.) grain fragment from a house in the Túrkeve-Terehalom Middle Bronze Age tell. Photograph by Tibor Kádas.

Fig. 177. Curled dock (*Rumex crispus*) glans from a house in the Túrkeve-Terehalom Middle Bronze Age tell. Photograph by Tibor Kádas.

Fig. 178. Common sorrel (*Rumex acetosa*) glans from a house in the Túrkeve-Terehalom Middle Bronze Age tell. Photograph by Tibor Kádas.

Fig. 179. Meadow-grass (*Poa* spec.) naked grain from the Túrkeve-Terehalom Middle Bronze Age tell settlement. Photograph by Tibor Kádas.

Fig. 180. Gipsywort (*Lycopus europaeus*) seeds from the Túrkeve-Terehalom Middle Bronze Age tell settlement. Photograph by Tibor Kádas.

Fig. 181. Field pansy (*Viola arvensis*) seed from the Túrkeve-Terehalom Middle Bronze Age tell settlement. Photograph by Tibor Kádas.

Fig. 182. Black-bindweed (*Fallopia convolvulus*) nutlets from the Túrkeve-Terehalom Middle Bronze Age tell settlement. Photograph by Tibor Kádas.

Fig. 183. Knotgrass (*Polygonum aviculare*) nutlet from the Túrkeve-Terehalom Middle Bronze Age tell settlement. Photograph by Tibor Kádas.

Fig. 184. Barley (*Hordeum vulgare*) grains hulled grains from a pit of Tiszaeszlár-Bashalom, Inventory of the Hungarian Agricultural Museum, Budapest.

Fig. 185. Two-rowed barley (*Hordeum vulgare* subsp. *distichum*) hulled grains from a pit of Tiszaeszlár-Bashalom, Inventory of the Hungarian Agricultural Museum, Budapest.

Fig. 186. Black-bindweed (*Fallopia convolvulus*) nutlets from a pit of Tiszaeszlár-Bashalom, Inventory of the Hungarian Agricultural Museum, Budapest.

Fig. 187. Spring wild-oat (*Avena fatua*) naked grains from a pit of Tiszaeszlár-Bashalom, Inventory of the Hungarian Agricultural Museum, Budapest.

Fig. 188. Spring wild-oat (*Avena fatua*) naked grain from a pit of Tiszaeszlár-Bashalom, Inventory of the Hungarian Agricultural Museum, Budapest.

Fig. 189. Two-rowed barley (*Hordeum vulgare* subsp. *distichum*) hulled grains from Poroszló-Aponhát, Inventory of the Hungarian Agricultural Museum, Budapest.

Fig. 190. Emmer (*Triticum turgidum* subsp. *dicoccum*) naked grains from Poroszló-Aponhát, Inventory of the Hungarian Agricultural Museum, Budapest.

Fig. 191. Spelta wheat (*Triticum aestivum* subsp. *spelta*) grains from Poroszló-Aponhát, Inventory of the Hungarian Agricultural Museum, Budapest.

Fig. 192. Spelta wheat (T*riticum aestivum* subsp. *spelta*) spicelets (*furca bicornis*) from Poroszló-Aponhát, Inventory of the Hungarian Agricultural Museum, Budapest.

Fig. 193. Common bread wheat (*Triticum aestivum* subsp. *vulgare*) grains from Poroszló-Aponhát, Inventory of the Hungarian Agricultural Museum, Budapest.

Fig. 194. Common millet (*Panicum miliaceum*) naked grains from Poroszló-Aponhát, Inventory of the Hungarian Agricultural Museum, Budapest.

Fig. 195. Carbonised seeds and fruits among inorganic residues (for the most part leguminous) waiting for selection. Floated sample from the Late Bronze Age pit in Balatonmagyaród-Hídvégpuszta. Photograph by the author.

Fig. 196. Small seed lentil (*Lens culinaris* subsp. *microsperma*) seeds from the waste pit of the Late Bronze Age settlement of Balatonmagyaród-Hídvégpuszta. Photograph by Tibor Kádas.

Fig. 197. Barley (*Hordeum vulgare*) grains hulled grain from the Late Bronze Age settlement of Balatonmagyaród-Hídvégpuszta. Photograph by Tibor Kádas.

Fig. 198. Bitter vetch (*Vicia ervilia*) seeds from a waste pit of the Late Bronze Age settlement of Balatonmagyaród-Hídvégpuszta. Photograph by Tibor Kádas.

Fig. 199. Pea (*Pisum sativum*) from a waste pit of the late Bronze Age settlement of Balatonmagyaród-Hídvégpuszta. Photograph by Tibor Kádas.

Fig. 200. Grass pea vine (*Lathyrus sativus*) seeds from a waste pit of the late Bronze Age settlement of Balatonmagyaród-Hídvégpuszta. Photograph by Tibor Kádas.

Fig. 201. Spring wild-oat (*Avena fatua*) naked grain from a waste pit of the Late Bronze Age settlement of Balatonmagyaród-Hídvégpuszta. Photograph by Tibor Kádas.

Fig. 202. Gruel fragment from a waste pit of the Late Bronze Age settlement of Balatonmagyaród-Hídvégpuszta. Photograph by Tibor Kádas.

Fig. 203. Common millet (*Panicum miliaceum*) naked grains from a waste pit of the Late Bronze Age settlement of Balatonmagyaród-Hídvégpuszta. Photograph by Tibor Kádas.

Fig. 204. Grape-vine (*Vitis vinifera* subsp. *vinifera*) seeds from La Tène settlement of Sopron-Krautäcker. After JEREM *et al.* 1985.

Fig. 205. Barley (*Hordeum vulgare*) grains from the Scythian settlement of Rákoskeresztúr-Újmajor. Photograph by the author.

Fig. 206. Common millet (*Panicum miliaceum*) naked grains from the Scythian settlement of Rákoskeresztúr-Újmajor. Photograph by the author.

Fig. 207. Late Bronze Age, Early Iron Age and Celtic botanical finds in Transdanubia.

Fig. 208. Barley (*Hordeum vulgare*) grains from Aggtelek cave (Szultán pamlagja), Hallstatt culture, Inventory of the Hungarian Agricultural Museum, Budapest.

Fig. 209. Grape-vine (*Vitis vinifera* subsp. *vinifera*) seeds from the late Urnfield settlement of Sopron-Krautäcker. After JEREM *et al.* 1984.

Fig. 210. Emmer (*Triticum turgidum* subsp. *dicoccum*) naked grains from Aggtelek cave, Early Iron Age, Inventory of the Hungarian Agricultural Museum, Budapest.

Fig. 211. Grape-vine (*Vitis vinifera* subsp. *vinifera*) seeds from the late Hallstatt settlement of Sopron-Krautäcker. After JEREM *et al.* 1985.

Fig. 212. Mahaleb cherry (*Prunus mahaleb*), plum (*Prunus* spec.) and cherry (*Prunus avium*) stones from a Roman house in Tác, Inventory of the Hungarian Agricultural Museum, Budapest.

Fig. 213. Sorvice tree (*Sorbus domestica*) seeds from a Roman house in Tác, Inventory of the Hungarian Agricultural Museum, Budapest.

Fig. 214. Two-rowed barley (*Hordeum vulgare* subsp. *distichum*) hulled grains from Sopron, Városház Str., Inventory of the Hungarian Agricultural Museum, Budapest.

Fig. 215. Wild sorvice (*Sorbus torminalis*) seeds, common bread wheat (*Triticum aestivum* subsp. *vulgare*) grains and grape-vine (*Vitis vinifera* subsp. *vinifera*) seeds from a Roman house in Tác, Inventory of the Hungarian Agricultural Museum, Budapest.

Fig. 216. Field cowwheat (*Melampyrum arvense*) seeds from Sopron, Városház Str., Inventory of the Hungarian Agricultural Museum, Budapest.

Fig. 217. Spring wild-oat (*Avena fatua*) naked grain from Sopron, Városház Str., Inventory of the Hungarian Agricultural Museum, Budapest.

Fig. 218. Two-rowed barley (*Hordeum vulgare* subsp. *distichum*) hulled grains from Sopron, Beloiannisz sq., Inventory of the Hungarian Agricultural Museum, Budapest.

Fig. 219. Charcoal cereal spikes from a Roman watchtower, Budakalász-Őrtorony, Lupa csárda, Inventory of the Hungarian Agricultural Museum, Budapest.

Fig. 220. Common millet (*Panicum miliaceum*) naked grains found on the peel of an oven in a Late Roman building, Budapest, Körte utca 29., Inventory of the Hungarian Agricultural Museum, Budapest.

Fig. 221. Cereal remains from a Roman watchtower, Budakalász-Őrtorony, Lupa csárda, Inventory of the Hungarian Agricultural Museum, Budapest.

Fig. 222. Barley (*Hordeum vulgare*) grain from a Roman watchtower, Budakalász-Őrtorony, Lupa csárda, Inventory of the Hungarian Agricultural Museum, Budapest.

Fig. 223. Barley (*Hordeum vulgare*) grain from a Roman watchtower, Budakalász-Őrtorony, Lupa csárda, Inventory of the Hungarian Agricultural Museum, Budapest.

Fig. 224. Late Roman common elder (*Sambucus nigra*) seeds from Budapest, Vörösvári utca 20., Inventory of the Hungarian Agricultural Museum, Budapest.

Fig. 225. Late Roman grape (*Vitis vinifera*) seeds from Budapest, Vörösvári utca 20., Inventory of the Hungarian Agricultural Museum, Budapest.

Fig. 226. Leafy spurge (*Euphorbia esula*) seed from a brickgrave found in a 4th-century cemetery in Intercisa-Papsziget, Inventory of the Hungarian Agricultural Museum, Budapest

Fig. 227. Field bindweed (*Convolvulus arvensis*) seeds from a brickgrave found in a 4th-century cemetery in Intercisa-Papsziget, Inventory of the Hungarian Agricultural Museum, Budapest.

Fig. 228. Common bread wheat (*Triticum aestivum* subsp. *vulgare*) grains from the Roman settlement in Tác, Inventory of the Hungarian Agricultural Museum, Budapest.

Fig. 229. Crab apple (*Malus silvestris*) seeds from a Roman house in Tác, Inventory of the Hungarian Agricultural Museum, Budapest.

Fig. 230. Einkorn (*Triticum monococcum*) grains from a building in the Roman settlement at Tác, Inventory of the Hungarian Agricultural Museum, Budapest.

Fig. 231. Roman castrum Fenékpuszta, reconstruction of Tivadar Vida with the location of archaeobotanical samples.

Fig. 232. Collected wheat grains in a matchbox – handwriting of Árpád Csák, 1905. Photograph by Gábor Gyulai.

Fig. 233. Rye (*Secale cereale*) grains from the Roman fortress at Fenékpuszta. Photograph by Miklós Füzes.

Fig. 256. Black-bindweed (*Fallopia convolvulus*) nutlets from a grave in Szőreg, Avar Period, Inventory of the Hungarian Agricultural Museum, Budapest.

Fig. 257. Two-rowed barley (*Hordeum vulgare* subsp. *distichum*), hulled grains with scars from burning from Fonyód-Bélatelep Late Migration Period lake dwelling. Photograph by the author.

Fig. 258. Wattle-and-daub wall, lake dwelling settlement at Fonyód-Bélatelep Late Migration Period lake dwelling. Photograph at the excavation.

Fig. 259. Six rowed barley (*Hordeum vulgare* subsp. *hexastichum*) hulled grains with straight axis from Fonyód-Bélatelep Late Migration Period lake dwelling. Photograph by the author.

Fig. 260. Small seed lentil (*Lens culinaris* subsp. *microsperma*) seed from a Fonyód-Bélatelep Late Migration Period lake dwelling. Photograph by the author.

Fig. 261. Einkorn (*Triticum monococcum*) grain from a Fonyód-Bélatelep Late Migration Period settlement. Photograph by the author.

Fig. 262. Common oat (*Avena sativa*) naked grain from the Fonyód-Bélatelep Late Migration Period site. Photograph by the author.

Fig. 263. Two-rowed naked barley (*Hordeum vulgare* subsp. *distichum* var. *nudum*) grains in dorsal and ventral view from a Fonyód-Bélatelep Late Migration Period lake dwelling. Photograph by the author.

Fig. 264. Six rowed barley (*Hordeum vulgare* subsp. *hexastichum*) hulled grains from the side with oblique axis from a Fonyód-Bélatelep Late Migration Period lake dwelling. Photograph by the author.

Fig. 265. Rye (*Secale cereale*) and barley (*Hordeum vulgare*) burnt grains from a Fonyód-Bélatelep Late Migration Period settlement. Photograph by the author.

Fig. 266. Common oat (*Avena sativa*) hulled grains from Fonyód-Bélatelep Late Migration Period site. Photograph by the author.

Fig. 267. Common millet (*Panicum miliaceum*) hulled grains (*cum caryopse corticata*) from a Fonyód-Bélatelep Late Migration Period lake dwelling. Photograph by the author.

Fig. 268. Common millet (*Panicum miliaceum*) naked (husked) grains (*caryopsis nuda*) with and without the germ disk from a Fonyód-Bélatelep Late Migration Period lake dwelling. Photograph by the author.

Fig. 269. Rye brome (*Bromus secalinus*) grain from the Fonyód-Bélatelep Late Migration Period settlement. Photograph by the author.

Fig. 270. Pea (*Pisum sativum*) seeds from a Fonyód-Bélatelep Late Migration Period lake dwelling. Photograph by the author.

Fig. 271. Korn-rade (*Agrostemma githago*) seeds from a Fonyód Bélatelep Late Migration Period lake dwelling. Photograph by the author.

Fig. 272. Korn-rade (*Agrostemma githago*) seed burned together with barley (*Hordeum vulgare*), from the Fonyód-Bélatelep Late Migration Period settlement. Photograph by the author.

Fig. 273. Spring wild-oat (*Avena fatua*) hulled grains from e Fonyód-Bélatelep Late Migration Period settlement. Photograph by the author.

Fig. 274. Field poppy or corn poppy (*Papaver rhoeas*) seed from the Fonyód-Bélatelep Late Migration Period settlement. Photograph by the author..

Fig. 275. Field cowwheat (*Melampyrum arvense*) seed from the Fonyód-Bélatelep Late Migration Period settlement. Photograph by the author.

Fig. 276. Wild rye (*Secale silvestre*) naked grain from the Fonyód-Bélatelep Late Migration Period settlement. Photograph by the author.

Fig. 277. Redshank (*Polygonum persicaria*) nutlets from the Fonyód-Bélatelep Late Migration Period settlement. Photograph by the author.

Fig. 278. Wild radish (*Raphanus raphanistrum*) seeds from the Fonyód-Bélatelep Late Migration Period settlement. Photograph by the author.

Fig. 279. Hemp (*Cannabis sativa*) achenes from the Fonyód-Bélatelep Late Migration Period settlement. Photograph by the aute hor.

Fig. 280. Hairy vetch (*Vicia villosa*) seed from the Fonyód-Bélatelep Late Migration Period settlement. Photograph by the author.

Fig. 281. Common elder (*Sambucus nigra*) seeds from the Fonyód-Bélatelep Late Migration Period settlement. Photograph by the author.

Fig. 282. White goosefoot or fan-hen (*Chenopodium album*) seeds from the Fonyód-Bélatelep Late Migration Period settlement. Photograph by the author.

Fig. 283. Sour cherry (*Prunus cerasus*) stone from the Fonyód-Bélatelep Late Migration Period settlement. Photograph by the author.

Fig. 284. Cherry (*Prunus avium*) stone from the Fonyód-Bélatelep Late Migration Period settlement. Photograph by the author.

Fig. 285. Crab apple (*Malus silvestris*) mummified fruit from the Fonyód-Bélatelep Late Migration Period settlement. Photograph by the author.

Fig. 286. Danewort (*Sambucus ebulus*) seeds from the Fonyód-Bélatelep Late Migration Period settlement. Photograph by the author.

Fig. 287. Mahaleb cherry (*Prunus mahaleb*) half stone from the Fonyód-Bélatelep Late Migration Period settlement. Photograph by the author.

Fig. 288. *Prunus* spec. stone fragment from the Fonyód-Bélatelep Late Migration Period settlement. Photograph by the author.

Fig. 289. *Prunus* spec. stone from the Fonyód-Bélatelep Late Migration Period settlement. Photograph by the author.

Fig. 290. Single-seed hawthorn (*Crataegus monogyna*) stone from the Fonyód-Bélatelep Late Migration Period settlement. Photograph by the author.

Fig. 291. European bird cherry (*Prunus padus*) stone from the Fonyód-Bélatelep Late Migration Period settlement. Photograph by the author.

Fig. 292. Hungarian(?) pear (*Pyrus* cf. *magyarica*) mummified fruit from the Fonyód-Bélatelep Late Migration Period settlement. Photograph by the author.

Fig. 293. Service tree (*Sorbus domestica*) seed from the Fonyód-Bélatelep Late Migration Period settlement. Photograph by the author.

Fig. 294. Apricot (*Prunus armeniaca*) fruit mummy with Monilia damage from the Fonyód-Bélatelep Late Migration Period settlement. Front view. Photograph by the author.

Fig. 295. Pear (*Pyrus* spec.) fruit-stalk (*pomum*) from the Fonyód-Bélatelep Late Migration Period settlement. Photograph by the author.

Fig. 319. Common bread wheat (*Triticum aestivum* subsp. *vulgare*) grains from the *ispán*'s castle at Hont, Early Árpádian Period, Inventory of the Hungarian Agricultural Museum, Budapest.

Fig. 320. Rye (*Secale cereale*) grains from the *ispán*'s castle at Hont, Early Árpádian Period, Inventory of the Hungarian Agricultural Museum, Budapest.

Fig. 321. Common bread wheat (*Triticum aestivum* subsp. *vulgare*) grains from Cegléd-Madarászhalom, Late Árpádian Period, Inventory of the Hungarian Agricultural Museum, Budapest.

Fig. 322. Spring wild-oat (*Avena fatua*) grains from the ispán's castle at Hont, Early Árpádian Period, Inventory of the Hungarian Agricultural Museum, Budapest.

Fig. 323. Knotgrass (*Polygonum aviculare*) nutlets from the floor of a house in Tiszaörvény (11th–13th centuries), Inventory of the Hungarian Agricultural Museum, Budapest.

Fig. 324. Flower-of-an-hour (*Hibiscus trionum*) seed from the floor of a house in Tiszaörvény (11th–13th centuries), Inventory of the Hungarian Agricultural Museum, Budapest.

Fig. 325. Redshank (*Polygonum persicaria*) nutlets from the floor of a house in Tiszaörvény (11th–13th centuries), Inventory of the Hungarian Agricultural Museum, Budapest.

Fig. 326. White goosefoot or fan-hen (*Chenopodium album*) seeds from the floor of a house in Tiszaörvény (11th–13th centuries), Inventory of the Hungarian Agricultural Museum, Budapest.

Fig. 327. Garlic (*Allium sativum*) shoot bulb clove from Nagyvázsony-Csepely (15th–16th centuries), Inventory of the Hungarian Agricultural Museum, Budapest.

Fig. 328. Korn-rade (*Agrostemma githago*) seeds from Nagyvázsony-Csepely (15th–16th centuries), Inventory of the Hungarian Agricultural Museum, Budapest.

Fig. 329. Almond (*Prunus amygdalus*) stone fragments from well No. 1., Budapest I., Dísz square 10. (Buda castle) (13th–14th centuries), Inventory of the Hungarian Agricultural Museum, Budapest.

Fig. 330. Sweet cherry (*Prunus avium*) stones from well No. 1., Budapest I., Dísz square 10. (Buda castle) (13th–14th centuries), Inventory of the Hungarian Agricultural Museum, Budapest.

Fig. 331. Sour cherry (*Prunus cerasus*) stones from well No. 1., Budapest I., Dísz square 10. (Buda castle) (13th–14th centuries), Inventory of the Hungarian Agricultural Museum, Budapest.

Fig. 332. Watermelon (*Citrullus lanatus*) seeds from well No. 1., Budapest I., Dísz square 10. (Buda castle) (13th–14th centuries), Inventory of the Hungarian Agricultural Museum, Budapest.

Fig. 333. Hazelnut (*Corylus avellana*) shell from well No. 1., Budapest I., Dísz square 10. (Buda castle) (13th–14th centuries), Inventory of the Hungarian Agricultural Museum, Budapest

Fig. 334. Common fig (*Ficus carica*) seeds from well No. 1., Budapest I., Dísz square 10. (Buda castle) (13th–14th centuries), Inventory of the Hungarian Agricultural Museum, Budapest.

Fig. 335. Walnut (*Juglans regia*) stones from well No. 1., Budapest I., Dísz square 10. (Buda castle) (13th–14th centuries), Inventory of the Hungarian Agricultural Museum, Budapest.

Fig. 336. Cherry plum (*Prunus* cf. *cerasifera*) stones from well No. 1., Budapest I., Dísz square 10. (Buda castle) (13th–14th centuries), Inventory of the Hungarian Agricultural Museum, Budapest.

Fig. 337. Garden plum (*Prunus domestica* subsp. *oeconomica*) stones from well No. 1., Budapest I., Dísz square 10. (Buda castle) (13th–14th centuries), Inventory of the Hungarian Agricultural Museum, Budapest.

Fig. 338. Vine grape (*Vitis vinifera* subsp. *vinifera*) seeds from well No. 1., Budapest I., Dísz square 10. (Buda castle) (13th–14th centuries), Inventory of the Hungarian Agricultural Museum, Budapest.

Fig. 339. Danewort (*Sambucus ebulus*) seeds from well No. 1., Budapest I., Dísz square 10. (Buda castle) (13th–14th centuries), Inventory of the Hungarian Agricultural Museum, Budapest.

Fig. 340. Muskmelon (*Cucumis melo*) seeds from a well, Budapest I., Úri Str. 40. (Buda castle) (13th–14th centuries), Inventory of the Hungarian Agricultural Museum, Budapest.

Fig. 341. Watermelon (*Citrullus lanatus*) seeds from a well, Budapest I., Úri Str. 40. (Buda castle) (13th–14th centuries), Inventory of the Hungarian Agricultural Museum, Budapest.

Fig. 342. Cornelian cherry or dogwood (*Cornus mas*) stone from a well, Budapest I., Úri Str. 40. (Buda castle) (13th–14th centuries), Inventory of the Hungarian Agricultural Museum, Budapest.

Fig. 343. Sloe, blackthorn (*Prunus spinosa*) stones from well No. 1., Budapest I., Dísz square 10. (Buda castle) (13th–14th centuries), Inventory of the Hungarian Agricultural Museum, Budapest.

Fig. 344. Common fig (*Ficus carica*) seeds from a well, Budapest I., Úri Str. 40. (Buda castle) (13th–14th centuries), Inventory of the Hungarian Agricultural Museum, Budapest.

Fig. 345. Walnut (*Juglans regia*) stones from a well, Budapest I., Úri Str. 40. (Buda castle) (13th–14th centuries), Inventory of the Hungarian Agricultural Museum, Budapest.

Fig. 346. Sloe, blackthorn (*Prunus spinosa*) stones from a well, Budapest I., Úri Str. 40. (Buda castle) (13th–14th centuries), Inventory of the Hungarian Agricultural Museum, Budapest.

Fig. 347. Vine grape (*Vitis vinifera* subsp. *vinifera*) seeds from a well, Budapest I., Úri Str. 40. (Buda castle) (13th–14th centuries), Inventory of the Hungarian Agricultural Museum, Budapest.

Fig. 348. Common millet (*Panicum miliaceum*) naked grains from Nagyvázsony-Csepely (15th–16th centuries), Inventory of the Hungarian Agricultural Museum, Budapest.

Fig. 349. Barley (*Hordeum vulgare*) grains from Nagyvázsony-Csepely (15th–16th centuries), Inventory of the Hungarian Agricultural Museum, Budapest.

Fig. 350. Common millet (*Panicum miliaceum*) naked grains from Nagyvázsony-Csepely (15th–16th centuries), Inventory of the Hungarian Agricultural Museum, Budapest.

Fig. 351. Rye (*Secale cereale*) grains from Nagyvázsony-Csepely (15th–16th centuries), Inventory of the Hungarian Agricultural Museum, Budapest.

Fig. 352. Common bread wheat (*Triticum aestivum* subsp. *vulgare*) grains from Nagyvázsony-Csepely (15th–16th centuries), Inventory of the Hungarian Agricultural Museum, Budapest.

Fig. 353. Big seed lentil (*Lens culinaris* subsp. *macrosperma*) from Nagyvázsony-Csepely (15th–16th centuries), Inventory of the Hungarian Agricultural Museum, Budapest.

Fig. 354. Horsebean (*Vicia faba*) seeds from Nagyvázsony-Csepely (15th–16th centuries), Inventory of the Hungarian Agricultural Museum, Budapest.

Fig. 355. Grape-vine (*Vitis vinifera* subsp. *vinifera*) seeds from Nagyvázsony-Csepely (15th–16th centuries), Inventory of the Hungarian Agricultural Museum, Budapest.

Fig. 356. Common flax (*Linum usitatissimum*) seeds from Nagyvázsony-Csepely (15th–16th centuries), Inventory of the Hungarian Agricultural Museum, Budapest.

Fig. 357. Catchweed or cleavers (*Galium aparine*) achenes from Nagyvázsony-Csepely (15th–16th centuries), Inventory of the Hungarian Agricultural Museum, Budapest.

Fig. 358. Grape-vine (*Vitis vinifera* subsp. *vinifera*) seeds and stalk fragments from Nagyvázsony-Csepely (15th–16th centuries), Inventory of the Hungarian Agricultural Museum, Budapest.

Fig. 359. Charlock (*Sinapis arvensis*) seed from Nagyvázsony-Csepely (15th–16th centuries), Inventory of the Hungarian Agricultural Museum, Budapest.

Fig. 360. Vetchling (*Lathyrus* spec.) seeds from Nagyvázsony-Csepely (15th–16th centuries), Inventory of the Hungarian Agricultural Museum, Budapest.

Fig. 361. Ball mustard (*Neslea paniculata*) siliqua from Nagyvázsony-Csepely (15th–16th centuries), Inventory of the Hungarian Agricultural Museum, Budapest.

Fig. 362. Common bread wheat (*Triticum aestivum* subsp. *vulgare*) grains from Pécs-Kórház Str. (16th–17th centuries), Inventory of the Hungarian Agricultural Museum, Budapest.

Fig. 363. Common millet (*Panicum miliaceum*) naked grains from the Benedictin monastery, Kaposvár-Kaposszentjakab, Turkish Period (16th century), Inventory of the Hungarian Agricultural Museum, Budapest.

Fig. 364. Common bread wheat (*Triticum aestivum* subsp. *vulgare*) grains fromthe Benedictin monastery, Kaposvár-Kaposszentjakab, Turkish Period (16th century), Inventory of the Hungarian Agricultural Museum, Budapest.

Fig. 365. Horsebean (*Vicia faba*) seeds from the Benedictin monastery, Kaposvár-Kaposszentjakab, Turkish Period (16th century), Inventory of the Hungarian Agricultural Museum, Budapest.

Fig. 366. Sweet cherry (*Prunus avium*) stones from well N° 15 of the former Military Headquarters in the Buda castle. Photograph by Ferenc Hegyi.

Fig. 367. Sour cherry (*Prunus cerasus*) stones from well N° 15 of the former Military Headquarters in the Buda castle. Photograph by Ferenc Hegyi.

Fig. 368. Bullace plum (*Prunus domestica* subsp. *insititia*) stones from well N° 15 of the former Military Headquarters in the Buda castle. Photograph by Ferenc Hegyi.

Fig. 369. Peach (*Prunus persica*) stone from well N° 15 of the former Military Headquarters in the Buda castle. Photograph by Ferenc Hegyi.

Fig. 370. Walnut (*Juglans regi*a) stone from well N° 15 of the former Military Headquarters in the Buda castle. Photograph by Ferenc Hegyi.

Fig. 371. Garden plum (*Prunus domestica* subsp. *oeconomica*) stones from well N° 15 of the former Military Headquarters in the Buda castle. Photograph by Ferenc Hegyi.

Fig. 372. Oblong hazelnut (*Corylus avellana*) shell fragment from well N° 15 of the former Military Headquarters in the Buda castle. Photograph by Ferenc Hegyi.

Fig. 373. Almond (*Prunus amygdalus*) stone fragment from well N° 15 of the former Military Headquarters in the Buda castle. Photograph by Ferenc Hegyi.

Fig. 374. Sloe or blackthorn (*Prunus spinosa*) stones from well N° 15 of the former Military Headquarters in the Buda castle. Photograph by Ferenc Hegyi.

Fig. 375. Vine grape (*Vitis vinifera* subsp. *vinifera*) seeds from well N° 15 of the former Military Headquarters in the Buda castle. Photograph by Ferenc Hegyi.

Fig. 376. Cornelian cherry or dogwood (*Cornus mas*) stones from well N° 15 of the former Military Headquarters in the Buda castle. Photograph by Ferenc Hegyi.

Fig. 377. Common fig (*Ficus carica*) seeds from well N° 15 of the former Military Headquarters in the Buda castle. Photograph by Ferenc Hegyi.

Fig. 378. Bramble (*Rubus caesius*) seeds from well N° 15 of the former Military Headquarters in the Buda castle. Photograph by Ferenc Hegyi.

Fig. 379. Muskmelon or cantaloupe (*Cucumis melo*) seeds from well N° 15 of the former Military Headquarters in the Buda castle. Photograph by Ferenc Hegyi. Type "A": oblong, large seeds (Cantaloupe type).

Fig. 380. Muskmelon (*Cucumis melo*) seeds from well N° 15 of the former Military Headquarters in the Buda castle. Photograph by Ferenc Hegyi. Type "B": short seeds (Turkestan type).

Fig. 381. Watermelon (*Citrullus lanatus*) seeds from well N° 15 of the former Military Headquarters in the Buda castle. Photograph by Ferenc Hegyi.

Fig. 382. Rock-well N° 8 of the Teleki mansion in the Buda castle, the beginning of the 15th century. Photograph Dorottya B. Nyékhelyi.

Fig. 383. Seed remains from rock-well N° 8 of the Teleki mansion in the Buda castle, the beginning of the 15th century. Photograph by Gábor Gyulai.

Fig. 384. Muskmelon (*Cucumis melo*) seeds from rock-well N° 8 of the Teleki mansion in the Buda castle, the beginning of the 15th century. Photograph by the author.

Fig. 385. Stratigraphy of rock-well N° 8 of the Teleki mansion in the Buda castle, the beginning of the 15th century. After Dorottya B. Nyékhelyi.

Fig. 386. Results of genetic examinations of medieval seeds coming from excavations in the Buda castle. Photographs by Gábor Gyulai.

Fig. 387. Well or sink from Budapest I., Kapucinusok Str. (Buda castle), end of the 15th century. Photograph by András Végh.

Fig. 388. Well or sink from Budapest I., Kapucinusok Str. (Buda castle), end of the 15th century. Photograph by András Végh

Fig. 389. Separation of floated organic remains from inorganic material in Budapest I., Kapucinusok Str. (Buda castle), end of the 15th century. Photograph by Renáta Bacsó.

Fig. 390. Separation of remains to taxa in Budapest I., Kapucinusok Str. (Buda castle), end of the 15th century. Photograph Renáta Bacsó.

Fig. 391. Identification of diaspores using a stereo-binocular microscope, from Budapest I., Kapucinusok Str. (Buda castle), end of the 15th century. Photograph Renáta Bacsó.

Fig. 392. Sweet cherry (*Prunus avium*) stones from Budapest I., Kapucinusok Str. (Buda castle), end of the 15th century. Photograph Renáta Bacsó.

Fig. 393. Sour cherry (*Prunus cerasus*) stones from Budapest I., Kapucinusok Str. (Buda castle), end of the 15th century. Photograph Renáta Bacsó.

Fig. 394. Garden plum (*Prunus domestica* subsp. *oeconomica*) stone from Budapest I., Kapucinusok Str. (Buda castle), end of the 15th century. Photograph Renáta Bacsó.

Fig. 395. Apple (*Malus domestica*) seed from Budapest I., Kapucinusok Str. (Buda castle), end of the 15th century. Photograph Renáta Bacsó.

Fig. 396. Watermelon (*Citrullus lanatus*) seed from Budapest I., Kapucinusok Str. (Buda castle), end of the 15th century. Photograph Renáta Bacsó.

Fig. 397. Sloe or blackthorn (*Prunus spinosa*) stone from Budapest I., Kapucinusok Str. (Buda castle), end of the 15th century. Photograph Renáta Bacsó.

Fig. 398. Almond (*Amygdalus communis*) stone from Budapest I., Kapucinusok Str. (Buda castle), end of the 15th century. Photograph Renáta Bacsó.

Fig. 399. Walnut (*Juglans regia*) stone fragment from Budapest I., Kapucinusok Str. (Buda castle), end of the 15th century. Photograph Renáta Bacsó.

Fig. 400. Haselnut (*Corylus avellana*) shell fragment from Budapest I., Kapucinusok Str. (Buda castle), end of the 15th century. Photograph Renáta Bacsó.

Fig. 401. Haselnut (*Corylus avellana*) shell fragment from Budapest I., Kapucinusok Str. (Buda castle), end of the 15th century. Photograph Renáta Bacsó.

Fig. 402. Endospermium of cornelian cherry or dogwood (*Cornus mas*) stone from Budapest I., Kapucinusok Str. (Buda castle), end of the 15th century. Photograph Renáta Bacsó.

Fig. 403. Cornelian cherry or dogwood (*Cornus mas*) stones from Budapest I., Kapucinusok Str. (Buda castle), end of the 15th century. Photograph Renáta Bacsó.

Fig. 404. "Spider" shaped grapevine (*Vitis vinifera* subsp. *vinifera*) seed from Budapest I., Kapucinusok Str. (Buda castle), end of the 15th century. Photograph Renáta Bacsó.

Fig. 405. Endospermium of vine grape (*Vitis vinifera* subsp. *vinifera*) seed from Budapest I., Kapucinusok Str. (Buda castle), end of the 15th century. Photograph Renáta Bacsó.

Fig. 406. Balkan type vine grape (*Vitis vinifera* subsp. *vinifera*) seed from Budapest I., Kapucinusok Str. (Buda castle), end of the 15th century. Leg. Géza Facsar. Photograph Renáta Bacsó.

Fig. 407. Balkan type vine grape (*Vitis vinifera* subsp. *vinifera*) seeds from Budapest I., Kapucinusok Str. (Buda castle), end of the 15th century. Leg. Géza Facsar. Photograph Renáta Bacsó.

Fig. 408. Goher type vine grape (*Vitis vinifera* subsp. *vinifera*) seeds from Budapest I., Kapucinusok Str. (Buda castle), end of the 15th century. Leg. Géza Facsar. Photograph Renáta Bacsó.

Fig. 409. Goher-Doblerab types vine grape (*Vitis vinifera* subsp. *vinifera*) seeds from Budapest I., Kapucinusok Str. (Buda castle), end of the 15th century. Leg. Géza Facsar. Photograph Renáta Bacsó.

Fig. 410. Morphology of ancient *Vitis* seeds excavated in Hungary. SEM micrographs of seeds excavated at a Roman villa at Aquincum, Budapest, (2nd–4th cent. AD) (1) at Fenékpuszta (2); at a vineyard site near Győr, Ece (11th–12th cent.) (3); at Debrecen (15th cent.) (4); at the King's Palace in the Buda castle, Budapest, (15th cent.) (5). The SEM micrograph of seeds of the contemporary *Vitis vinifera* cultivar 'Kék Bakator' is also shown (6). Upper and middle rows show seed morphology at 20x magnification. Bottom row shows seed coat detail at 500x magnification. GYULAI *et. al.* 2009.

Fig. 411. Common bread wheat (*Triticum aestivum* subsp. *vulgare*) grains from the 16th-century settlement layer at Vác. Photograph by Tibor Kádas.

Fig. 412. Club wheat (*Triticum aestivum* subsp. *compactum*) grains from the 16th-century settlement layer at Vác. Photograph by Tibor Kádas.

Fig. 413. Bread wheat (*Triticum aestivum*) ear spindle fragments from the 16th-century settlement layer at Vác. Photograph by Tibor Kádas.

Fig. 414. Rye (*Secale cereale*) grains from the 16th-century settlement layer at Vác. Photograph by Tibor Kádas.

Fig. 415. Jimsonweed (*Datura stramonium*) seeds from the 16th-century settlement layer at Vác. Photograph by Tibor Kádas.

Fig. 416. Opium poppy (*Papaver somniferum*) seed from the 16th-century settlement layer at Vác. Photograph by Tibor Kádas.

Fig. 417. Barley (*Hordeum vulgare*) hulled grain from Dunaföldvár-Öregtorony, beginning of the 17th century, Inventory of the Hungarian Agricultural Museum, Budapest.

Fig. 418. Germinated rye (*Secale cereale*) grains from Dunaföldvár-Öregtorony, beginning of the 17th century, Inventory of the Hungarian Agricultural Museum, Budapest.

Fig. 419. Common millet (*Panicum miliaceum*) naked grains from Dunaföldvár-Öregtorony, beginning of the 17th century, Inventory of the Hungarian Agricultural Museum, Budapest.

Fig. 420. Hare's-ear mustard (*Conringia orientalis*) seeds from Dunaföldvár-Öregtorony, beginning of the 17th century, Inventory of the Hungarian Agricultural Museum, Budapest.

Fig. 421. Steppen-sesel (*Seseli annuum*) achenes from Dunaföldvár-Öregtorony, Turkish Period, Inventory of the Hungarian Agricultural Museum, Budapest.

Fig. 422. Common wheat (*Triticum aestivum* subsp. *vulgare*) grains from Dunaföldvár-Öregtorony, beginning of the 17th century, Inventory of the Hungarian Agricultural Museum, Budapest.

Fig. 423. Common wheat (*Triticum aestivum* subsp. *vulgare*) grains from Dunaföldvár-Öregtorony, beginning of the 17th century, Inventory of the Hungarian Agricultural Museum, Budapest.

Fig. 424. Charlock (*Sinapis arvensis*) seeds from Dunaföldvár-Öregtorony, beginning of the 17th century, Inventory of the Hungarian Agricultural Museum, Budapest.

Fig. 425. Rye (*Secale cereale*) grains from the gate of Hollókő castle, 17th century, Inventory of the Hungarian Agricultural Museum, Budapest.

Fig. 426. Rye brome (*Bromus secalinus*) naked grains from the gate of Hollókő castle, 17th century, Inventory of the Hungarian Agricultural Museum, Budapest.

Fig. 427. Hoary cress or hoary pepperwort (*Lepidium draba*) seed from the gate of Hollokő castle, 17th century, Inventory of the Hungarian Agricultural Museum, Budapest.

Fig. 428. Thorn-apple (*Datura stramonium*) seeds from the ditch of Hollokő castle, 17th century, Inventory of the Hungarian Agricultural Museum, Budapest.

Fig. 429. White goosefoot or fan-hen (*Chenopodium album*) seeds from the cellar of the barrack-room in Esztergom castle, 16th–18th centuries, Inventory of the Hungarian Agricultural Museum, Budapest.

Fig. 430. Yellow bristle-grass (*Setaria lutescens*) hulled grains from the cellar of a barrack-room in Esztergom castle, 16th–18th centuries, Inventory of the Hungarian Agricultural Museum, Budapest.

Fig. 431. Curltop lady's-thumb (*Polygonum lapathifolium*) seeds from the cellar of the barrack-room in Esztergom castle, 16th–18th centuries, Inventory of the Hungarian Agricultural Museum, Budapest.

Fig. 432. Annual woundwort (*Stachys annua*) nutlets from the cellar of a barrack-room in Esztergom castle, 16th–18th centuries, Inventory of the Hungarian Agricultural Museum, Budapest.

Fig. 433. Common wheat (*Triticum aestivum* subsp. *vulgare*) grain from Feldebrő Roman Catholic church, 18th-century, Inventory of the Hungarian Agricultural Museum, Budapest.

Fig. 434. Common wheat (*Triticum aestivum* subsp. *vulgare*) grains from Feldebrő Roman Catholic church, 18th-century cereal storage pit, Inventory of the Hungarian Agricultural Museum, Budapest.

Fig. 435. Spring wild-oat (*Avena fatua*) naked grains from Feldebrő Roman Catholic church, 18th-century cereal storage pit, Inventory of the Hungarian Agricultural Museum, Budapest.

Fig. 436. Common millet (*Panicum miliaceum*) naked grains from Ebes, 18th-century pit, Inventory of the Hungarian Agricultural Museum, Budapest

Fig. 437. Yellow bristle-grass (*Setaria lutescens*) grains from Ebes, 18th-century pit, Inventory of the Hungarian Agricultural Museum, Budapest.

Fig. 438. Black-bindweed (*Fallopia convolvulus*) nutlets from Ebes, 18th-century pit, Inventory of the Hungarian Agricultural Museum, Budapest.

Fig. 439. Spring wild-oat (*Avena fatua*) hulled grain from Tószeg, prehistoric site. Inventory of the Hungarian Agricultural Museum, Budapest.

Fig. 440. Field pennycress (*Thlaspi arvense*) seed from Ebes, 18th-century pit, Inventory of the Hungarian Agricultural Museum, Budapest.

Fig. 441. Two-rowed barley (*Hordeum vulgare* subsp. *distichum*) hulled grain from Tószeg, prehistoric site. Inventory of the Hungarian Agricultural Museum, Budapest.

Fig. 442. Club wheat (*Triticum aestivum* subsp. *compactum*) grains from Szentes Tűzköves, Neolithic or Bronze Age. Inventory of the Hungarian Agricultural Museum, Budapest.

Fig. 443. Emmer (*Triticum turgidum* subsp. *dicoccum*) naked grains from Szob-Ipolypart, prehistoric site. Inventory of the Hungarian Agricultural Museum, Budapest.

Fig. 444. Einkorn (*Triticum monococcum*) spicelets (furca bicornis) from Szob-Ipolypart, prehistoric site. Inventory of the Hungarian Agricultural Museum, Budapest.

Fig. 445. Einkorn (*Triticum monococcum*) naked grains from Szob-Ipolypart, prehistoric site. Inventory of the Hungarian Agricultural Museum, Budapest.

Fig. 446. Einkorn (*Triticum monococcum*) hulled grain from Szob-Ipolypart, prehistoric site. Inventory of the Hungarian Agricultural Museum, Budapest.

Fig. 447. Barley (*Hordeum vulgare*) hulled grain from Tószeg, Roman or medieval site. Inventory of the Hungarian Agricultural Museum, Budapest.

Fig. 448. Pot fragment with burnt food remains from the Copper Age site on the Zalaszentbalázs-Szőlőhegy meadow. Photograph by the author.

Fig. 449. Pot fragment with burnt food remains from the Albertfalva site of the Bell-Beaker Csepel-group. Photograph by Anna Endrődi.

Fig. 450. Carbonised crab apple (*Malus silvestris*) fruits from the Albertfalva site of the Bell-Beaker Csepel-group. Photograph by Anna Endrődi.

Fig. 451. Bread remains from the floor level in a house in the Túrkeve-Terehalom Middle Bronze Age tell. Rear view. Photograph by Tibor Kádas.

Fig. 452. Bread remains from the floor level in a house in the Túrkeve-Terehalom Middle Bronze Age tell. Front view. Photograph by Tibor Kádas.

Fig. 453. Millet grains (*Panicum miliaceum*) cooked for mush from the Late Bronze Age waste pit of Balatonmagyaród-Hídvégpuszta. Photograph by the author.

Fig. 454. Grain mush remains from the Bölcske-Vörösgyír Middle Bronze Age tell settlement. Photograph by Tibor Kádas.

Fig. 455. Millet mush remains from the Late Bronze Age settlement of Balatonmagyaród-Hídvégpuszta. Photograph by Tibor Kádas.

Fig. 476. Millet (*Panicum miliaceum*) baked in cereal-gruel (fine milling product) found in 1992 at the Edelény-Borsod motte settlement from the Conquest Period. Enlargement. Photograph by the author.

Fig. 477. Millet (*Panicum miliaceum*) baked in cereal-gruel (fine milling product) found in 1992 at the Edelény-Borsod motte settlement from the Conquest Period. Photograph by Endre Tóth.

Fig. 478. Remains of the one-course dish (meat-grual) found in 1998 at the Edelény-Borsod motte settlement from the Conquest Period. Enlargement. Photograph by Endre Tóth.

Fig. 479. Wild berries and seeds of lictarium (thin jam) found in 1998 at the Edelény-Borsod motte settlement from the Conquest Period. Photograph Anna Endrődi.

Fig. 480. Remains of a dish: in flour baked millet (*Panicum miliaceum*), foxtail millet (*Setaria italica*), field pea (*Pisum sativum* subsp. *arvense*) using fat by cooking found in 1992 at the Edelény-Borsod motte settlement from the Conquest Period, including millet, foxtail millet and pea embedded in grain meal. Photograph by the author.

Fig. 481. Lictarium (thin jam) made from sloe (*Prunus spinosa*), wild pear (*Pyrus achras*), crab apple (*Malus silvestris*), field rose (*Rosa silvestris*) and cornelian cherry (*Cornus mas*) found in 1998 at the Edelény-Borsod motte settlement from the Conquest Period. Photograph by Endre Tóth.

Fig. 482. Nematoda cocoon from the dish remains samples (in flour baked millet (*Panicum miliaceum*), foxtail millet (*Setaria italica*), field pea (*Pisum sativum* subsp. *arvense*) used fat by cooking) found at the 1992 excavation of the Edelény-Borsod motte, Hungarian Conquest Period. Magnification: 400x. Photo by the Author.

Fig. 483. Phytolith of millet (*Panicum miliaceum*) from lamb-gruel samples found at the 1992 excavation of the Edelény-Borsod motte, Hungarian Conquest Period. Magnification: 400x. Photo by the Author.

Fig. 484. Exposed flour particles from the cereals' aleuron layer from the meat-gruel (goulash soup made from dough, vegetables, onion/garlic and lamb) samples found at the 1998 excavation of the Edelény-Borsod motte, Hungarian Conquest Period. Magnification: 200x. Photo by the Author.

Fig. 485. Exposed flour particles from the cereals' aleuron layer from lamb-gruel samples found at the 1992 excavation of the Edelény-Borsod motte, Hungarian Conquest Period. Magnification: 400x. Photo by the Author.

Fig. 486. Rye (*Secale cereale*) phytoliths from meat-gruel (goulash soup made from dough, vegetables, onion/barlic and lamb) samples found at the 1998 excavation of the Edelény-Borsod motte, Hungarian Conquest Period. Magnification: 400x. Photo by the Author.

Fig. 487. Trapezoid phytolith of common bread wheat (*Triticum aestivum* subsp. *vulgare)* from meat-gruel (goulash soup made from dough, vegetables, onion/garlic and lamb) samples found at the 1998 excavation of the Edelény-Borsod motte, Hungarian Conquest Period. Magnification: 400x. Photo by the Author.

Fig. 488. Plant hair from meat-gruel (goulash soup made from dough, vegetables, onion/garlic and lamb) samples found at the 1998 excavation of the Edelény-Borsod motte, Hungarian Conquest Period. Magnification: 400x. Photo by the Author.

14. Tables

Table 1. Seed and fruit remains in Hungary, 1860–2006. Completed by F. Gyulai, 2006.

Latin name	English name	Early Neolithic (6000–5300 BC)	Middle Neolithic (5300–4700 BC)	Late Neolithic (4700–4300 BC)	Neolithic without epoch	Early Copper age (4300–3500 BC)	Late Copper age (3500–3000 BC)	Copper age without epoch	Early Bronze age (3000–1600 BC)	Middle Bronze age (1600–1300 BC)	Late Bronze age (1300–900 BC)	Bronze age without epoch	Early Iron age (Hallstatt) (900–500 BC)	Early Iron age (Skythian) (700–500 BC)
Number of archaeological sites		12	47	26	5	12	12	3	12	26	16	5	10	2
Number of plant taxons		53	104	119	27	19	38	10	99	126	202	20	57	69
Number of seed and fruits		4456	163725	277944	620	5873	4564	1221	5449	382991	91909	4986	3452	14123
Acer campestre L.	hedge maple													
Acer spec.	maple													
Achillea millefolium L.	common yarrow									1	1			
Adonis aestivalis L.	summer adonis													
Adonis cf. aestivalis L.	summer adonis													
Adonis spec.	adonis													
cf. Adonis spec.	adonis													
Aegilops cf. cylindrica Host	jointed goatgrass								1					
Aegilops spec.	goatgrass						2							
Aesculus cf. pavia L.	horse chestnut													
Aethusa cynapium L.	fool's parsley			1							13			
Agrimonia eupatoria L.	common agrimony		1						x					
Agropyron repens (L.) P. B.	twitch quackgrass								x	7			1	
Agropyron cf. repens (L.) P. B.	twitch quackgrass													
Agropyron spec.	quackgrass													
Agrostemma githago L.	korn-rade		4	x			18	134	22	xxx	91		2	1
cf. Agrostemma githago L.	korn-rade		2						1					
Agrostemma githago L. subsp. linicola	korn-rade													
Ajuga chamaepitys (L.) Schreb.	ground pine									1	2			
Ajuga genevensis L.	blue bugle										2			
Ajuga geneviensis L./reptans L.	blue/carpet bugle													
Ajuga reptans L.	carpet bugle									1	1			
Ajuga cf. reptans L.	carpet bugle													
Ajuga spec.	bugle									1				
Alchemilla vulgaris agg.	common lady's-mantle			3										
Alisma plantago-aquatica L.	common water-plantain										22			
Allium sativum L.	garlic													
cf. Alnus glutinosa (L.) Gaertn.	sticky alder										2			
Alnus incana (L.) Mönch	speckled alder													
Alopecurus cf. myosuroides Huds.	slender foxtail													
Althaea officinalis L.	marshmallow													
Amaranthus cf. blitoides S. Watson	prostrate amaranth										3			
Amaranthus gracizans L.	tumbleweed amaranth													
Amaranthus lividus L.	livid amaranth			1						2	3			
Amaranthus cf. lividus L.	livid amaranth													
Amaranthus lividus L./ paniculatus L.	livid amaranth/ love-lies-bleeding													
Amaranthus retroflexus L.	green amaranth												x?	
Amaranthus spec.	amaranth			1										
Amygdalus communis L.	almond													
Anagallis arvensis L.	scarlet pimpernel								2	1	1			7
cf. Anagallis arvensis L.	scarlet pimpernel		1											
Anchusa arvensis L.	alkanet													

Late Iron age (La Tène) (5th century BC–1st century AD)	Roman age (1st–middle 5th century AD)	Barbaricum (1st–5th century AD)	Early Migration period (476–568 AD)	Late Migration period (Avar age) (568–895 AD)	Late Migration periode (Caroling age) (811–895 AD)	Late Migration period (German, Slavonic) (6th–10th century AD)	Hungarian conquest time (895 AD)	Early Árpád age (1001–12th century AD)	Late Árpád age (13th century–1301 AD)	Árpád age without epoch	Hungarian Kingdom (14th–16th century AD)	Cumanian (15th–16th century AD)	Turkish occupation (16th–17th century AD)	17th–18th century–Total	19th century–Total	Uncertain age (prehistoric – historic)	Number of remains
16	58	24	4	19	5	2	5	15	13	13	32	1	17	10	6	27	455
98	163	193	21	120	150	10	40	124	262	121	319	165	180	60	207	139	636 (overlapping)
7219	239651	99918	57017	5828	186176	32	822	183777	165520	25055	5469513	11880	1211768	112566	15004	15121	8768180
															1		1
											x						x
																	2
	1	5							2		4		2		4	1	19
1																	1
											1						1
										1							1
								1								1	3
																	2
														x			x
3		2									117				4		140
		2							1								x
4	2	2							1		1						xx
																1	1
1																	1
26	xx	28	13		263		3	417	41	8	7030	15	xxxx	x + 108		18	xxxx
																	3
									36		2						38
	6	11							11	2	341	2	4		2	2	384
									279				2		8	13	304
									1								1
							2		29		69	19	1		4		126
									8						3		11
																	1
2		4						1				6			2		18
																	22
					2						1		3				6
																	2
															3		3
										1							1
												4				2	6
											26						29
7											2						9
	17			7					7		1584	14	37				1672
									4				1				5
											207						207
														16	226		xxx
											128						129
	xx										80			10			xx + 90
	1	3			2			1	97	10	2				6	9	142
		1															2
									1								1

Latin name	English name	Early Neolithic (6000–5300 BC)	Middle Neolithic (5300–4700 BC)	Late Neolithic (4700–4300 BC)	Neolithic without epoch	Early Copper age (4300–3500 BC)	Late Copper age (3500–3000 BC)	Copper age without epoch	Early Bronze age (3000–1600 BC)	Middle Bronze age (1600–1300 BC)	Late Bronze age (1300–900 BC)	Bronze age without epoch	Early Iron age (Hallstatt) (900–500 BC)	Early Iron age (Skythian) (700–500 BC)
cf. Anemona nemorosa L.	European wood anemone		1											
Anethum graveolens L.	common dill										3			
Anthemis arvensis L.	field camomile													
Anthemis tinctoria L.	golden camomile/dyer's camomile													
Anthemis cf. tinctoria L.	golden camomile/dyer's camomile													
Anthyllis vulneraria L.	kidney-vetch				10									
Antirrhinum orontium L.	corn snapdragon													
Aphanes arvensis L.	parsley-piert			3							1			
Apium graveolens L.	wild celery													
cf. Apium graveolens L.	wild celery										1			
Apiaceae	apium family										4			
Arctium lappa L.	great burdock													
Arctium minus (Hill.) Bernh.	smaller burdock													
Arctium tomentosum Mill.	woolly burdock										4			
Arctium spec.	burdock													
Arenaria serpyllifolia L.	thyme-leaved sandwort									3				
Aristolochia clematitis L.	birthwort													
Asperula arvensis L.	blue woodruff													
Aster amellus L.	summer Michaelmas-daisy													
Aster spec.	sea-starwort										2			
Asteraceae	aster family	2												
Astragalus glycyphyllus L.	common milkvetch			x										
Atriplex cf. hortensis L.	garden orache													
Atriplex patula L.	spreading/cammon orache		6	7						x	84			
cf. Atriplex patula	spreading/cammon orache									1				
Atriplex spec.	orache													
Atropa bella-donna L.	banewort/deadly nightshade													
Avena fatua L.	spring wild-oat	1	29	xx		2	15		26	153	14	20	3	51
Avena cf. fatua L.	spring wild-oat		1											
Avena fatua L./sativa L.	spring wild-oat/common oat													13
Avena fatua L./sterilis L.	spring wild/steril oat										2			
Avena cf. nuda Höjer	naked oat			2										
Avena sativa L.	common oat		1?											7
Avena cf. sativa L.	common oat													2
Avena cf. sterilis L.	steril oat													
cf. Avena sp. con cultiv.	cultivated oat		6											
Avena spec.	oat		1						1	1			1	
Avena/Bromus spec.	oat/brome													
Ballota nigra L.	black horehound										1			
Bellis perennis L.	English daisy/garden daisy										1			
cf. Bellis perennis L.	English daisy/garden daisy													
Berberis vulgaris L.	European barberry			x										
Berteroa incana (L.) DC.	hoary alyssum													
Betula pendula	common silver birch													
Bidens cernua L.	nodding bur marigold													

Late Iron age (La Tène) (5th century BC–1st century AD)	Roman age (1st–middle 5th century AD)	Barbaricum (1st–5th century AD)	Early Migration period (476–568 AD)	Late Migration period (Avar age) (568–895 AD)	Late Migration periode (Caroling age) (811–895 AD)	Late Migration period (German, Slavonic) (6th–10th century AD)	Hungarian conquest time (895 AD)	Early Árpád age (1001–12th century AD)	Late Árpád age (13th century–1301 AD)	Árpád age without epoch	Hungarian Kingdom (14th–16th century AD)	Cumanian (15th–16th century AD)	Turkish occupation (16th–17th century AD)	17th–18th century–Total	19th century–Total	Uncertain age (prehistoric – historic)	Number of remains
																	1
									2		3943						3948
		6											3				9
									2								2
											2		1				3
																	10
		1															1
										3		11					18
											151						151
				1													2
											68	1			1		74
	1	1							1		3						6
	2								2		2				1		7
		1		1	4						1	1698					1709
												1					1
		1							255								259
					1				2								3
								2				182					184
		1															1
																	2
		1							7		4	32			1		47
					3				1		1						x
									1								1
	48	13		63	2			2	573	12	11	5	35		1		xxx
										1							2
													2				2
															43		43
9	87	xxx	4	1	24		44	29	42	40	26	7		x + 4	3	22	xxx
		3														1	5
5		15							1		1						35
																	2
																	2
xx	xxx	255			318			65	23	63	1686		97	x		3	xxxx
		6								2							10
		11															11
				1													6
x		xxxx						38	2								xxxx
															2		2
				1	1				38		3	3					47
															1		2
											1						1
																	x
									36								36
									2						1273		1275
				16	1												17

Latin name	English name	Early Neolithic (6000–5300 BC)	Middle Neolithic (5300–4700 BC)	Late Neolithic (4700–4300 BC)	Neolithic without epoch	Early Copper age (4300–3500 BC)	Late Copper age (3500–3000 BC)	Copper age without epoch	Early Bronze age (3000–1600 BC)	Middle Bronze age (1600–1300 BC)	Late Bronze age (1300–900 BC)	Bronze age without epoch	Early Iron age (Hallstatt) (900–500 BC)	Early Iron age (Skythian) (700–500 BC)
Blysmus compressus (L.) Panzer	flat-rush								1					
Bolboschoenus (syn. Schoenoplectus) maritimus (L.) Palla	sea club rush			1							1			
Brachypodium silvaticum (Huds.) R. et Sch.	slender fase brome													
Brassica campestris L. (syn. Brassica rapa L. subsp. campestris)	wild turnip		1						1	1	6			
Brassica campestris L. var. oleifera DC.	naven													
Brassica cf. juncea (L.) Czern. (B. nigra x rapa)	indian/chinese mustard													
Brassica cf. napus L. (oleracea?) L.	rape (cabbage?)													
Brassica nigra (L.) Koch	black mustard									5				
Brassica cf. oleracea L.	headed cabbage													
Brassica rapa L. em. Metzger	cabbage													
Brassica cf. rapa L. em. Metzger	turnip													
Brassica spec.	cabbage										1			
Brassica/Sinapis spec.	cabbage/mustard	1												
Brassicaceae	cabbage family	1								2				1
Bromus arvensis L.	field brome	64	8	xx		1	3		11	542	129		1	1
Bromus cf. arvensis L.	field brome													
Bromus arvensis L./secalinus L.	field/rye brome									2				
Bromus cf. commutatus Schrad	hairy chess/meadow brome									4				
Bromus erectus Huds.	upright brome									1	1			
Bromus inermis Leyss.	smooth brome/awnless brome			1							23			
Bromus mollis L. (syn. B. hordeaceus)	soft brome			1			2		2	8	4	1		2
Bromus secalinus L.	rye brome		4	xx	1		16	32	24	670	385		17	31
Bromus cf. secalinus L.	rye brome													
Bromus sterilis L.	barren brome	6	1					1	1	3				
Bromus cf. sterilis L.	barren brome													
Bromus sterilis L./tectorum L.	barren/downy brome													
Bromus tectorum L.	downy brome								x	1				
Bromus cf. tectorum L.	downy brome		2											
Bromus spec.	brome	17	5					21		xxx	355			2
Bryonia alba L.	white bryony													
cf. Bryonia alba L.	white bryony													
Bupleurum rotundifolium L.	thorow-wax										1			
Calamintha acinos (L.) Clairv. (syn. Satureja acinos (L.) Schleehe	basil thyme								1		1			
Calamintha nepeta (C. silvatica)	greater calamint													
Calystegia sepium (L.) R. Br.	hedge bindweed													
Calystegia spec.	bindweed									x				
Camelina microcarpa Andrz.	hairy gold-of-pleasure													
Camelina sativa (L.) Cr.	gold-of-pleasure			1					x		359		x	
Camelina spec.	gold-of-pleasure													
cf. Camelina spec.	gold-of-pleasure													
Campanula spec.	bellflower													

Late Iron age (La Tène) (5th century BC–1st century AD)	Roman age (1st–middle 5th century AD)	Barbaricum (1st–5th century AD)	Early Migration period (476–568 AD)	Late Migration period (Avar age) (568–895 AD)	Late Migration periode (Caroling age) (811–895 AD)	Late Migration period (German, Slavonic) (6th–10th century AD)	Hungarian conquest time (895 AD)	Early Árpád age (1001–12th century AD)	Late Árpád age (13th century–1301 AD)	Árpád age without epoch	Hungarian Kingdom (14th–16th century AD)	Cumanian (15th–16th century AD)	Turkish occupation (16th–17th century AD)	17th–18th century–Total	19th century–Total	Uncertain age (prehistoric – historic)	Number of remains
				3								2			2		8
		x															x
1																	1
		8			1	3	1	3	59		84	2	7		13	3	193
																1	1
															24		24
									9				19		37		65
													1				6
									20		55	4					79
													5				5
									2								2
		1							108		1						111
																	1
		1							2	5	11		5			2	30
1	6	32	3	1	3			23	2	66	15	13	76	x	3	5	xxxx
		1							1						6		8
																	2
																	4
																	2
											1						25
1	3	16			7				5	2		12			1	15	82
27	4	43			4			48	1	3	25	14	30	x + 1	2	9	xxxx
										1							1
	2		48													1	63
																1	1
1																	1
																4	5
								11									13
1	x	x							2	8		7	1				xxx
									1	1							2
		2									1						3
							1	15	179		1158	4	5				1363
	5								31	1	9						48
															1		1
		1		10							1	2					14
																	x
									1								1
3		18							3					x		2	xxx
									6		99						105
									15								15
											1						1

Latin name	English name	Early Neolithic (6000–5300 BC)	Middle Neolithic (5300–4700 BC)	Late Neolithic (4700–4300 BC)	Neolithic without epoch	Early Copper age (4300–3500 BC)	Late Copper age (3500–3000 BC)	Copper age without epoch	Early Bronze age (3000–1600 BC)	Middle Bronze age (1600–1300 BC)	Late Bronze age (1300–900 BC)	Bronze age without epoch	Early Iron age (Hallstatt) (900–500 BC)	Early Iron age (Skythian) (700–500 BC)
Cannabis sativa L.	hemp												1	
Capsella bursa-pastoris (L.) Medic.	shepherd's-purse									1				
Carduus acanthoides L.	welted thistle									1	12			
Carduus crispus L.	curly welted-thistle													
Carduus nutans L.	musk thistle										1			
Carduus spec.	thistle													
Carduus/Cirsium spec.	thistle/thistle													
Carex cf. brizoides L.	alpine grass													
Carex caryophyllea Latour	spring sedge													
Carex distans L./digitata L.	distant-flowered/fingered sedge													
Carex cf. disticha Huds.	brown sedge													
Carex cf. divulsa Stokes	grassland sedge													
Carex elata All.	tussock sedge										1			
Carex elata All./gracilis Curt. (-Typ)	tussock/slender tufted sedge										1		1	
Carex elongata L.	elongated sedge													
Carex cf. elongata L.	elongated sedge													
Carex flacca Schreb.	heath sedge													
Carex flava L.	yellow sedge													
Carex flava L./pendula Huds.	yellow/pendulous sedge								4		1			
Carex cf. gracilis Curt.	slender tufted-sedge													
Carex hirta L.	hairy sedge			2					2	1	29			
Carex hordeistichos Vill.	barley sedge									5				
Carex cf. lasiocarpa Ehrh.	woollyfruit sedge													
Carex cf. leporina L.	common hare sedge													
Carex muricata L.	greater starry sedge													
Carex pallescens L.	pale sedge								2		29			
Carex panicea L.	pink-leaved sedge													
Carex cf. paniculata Jusl.	greater Tussock-sedge													
Carex pendula Huds./strigosa Huds.	pendulous sedge/thin-spiked Wood-sedge													
Carex pseudocyperus L.	cyperous sedge										1			
Carex remota L.	greater Tussock-sedge													
Carex riparia Curt.	stream-bank sedge										3			
Carex cf. riparia Curt.	great pond sedge													
Carex riparia Curt./silvatica Huds.	great pond/wood sedge													
Carex rostrata Stokes ex W.	bottle sedge													
Carex cf. serotina Mérat	small-fruited yellow-sedge													
Carex silvatica Huds.	wood sedge										4			
Carex cf. stellulata Good.	rough sedge/prickly sedge													
Carex cf. vesicaria L.	bladder sedge													
Carex vulpina L.	fox sedge			2										
Carex vulpina L./muricata L. Typ.	fox sedge/prickly sedge		1								1			
Carex bicarpellat	sedge bicarpellat										2			
Carex tricarpellat	sedge tricarpellat								1		21			
Carex spec.	sedge			x							5			
Carpinus betulus L.	hornbeam													

Late Iron age (La Tène) (5th century BC–1st century AD)	Roman age (1st–middle 5th century AD)	Barbaricum (1st–5th century AD)	Early Migration period (476–568 AD)	Late Migration period (Avar age) (568–895 AD)	Late Migration periode (Caroling age) (811–895 AD)	Late Migration period (German, Slavonic) (6th–10th century AD)	Hungarian conquest time (895 AD)	Early Árpád age (1001–12th century AD)	Late Árpád age (13th century–1301 AD)	Árpád age without epoch	Hungarian Kingdom (14th–16th century AD)	Cumanian (15th–16th century AD)	Turkish occupation (16th–17th century AD)	17th–18th century–Total	19th century–Total	Uncertain age (prehistoric – historic)	Number of remains
		23		25	103		61	2	178		2982	240	9	x	88	34	xxxx
									3		2						6
									3857		6	3			10		3889
		39							6		2	1			25	3	76
	4	1													1		7
	2												1				3
				2							1						3
												2					2
												4					4
									1								1
				1													1
											25						25
																	1
				210	3			2	22		2	25			8		274
									64		9				1		74
				2						1							3
				11							1	6					18
								1			4		3				8
	2			48				1	2	1		2	2		3		66
		1						3									4
	2	1		25	6				16		49	25	37		58	13	266
																1	6
	1								40		4				1		46
													1				1
		2		9					36	1	3				2		53
	5			610					16		39	9	7		14	15	746
		x															x
											2						2
		1															1
																	1
				41				1	46		2	2	1		11		104
		x		4								2			2	2	xx
									1								1
															3		3
	4								14		11		4		1	3	37
												2					2
		1			1						1	2	5		9		23
		1															1
									1				1				2
				2	2								1				7
		1		14					7	1	31	16	14		11	2	99
		2		3				25	1		2	7	2				44
	5				1				3		9	2	15		6		63
													1				x
1											36				158		195

Latin name	English name	Early Neolithic (6000–5300 BC)	Middle Neolithic (5300–4700 BC)	Late Neolithic (4700–4300 BC)	Neolithic without epoch	Early Copper age (4300–3500 BC)	Late Copper age (3500–3000 BC)	Copper age without epoch	Early Bronze age (3000–1600 BC)	Middle Bronze age (1600–1300 BC)	Late Bronze age (1300–900 BC)	Bronze age without epoch	Early Iron age (Hallstatt) (900–500 BC)	Early Iron age (Skythian) (700–500 BC)
Carpinus cf. orientalis Mill.	orientalis hornbeam									1				
Carthamus tinctorius L.	safflower									1246				
Carum carvi L.	caraway													
Caryophyllaceae	caryophyllaceous plants	2								5	2			
Castanea sativa Mill.	European chestnut													
cf. Castanea sativa Mill.	European chestnut													
Caucalis platycarpos L. (syn. C. lappula (Web.) Grande	small bur-parsley													
Centaurea cyanus L.	cornflower													
Centaurea jacea L.	brown knapweed								1					1
Centaurea pannonica (Heuff.) Simk.	hungarian knapweed													
Centaurea scabiosa L.	greater knapweed													
Centaurea spec.	knapweed								x	1				
cf. Centaurea sp.	knapweed													
Centaurea/Cirsium spec.	knapweed/thistle													
Cerastium arvense L.	field mouse-ear													
Ceratophyllum submersum L.	soft hornwort										21			
Ceratophyllum cf. submersum L.	soft hornwort													
Ceratophyllum spec.	hornwort													
Cerealia	cereals	xxx	xxxx	xxxxx	36	174	711	1	1615	xxxx	1076	3082	744	6773
Cerinthe minor L.	lesser honeywort										1			
Chelidonium majus L.	greater celandine										1			
Chenopodium album agg.	white goosefoot/ fan-hen	31	3055	251	8	8	11	15	20	36730	1649	623	6	2424
Chenopodium cf. album agg.	white goosefoot/ fan-hen													
Chenopodium ficifolium Sm.	fig-leaved goosefoot										2			
Chenopodium cf. ficifolium Sm.	fig-leaved goosefoot													
Chenopodium glaucum L.	oak-leaved goosefoot													
Chenopodium hybridum L.	maple-leaved goosefoot	8	120	3			1		1	2	273	2	x	2
Chenopodium murale L.	nettle-leaved goosefoot								1					
Chenopodium polyspermum L.	many-seeded goosefoot								7					
Chenopodium rubrum L.	red goosefoot													
Chenopodium cf. vulvaria L.	stinking goosefoot													
Chenopodium spec.	goosefoot	10	2	4						2	10			
Chenopodiaceae	goosefoot family													
Chrysanthemum leucanthemum L.	exeye daisy/woundwort													
Chrysanthemum segetum L.	corn marigold													
Cicer arietinum L.	chick-pea									x				
Cichorium intybus L.	chicory		1	1							2			
Circaea lutetiana L.	enchanter's nightshade													
Cirsium arvense (L.) Scop.	creeping thistle								1	5				1
Cirsium oleraceum (L.) Scop.	cabbage thistle										1			
Cirsium palustre (L.) Scop.	marsh thistle													
Cirsium vulgare (Savi) Ten.	spear thistle													
Cirsium spec.	thistle									1				
Citrullus lanatus (Thunb.) Mansfeld	watermelon													
Cladium mariscus (L.) Pohl.	European cut-sedge													

Late Iron age (La Tène) (5th century BC–1st century AD)	Roman age (1st–middle 5th century AD)	Barbaricum (1st–5th century AD)	Early Migration period (476–568 AD)	Late Migration period (Avar age) (568–895 AD)	Late Migration periode (Caroling age) (811–895 AD)	Late Migration period (German, Slavonic) (6th–10th century AD)	Hungarian conquest time (895 AD)	Early Árpád age (1001–12th century AD)	Late Árpád age (13th century–1301 AD)	Árpád age without epoch	Hungarian Kingdom (14th–16th century AD)	Cumanian (15th–16th century AD)	Turkish occupation (16th–17th century AD)	17th–18th century–Total	19th century–Total	Uncertain age (prehistoric – historic)	Number of remains
																	1
																	1246
								1									1
	1									2	9		1		1	5	28
											361						361
									3								3
											1						1
									7	4	6	2	8	x			xx
	1	5						2		1	557						568
1																	1
											4						4
											3						x
										1							1
											2						2
									3	1							4
									68							260	349
	2																2
															1		1
2080	xxxxx	21866		413	xxx	1	32	xxxxx	6985	12556	4329	378	54		18	xxxx	xxxxxx
		2			2		4		2	1	20		1		4		37
											1036						1037
628	907	2140		1632	137783		12	153	35852	499	40298	3047	1154	179	1091	685	xxxxxx
																1	1
		1		2	1						42	3	42				93
	3										2		18			1	24
		1		22													23
9	9	5			37		23	15	490	6	1102	10	17	17	19	17	xxxx
		1						1				1	15		9		28
2		2		15	2						1		1				30
				103													103
6																	6
12	x	2			2		2				1		1			1	xx
		1							3		23	2					29
									37	1	3	1					42
									21		3					1	25
											1						1
				1					2	1	48	3	1		1		61
											1						1
1		1		45				2	5	49	9	26			4		149
									1	1	5				9	3	20
				2													2
												14	1				15
				2					1								4
									251		97374	63	81		22		97791
									1								1

Latin name	**English name**	Early Neolithic (6000–5300 BC)	Middle Neolithic (5300–4700 BC)	Late Neolithic (4700–4300 BC)	Neolithic without epoch	Early Copper age (4300–3500 BC)	Late Copper age (3500–3000 BC)	Copper age without epoch	Early Bronze age (3000–1600 BC)	Middle Bronze age (1600–1300 BC)	Late Bronze age (1300–900 BC)	Bronze age without epoch	Early Iron age (Hallstatt) (900–500 BC)	Early Iron age (Skythian) (700–500 BC)
Clinopodium vulgare L. (syn. Calamintha clinopodium Spenner, Satureje vulgaris Fritsch.)	wild basil								1		1			1
Coffea arabica L.?	coffee													
Conium maculatum L.	hemlock										5			
Conringia orientalis (L.) Andrz.	hare's-ear mustard									1				
Convolvulus arvensis L.	field bindweed			5		1			x	9	8			
Coriandrum sativum L.	coriander													
Cornus mas L.	cornelian cherry/dogwood					1	220			14	4	1		
Cornus sanguinea L.	blood-twig dogwood/red dogwood	3	1	xx	5									
Cornus spec.	dogwood													
Coronilla varia L.	crownvetch		2											
Corylus avellana L.	European filbert (haselnut)	xxx	1				125				2		1	
Crataegus monogyna Jacq.	single-seed hawthorn			1			2			1				
Crataegus spec.	hawthorn													
Cucumus melo L.	muskmelon													
Cucumus melo L. (Cantaloupe Typ)	muskmelon/cantaloupe "longseed" Typ													
Cucumus melo L. (Togo Typ)	muskmelon/cantaloupe "shortseed" Typ													
Cucumis sativus L.	cucumber													
Cucumis cf. sativus L.	cucumber										3			
Cucumis spec.	muskmelon/cucumber									x				
Cucurbita pepo L.	pumpkin													
Cucurbita pepo L./maxima Duch.	pumpkin/winter squash													
Cucurbita/Lagenaria siceraria (Mol.) Standley	squash/calabash or bottle gourd													
Cucurbita spec.	squash													
Cuscuta cf. campestris Yuncker	field dodder													
Cuscuta europaea L.	greater dodder			1										
Cuscuta spec.	dodder			x										
cf. Cydonia oblonga Mill. emend Beck	quince													
Cyperus fuscus L.	brown galingale										6			
Cyperus cf. fuscus L.	brown galingale													
Cyperus longus L.	common galingale													
Cyperus cf. longus L.	common galingale													
Cyperaceae	cyperaceous plants			x							2			
Dactylis glomerata L.	cocksfoot													
cf. Dactylis glomerata L.	cocksfoot													
Datura stramonium L.	thorn-apple								x					
Daucus carota L.	wild carrot									1	22			
Daucus spec.	carrot									1				
Descurainia sophia (L.) Webb	flixweed													
Dianthus deltoides L.	maiden pink													
Dianthus cf. suberbus L.	large pink													
Dianthus spec.	pink			x										
Digitaria ischaemum (Schreb.) Mühlenb.	smooth finger-grass		1	57	1		1		4	1	27		2	2
Digitaria sanguinalis (L.) Scop.	red finger-grass			22							27			1

Late Iron age (La Tène) (5th century BC–1st century AD)	Roman age (1st–middle 5th century AD)	Barbaricum (1st–5th century AD)	Early Migration period (476–568 AD)	Late Migration period (Avar age) (568–895 AD)	Late Migration periode (Caroling age) (811–895 AD)	Late Migration period (German, Slavonic) (6th–10th century AD)	Hungarian conquest time (895 AD)	Early Árpád age (1001–12th century AD)	Late Árpád age (13th century–1301 AD)	Árpád age without epoch	Hungarian Kingdom (14th–16th century AD)	Cumanian (15th–16th century AD)	Turkish occupation (16th–17th century AD)	17th–18th century–Total	19th century–Total	Uncertain age (prehistoric – historic)	Number of remains
								1				1			1	6	12
	x																x
		2			2			2	50	3	1						65
								5		1	1		19			3	30
	xx	7		118	14	1	3				2	1	29	3		2	xxx
											53	5					58
x	1				2			2	2		4236		27		1	30	xxxx
					1						5				39		xx
													4				4
		2			1				1	2	1	18					27
	1		11	9	426				xx		728	2	18	12	9	4	xxxx
					4							1			2		11
x									x								x
	2								x		5853	30	7		1		xxxx
											23341						23341
									69		9734						9803
3									1		15136	1	1		1		15143
	4								4		3						14
									2		2371		1				xxx
													1				1
															2		2
															2		2
													22				22
						6											6
	29	1				3					7	1					42
																	x
											498	1					499
				1													7
											2						2
				1													1
		1															1
	4	1		3									2		5		xx
				1											2		3
									1								1
													xxxxxx		6		xxxx
				5					269		195		2		3		497
																	1
										1							1
															1		1
									36								36
				1							3						x
1	5	12			50			1	6	4	2	10				2	189
1		44			4				1	1	1	1					103

Latin name	English name	Early Neolithic (6000–5300 BC)	Middle Neolithic (5300–4700 BC)	Late Neolithic (4700–4300 BC)	Neolithic without epoch	Early Copper age (4300–3500 BC)	Late Copper age (3500–3000 BC)	Copper age without epoch	Early Bronze age (3000–1600 BC)	Middle Bronze age (1600–1300 BC)	Late Bronze age (1300–900 BC)	Bronze age without epoch	Early Iron age (Hallstatt) (900–500 BC)	Early Iron age (Skythian) (700–500 BC)
Diplotaxis muralis (L.) DC.	annual wall-rocket													
Echinochloa crus-galli (L.) P. B.	barnyard grass	1	2	xx	2				1		31		2	22
cf. Echinochloa crus-galli (L.) P. B.	barnyard grass													
cf. Echinops sphaerocephalus L.	pale globe-thistle													
Echium vulgare L.	viper's-bugloss													
Eleocharis palustris agg.	common spike-rush		6	115					1		5			
Equisetum arvense L.	field horsetail									x				
Eriophorum vaginatum L.	cottongrass													
Euphorbia amygdaloides L.	wood spurge										1			
Euphorbia cf. amygdaloides L.	wood spurge													
Euphorbia amygdaloides L./ platyphyllos L.	wood/broad-leaved spurge													
Euphorbia cyparissias L.	cypress spurge		1							1	1			
Euphorbia cf. cyparissias L.	cypress spurge													
Euphorbia esula L.	leafy spurge													
Euphorbia exigua L.	dwarf spurge													
cf. Euphorbia exigua L.	dwarf spurge													
Euphorbia helioscopia L.	sun spurge													
Euphorbia cf. helioscopia L.	sun spurge													
Euphorbia lathyris L.	caper spurge													
Euphorbia palustris agg.	marsh spurge								1					
Euphorbia platyphyllos L.	broad-leaved spurge													
Euphorbia segetalis L.	spurge													
Euphorbia spec.	spurge													
Euphorbiaceae	spurge family													
Fagus silvatica L.	European beech						2							
Fallopia convolvulus (L.) A. Löve	black-bindweed	28	89	25	12	6	22		6	91	116	2	x	568
cf. Fallopia convolvulus (L.) A. Löve	black-bindweed	5												
Fallopia dumetorum L. Holub (syn. Bilderdykia dumetorum (L.) Dum.)	hedge buckwheat	1	2							5				
Festuca pratensis Huds.	meadow fescue			1										
Festuca cf. pratensis Huds.	meadow fescue		2							2	1			
Festuca spec.	fescue													
Ficus carica L.	common fig													
Foeniculum vulgare Mill.	common fennel													
Fragaria vesca L.	wild strawberry	2	4											1
Fraxinus spec.	ash													
Fumaria officinalis L.	common fumitory										1			
Fumaria schleicheri Soy.-Will.	small-flowered furmitory								2		4			
Fumaria spec.	fumitory													
Galeopsis ladanum L.	red hemp-nettle													
Galeopsis tetrahit L.	common hemp-nettle													
Galium aparine agg.	catchweed/cleavers	12			1				1	3	3	1	x	1
Galium cruciata L.	cross-wort													
Galium mollugo L.	hedge bedstraw			2							2			
Galium palustre L.	common marsh-bedstraw												x	
Galium spurium L.	false cleavers	64	9	9		1	1		6	4	41			2

Late Iron age (La Tène) (5th century BC–1st century AD)	Roman age (1st–middle 5th century AD)	Barbaricum (1st–5th century AD)	Early Migration period (476–568 AD)	Late Migration period (Avar age) (568–895 AD)	Late Migration periode (Caroling age) (811–895 AD)	Late Migration period (German, Slavonic) (6th–10th century AD)	Hungarian conquest time (895 AD)	Early Árpád age (1001–12th century AD)	Late Árpád age (13th century–1301 AD)	Árpád age without epoch	Hungarian Kingdom (14th–16th century AD)	Cumanian (15th–16th century AD)	Turkish occupation (16th–17th century AD)	17th–18th century–Total	19th century–Total	Uncertain age (prehistoric – historic)	Number of remains
									10	4	3						17
4		48		1	4			5	17	3	6	15				4	xxx
										1	5						6
											2						2
													2		5		7
	56	41		478	2			15	3	3	24	28	5		25	49	856
																	x
									2		8				3		13
																	1
															1		1
											1						1
		2		11			2	4	5	2	139	2			1	2	173
															1		1
	1														3		4
											1						1
		1															1
									54		1						55
											1						1
							398										398
				20	2				16		20		19		13		91
									10		7				25	7	49
															2		2
1		7							3		2		2		33		48
											2						2
																	2
63	96	887		13	311		3	57	1986	67	4674	351	43	x + 5	25	25	xxxx
																	5
	25			1	1				93								128
																	1
		7							1	1	1					1	16
											5	16					21
	13								1		351668	1	445	1	7		352136
											2						2
	5	10							29	2	1409127	3	1455		3	4	1410645
		x															x
									1				1		1	1	5
	8	15	207					216	242	1	22	1	1				719
	1								5				1				7
		1		4							5		2				12
	1										1		1				3
	4	5	5		163			40	2		326	15				1	xxx
	3												24				27
2					1						43	99					149
																	x
32	113	299	44	1	10	1	1	28	13	11	223	6	2	1		20	942

Latin name	English name	Early Neolithic (6000–5300 BC)	Middle Neolithic (5300–4700 BC)	Late Neolithic (4700–4300 BC)	Neolithic without epoch	Early Copper age (4300–3500 BC)	Late Copper age (3500–3000 BC)	Copper age without epoch	Early Bronze age (3000–1600 BC)	Middle Bronze age (1600–1300 BC)	Late Bronze age (1300–900 BC)	Bronze age without epoch	Early Iron age (Hallstatt) (900–500 BC)	Early Iron age (Skythian) (700–500 BC)
Galium tricornutum Dandy (syn. G. tricorne auct.)	corn bedstraw													
Galium verum agg.	lady's bedstraw			3							2		x	
Galium spec.	bedstraw	25	1							2				
Genista cf. ovata W. et K.	greenweed													
Geranium cf. dissectum Jusl.	cut-leaved cranesbill													
Geranium spec.	cranesbill													
Glaucium corniculatum (L.) Rudolph	red horned-poppy		2						4	4				4
Glechoma hederacum L.	ground-ivy										1			
Glyceria maxima (Hartm.) Holmbg.	reed sweet-grass			1										
Gossypium spec.	cotton													
Helianthus annuus L.	common sunflower													
Heliotropium europaeum L.	common heliotrope		1						2		3			1
Hieracium cf. umbellatum L.	leafy hawkweed													
Hibiscus trionum L.	flower-of-an-hour										1		x	
Holosteum umbellatum L.	jagged chickweed													
Holoschoenus vulgaris Lk.	round-headed club-rush													
Hordeum murinum L.	wild barley-grass			3					1	6	1			
Hordeum cf. murinum Huds.	wild barley-grass													
Hordeum vulgare L. subsp. distichum Zoh.	two-rowed barley	8	21	xxx	5					xxxx	275	xxx	15	48
Hordeum vulgare cf. subsp. distichum Zoh.	two-rowed barley													
Hordeum vulgare L. subsp. hexastichum Zoh.	six-rowed barley		25	1213		34	1		102	xxxxx	254	9		
Hordeum vulgare L. polystichum (H. tetrastichum + H. hexastichum)	many-rowed barley	2	380	285			82		1089	xxx	33		300	509
Hordeum vulgare L. tetrastichon	four-rowed barley		1							xxx				
Hordeum vulgare L. var. nudum	naked barley	22	30	xxxx	2		9		8	xxx	20		xx	1
Hordeum vulgare L. cf. var. nudum	naked barley	23												
Hordeum vulgare L. subsp. distichum Zoh. var. nudum	two-rowed naked barley	13	55	71	2		7	1	9	xxxx				17
Hordeum vulgare subsp. polystichum var. nudum	many-rowed naked barley										1			
Hordeum vulgare L. subsp. hexastichum Zoh. var. nudum	six-rowed naked barley		x											
Hordeum vulgare L.	barley	xxx	xxxxx	xx	14	24	231		xxx	xxxxx	882	22	xx	1433
cf. Hordeum vulgare L.	barley	4	x		25					1				
Hordeum/Triticum spec.	barley/wheat	xxx	1102	838	320	11	745		289	282	6034	10	84	48
Hordeum/Triticum/Secale spec.	barley/wheat/rye													
Humulus lupulus L	common hop													
Hyoscyamus niger L.	common henbane		2						1		9		41	
Hypericum perforatum L.	common Saint John's-wort										5			
Hypericum hirsutum L.	hairy Saint John's-wort													
Iris pseudacorus L.	yellow iris										1			
Juglans regia L.	walnut													
Juglans regia L.	walnut lonshaped "papershell"													

Late Iron age (La Tène) (5th century BC–1st century AD)	Roman age (1st–middle 5th century AD)	Barbaricum (1st–5th century AD)	Early Migration period (476–568 AD)	Late Migration period (Avar age) (568–895 AD)	Late Migration periode (Caroling age) (811–895 AD)	Late Migration period (German, Slavonic) (6th–10th century AD)	Hungarian conquest time (895 AD)	Early Árpád age (1001–12th century AD)	Late Árpád age (13th century–1301 AD)	Árpád age without epoch	Hungarian Kingdom (14th–16th century AD)	Cumanian (15th–16th century AD)	Turkish occupation (16th–17th century AD)	17th–18th century–Total	19th century–Total	Uncertain age (prehistoric – historic)	Number of remains
2													1				3
1	36	7			3				7	2	1	36				1	xxx
		1								1	4	1	5				40
									1								1
			1						1								2
	1	2															3
	65	2		4				2	43		35	4	282	x	486	18	xxx
		4		1							3				3		12
								1									2
			2	x													x
															3		3
	5	8								1	4				1		26
										1							1
		2			1			3	59	1	13	3	6	3	7	5	104
															11		11
				11					72								83
	1	1						1		6	2		1			1	24
										2							2
4	xxx	116	22		3864		2	141	69	21	134		6		1	xxx	xxxxx
		2															2
	669										127	659	12				xxxxx
1263	520	48614					19	296	5275	1829	1284		3			128	xxxxx
																	xxx
5	xxx	16		1			1	5		3	14	1				3	xxxx
																	23
132	27	1			1558			98	16	4							xxxx
	107			9					9						8		134
			3	2	22473												22478
xxx	xxx	2434		44	13	6	15	xxx	xx	2191	308	6	13	x +10	1	256	xxxxxx
		57															xx
234	585	129		27				27	14	69	233				7	299	xxxx
									31		49		7		7	71	165
				1					2		247	1					251
1	80	9		8					647		218	8	208		2687	26	3945
				2	1		2		199		14						223
		1															1
					2												3
	xx				1703			x	x	1	xxx + 2169	7	xxx	1	20	6	xxxx
											41				1		1

Latin name	English name	Early Neolithic (6000–5300 BC)	Middle Neolithic (5300–4700 BC)	Late Neolithic (4700–4300 BC)	Neolithic without epoch	Early Copper age (4300–3500 BC)	Late Copper age (3500–3000 BC)	Copper age without epoch	Early Bronze age (3000–1600 BC)	Middle Bronze age (1600–1300 BC)	Late Bronze age (1300–900 BC)	Bronze age without epoch	Early Iron age (Hallstatt) (900–500 BC)	Early Iron age (Skythian) (700–500 BC)
Juglans regia L.	walnut middle large "stoneshell"													
Juglans regia L.	walnut "long shaped/ middle big"													
Juglans regia L. cf. var. connata	walnut "stone"													
Juglans regia L. cf. var. oblonga Hort.	walnut "long shaped"													
Juglans regia L. cf. var. durissima	walnut "hard nutshall"													
Juglans regia L. cf. var. rotunda	walnut "common round"													
Juncus articulatus L.	jointed rush													
Juncus bufonius L.	toad rush													
Juncus spec.	rush													
Knautia arvensis (L.) Coult.	field scabious													
Lactuca serriola Torn.	prickly lettuce													
Lamium amplexicaule L.	henbit dead-nettle										2			
Lamium cf. amplexicaule L.	henbit dead-nettle													
Lamium galeobdolon L. (syn. Galeobdolon luteum Huds.)	yellow dead-nettle													
Lamium purpureum L.	red dead-nettle										11			
Lamiaceae	lamium family										1			
Lapsana communis L.	nipplewort													
Laserpitium latifolium L.	sermountain										1			
Laserpitium pruthenicum L.	gentian													
Lathyrus cf. aphaca L.	yellow vetchling													
Lathyrus cicera L.	red vetchling						1							
Lathyrus cf. hirsutus L.	hairy vetchling													
Lathyrus pratensis L.	meadow vetchling													
Lathyrus sativus L.	grass pea wine		1	xxxx						xxxx	37		2	
Lathyrus tuberosus L.	tuberous pea													
Lathyrus spec.	vetchling	1											1	
Lens culinaris Medic. subsp. microsperma Bar.	small seed lentil	4	53	34	5	4			1	4873	43		7	3
Lens culinaris Medic. subsp. macrosperma Bar.	big seed lentil													
Lens culinaris Medic.	common lentil		5	2180					x	xxxx	98			
Leontodon autumnalis L.	autumn hawkbit													
Leontodon hispidus L.	rough hawkbit													
cf. Leonurus marrubiastrum L.	false motherwort													
Lepidium campestre (L.) R. Br.	common field pepperwort													
Lepidium draba L.	hoary cress/hoary pepperwort													
Lepidium perfoliatum L. R.Br.	pepperwort													
Lepidium spec.	pepperwort													
Ligustrum vulgare L.	common privet													
Linaria vulgaris Mill.	common toadflax													
Linum cf. austriacum L.	Austrian flax													
Linum catharticum L.	purging flax													
Linum cf. catharticum L.	purging flax													
Linum perenne L.	perennial flax													
Linum perenne L./austriacum L.	perennial/Austrian flax										1			
Linum cf. tenuifolium L.	narrow-leaved flax						1							
Linum usitatissimum L.	common flax		292	245	6	4				15				

Late Iron age (La Tène) (5th century BC–1st century AD)	Roman age (1st–middle 5th century AD)	Barbaricum (1st–5th century AD)	Early Migration period (476–568 AD)	Late Migration period (Avar age) (568–895 AD)	Late Migration periode (Caroling age) (811–895 AD)	Late Migration period (German, Slavonic) (6th–10th century AD)	Hungarian conquest time (895 AD)	Early Árpád age (1001–12th century AD)	Late Árpád age (13th century–1301 AD)	Árpád age without epoch	Hungarian Kingdom (14th–16th century AD)	Cumanian (15th–16th century AD)	Turkish occupation (16th–17th century AD)	17th–18th century–Total	19th century–Total	Uncertain age (prehistoric – historic)	Number of remains
															3		3
					9						15		5				9
					x												x
					x												x
					x												x
					x												x
		x															x
		x															x
	1												2				3
									8								8
												7					7
	16	1		2	1				5		27	56	3		4		117
															1		1
															4		4
	28							3			22	6					70
				4					40		6	2	2		1		56
									36			3					39
											1				2		4
															2		2
2											1						3
					2												3
					4							1					5
1		1							1	8							11
	1																xxxx
1																	1
	1				1	1					13					2	20
54	xx	25			21	1			8	1	2	1				21	xxxx
	156						1	1	13	16	xxx		2		1	4	xxx
15	7						62	7		1	xxx		10				xxxx
											1		3		1		5
													1				1
											34						34
38		83						1	6	18	7		213				366
		1							6	1	2		3			1	14
											1						1
			1						1								2
									1						1		2
				1					144								145
											3						3
	1																1
											1						1
				1				1									2
																	1
																	1
52		1250			15					2	21	29				16	1947

Latin name	English name	Early Neolithic (6000–5300 BC)	Middle Neolithic (5300–4700 BC)	Late Neolithic (4700–4300 BC)	Neolithic without epoch	Early Copper age (4300–3500 BC)	Late Copper age (3500–3000 BC)	Copper age without epoch	Early Bronze age (3000–1600 BC)	Middle Bronze age (1600–1300 BC)	Late Bronze age (1300–900 BC)	Bronze age without epoch	Early Iron age (Hallstatt) (900–500 BC)	Early Iron age (Skythian) (700–500 BC)
cf. Linum usitatissimum L.	common flax													
Linum spec.	flax			x					1					
Linaceae	linum family													
Linum/Cannabis spec.	flax/hemp	4	x	x										
Lithospermum arvense L.	field gromwell									xxx	1			
Lithospermum offinale L.	common gromwell									7				
Lolium perenne L.	perenniel ryegrass										1			1
Lolium cf. perenne L.	perenniel ryegrass													
Lolium remotum Schrank	hardy ryegrass		2											
Lolium cf. remotum Schrank	hardy ryegrass		1											
Lolium remotum Schrk./ temulentum L.	hardy ryegrass/darnel													
Lolium temulentum L.	darnel	1									400			
Lolium cf. temulentum L.	darnel												1	1
Lolium spec.	ryegrass								x					
cf. Lolium spec.	ryegrass	1												
Lotus corniculatus agg.	common birdsfoot-trefoil		4							1	3		1	
cf. Luzula spec.	wood-rush													
Lychnis flos-cuculi L.	ragged-robin			26						1	3			
Lycopus europaeus L.	gipsywort									1	27			
Malus domestica Borkh.	apple													
Malus cf. domestica Borkh.	apple													
Malus silvestris agg.	crab apple								8	7	16	1		
Malus cf. silvestris agg.	crab apple													
Malus spec.	apple									1				
Malva alcea L.	large musk-mallow													
Malva cf. alcea L.	large musk-mallow													
Malva moschata L.	musk mallow													
Malva cf. moschata L.	musk mallow													
Malva neglecta Wallr.	dwarf mallow			3					2					
Malva cf. neglecta Wallr.	dwarf mallow													1
Malva pusilla Sm.	northern dwarf mallow		1											
Malva silvestris L.	cammon mallow			11			2		6	1	8	1		
Malva spec.	mallow			1					x				1	
Malvaceae	mallow family													
Marrubium peregrinum L.	branched horehound													
Marrubium peregrinum L./ Ballota nigra L.	branched/black horehound													
Marrubium vulgare L.	white horehound										3			
Matricaria cf. chamomilla L.	wild chamomile													
Matricaria inodora L.	scentless mayweed				2						1			
Medicago cf. falcata L.	yellow sickle medic		1											
Medicago lupulina L.	black medic/hop clover/ yellow trefoil		7	1	1					3	17		5	11
cf. Medicago lupulina L.	black medic/hop clover/ yellow trefoil		1											
Medicago minima (L.) Desr.	bur medic		1	1					1	2	2			
Medicago cf. minima (L.) Desr.	bur medic													
Medicago cf. orbicularis (L.) All.	button medic													
Medicago spec.	medic								1	1	1			1
cf. Medicago spec.	medic	1												

Late Iron age (La Tène) (5th century BC–1st century AD)	Roman age (1st–middle 5th century AD)	Barbaricum (1st–5th century AD)	Early Migration period (476–568 AD)	Late Migration period (Avar age) (568–895 AD)	Late Migration periode (Caroling age) (811–895 AD)	Late Migration period (German, Slavonic) (6th–10th century AD)	Hungarian conquest time (895 AD)	Early Árpád age (1001–12th century AD)	Late Árpád age (13th century–1301 AD)	Árpád age without epoch	Hungarian Kingdom (14th–16th century AD)	Cumanian (15th–16th century AD)	Turkish occupation (16th–17th century AD)	17th–18th century–Total	19th century–Total	Uncertain age (prehistoric – historic)	Number of remains
													1				1
		1	101	4							1				1		xx
									1								1
			3	x													x
	xx	33						4	3		2	159	156		1		xxx
5											58						70
		1							2			51					56
		2								3						1	6
																	2
																	1
																1	1
		5								1				x			xxx
																	2
									2	1						1	4
																	1
	1	679			3			24	5	5	2	7			1		736
				1													1
		1		13					5	1	2		1				53
											27	2	41		8		106
									7		54145		207		3		54362
								1									1
	8				4		1	2	3	1	21	1					73
																1	1
	2								2		27322		2				27329
				1													1
											2		1				3
									2								2
				10													10
		18			2			1	5	7	1	228					267
																	1
		2								3	1					5	12
7	1	30		1	14			9	23	56	213	83	1			6	473
					2	1				1	2		11		1		xx
		2										17					19
									380		1681		5		2		2068
				1													1
		1		2					126		171	6			4		313
															3		3
											12						15
																	1
18	9	90			2			15	10	60	18	20	2		2	16	307
																	1
	2	9														1	19
									2								2
										198							198
		25		1				5	3	1					1	3	43
																	1

Latin name	English name	Early Neolithic (6000–5300 BC)	Middle Neolithic (5300–4700 BC)	Late Neolithic (4700–4300 BC)	Neolithic without epoch	Early Copper age (4300–3500 BC)	Late Copper age (3500–3000 BC)	Copper age without epoch	Early Bronze age (3000–1600 BC)	Middle Bronze age (1600–1300 BC)	Late Bronze age (1300–900 BC)	Bronze age without epoch	Early Iron age (Hallstatt) (900–500 BC)	Early Iron age (Skythian) (700–500 BC)
Melampyrum arvense L.	field cowwheat			8					2	2				1
Melampyrum barbatum W. et K.	bearded cow-wheat													
Melampyrum cf. cristatum L.	crested cow-wheat													
Melampyrum spec.	cow-wheat									x				
cf. Melica nutans L.	mountain melick													
Melilotus albus Desr.	white melilot		2						1		7			2
Melilotus cf. albus Desr.	white melilot		1											
Melilotus albus Desr./altissimus Thuill.	white/tall melilot												1	
Melilotus albus Desr./officinalis (L.) Pall.	white/ribbed melilot								1					
Melilotus altissimus Thuill.	tall melilot													
Melilotus dentatus (W. et K.) Pers.	melilot									1				1
Melilotus officinalis (L.) Pall.	ribbed melilot										1			
Melilotus cf. officinalis (L.) Pall.	ribbed melilot													
Melilotus spec.	melilot													
Melilotus/Trifolium spec.	melilot/clover										1			
Melissa officinalis L.	common balm													
Mentha arvensis L.	corn mint													
Mentha cf. longiflora (L.) Nath.	horse mint										1			
Mentha spec.	mint													
Mercurialis annua L.	annual mercury													1
Mespilus germanica L.	medlar													
Moehringia trinervia (L.) Clairv.	three-nerved sandwort													
Molinia coerulea (L.) Mönch.	purple moor-grass			1										
Molinia cf. coerulea (L.) Mönch.	purple moor-grass													
Morus cf. alba	white mulberry													
Morus nigra L.	black mulberry													
Muscari comosum (L.) Mill.	tassel-hyacinth		2							8				
Myagrum perfoliatum L.	mitre cress													
Myosotis arvensis (L.) Hill.	field forget-me-not/field scorpion grass													
Myosotis cf. arvensis (L.) Hill.	field forget-me-not/field scorpion grass	1												
Myosotis palustris agg.	forget-me-not													
Myosoton aquaticum (L.) Mönch	water chickweed										1			
Nepeta cataria L. (syn. Calamintha nepeta)	cat-mint								1		2			
Neslea paniculata (L.) Desv.	ball mustard								1					1
Nigella arvensis L.	field nigella													
cf. Nuphar lutea (L.) Sm.	yellow water-lily										9			
cf. Ocinum basilicum L.	basil													
Oenanthe aquatica Poiret	water fennel										177			
Oenanthe fistulosa L.	common water-dropwort													
Olea europaea L.	common olive													
Onobrychis viciifolia Scop.	common sainfoin									2				
Ononis spinosa L.	spring restharrow													
Onopordum acanthium L.	cotton thistle										5			
Origanum vulgare L.	wild marjoran										1			
Orlaya grandiflora (L.) Hofm.	white laceflower										1			

Late Iron age (La Tène) (5th century BC–1st century AD)	Roman age (1st–middle 5th century AD)	Barbaricum (1st–5th century AD)	Early Migration period (476–568 AD)	Late Migration period (Avar age) (568–895 AD)	Late Migration periode (Caroling age) (811–895 AD)	Late Migration period (German, Slavonic) (6th–10th century AD)	Hungarian conquest time (895 AD)	Early Árpád age (1001–12th century AD)	Late Árpád age (13th century–1301 AD)	Árpád age without epoch	Hungarian Kingdom (14th–16th century AD)	Cumanian (15th–16th century AD)	Turkish occupation (16th–17th century AD)	17th–18th century–Total	19th century–Total	Uncertain age (prehistoric – historic)	Number of remains
7	1	2			6			5			5		9	x			xx
					25									x			xx
											3						3
													5				x
		x															x
	2	15			2		8	2	5	3	5						54
																1	2
																1	2
																1	2
1															1		2
4																2	8
1									1	1	1			x		1	6
	1																1
		6															6
																	1
											1						1
											2				1	2	5
											1						2
										1							1
1	6	2			1						2		4				17
	1										3514						3515
															1		1
								5									6
									2								2
													287				287
											61345		72	3	3		61423
							1		22		2		4	18			57
	x																x
											3						3
																	1
					2												2
																	1
									24		4		3		1		35
	x	17							4		60					1	xx
	1																1
																	9
											1						1
		24								17					1		219
									2								2
	x																x
																	2
		1							1				1	xxxxx	1		xxxxx
	6	1							10			1				1	24
		8		1							4						14
											75						76

Latin name	English name	Early Neolithic (6000–5300 BC)	Middle Neolithic (5300–4700 BC)	Late Neolithic (4700–4300 BC)	Neolithic without epoch	Early Copper age (4300–3500 BC)	Late Copper age (3500–3000 BC)	Copper age without epoch	Early Bronze age (3000–1600 BC)	Middle Bronze age (1600–1300 BC)	Late Bronze age (1300–900 BC)	Bronze age without epoch	Early Iron age (Hallstatt) (900–500 BC)	Early Iron age (Skythian) (700–500 BC)
Ornithogalum pyramidale L.	spiked star of Betlehem								8	9	1			
cf. Ornitogalum pyramidale L.	spiked star of Betlehem													1
Oxalis acetosella L.	wood-sorrel													
Oxalis cf. corniculata L.	procumbent yellow-sorrel		1											
Panicum miliaceum L.	common millet	1	xx	xxx	5	5	12		10	xxx	xxxx		xxx	958
cf. Panicum miliaceum L.	common millet					1								
Panicum spec.	millet													
Panicum /Echninochloa spec.	millet/barnyard grass													
Papaver dubium L.	long-pod poppy							1						
Papaver cf. dubium L.	long-pod poppy													
Papaver rhoeas L.	field/corn poppy													
Papaver cf. rhoeas L.	field/corn poppy													
Papaver somniferum L.	opium poppy													
Papaver somniferum L. (subsp. setigerum)	opium poppy										7			
Papaver spec.	poppy									1				
Papilionaceae	papilionaceous plants									21	2618		2	1
Papilionaceae cultiv.	papilionaceous plants cultivated	7												
Papilionaceae non cultiv.	papilionaceous plants non cultivated	50	6	7			1							
Pastinaca sativa L.	garden parsnip													
Peucedanum spec.	haarstrang													
Phalaris arundinacea (L.) Rauschert (syn. Baldingera arundinacea Dum., Typhoides arundinacea (L.) Dum.)	reed canary-grass/green reedgrass													
Phleum pratense L.	timothy/herd's-grass			2										
Phoenix dactylifera L.	date													
Phragmites australis (Cav.) Trin.	common reed		xxx	1							2			
cf. Phragmites australis (Cav.) Trin.	common reed										1			
Physalis alkakengi L.	strawberry ground cherry/ alkekengi										53			
Picea abies (L.) Karst.	Norway spruce													
Picris hieracioides L.	hawkweed extongue													
Pimpinella saxifraga L.	burnet saxifraga			11						1				
Pinus silvestris L.	Scotch pine			x										
Pinaceae	pine family													11
cf. Piper nigrum L	black pepper													
Pisum elatius Stev.	wild pea									4			1	
Pisum sativum L.	pea		2	126			1		3	xxxx	686		x	
Pisum sativum L. subsp. arvense	field pea		5	1774					1	385	135			3
cf. Pisum sativum L. subsp. arvense	field pea	3							x					
Pisum sativum L. var. microspermum	small seed pea								x					
Pisum spec.	pea		1		1					6			2	
Plantago lanceolata L.	ribwort-plantain		3	x							1	1	x	4
Plantago major L.	greater plantain													
Plantago media L.	hoary plantain									1				

Late Iron age (La Tène) (5th century BC–1st century AD)	Roman age (1st–middle 5th century AD)	Barbaricum (1st–5th century AD)	Early Migration period (476–568 AD)	Late Migration period (Avar age) (568–895 AD)	Late Migration periode (Caroling age) (811–895 AD)	Late Migration period (German, Slavonic) (6th–10th century AD)	Hungarian conquest time (895 AD)	Early Árpád age (1001–12th century AD)	Late Árpád age (13th century–1301 AD)	Árpád age without epoch	Hungarian Kingdom (14th–16th century AD)	Cumanian (15th–16th century AD)	Turkish occupation (16th–17th century AD)	17th–18th century–Total	19th century–Total	Uncertain age (prehistoric – historic)	Number of remains
																	18
																1	2
					3												3
																	1
xxx	xxxxx	xxxx	1493	x	6648		2	xxx	13516	99	xxxxxx + 1128201	912	xxxx	xxxx	173	xxxx	xxxxxx
																	1
									1								1
									3								3
																	1
											3						3
									2								2
					1						57			x			xxx
2	2	1									457587	18	38				457648
																	7
																	1
3			4						6		7	5				8	2675
	3	1						1									12
		x								1							xx
		29									85		1				115
															1		1
		x															x
					1							1					4
	7																7
			23		1				1					xxx			xxx
																	1
											5558		54				5665
											9						9
											5	2					7
																	12
																	x
																	11
											594						594
	1						4										10
6	41	1		2	5			2	11	1	xx	5	1				xxxx
2	2	12							1							5	2325
																	x
																	x
																	10
	4	24			2	1		3	5	11		335			3	1	xxx
											1						1
																	1

Latin name	English name	Early Neolithic (6000–5300 BC)	Middle Neolithic (5300–4700 BC)	Late Neolithic (4700–4300 BC)	Neolithic without epoch	Early Copper age (4300–3500 BC)	Late Copper age (3500–3000 BC)	Copper age without epoch	Early Bronze age (3000–1600 BC)	Middle Bronze age (1600–1300 BC)	Late Bronze age (1300–900 BC)	Bronze age without epoch	Early Iron age (Hallstatt) (900–500 BC)	Early Iron age (Skythian) (700–500 BC)
Plantago cf. media L.	hoary plantain													
Poa annua L.	annual meadow-grass		2		1									3
Poa cf. annua L.	annual meadow-grass		1		1				1		1			
Poa annua L./pratensis agg.	annual/smooth meadow-grass													1
Poa palustris L.	swamp meadow-grass													
Poa pratensis agg.	smooth meadow-grass			2										1
Poa cf. pratensis agg.	smooth meadow-grass		1							2	2		1	
Poa spec.	meadow-grass	6	4				1						1	2
Poa/Alopecurus spec.	meadow-grass/foxtail								1					
Poaceae non cultiv.	grasses non cultivated	43	xx	xx	1	1	10		23	xxx	278	1	15	74
Poaceae cultiv. (Cerealia)	grasses cultivated (cereals)													1
Polygonum amphybium L.	amphibious bistort										1			
Polygonum aviculare agg.	knotgrass		31	24			2		1	64	418		4	23
Polygonum cf. aviculare agg.	knotgrass													
Polygonum hydropiper L.	water-pepper								1					
Polygonum lapathifolium L.	curltop lady's-thumb		2								5		x	
Polygonum minus Huds.	small water-pepper		5	3							1			4
Polygonum minus Huds./mite Schrank	small/tasteless water-pepper													
Polygonum cf. minus Huds.	small water-pepper													
Polygonum mite Schrank	tasteless water-pepper			3					2		5			1
Polygonum persicaria L.	redshank										1			
Polygonum cf. persicaria L.	redshank	1												
Polygonum spec.	water-pepper	2	5	5										
Polygonaceae	polygonaceous plants		1	2						1			2	3
Portulaca oleracea L.	redshank/purslane									1	1			1
Potamogeton natans L.	broad-leaved pondweed										1			
Potamogeton perfoliatus L.	perfoliate pondweed													
Potamogeton cf. perfoliatus L.	perfoliate pondweed										3			
Potamogeton spec.	pondweed										1			
Potentilla alba L.	white cinsquefoil													
Potentilla anserina L.	silverweed cinquefoil													
Potentilla arenaria Borkh.	cinquefoil										16			
Potentilla argentea agg.	silver cinquefoil										5			
Potentilla erecta (L.) Räuschel	tormentilla cinquefoil								1		6			
Potentilla cf. palustris (L.) Scop.	cinquefoil													
Potentilla reptans L.	creeping cinquefoil		3								46			
Potentilla cf. reptans L.	creeping cinquefoil													
Potentilla rupestris L.	cinquefoil													
Potentilla supina L.	carpet cinquefoil										1			
Potentilla cf. supina L.	carpet cinquefoil										1			
Potentilla spec.	cinquefoil									3				
Primula elatior (L.) Schreb.	oxlip													
Prunella vulgaris L.	common self-heal			15							3			1
Prunus armeniaca L.	apricot													
Prunus avium L.	cherry													
Prunus avium L. subsp. silvestris	wild cherry			x										
Prunus avium L. cultae	sweet cherry													

Late Iron age (La Tène) (5th century BC–1st century AD)	Roman age (1st–middle 5th century AD)	Barbaricum (1st–5th century AD)	Early Migration period (476–568 AD)	Late Migration period (Avar age) (568–895 AD)	Late Migration periode (Caroling age) (811–895 AD)	Late Migration period (German, Slavonic) (6th–10th century AD)	Hungarian conquest time (895 AD)	Early Árpád age (1001–12th century AD)	Late Árpád age (13th century–1301 AD)	Árpád age without epoch	Hungarian Kingdom (14th–16th century AD)	Cumanian (15th–16th century AD)	Turkish occupation (16th–17th century AD)	17th–18th century–Total	19th century–Total	Uncertain age (prehistoric – historic)	Number of remains
											2						2
		5								4					1	2	18
	1	9			4			1		3	3		1				26
		1															2
		x															x
		2								16						5	26
		1								4		6					17
		3		6			1	1	3	4	1	1	1				35
																	1
4	xx	142	5	7	12		2	xx	63	62	5464	55	2		14	21	xxxx
		40							222		85						348
					2											43	46
7	21	141		26	97	1	12	28	15833	172	766	145	104		45	11	17976
	1																1
	1	39										22					63
	1	4			5				2	4	4		1	5	10	1	xx
2		22		1	1				1	3	2				3	3	51
											1						1
																1	1
		7		3	3				1	4		5				1	35
1	2	2	20	1	13		4	7	6		49					1	107
											1						2
	1	10						1		2			1			2	29
										1		22			6	1	39
6	2	3		2					738	2	1						757
																	1
				16												1	17
																	3
																	1
											1				2		3
	1											1			1		3
																	16
									5		1	2			1		14
				30					148		3	4	1			1	194
									1								1
2	2			18	1		1		485		9	3	3		2		575
											1						1
															1		1
											2		1				4
									36								37
									4						1	1	9
											1						1
				1				1		2	1				8		32
	2				x						17	1		8			xx
	2				1										25		28
																	x
	1				1				1		50543	2	1235	109		1	51893

Latin name	English name	Early Neolithic (6000–5300 BC)	Middle Neolithic (5300–4700 BC)	Late Neolithic (4700–4300 BC)	Neolithic without epoch	Early Copper age (4300–3500 BC)	Late Copper age (3500–3000 BC)	Copper age without epoch	Early Bronze age (3000–1600 BC)	Middle Bronze age (1600–1300 BC)	Late Bronze age (1300–900 BC)	Bronze age without epoch	Early Iron age (Hallstatt) (900–500 BC)	Early Iron age (Skythian) (700–500 BC)
Prunus avium L. cultae	sweet cherry													
Prunus avium L./cerasus L.	cherry/sour cherry													
Prunus avium L. cultae/cerasus L.	sweet cherry/sour cherry													
Prunus cerasifera Ehrh.	cherry plum			23										
Prunus cerasus L.	sour cherry													
Prunus cerasus L.	sour cherry (small stone)													
Prunus cerasus L.	sour cherry (large stone)													
Prunus cerasus L. subsp. acida (Dum.) Dost.	sour cherry													
Prunus domestica L.	garden plum			x										
Prunus domestica L. subsp. oeconomica	garden plum													
Prunus domestica L. subsp. oeconomica cv. Besztercei	garden plum convar. Besztercei													
Prunus domestica L. subsp. oeconomica cv. Dobzó	garden plum convar. Dobzo													
Prunus domestica L. subsp. oeconomica cv. Lószemű	garden plum (“horseye”)													
Prunus domestica L. subsp. oeconomica	garden plum (“halfround”)													
Prunus domestica subsp. insititia Jusl.	bullace plum													
Prunus domestica subsp. insititia Jusl. var. Juliana	bullace plum “Juliana”													
Prunus domestica subsp. insititia Jusl. var.subrotunda	bullace plum “subrotunda”													
Prunus domestica subsp. insititia Jusl. var.oxicarpa	bullace plum “oxycarpa”													
Prunus domestica L. subsp. italica (Borkh.) Gams ap. Hegi	greengage													
Prunus domestica subsp. syriaca (Borkh.) Janchen	Mirabella plum													
Prunus fruticosus Pall.	ground cherry/dwarf cherry													
Prunus mahaleb Mill.	mahaleb cherry/St. Lucy cherry													
Prunus padus L.	European bird cherry													
Prunus persica (L.) Batsch	peach													
Prunus persica (L.) Batsch	clingstone peach middle large stone													
Prunus persica (L.) Batsch	peach round shaped stone													
Prunus persica (L.) Batsch	peach egg-shaped stone													
Prunus persica (L.) Batsch	peach oboval shaped stone													
Prunus persica (L.) Batsch	peach big stone													
Prunus spinosa agg.	sloe/blackthorn		2	135						7				
Prunus spinosa agg.	blackthorn (small round)													
Prunus spinosa agg.	blackthorn (small longshaped)													
Prunus spinosa agg. (macrocarpa)	blackthorn (large/round)													
Prunus spinosa agg. (macrocarpa)	blackthorn (large/long)													
Prunus cf. spinosa agg.	sloe/blackthorn				1									
Prunus spec.	plum		1	x										
Punica granatum L.	pomegranate													

Late Iron age (La Tène) (5th century BC–1st century AD)	Roman age (1st–middle 5th century AD)	Barbaricum (1st–5th century AD)	Early Migration period (476–568 AD)	Late Migration period (Avar age) (568–895 AD)	Late Migration periode (Caroling age) (811–895 AD)	Late Migration period (German, Slavonic) (6th–10th century AD)	Hungarian conquest time (895 AD)	Early Árpád age (1001–12th century AD)	Late Árpád age (13th century–1301 AD)	Árpád age without epoch	Hungarian Kingdom (14th–16th century AD)	Cumanian (15th–16th century AD)	Turkish occupation (16th–17th century AD)	17th–18th century–Total	19th century–Total	Uncertain age (prehistoric – historic)	Number of remains
											13253				7	45	13305
	1										5192		136		9		5338
									2								2
									2		605		67				697
									xxx		54640	1	507		4		xxxxx
											14317						14317
											1389						1389
					14												14
	x				7				x		11		xxxx			3	xx
											1209			29	7		1245
											54				1		xx
											251	5			3		xxx
	1										4						5
											26						26
									4		1908	2	63		1		1978
									3		1274						1277
											65						65
											233						233
											17						17
											15						15
1											1090						1091
	1				5						4931						4937
					1						7		2				10
1	5			161				1	x		368	1	8	41	4	1	xxx
											70		2		5		5
				x				1									x
				x				1									x
				x													x
									11						1		12
x					10			1	x		574	2	3224	1	1	2	xxxx
					6						2243					1	2250
								1			588					1	590
											11						11
											563						563
																	1
	1				27				4		6876		35		3		xxxx
											183						183

Latin name	English name	Early Neolithic (6000–5300 BC)	Middle Neolithic (5300–4700 BC)	Late Neolithic (4700–4300 BC)	Neolithic without epoch	Early Copper age (4300–3500 BC)	Late Copper age (3500–3000 BC)	Copper age without epoch	Early Bronze age (3000–1600 BC)	Middle Bronze age (1600–1300 BC)	Late Bronze age (1300–900 BC)	Bronze age without epoch	Early Iron age (Hallstatt) (900–500 BC)	Early Iron age (Skythian) (700–500 BC)
Pyrus achras (Gaertn.) Stohr (syn. P. pyraster (L.) Burgsdorf	wild pear													
Pyrus cf. magyarica Terpo	Hungarian-pear													
Pyrus communis L.	common pear													
Pyrus cf. communis L.	common pear													
Pyrus spec.	pear									x				
Pyrus/Malus spec.	pear/apple													
Quercus robur L./petraea (Matt.) Lieb.	cammon/chestnut oak						1344				2993			
Quercus cf. pubescens Willd.	pubescent oak			xxx										
Quercus spec.	oak		58	3						xx			x	
cf. Quercus spec.	oak													
Ranunculus cf. acris L.	meadow buttercup										1			
Ranunculus aquatilis L.	common water-crowfoot										7			
Ranunculus bulbosus L.	bulbous buttercup									2				
Ranunculus flammula L.	lesser spearwort													
Ranunculus repens L.	creeping buttercup		3	2						1	13			
Ranunculus cf. repens L.	creeping buttercup	1												
Ranunculus sardous Crau.	hairy buttercup													
Ranunculus sceleratus L.	blister buttercup										2			
Ranunculus spec.	buttercup									1				
Raphanus raphanistrum L.	wild radish													
Raphanus cf. raphanistrum L.	wild radish													
Raphanus cf. sativus L.	garden radish													
Raphanus spec.	radish													
Rapistrum perenne L.	steppe cabbage													
Reseda lutea L.	yellow mignonette								2					
Reseda luteola L.	weld mignonette								1					
Rhamnus catharticus L.	common buckthorn										1			
Rhynchospora cf. alba (L.) Vahl.	white beak-sedge													
Ribes uva-crispa L.	gooseberry													
Ribes spec.	gooseberry/red currant													
Rosa cf. arvensis L.	field rose													
Rosa cf. canina L.	dog-rose													
Rosa spec.	rose													
Rubiaceae	madder family	14												
cf. Rubiaceae	madder family	1												
Rubus caesius L.	blackberry									1	15			
Rubus caesius L./fruticosus agg.	balckberry/bramble									3				
Rubus discolor Wh. Et N. (syn. R. procerus P. J. Muell.)	blackberry													
Rubus fruticosus agg.	bramble													
Rubus idaeus L.	wild raspberry													
Rubus spec.	dewberry/raspberry									16				
Rumex acetosa L.	common sorrel		3							6	17			4
Rumex acetosella agg.	sheep's sorrel		1	3					x	1	1			1
Rumex aquaticus L.	Scottish dock										1			
Rumex conglomeratus Murr.	clustered dock					2								
cf. Rumex conglomeratus Murr.	clustered dock													
Rumex conglomeratus Murr./ crispus L.	clustered/curled dock													

Late Iron age (La Tène) (5th century BC–1st century AD)	Roman age (1st–middle 5th century AD)	Barbaricum (1st–5th century AD)	Early Migration period (476–568 AD)	Late Migration period (Avar age) (568–895 AD)	Late Migration periode (Caroling age) (811–895 AD)	Late Migration period (German, Slavonic) (6th–10th century AD)	Hungarian conquest time (895 AD)	Early Árpád age (1001–12th century AD)	Late Árpád age (13th century–1301 AD)	Árpád age without epoch	Hungarian Kingdom (14th–16th century AD)	Cumanian (15th–16th century AD)	Turkish occupation (16th–17th century AD)	17th–18th century–Total	19th century–Total	Uncertain age (prehistoric – historic)	Number of remains
					2												2
					1												1
	2										8320		217				8539
	8								5								13
	1				1					1	11599					1	xxxxx
	1										4375		121				4497
																	4337
											x						xxx
	2			x							2	2	xx			x	xx
																11	11
									2								3
																	7
																	2
					2												2
1	6	5		6	4		3	1	13	2	163	6	11		54	11	305
																	1
											2						2
1				7				1	3						8		22
									4				1				6
2				1	1				2	3	23						32
																2	2
									215		52						267
											3						3
									2				1				3
	2	2					2		77	1	1131	5	3		7		1232
											1						2
		1			5			1									8
									19								19
														2			2
															1		1
													1				1
											36						36
											3384				1		3385
																	14
																	1
				3					41		46011		8		3	1	46083
											1		2				6
													270				270
1	1				4				2		7928		186	1	15		8138
											2387		xxx		15	2	2404
								1			618		1				636
		55			6			8	4	3	1	10			1	1	119
3	1	3						1	147	6	12	3	5		5	1	xxx
												109					110
		x			11												xx
															2		2
									1								1

Latin name	**English name**	Early Neolithic (6000–5300 BC)	Middle Neolithic (5300–4700 BC)	Late Neolithic (4700–4300 BC)	Neolithic without epoch	Early Copper age (4300–3500 BC)	Late Copper age (3500–3000 BC)	Copper age without epoch	Early Bronze age (3000–1600 BC)	Middle Bronze age (1600–1300 BC)	Late Bronze age (1300–900 BC)	Bronze age without epoch	Early Iron age (Hallstatt) (900–500 BC)	Early Iron age (Skythian) (700–500 BC)
Rumex conglomeratus Murr./ sanguineus L.	clustered/wood dock													
Rumex crispus L.	curled dock			1					x	2	1			
Rumex cf. crispus L.	curled dock													1
Rumex hydrolapathum Huds.	water dock			1										
Rumex obtusifolius L.	broad-leaved dock		3	9	1				1		1		x	
Rumex patientia L.	spinach dock													
Rumex cf. pulcher L.	fiddle dock													
Rumex sanguineus L.	wood dock			16										
Rumex cf. scutatus L.	French sorrel													
Rumex cf. thyrsiflorus Finger.	dock													
Rumex spec.	dock								x					
cf. Sagittaria sagittifolia L.	arrowhead										1			
Salix fragilis L.	willow		2											
Salvia nemorosa L.	wild sage													
Salvia cf. nemorosa L.	wild sage													
Salvia pratensis L.	meadow clary									1			x	
Salvia verticillata L.	whorled clary													
Salvia cf. verticillata L.	whorled clary													
Salvia spec.	clary													
Sambucus ebulus L.	danewort	32	24	2		1	21		1	26	293	1	x	
Sambucus nigra L.	common elder									1	4			
Sambucus nigra L./racemosa L.	common elder/red elder	6									19			
Sambucus spec.	elder	1	2								11			
Sanguisorba officinalis L.	great burnet	1	45	2										
Saponaria officinalis L.	soapwort		1	xx							1			
Satureja hortensis L.	summer-savory										1			
Schoenoplectus lacustris (L.) Palla	common bulrush		4	5					55		108			16
Schoenoplectus lacustris (L.) Palla/Scirpus maritimus (L.) Palla	common bulrush/sea club-rush	3				xxxx								
Schoenoplectus mucronatus (L.) Palla	roughseed club-rush		2											
Schoenoplectus (syn. Scirpus) setaceus (L.) Palla	bristle club-rush										67			
Schoenoplectus tabernaemontani (Gmelin) Palla	tabernaemontanus bulrush													
Schoenoplectus cf. tabernaemontani (Gmelin) Palla	tabernaemontanus bulrush										2			
Schoenoplectus spec.	bulrush													
Schoenus nigricans L.	bog-rush								1		2			
Scirpus (Bolboschoenus) maritimus (L.) Palla	sea club-rush													
Scleranthus annuus L.	annual knawel		2	6					12	3	2		1	
Scorzonera humilis L.	common viper's-grass													
Scutellaria galericulata L.	common skullcap													
Secale cereale L.	rye		74	27				2	1	5	3			1
Secale cereale L.	rye "small"										23			
Secale cereale L.	rye "long shaped/small"													
Secale cereale L.	rye "long and thin shaped"									xx				

Late Iron age (La Tène) (5th century BC–1st century AD)	Roman age (1st–middle 5th century AD)	Barbaricum (1st–5th century AD)	Early Migration period (476–568 AD)	Late Migration period (Avar age) (568–895 AD)	Late Migration periode (Caroling age) (811–895 AD)	Late Migration period (German, Slavonic) (6th–10th century AD)	Hungarian conquest time (895 AD)	Early Árpád age (1001–12th century AD)	Late Árpád age (13th century–1301 AD)	Árpád age without epoch	Hungarian Kingdom (14th–16th century AD)	Cumanian (15th–16th century AD)	Turkish occupation (16th–17th century AD)	17th–18th century–Total	19th century–Total	Uncertain age (prehistoric – historic)	Number of remains
		1															1
	1	3	1						4				1		1	1	xx
																	1
	1			1	9				22		15	103			3		155
	1	6			2			3	8	4	1	13	2		4		xx
											4						4
	2										1						3
					11						1				33		61
															1		1
								1					4				5
					2					1	x		1				x
																	1
																	2
		1								1							2
				1							2						3
		3			1						4	1	1		1		12
									3		118		1		91		213
									1						1		2
									2				1		2		5
43	xxx	8			12		5		27340	8	xxx	5	58971		131	460	xxxxx
1	xxxx	9						xxx	4610		442	32	xxxxxx		35	74	xxxxx
									4				1		3		33
	23	6						xxx	6		37		1		5	1	xxx
	1			35	2				760				2				848
	9	14									37	1	1		1		xx
													1				2
1	230	17		529	4		2	3	15	49	38	140	22		34	50	1322
																	xxxx
		1							1			15	2				21
																	67
		7		25				40									72
		1									2		1		1		7
		2															2
		8		2					6		2	8					29
											1						1
		8						1	1978	2	4		1		1		2021
					1												1
				7					2								9
xx	xxxxx	141		8	130		4	xxxx	1086	430	xxxxx	272	xxxxx	x	18	xx	xxxxx
	215	13															xxxx
	xx				278			160									438
		2			154				61		36		xxx	xxx		50	xxx

Latin name	**English name**	Early Neolithic (6000–5300 BC)	Middle Neolithic (5300–4700 BC)	Late Neolithic (4700–4300 BC)	Neolithic without epoch	Early Copper age (4300–3500 BC)	Late Copper age (3500–3000 BC)	Copper age without epoch	Early Bronze age (3000–1600 BC)	Middle Bronze age (1600–1300 BC)	Late Bronze age (1300–900 BC)	Bronze age without epoch	Early Iron age (Hallstatt) (900–500 BC)	Early Iron age (Skythian) (700–500 BC)
Secale cereale L.	rye “long and plump shaped”													
Secale cereale L.	rye “small and squat shaped”									xx				
Secale cereale L.	rye “large and and squat shaped”													
cf. Secale cereale L.	rye													1
Secale silvestre Host	wild rye													
Secale cf. silvestre Host	wild rye													
Senecio aquaticus agg.	marsh ragwort													
Senecio spec.	ragwort													
Seseli annuum L.	steppen-sesel													
Setaria italica (L.) P. B.	foxtail millet			x										
Setaria cf. italica (L.) P. B.	foxtail millet		1											
Setaria lutescens (Weigel) Hubbard	yellow bristle-grass		2								43			2
Setaria verticillata (L.) R. et Sch.	rough bristle-grass		1											
Setaria viridis (L.) P. B.	green bristle-grass									2	1		x	
Setaria viridis (L.) PB./ verticillata (L.) R. et Sch.	green/rough bristle-grass		32	78			2			2	32		4	13
Setaria spec.	bristle-grass						2						5	
Setaria/Echinochloa spec.	bristle/barnyard grass													
Sherardia arvensis L.	field madder			1					1					4
Sideritis montana L.	iron woundwort/ironwort												x	
Silene alba (Mill.) E.H.L. Krause	white campion		1	2						1	77			1
Silene cf. armeria L.	sweet-William catchfly													
Silene dioica (L.) Clairv.	red campion													
Silene cf. dioica (L.) Clairv.	red campion													
Silene noctiflora L. (syn. Melandrium noctiflorum (L.) Fa)	night-flowering catchfly													
Silene nutans L.	nodding catchfly										1			
Silene vulgaris (Mönch) Garcke	bladder campion								1					
Silene spec.	campion										2			
Sinapis alba L.	white mustard													
Sinapis arvensis L.	charlock			1		xxx			1	7	5			
Sinapis cf. arvensis L.	charlock													
Solanum dulcamara L.	woody nightshade		1				4				6			
Solanum cf. dulcamara L.	woody nightshade													
Solanum nigrum L.	black nightshade	2	5	5					2		173			1
cf. Solanum nigrum L.	black nightshade													
Solanaceae	solanum family													
Sonchus arvensis L.	perennial sow-thistle													
Sonchus asper (L.) Hill.	prickly sow-thistle										1			
Sonchus oleraceus L.	smooth sow-thistle													
Sorbus domestica L.	sorvice tree													
Sorbus torminalis (L.) Cr.	wild service													
Sorbus spec.	sorvice													
Sparganium erectum L.	branched bur-reed			16										
cf. Sparganium minimum Wallr.	least bur-reed													
Sparganium minimum Wallr./ erectum L.	least/common bur-reed													
Spergula arvensis L.	field woundwort													

Late Iron age (La Tène) (5th century BC–1st century AD)	Roman age (1st–middle 5th century AD)	Barbaricum (1st–5th century AD)	Early Migration period (476–568 AD)	Late Migration period (Avar age) (568–895 AD)	Late Migration periode (Caroling age) (811–895 AD)	Late Migration period (German, Slavonic) (6th–10th century AD)	Hungarian conquest time (895 AD)	Early Árpád age (1001–12th century AD)	Late Árpád age (13th century–1301 AD)	Árpád age without epoch	Hungarian Kingdom (14th–16th century AD)	Cumanian (15th–16th century AD)	Turkish occupation (16th–17th century AD)	17th–18th century–Total	19th century–Total	Uncertain age (prehistoric – historic)	Number of remains
	xx													xxxx			xxxx
					90				10				xxx	xxx		94	xxx
					1710												1710
																	1
		x												3			3
	xxxx	1			1												xxxx
				3								3				1	7
									1								1
													16				16
					1			1					1				x
		x							1								x
	2	3		2	2			1	353		914	15	8	15	22	1	1385
		7			56			1	1		4						70
3	5	7			18			6			29		5		1		xx
24		155		6			2		62	34	43289	37			1	17	43790
	1	1									69						78
									8								8
		18						22	1	22	7	1				1	78
																	x
	5	1						1	779	6	191	83	11		10		1169
									72								72
											3						3
											1	5					6
											4						4
																	1
								1	1		3						6
																	2
										5	1						6
2	2	47			4		3	12	218	2785	959	384	30	x + 1	35	44	xxxx
	1																1
1				4	22				147		67164		4		9		67362
									1								1
16	1	12			123				132		1163	95	301		280	5	2316
																1	1
															1		1
		1										4					5
											1		1		1		4
	2			1							23	1					27
	2				1						1202						1205
	1																1
															1		1
										1							17
												3					3
	1																1
									2								2

Latin name	English name	Early Neolithic (6000–5300 BC)	Middle Neolithic (5300–4700 BC)	Late Neolithic (4700–4300 BC)	Neolithic without epoch	Early Copper age (4300–3500 BC)	Late Copper age (3500–3000 BC)	Copper age without epoch	Early Bronze age (3000–1600 BC)	Middle Bronze age (1600–1300 BC)	Late Bronze age (1300–900 BC)	Bronze age without epoch	Early Iron age (Hallstatt) (900–500 BC)	Early Iron age (Skythian) (700–500 BC)
Staphylea pinnata L.	shrubberry													
Stachys annua L.	annual woundwort	1		3						5	56			1
Stachys annua L./arvensis L.	annual/field woundwort										28			
Stachys arvensis (L.) L.	field woundwort								4		27			
Stachys palustris L.	marsh woundwort													
Stachys cf. palustris L.	marsh woundwort										2			
Stachys recta agg.	perennial yellow-woundwort													
Stachys spec.	woundwort										2			
Stellaria graminea L.	lesser stitchwort									1				
Stellaria holostea L.	greater stichwort													
Stellaria media agg.	common chickweed		3	1	1				3	2	23			
Stipa spec.	feather-grass					x								
Tanacetum vulgare L. (syn. Chrysanthemum vulgare (L.) Bernh.)	common tansy													
Taraxacum officinale agg.	common dandelion										3			
Tetragonolobus maritimus (L.) Roth (syn. Lotus siliquosus L.)	asparagus pea													
Teucrium chamaedrys L.	wall germander						55		10	14	29		1	2
Teucrium cf. chamaedrys L.	wall germander	1												
Teucrium scordium agg.	water germander										9			
Teucrium spec.	germander		1											
Thalicrum flavum agg.	common meadow-rue		1	1		1			1					10
Thalictrum minus L.	lesser meadow-rue	1												
Thalictrum spec.	meadow-rue													
Thlaspi arvense L.	field pennycress										1			
Thymeleae passerina (L.) Coss. Et Germ.	En (-)/Gewöhnlicher Vogelkopf (D)													2
Thymus spec.	thyme													
Tilia cordata	small leaved lime/linden													
Tilia platyphyllos Scop.	large leaved lime/linden													
Tilia cf. platyphyllos Scop.	large leaved lime/linden													
Tilia spec.	lime/linden													
Torilis japonica (Houtt.) DC	upright hedge-parsley													
Trapa natans	water-chestnut	14												
cf. Trapa natans	water-chestnut	306	49											
Trifolium arvense L.	hare's-foot clover		2	164					22	2	13	1	8	
Trifolium cf. arvense L.	hare's-foot clover	2	1								6			
Trifolium arvense L./campestre Schreb.	hare's-foot/large hop clover													
Trifolium campestre Schreb.	large hop clover													
Trifolium cf. campestre Schreb.	large hop clover												3	
Trifolium hybridum L.	alsike clover													
Trifolium cf. hybridum	alsike clover													1
Trifolium incarnatum L.	crimson clover													
Trifolium pratense (L.) Kelch	red clover			6							1	3		
Trifolium repens (L.) Kelch	white clover			8									1	
Trifolium spec.	clover			10					4	1	5			
Triticum aestivum L. subsp. compactum (Host.) MacKey	club wheat	1	24	5	1		3		x	xxx	12		6	1

Late Iron age (La Tène) (5th century BC–1st century AD)	Roman age (1st–middle 5th century AD)	Barbaricum (1st–5th century AD)	Early Migration period (476–568 AD)	Late Migration period (Avar age) (568–895 AD)	Late Migration periode (Caroling age) (811–895 AD)	Late Migration period (German, Slavonic) (6th–10th century AD)	Hungarian conquest time (895 AD)	Early Árpád age (1001–12th century AD)	Late Árpád age (13th century–1301 AD)	Árpád age without epoch	Hungarian Kingdom (14th–16th century AD)	Cumanian (15th–16th century AD)	Turkish occupation (16th–17th century AD)	17th–18th century–Total	19th century–Total	Uncertain age (prehistoric – historic)	Number of remains
																	x
4	389	1		1	2			34	29530	9	870	145	7	7	8	25	31098
		14							1861	8	307		140		1229		3587
1	14			1			2		10	1	276						336
				1													1
									48		2	23	1		2		78
									1						3		4
									6		x					1	9
					1							1	4		1		8
					1						1				12		14
	34	5		1	2			8	112		147	2	7		77		428
																	x
		2		2													4
								2	4		7	1	1		107		125
					1												1
1	6	12		15	17		1	2	50	4	7	3	1		4	6	240
																	1
				4								1					14
		1															2
	1	93							2	14	3	5			1	4	137
		16		2							17						36
				1													1
		1						1	342	4	422	8	123		4		906
		18							22	9	3		1		2	1	58
													1				1
															1		1
					1												1
									1								1
											2				5		7
															1		1
					1				x								xx
																	355
1	1	47			5			25	5	15	22	1				12	346
		5								1						2	17
																5	5
14	2	10							3	2	1				1		33
										1							4
					1			1				2					4
																	1
															1		1
2	3	2			11			3	10		4	13			1	1	60
	1	5							1		1	235	2		2		256
		1						1					26				48
76	48	25						9	53	101	xxxxx		1		3	3	xxxx

Latin name	English name	Early Neolithic (6000–5300 BC)	Middle Neolithic (5300–4700 BC)	Late Neolithic (4700–4300 BC)	Neolithic without epoch	Early Copper age (4300–3500 BC)	Late Copper age (3500–3000 BC)	Copper age without epoch	Early Bronze age (3000–1600 BC)	Middle Bronze age (1600–1300 BC)	Late Bronze age (1300–900 BC)	Bronze age without epoch	Early Iron age (Hallstatt) (900–500 BC)	Early Iron age (Skythian) (700–500 BC)
Triticum aestivum L. subsp. compactum (Host.) MacKey	club wheat strong compactoid typ													
Triticum aestivum L. subsp. compactum (Host.) MacKey	club wheat round on the basal site													
Triticum aestivum L. subsp. vulgare (Vill.) MacKey	cammon wheat	3	14	225			1		18	xxx	52	2	x	
Triticum aestivum L. subsp. vulgare (Vill.) MacKey	common wheat "short and squat shaped"												x	
Triticum aestivum L. subsp. vulgare (Vill.) MacKey	common wheat "squat and column shaped"													
Triticum aestivum L. subsp. vulgare (Vill.) MacKey	common wheat "squat/ elliptical shaped with convex lateral site													
Triticum aestivum L. subsp. vulgare (Vill.) MacKey	common wheat "long shaped"			1										2
Triticum aestivum L. subsp. vulgare (Vill.) MacKey	common wheat "long and narrow shaped"													
Triticum aestivum L. subsp. vulgare (Vill.) MacKey	common wheat "long and egg shaped"													
Triticum aestivum L. subsp. vulgare (Vill.) MacKey	common wheat "oval and flat shaped"													
Triticum aestivum L. subsp. vulgare (Vill.) MacKey	common wheat "funnel shaped towards scutellum"													
Triticum aestivum L. subsp. vulgare (Vill.) MacKey	common wheat "funnel shaped towards scutellum/ small"													
Triticum aestivum L. subsp. vulgare (Vill.) MacKey var. barbarossa	cammon wheat													
Triticum aestivum L. subsp. vulgare (Vill.) MacKey var. hostianum compactoides	cammon wheat													
Triticum aestivum L. subsp. vulgare (Vill.) MacKey var. erythrospermum	cammon wheat													
Triticum aestivum L. subsp vulgare (Vill.) Mackey/T. turgidum cv. durum (Desf.) Mackey	common/durum wheat	13	7	619						7	4			11
Triticum aestivum L. subsp vulgare (Vill.) Mackey inc. compactum/T. turgidum cv. durum (Desf.) Mackey	common/club/durum wheat													
Triticum aestivum L. subsp. spelta (L.) Thell.	spelta wheat		26	xxx	2		4	2	2	51	xxxx	91		19
Triticum aestivum L. cf. subsp. spelta (L.) Thell.	spelta wheat	6	4											3
Triticum aestivum L.	bread wheat		1	xxx					1		1			
Triticum aestivum L./Secale cereale L.	bread wheat/rye													
Triticum monococcum L.	einkorn	67	xxx	xxxxx	15	14	510	508	xxx	xxxxx	xxxx	4	xx	394
Triticum monococcum L.	oneseed einkorn	18												
Triticum monococcum L.	twoseeds einkorn	21	1											
Triticum monococcum L.	einkorn "long and flat shaped"									425				
Triticum monococcum L.	einkorn "squat and plump shaped"									144				
Triticum cf. monococcum L.	einkorn	9	3											
Triticum monococcum/dicoccum	einkorn/emmer	69	144	253		51			19		25		2	9

Late Iron age (La Tène) (5th century BC–1st century AD)	Roman age (1st–middle 5th century AD)	Barbaricum (1st–5th century AD)	Early Migration period (476–568 AD)	Late Migration period (Avar age) (568–895 AD)	Late Migration periode (Caroling age) (811–895 AD)	Late Migration period (German, Slavonic) (6th–10th century AD)	Hungarian conquest time (895 AD)	Early Árpád age (1001–12th century AD)	Late Árpád age (13th century–1301 AD)	Árpád age without epoch	Hungarian Kingdom (14th–16th century AD)	Cumanian (15th–16th century AD)	Turkish occupation (16th–17th century AD)	17th–18th century–Total	19th century–Total	Uncertain age (prehistoric – historic)	Number of remains
					2524											xx	xxxx
					280												280
128	xxxx	55	2	xx	89		11	xxxx	xxx	67	xxxxx	240	xxx + 303	x + 4	11	50	xxxxx
	xxx	5	xxxxx		xxx			xxxxx	284	1115			xxx	xx		xx	xxxxx
														xxxxx		xx	xxxx
											49		xxxx				xxxx
		19															22
	xxx		xxxx		215			xxxx	252	356	290			x			xxxx
											22						22
	240				1597												1837
	xxx				1017												1017
	562																562
															1		1
															1		1
														x	xxxx		xxxx
		5															666
x																	x
xx	9	12							2		8						xxxx
																34	47
	x				1			xxx	9	897					1		xxxx
											222						222
xx		376			2			1	3	4	3			78		80	xxxxx
	xxxx																xxxx
																	22
																	425
																	144
		1								1						3	17
5	1	429			1			1		2						188	1199

Latin name	**English name**	Early Neolithic (6000–5300 BC)	Middle Neolithic (5300–4700 BC)	Late Neolithic (4700–4300 BC)	Neolithic without epoch	Early Copper age (4300–3500 BC)	Late Copper age (3500–3000 BC)	Copper age without epoch	Early Bronze age (3000–1600 BC)	Middle Bronze age (1600–1300 BC)	Late Bronze age (1300–900 BC)	Bronze age without epoch	Early Iron age (Hallstatt) (900–500 BC)	Early Iron age (Skythian) (700–500 BC)
Triticum monococcum/ dicoccum/spelta	einkorn/emmer/spelt	490												
Triticum dicoccum/durum/ aestivum	emmer/durum/common wheat	3												
Triticum turgidum L. subsp. dicoccum (Scrank) Thell.	emmer	143	xxxxx	xxxxx	70	11	226	xxx	329	xxxxx	xxxxx	559	382	214
Triticum turgidum L. subsp. dicoccum (Scrank) Thell.	emmer "long shaped"			xxxx						460				
Triticum turgidum L. subsp. dicoccum (Scrank) Thell.	emmer "wide and flat shaped"		xxxxx	xx										
Triticum turgidum L. subsp. dicoccum (Scrank) Thell.	emmer "rectangular shaped"			237						494				
Triticum turgidum L. subsp. dicoccum (Scrank) Thell.	emmer "einkorn charakter"		xxx	xxx							54			
Triticum cf. turgidum L. subsp. dicoccum (Schrank) Thell.	emmer	42	12								1			
Triticum turgidum L. subsp. dicoccum/spelta	emmer/spelt		2											
Triticum cf. turgidum L. subsp. dicoccum (Schrank) Thell.	new typ emmer	117												
Triticum vulgare antiquorum Heer	wheat			x						x				
Triticum tetraploid (cf. Triticum parvicoccum Kislev)	tetraploid free treshing (naked) wheat	1	1											
Triticum cf. boeoticum Boiss.	wild einkorn		16						1					
Triticum spec.	wheat	xxx	xxx	xxxxx	27		3		5	xxxxx	2614	x	5	3
Triticum spec.	hulled wheat									7		13		
Triticum spec.	new typ hulled wheat	33												
cf. Triticum spec.	new typ hulled wheat	8												
cf. Triticum spec.	wheat	3												
Typha cf. angustifolia L.	narrow-leaved cattail		8	x										
Typha latifolia L.	common cattail		4	x										
Typha spec.	cattail		2											
Ulmus scabra Mill.	elm													
Urtica dioica L.	common nettle										4			
Urtica urens L.	annual nettle			14										
Vaccaria pyramidata Medic.	cow soapwort													
Valerianella cf. carinata Lois.	keeled-fruited cornsalad													
Valerianella dentata Poll.	narrow-fruited cornsalad										5			
Valerianella locusta (l.) Latterade	common cornsalad													
Verbascum spec.	mullein								x					
Verbena officinalis L.	common vervain										11			
Veronica hederifolia agg.	ivy-leaved speedwell								1		3			
Veronica persica Poir.	common field-speedwell				1									
Veronika spec.	speedwell													
Viburnum lantana L.	wayfaring tree													
Viburnum opulus L.	European cranberry bush													
Vicia angustifolia L.	narrow-leaved vetch/ summer vetch			8					3	x	5		3	
Vicia cf. angustifolia L.	narrow-leaved vetch/ summer vetch	1					1							
Vicia angustifolia L./tetrasperma (L.) Schreb.	narrow-leaved vetch/ summer vetch/smooth tare		1											

Late Iron age (La Tène) (5th century BC–1st century AD)	Roman age (1st–middle 5th century AD)	Barbaricum (1st–5th century AD)	Early Migration period (476–568 AD)	Late Migration period (Avar age) (568–895 AD)	Late Migration periode (Caroling age) (811–895 AD)	Late Migration period (German, Slavonic) (6th–10th century AD)	Hungarian conquest time (895 AD)	Early Árpád age (1001–12th century AD)	Late Árpád age (13th century–1301 AD)	Árpád age without epoch	Hungarian Kingdom (14th–16th century AD)	Cumanian (15th–16th century AD)	Turkish occupation (16th–17th century AD)	17th–18th century–Total	19th century–Total	Uncertain age (prehistoric – historic)	Number of remains
																	490
																	3
xxx	xxxx	1236		1	6			49		12	x					109	xxxxx
																	xxxx
																	xxxxx
																	731
																	xxx
																	55
																	2
																	117
																	x
																	2
	38																55
17	xxxxx	xxxx	1					14	6	1	xxx		xxxx	3		xx	xxxxxx
										1							21
																	33
																	8
																	3
																	x
																	x
	1							1	x								x
														x			x
	2			16				1	660		1951		4		57		2695
				6					397		1	1	16				435
											328			x			328
											2						2
	4	4		1				8	236	13	124	3	1		9	55	463
				2													2
																	x
									40		3	1			5	10	70
14									2			1					21
																	1
											1						1
											40		4				44
	1										57	10					68
8	2	12			12			3		9	6	2	1		2	5	xx
																	2
																	1

Latin name	English name	Early Neolithic (6000–5300 BC)	Middle Neolithic (5300–4700 BC)	Late Neolithic (4700–4300 BC)	Neolithic without epoch	Early Copper age (4300–3500 BC)	Late Copper age (3500–3000 BC)	Copper age without epoch	Early Bronze age (3000–1600 BC)	Middle Bronze age (1600–1300 BC)	Late Bronze age (1300–900 BC)	Bronze age without epoch	Early Iron age (Hallstatt) (900–500 BC)	Early Iron age (Skythian) (700–500 BC)
Vicia cracca L.	tufted vetch			x										
Vicia ervilia (L.) Willd.	bitter vetch			2						769	1251		2	
Vicia faba L.	horsebean			x										
Vicia faba L. var. minor Beck	small seed horsebean	1							1	22	2		xxx	
Vicia grandiflora Scop.	large yellow-vetch													
Vicia hirsuta (L.) Gray	hairy tare									1	1			
Vicia cf. hirsuta (L.) Gray	hairy tare	1												
Vicia sativa L.	common vetch			1						4				
Vicia cf. sativa L.	common vetch												1	
Vicia sativa L. subsp.sativa var. platysperma (syn. V. s. var. lentisperma Rpcs.)	common vetch (Linsenwicke - D)													
Vicia cf. sepium L.	bush vetch													
Vicia tetrasperma (L.) Schreb.	smooth tare		1							2				
Vicia cf. tetrasperma (L.) Schreb.	smooth tare													
Vicia villosa Roth.	fodder vetch													
Vicia spec.	vetch		1	x						5	2		x	
Viola arvensis agg.	field pansy		1	1						2				
Viola cf. arvensis agg.	field pansy													
Viola arvensis/tricolor agg.	field pansy/wild pansy													
Viola cf. elatior Fr.	tall violet										1			
Viola odorata L.	sweet violet													
Viola cf. odorata L.	sweet violet													
Viola reichenbachiana Jord. ex Boreau (syn. Viola cf. silvestris Lam.)	pale wood-violet/Sylvan										4			
Viola reichenbachiana Jord. ex Boreau/riviniana Rchb.	Sylvan/Rivini violet													
Viola tricolor agg.	wild pansy													
Viola spec.	violet										1			
Violaceae	violet family													
Vitis vinifera L. subsp. silvestris (Gmel.) Hegi	woodland European grape			1			18				7			
Vitis vinifera L. subsp. vinifera (Gmel.) Hegi	grape vine (wine and delicacy)										1			
Vitis spec.	grape		1											
Zannichellia palustris L.	horned pondweed													
Xanthium italicum Mor.	Italian cocklebur													
Xanthium cf. italicum Mor.	Italian cocklebur													
Xanthium strumarium L.	rough cocklebur			2										8
Xanthium spec.	cocklebur													
Indet.	non identified	25	547	xxxxx	34	10	103		xxx	xxxxx	956	30	xxx	300
Fruit stalk (pomum)	fragment		1											
Fruit-meat	dried fruit fragment													
Fruit-meat?	dried fruit fragment	x												
Acornshell/Stone	fragment	x												
Moss	fragment										2			
Flower bud	fragment			1						2				
Stem	fragment													

Late Iron age (La Tène) (5th century BC–1st century AD)	Roman age (1st–middle 5th century AD)	Barbaricum (1st–5th century AD)	Early Migration period (476–568 AD)	Late Migration period (Avar age) (568–895 AD)	Late Migration periode (Caroling age) (811–895 AD)	Late Migration period (German, Slavonic) (6th–10th century AD)	Hungarian conquest time (895 AD)	Early Árpád age (1001–12th century AD)	Late Árpád age (13th century–1301 AD)	Árpád age without epoch	Hungarian Kingdom (14th–16th century AD)	Cumanian (15th–16th century AD)	Turkish occupation (16th–17th century AD)	17th–18th century–Total	19th century–Total	Uncertain age (prehistoric – historic)	Number of remains
											1						x
	213															13	2250
7	8										xxx		209	xx	xx		xxx
	10							2	1								xx
					8												8
1																	3
											16						17
4											2			x			xx
																	1
							41										41
																6	6
		x			1			1			62						xx
				1											1		2
					10												10
	x	5						1		9	7		16		2	2	xx
	5	3						2	1	2	5				3	2	27
1											1						2
												4					4
																	1
															1		1
															6		6
									59				1				64
												1					1
		3						2	64						3		72
									2						4		7
																2	2
1		1						1			2						31
11	xxxx	10		2	394			28	27		699687	5	61208	xx + 90	32	3	xxxxxx
	xxx				x				1		xx		150				xxx
				98													98
									1								1
									3		1				1		5
		4			2			3			2	1	1		2		25
3		1							6	1		8	1				20
269	xx	1060		xxx		5	16	63801	244	523	590	174	40	1	11	xxx	68739
										1	17	x	11				xx
	7												87				94
																	x
																	x
				1							83	x			2		xx
					1							x	1				5
										67							67

Latin name	English name	Early Neolithic (6000–5300 BC)	Middle Neolithic (5300–4700 BC)	Late Neolithic (4700–4300 BC)	Neolithic without epoch	Early Copper age (4300–3500 BC)	Late Copper age (3500–3000 BC)	Copper age without epoch	Early Bronze age (3000–1600 BC)	Middle Bronze age (1600–1300 BC)	Late Bronze age (1300–900 BC)	Bronze age without epoch	Early Iron age (Hallstatt) (900–500 BC)	Early Iron age (Skythian) (700–500 BC)
Berry									1	x				
Rind of tree	fragment													
Carnivora excrement											2		1	
Carnivora excrement		1	8											
Earthworm-mumy	Lumbricidae													
Fruit-mumy														
Bud							5	3		1	3		8	
Bud-basic				37						2				
Bud-scale											7			
Flower/bud		2												
Leaves of tree										1				
Inflorescence	fragment													
Flower–stalk														
Ergot	Claviceps purpurea (Fries) Tulasne		1								6			2
Monilia on apricot (Prunus armeniaca L.)	fruit mumy with Monila sporodochiums													
Mushrum screrotium	gomba sclerotium		1	23			4				42			1
Bunt sporodochium										1				
Mole/mildew	Tilletia spec.													
Nematode	Tilenchus tritici								5		3			
Nematode on wheat grain	Tilenchus tritici									7				
Nematode on emmer	Tilenchus tritici													
Nematode on barley	Tilenchus tritici													
Nematode on millet	Tilenchus tritici													
Carnivora excrement											2		1	
Carnivora excrement		1	8											
Pestprint on sweet cherry drupa (stone)	pestprint on Prunus avium L. cultae													
Pestprint on sour cherry drupa (stone)	pestprint on Prunus cerasus L.													
Pestprint on ground/dwarf cherry drupa (stone)	pestprint on Prunus fruticosus Pall.													
Pestprint on charlock seed	pestprint on Sinapis arvensis L.													
Weevil prints on peas	pestprint of Bruchus pisorum													
Weevil prints on grains	pestprint of Calandra granaria													
Gall (Quercus)										x				
Web														
Colored (yellowish) cabvas web	fragment			x										
Textile	fragment		x											
Textile made from animal hair?	fragment		x											

Legend (x = in case not mentioned number of seed):
x = sporadic (1–10 pieces, mean value 5),
xx = little (10–100 pieces, mean value 50),
xxx = much (100–1000 pieces, mean value 500),
xxxx = large (1000–10,000 pieces, mean value 5000),
xxxxx = extra large (10,000–100,000 pieces, mean value 50,000),
xxxxxx = extremely large (more than 100,000 pieces, mean value 500,000)

Late Iron age (La Tène) (5th century BC–1st century AD)	Roman age (1st–middle 5th century AD)	Barbaricum (1st–5th century AD)	Early Migration period (476–568 AD)	Late Migration period (Avar age) (568–895 AD)	Late Migration periode (Caroling age) (811–895 AD)	Late Migration period (German, Slavonic) (6th–10th century AD)	Hungarian conquest time (895 AD)	Early Árpád age (1001–12th century AD)	Late Árpád age (13th century–1301 AD)	Árpád age without epoch	Hungarian Kingdom (14th–16th century AD)	Cumanian (15th–16th century AD)	Turkish occupation (16th–17th century AD)	17th–18th century–Total	19th century–Total	Uncertain age (prehistoric – historic)	Number of remains
		2		1													x
										3							3
	1				1					37		x					xx
																	9
													1				1
											1	1					2
1	8			1	1				358	1	26	xx	7		16	18	xxx
					76										1		116
											2		1		266		276
																	2
																	1
											6	24					30
									1								1
																2	11
					1												1
					1	1			2		135				80	3	293
												x					x
											7						7
		1									2						11
																	7
		33															33
		7															7
		1															1
	1				1					37		x					xx
																	9
											903						903
									1		53						54
											7						7
											32						32
					x												x
					x												x
																	x
													1				1
																	x
																	x
																	x

Table 2. Seed, fruit, food and drink remains in Hungary, 1860–2006, without uncerten cultures/ages. Completed by F. Gyulai, 2006

	Early Neolithic (6000–5300 BC)	Middle Neolithic (5300–4700 BC)	Late Neolithic (4700–4300 BC)	Early Copper age (4300–3500 BC)	Late Copper age (3500–3000 BC)	Early Bronze age (3000–1600 BC)	Middle Bronze age (1600–1300 BC)	Late Bronze age (1300–900 BC)	Early Iron age (Hallstatt) (900–500 BC)	Early Iron age (Skythian) (700–500 BC)	Late Iron age (La Tène) (5th century BC–1st century AD)	Roman age (1st–middle 5th century AD)
Piece of seed and fruitremains	4456	163725	277944	5873	4564	5449	382991	91909	3452	14123	7219	239651
Piece of food and drinkremains	40	86	53	110	26	925	90	753	1466	6	653	169
Total macro remains	4496	163811	277997	5983	4590	6374	383081	92662	4918	14129	7872	239820
Number of plant taxons	53	104	119	19	38	99	126	202	57	69	98	163
Cereal species	12	15	14	4	8	9	13	11	8	12	11	11
Cereal (ecotyps, varietas, etc.)	2	4	6	0	1	1	5	3	1	2	1	10
Cereal grains	1578	156763	165295	264	1797	4190	219692	67691	2107	10394	5388	171040
Leguminous species	3	4	3	1	2	3	8	5	6	2	4	6
Leguminous (ecotyps, varietas, etc.)	0	0	0	0	0	2	1	1	1	0	0	1
Leguminous seeds	8	66	9122	4	2	21	21060	2252	517	6	88	488
Vegetables species	0	0	0	0	0	0	0	2	0	0	1	2+1
Vegetables seeds	0	0	0	0	0	0	0	4	0	0	3	6
Import plant species	0	0	0	0	0	0	0	0	0	0	0	4
Import plant remains	0	0	0	0	0	0	0	0	0	0	0	30
Oil and fibre crops species	0	1	2	1	0	1	1	2	2	0	3	1
Oil and fibre (ecotyps, varietas, etc.)	0	0	0	0	0	0	0	1	0	0	0	0
Oil and fibre crops seeds	0	292	246	4	0	5	15	366	6	0	57	2
Fruit and grape species	0	0	1	0	0	0	1	1	0	0	2	11
Fruit and grape (ecotyps, varietas, etc.)	0	0	0	0	0	0	0	0	0	0	0	6
Fruit and grape seeds	0	0	5	0	0	0	5	1	0	0	12	5133
Spice species	0	0	0	0	0	0	2	3	0	0	0	0
Spice seeds	0	0	0	0	0	0	1251	5	0	0	0	0
Cereal replace plant species	2	3	3	2	2	3	2	3	2	3	2	3
Cereal replace plant seeds	34	3063	306	5008	27	99	37400	2142	23	2471	656	1141
Wild (gathered) fruit species	5	5	6	1	6	1	7	7	2	1	6	8
Forest and forestside plants (ecotyps, varietas, etc.)	0	0	0	0	0	0	0	0	0	0	0	0
Wild (gathered) fruit seeds	828	114	668	1	1711	8	84	3060	6	1	14	5019
Gathered wild useful plant species	0	2	4	1	0	3	3	3	0	3	2	2
Gathered wild useful plant seeds	0	4	8	500	0	8	14	23	0	6	5	4
Herbs species	0	0	0	0	0	0	0	3	0	0	0	0
Herbs seeds	0	0	0	0	0	0	0	9	0	0	0	0
Poisonous pant species	2	3	3	1	2	5	3	5	4	3	5	5
Poisonous pant seeds	33	32	15	1	39	32	533	970	49	3	78	632
Cultics/religious plant species	0	0	1	1	0	1	1	1	0	0	0	1
Cultics/religious plant seeds	0	0	5	1	0	5	9	8	0	0	0	50
Fodder plant species	0	0	1	0	0	0	0	1	1	1	1	1
Fodder plant seeds	0	0	6	0	0	0	0	1	0	0	2	3
Building material plant taxons	0	3	3	0	0	1	1	2	1	0	0	2
Building material plant seeds	0	515	15	0	0	9	6	101	1	0	0	20

Barbaricum (1st–5th century AD)	Early Migration period (476–568 AD)	Late Migration period (Avar age) (568–895 AD)	Late Migration period (Caroling age) (811–895 AD)	Late Migration period (German, Slavonic) (6th–10th century AD)	Hungarian conquest time (895 AD)	Early Árpád age (1001–12th century AD)	Late Árpád age (13th century–1301 AD)	Hungarian Kingdom (13th–16th century AD)	Cumanian (15th–16th century AD)	Turkish occupation (16th–17th century AD)	17th–18th century–Total	19th century–Total	Total
99918	57017	5828	186176	32	822	183777	165520	5469513	11880	1211768	112566	15004	8721177
5552	12	1832	8	1	2005	64	84	883	28	37	0	82	14965
105470	57029	7660	186184	33	2827	183841	165604	5470396	11908	1211805	112566	15086	8736142
193	21	120	150	10	40	124	262	319	165	180	60	207	636 (overlapping)
13	3	6	10	1	6	11	11	11	5	9	6	6	21 (overlapping)
5	3	2	12	0	0	4	6	4	0	4	7	4	27 (overlapping)
80262	56520	534	43968	7	86	117325	28204	1786496	2468	62495	61167	234	3045965
2	0	1	3	1	2	3	4	6	2	3	2	2	12 (overlapping)
0	0		0	0	1	1	1	0	0	0	0	0	3 (overlapping)
38	0	2	28	1	104	12	34	1555	6	222	55	51	35742
1	0	1	1	0	0	0	8	9	1	8	0	6	16 (overlapping)
29	0	1	2	0	0	0	568	151785	98	121	0	29	152646
0	1	1	0	0	0	0	2	3	1	2	1	1	9 (overlapping)
0	2	5	0	0	0	0	4	352806	1	732	1	7	353588
4	0	1	2	0	1	1	3	4	3	4	2	3	7 (overlapping)
0	0	0	0	0	0	0	0	0	0	0	0	0	1 (overlapping)
1292	0	25	118	0	61	2	211	460689	287	67	10	128	463883
1	0	2	8	0	0	4	9	15	9	10	9	10	18 (overlapping)
0	0	3	6	0	0	2	4	21	1	2	1	8	27 (overlapping)
10	0	178	2155	0	0	37	573	983079	25	69024	343	96	1060676
1	0	2	0	0	0	1	1	7	2	2	0	1	12 (overlapping)
8	0	2	0	0	0	1	4	4251	6	2	0	24	5554
3	0	2	3	0	2	3	3	3	3	3	2	3	3 (overlapping)
2200	0	2161	137791	0	14	204	35868	40361	3201	1206	185	1127	276688
3	1	3	11	0	1	5	10	15	7	11	3	9	21 (overlapping)
0	0	0	1	0	0	1	0	4	0	0	0	0	4 (overlaping)
20	11	17	466	0	1	507	4753	1484922	43	505541	14	88	2007897
4	0	0	3	0	1	5	4	4	4	3	1	3	6 (overlapping)
123	0	0	12	0	3	23	374	973	625	36	6	41	2788
1	0	2	3	0	2	1	5	5	1	0	0	4	10 (overlapping)
1	0	4	3	0	400	0	329	191	6	0	0	14	957
7	1	1	3	0	2	3	4	5	3	6	3	6	9 (overlapping)
81	13	8	306	0	8	439	28066	7762	28	564195	128	2872	606323
2	1	2	1	1	1	0	0	2	1	2	1	0	2 (overlapping)
36	0	118	14	1	3	0	0	87	1	30	3	0	371
1	0	0	1	0	0	1	2	1	1	0	0	1	2 (overlapping)
2	0	0	11	0	0	3	10	4	13	0	0	1	56
1	1	1	2	0	0	2	3	1	1	1	1	1	3 (overlapping)
10	23	980	14	0	0	34	276	194	106	94	500	130	3028

	Early Neolithic (6000–5300 BC)	Middle Neolithic (5300–4700 BC)	Late Neolithic (4700–4300 BC)	Early Copper age (4300–3500 BC)	Late Copper age (3500–3000 BC)	Early Bronze age (3000–1600 BC)	Middle Bronze age (1600–1300 BC)	Late Bronze age (1300–900 BC)	Early Iron age (Hallstatt) (900–500 BC)	Early Iron age (Skythian) (700–500 BC)	Late Iron age (La Tène) (5th century BC–1st century AD)	Roman age (1st–middle 5th century AD)
Ornamental plant species	0	0	0	0	0	0	0	0	0	0	0	0
Ornamental plant seeds	0	0	0	0	0	0	0	0	0	0	0	0
Trace of plant protection species	1	2	1	0	1	1	2	4	1	2	0	1
Trace of plant protection remains	1	10	23	0	4	5	8	53	1	3	0	1
Species of cereal-crop weeds	7	13	15	5	6	20	21	25	10	13	19	24
Cereal-crop weeds (ecotyps, varietas, etc.)	0	0	0	0	0	0	0	0	0	0	0	0
Seeds of cereal-crop weeds	103	151	358	510	73	159	1476	748	46	673	199	507
Species of root-crop weeds	3	6	4	0	2	4	4	8	4	6	6	7
Seeds of root-crop weeds	10	162	139	0	4	15	9	803	17	21	41	63
Species of ruderal weeds	7	16	21	2	4	18	18	31	8	11	15	30
Seeds of ruderal weeds	89	3136	442	9	36	53	37357	2671	73	2486	711	6745
Species of aquaceous (submerse, floating pondweed etc.) plants	1	1	0	0	0	0	0	5	0	0	0	1
Seeds of aquaceous (submerse, floating pondweed etc.) plants	320	49	0	0	0	0	0	35	0	0	0	2
Species of watershore (reed, high sedge) and marshland plants	1	13	14	2	1	9	2	31	3	4	4	13
Seeds of watershore (reed, high sedge) and marshland plants	3	539	160	5001	4	67	6	475	11	31	5	303
Species of forest plants	1	2	7	1	4	2	5	13	2	0	5	5
Seeds of forest plants	500	59	525	2	1489	9	66	3138	6	0	22	13
Species of cleared forest and forestside plants	4	5	5	1	2	2	5	9	0	3	5	6
Forest and forestside plants (ecotyps, varietas, etc.)	0	0	0	0	0	0	0	0	0	0	0	0
Seeds of cleared forest and forestside plants	8	14	207	1	222	3	30	188	0	3	28	34
Species of meadow and pasture plants	5	19	21	2	5	17	31	47	10	16	17	25
Seeds of meadow and pasture plants	68	91	151	6	63	50	625	512	28	36	95	5517
Taxons of diverse (non classifiable) plants	19	16	15	1	4	13	25	21	11	10	9	22
Seeds of diverse (non classifiable) plants	214	634	50119	11	117	558	51088	4282	536	409	308	666

Barbaricum (1st–5th century AD)	Early Migration period (476–568 AD)	Late Migration period (Avar age) (568–895 AD)	Late Migration period (Caroling age) (811–895 AD)	Late Migration period (German, Slavonic) (6th–10th century AD)	Hungarian conquest time (895 AD)	Early Árpád age (1001–12th century AD)	Late Árpád age (13th century–1301 AD)	Hungarian Kingdom (13th–16th century AD)	Cumanian (15th–16th century AD)	Turkish occupation (16th–17th century AD)	17th–18th century–Total	19th century–Total	Total
0	0	1	0	0	0	0	2	1	0	0	1	0	6 (overlapping)
0	0	10	0	0	0	0	74	1	0	0	5	0	90
1	0	0	5	1	0	0	2	4	2	0	0	1	10 (overlapping)
42	0	0	12	1	0	0	3	1139	10	0	0	80	1396
30	3	8	16	3	9	19	24	45	24	22	14	18	71 (overlapping)
0	0	0	0	0	0	0	1	1	0	0	0	0	1 (overlapping)
2031	61	144	708	5	63	687	5080	15689	1015	5626	183	628	36923
10	2	5	9	1	3	7	8	13	5	6	3	7	18 (overlapping)
205	21	25	181	6	29	39	1085	45570	74	42	37	147	48745
39	4	16	17	2	5	26	46	47	32	34	3	30	69 (overlapping)
2591	261	1818	138099	4	34	992	88606	48482	5710	1060761	198	4433	1405797
0	0	2	2	0	0	0	2	0	0	0	0	0	7 (overlapping)
0	0	114	3	0	0	0	73	0	0	0	0	1	597
26	1	26	17	0	1	10	34	32	22	15	2	25	89 (overlapping)
310	23	1543	63	0	2	73	527	67473	526	183	505	167	78000
8	1	5	12	0	1	4	13	24	12	9	2	18	61 (overlapping)
20	11	59	459	0	1	5	210	52786	42	140	17	1495	61074
6	0	5	10	0	1	3	7	21	6	11	2	15	37 (overlapping)
0	0	0	1	0	0	1	0	4	0	0	0	0	4 (overlaping)
36	0	6	168	0	2	5	462	1435992	107	5971	2	440	1443929
40	1	25	26	1	8	22	65	59	38	33	5	49	116 (overlapping)
1147	3	794	80	1	21	133	38592	4881	1063	550	50033	1622	106162
29	5	13	10	2	3	11	38	43	16	33	1	23	107 (overlapping)
6310	114	540	54	7	21	64425	578	60688	353	463	1	131	242627

Table 3. Frequency of cereals in Hungary based on macroremains, 1860–2006. Completed by F. Gyulai, 2006.

	Neolithic			Copper age		Bronze age			Iron age			Roman Period		Migration Period				Hungarian conquest time and Árpád age			Late Medieval Period			Modern age	
Cereals	Early 6000 BC	Middle 5300 BC	Late 4700 BC	Early 4300 BC	Late 3500 BC	Early 3000 BC	Middle 1600 BC	Late 1300 BC	Hallstatt 900 BC	Skythian 700 BC	La Tène 500 BC	Pannonia 1st cent.	Barbaricum 1st cent.	Early 476 AD	Avar age 568 AD	Caroling age 811 AD	German, Slavonic 6–9th cent.	Hungarian conquest 895 AD	Early Árpád age 11th cent.	Late Árpád age 13th cent.	Hungarian Kingdom 1301–16th cent.	Cumanian 15th–16th cent.	Turkish occupation 16th–17th cent.	17th–18th cent.	19th century
Avena sativa common oat																									
Avena spec. cultiv. cultivated oat																									
Avena nuda naked oat																									
Cerealia cereals																									
Fagopyrum esculentum buckwheat																									
Hordeum distichon two-rowed barley																									
Hordeum hexastichon six-rowed barley																									
Hordeum polystichum many-rowed barley																									
Hordeum tetrastichon four-rowed barley																									
Hordeum nudum naked barley																									
Hordeum vulgare L. barley																									
Panicum miliaceum common millet																									
Secale cereale rye																									
Setaria italica foxtail millet																									

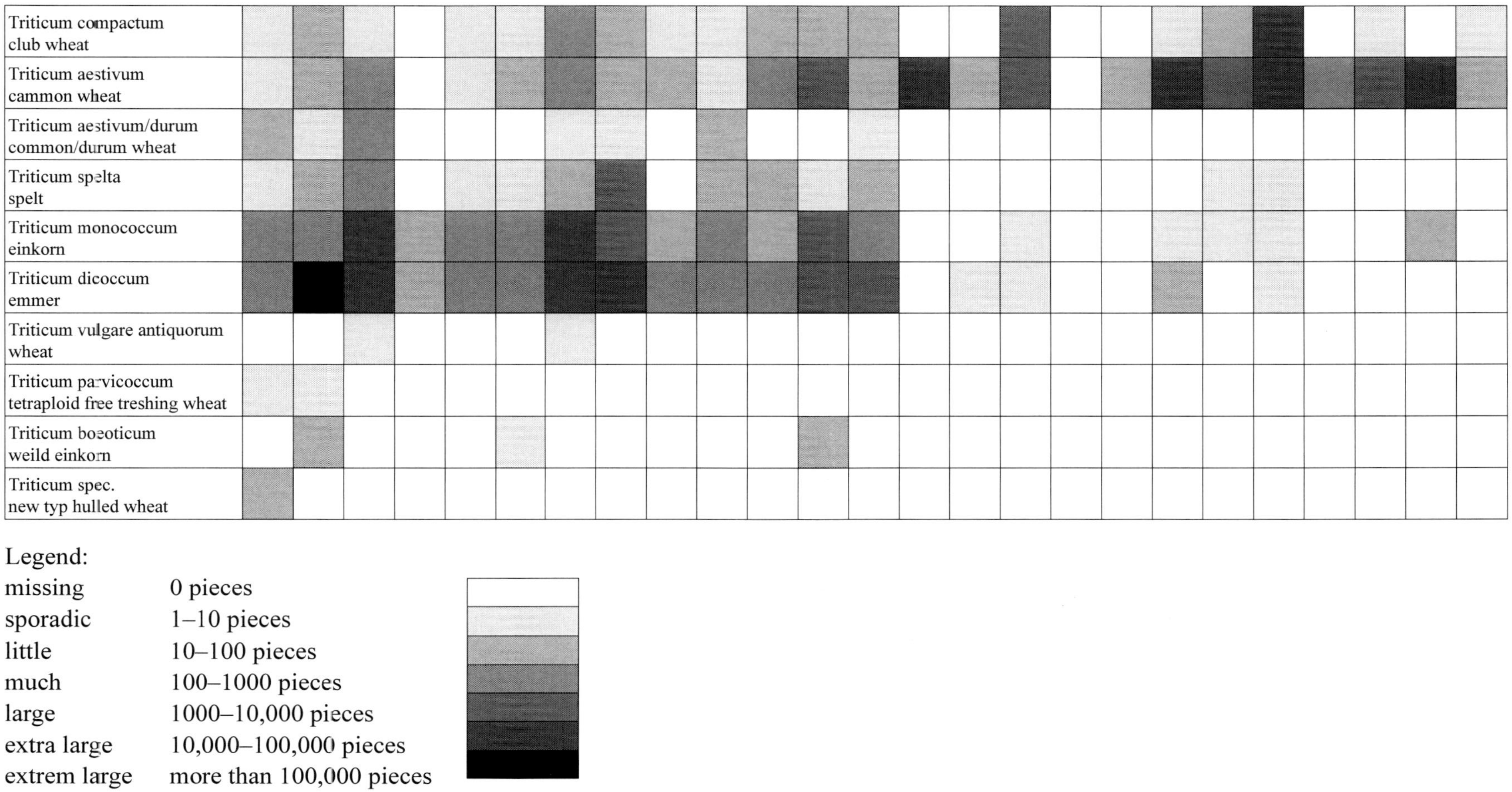
Triticum compactum
club wheat
Triticum aestivum
cammon wheat
Triticum aestivum/durum
common/durum wheat
Triticum spelta
spelt
Triticum monococcum
einkorn
Triticum dicoccum
emmer
Triticum vulgare antiquorum
wheat
Triticum parvicoccum
tetraploid free treshing wheat
Triticum boeoticum
weild einkorn
Triticum spec.
new typ hulled wheat
Legend:
missing 0 pieces
sporadic 1–10 pieces
little 10–100 pieces
much 100–1000 pieces
large 1000–10,000 pieces
extra large 10,000–100,000 pieces
extrem large more than 100,000 pieces

Table 4. Frequency of leguminous in Hungary based on macroremains, 1860–2006. Completed by F. Gyulai, 2006.

	Neolithic			Copper age		Bronze age			Iron age			Roman Period		Migration Period				Hungarian conquest time and Árpád age			Late Medieval Period			Modern age	
Leguminous	Early 6000 BC	Middle 5300 BC	Late 4700 BC	Early 4300 BC	Late 3500 BC	Early 3000 BC	Middle 1600 BC	Late 1300 BC	Hallstatt 900 BC	Skythian 700 BC	La Tène 500 BC	Pannonia 1st cent.	Barbaricum 1st cent.	Early 476 AD	Avar age 568 AD	Caroling age 811 AD	German, Slavonic 6–9th cent.	Hungarian conquest 895 AD	Early Árpád age 11th cent.	Late Árpád age 13th cent.	Hungarian Kingdom 1301–16th cent.	Cumanian 15th–16th cent.	Turkish occupation 16th–17th cent.	17th–18th cent.	19th century
Cicer arietinum Chickpea																									
Lathyrus cicera L. Red vetchling																									
Lathyrus sativus Grass pea																									
Lens microsperma Small seed lentil																									
Lens macrosperma Big seed lentil																									
Lens culinaris Lentil																									
Onobrychis viciifolia Common sainfoin																									
Pisum sativum Pea																									
Pisum arvense Field pea																									
Vicia ervilia Bitter vetch																									
Vicia faba Horse bean																									

Vicia sativa var. sativa Common vetch																									
Vicia sativa var. lentisperma Lentilvetch																									

Legend:

missing	0 piece
sporadic	1–10 pieces
little	10–100 pieces
much	100–1000 pieces
large	1000–10,000 pieces
extra large	10,000–100,000 pieces
extrem large	more than 100,000 pieces

Table 5. Frequency of vegetables in Hungary based on macroremains, 1860–2006. Completed by F. Gyulai, 2006.

Vegetables	Neolithic			Copper age		Bronze age			Iron age			Roman Period		Migration Period				Hungarian conquest time and Árpád age			Late Medieval Period			Modern age	
	Early 6000 BC	Middle 5300 BC	Late 4700 BC	Early 4300 BC	Late 3500 BC	Early 3000 BC	Middle 1600 BC	Late 1300 BC	Hallstatt 900 BC	Skythian 700 BC	La Tène 500 BC	Pannonia 1st cent.	Barbaricum 1st cent.	Early 476 AD	Avar age 568 AD	Caroling age 811 AD	German, Slavonic 6–9th cent.	Hungarian conquest 895 AD	Early Árpád age 11th cent.	Late Árpád age 13th cent.	Hungarian Kingdom 1301–16th cent.	Cumanian 15th–16th cent.	Turkish occupation 16th–17th cent.	17th–18th cent.	19th century
Allium sativum garlic																									
Apium graveolens (wild) celery																									
Atriplex cf. hortensis garden orache																									
Brassica cf. oleracea headed cabbage																									
Brassica rapa cabbage																									
Brassica cf. rapa turnip																									
Citrullus lanatus watermelon																									
Cucumus melo muskmelon																									
Cucumus melo (Cantaloupe Typ) cantaloupe																									
Cucumus melo (Togo Typ) muskmelon																									
Cucumis sativus cucumber																									
Cucurbita pepo pumpkin																									
Cucurbita pepo/maxima pumpkin/winter squash																									

Cucurbita/Lagenaria siceraria squash/calabash or bottle gourd																									sporadic
Cucurbita spec. squash																							little		
Pastinaca sativa garden parsnip													little								little		sporadic		
Raphanus cf. sativus garden radish																				much	little				
Rumex cf. scutatus French sorrel																									sporadic

Legend:

missing	0 piece	(white)
sporadic	1–10 pieces	(light grey)
little	10–100 pieces	(grey)
much	100–1000 pieces	(mid grey)
large	1000–10,000 pieces	(dark grey)
extra large	10,000–100,000 pieces	(darker grey)
extrem large	more than 100,000 pieces	(black)

Table 6. Frequency of import plants in Hungary based on macroremains, 1860–2006. Completed by F. Gyulai, 2006.

	Neolithic			Copper age		Bronze age			Iron age			Roman Period		Migration Period				Hungarian conquest time and Árpád age			Late Medieval Period			Modern age	
Import plants	Early 6000 BC	Middle 5300 BC	Late 4700 BC	Early 4300 BC	Late 3500 BC	Early 3000 BC	Middle 1600 BC	Late 1300 BC	Hallstatt 900 BC	Skythian 700 BC	La Tène 500 BC	Pannonia 1st cent.	Barbaricum 1st cent.	Early 476 AD	Avar age 568 AD	Caroling age 811 AD	German, Slavonic 6–9th cent.	Hungarian conquest 895 AD	Early Árpád age 11th cent.	Late Árpád age 13th cent.	Hungarian Kingdom 1301–16th cent.	Cumanian 15th–16th cent.	Turkish occupation 16th–17th cent.	17th–18th cent.	19th century
Castanea sativa European chestnut																									
Coffea arabica? coffee																									
Ficus carica fig																									
Gossypium spec. cotton																									
Morus cf. alba white mulberry																									
Olea europaea common olive																									
Phoenix dactylifera date																									
Peper nigrum pepper																									
Punica granatum pomegranate																									

Legend:

missing	0 piece
sporadic	1–10 pieces
little	10–100 pieces
much	100–1000 pieces
large	1000–10,000 pieces
extra large	10,000–100,000 pieces
extrem large	more than 100,000 pieces

Table 7. Frequency of oil and fibre crops in Hungary based on macroremains, 1860–2006. Completed by F. Gyulai, 2006.

	Neolithic			**Copper age**		**Bronze age**			**Iron age**			**Roman Period**		**Migration Period**				**Hungarian conquest time and Árpád age**			**Late Medieval period**			**Modern age**	
Oil and fibre crops	**Early 6000 BC**	**Middle 5300 BC**	**Late 4700 BC**	**Early 4300 BC**	**Late 3500 BC**	**Early 3000 BC**	**Middle 1600 BC**	**Late 1300 BC**	**Hallstatt 900 BC**	**Skythian 700 BC**	**La Tène 500 BC**	**Pannonia 1st cent.**	**Barbaricum 1st cent.**	**Early 476 AD**	**Avar age 568 AD**	**Caroling age 811 AD**	**German, Slavonic 6–9th cent.**	**Hungarian conquest 895 AD**	**Early Árpád age 11th cent.**	**Late Árpád age 13th cent.**	**Hungarian Kingdom 1301–16th cent.**	**Cumanian 15th–16th cent.**	**Turkish occupation 16th–17th cent.**	**17th–18th cent.**	**19th century**
Brassica campestris var. oleifera naven																									
Brassica cf. napus (oleracea?) rape (cabbage?)																									
Camelina sativa gold-of-pleasure																									
Cannabis sativa hemp																									
Helianthus annuus common sunflower																									
Linum usitatissimum flax																									
Papaver somniferum poppy																									

Legend:

missing	0 piece	
sporadic	1–10 pieces	
little	10–100 pieces	
much	100–1000 pieces	
large	1000–10,000 pieces	
extra large	10,000–100,000 pieces	
extrem large	more than 100,000 pieces	

Table 8. Frequency of fruits and grapes in Hungary based on macroremains, 1860–2006. Completed by F. Gyulai, 2006.

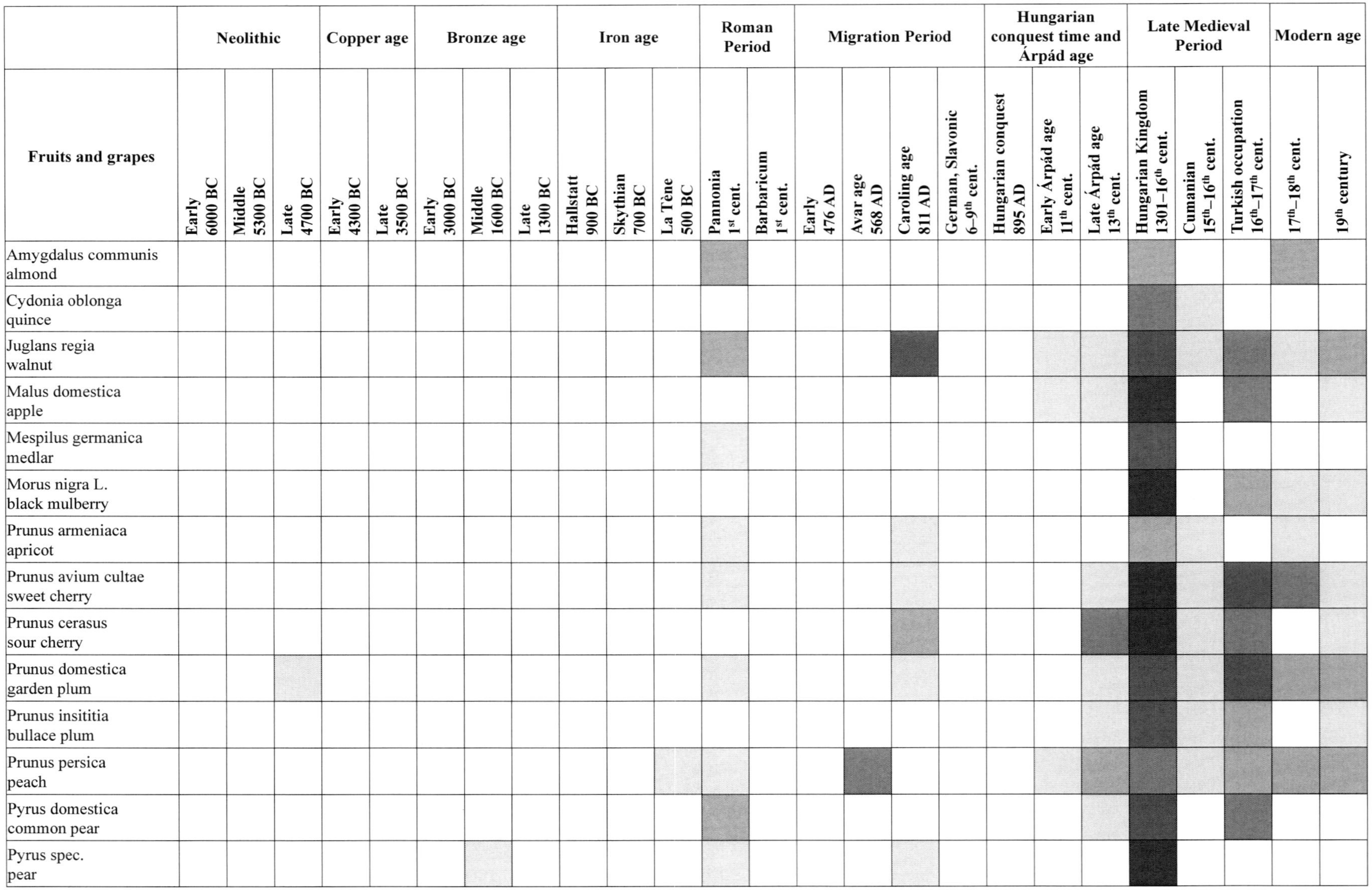

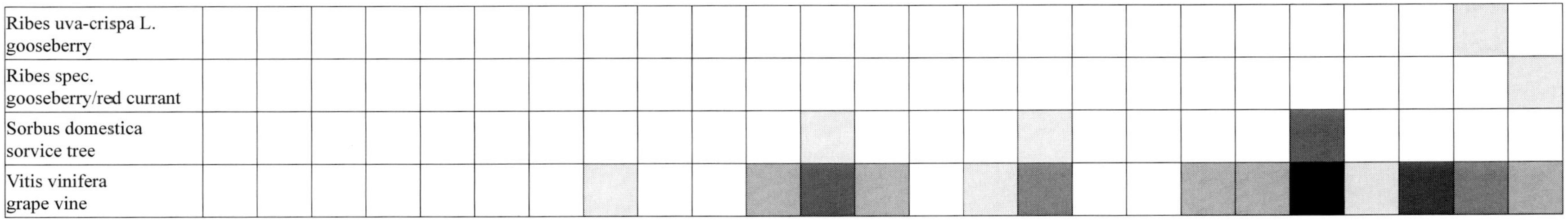

Legend:

missing	0 piece	
sporadic	1–10 pieces	
little	10–100 pieces	
much	100–1000 pieces	
large	1000–10,000 pieces	
extra large	10,000–100,000 pieces	
extrem large	more than 100,000 pieces	

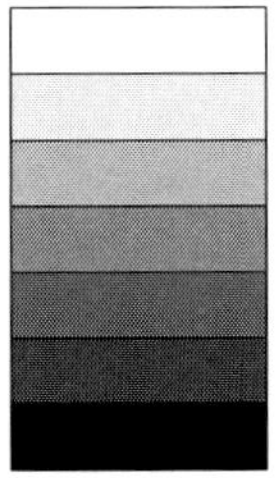

Table 9. Frequency of spice plants in Hungary based on macroremains, 1860–2006. Completed by F. Gyulai, 2006.

	Neolithic			Copper age		Bronze age			Iron age			Roman Period		Migration Period				Hungarian conquest time and Árpád age			Late Medieval Period			Modern age	
Spice plants	**Early 6000 BC**	**Middle 5300 BC**	**Late 4700 BC**	**Early 4300 BC**	**Late 3500 BC**	**Early 3000 BC**	**Middle 1600 BC**	**Late 1300 BC**	**Hallstatt 900 BC**	**Skythian 700 BC**	**La Tène 500 BC**	**Pannonia 1st cent.**	**Barbaricum 1st cent.**	**Early 476 AD**	**Avar age 568 AD**	**Caroling age 811 AD**	**German, Slavonic 6–9th cent.**	**Hungarian conquest 895 AD**	**Early Árpád age 11th cent.**	**Late Árpád age 13th cent.**	**Hungarian Kingdom 1301–16th cent.**	**Cumanian 15th–16th cent.**	**Turkish occupation 16th–17th cent.**	**17th–18th cent.**	**19th century**
Anethum graveolens common dill																									
Brassica cf. juncea indian/chinese mustard																									
Brassica nigra black mustard																									
Carthamus tinctorius safflower																									
Carum carvi caraway																									
Coriandrum sativum coriander																									
Foeniculum vulgare common fennel																									
Humulus lupulus common hop																									
Ocinum basilicum basil																									
Origanum vulgare wild marjoran																									

Satureja hortensis summer-savory																									
Sinapis alba white mustard																									

Legend:

missing	0 piece
sporadic	1–10 pieces
little	10–100 pieces
much	100–1000 pieces
large	1000–10,000 pieces
extra large	10,000–100,000 pieces
extrem large	more than 100,000 pieces

Table 10. Frequency of cereal replace plants in Hungary based on macroremains, 1860–2006. Completed by F. Gyulai, 2006.

	Neolithic			Copper age		Bronze age			Iron age			Roman Period		Migration Period				Hungarian conquest time and Árpád age			Late Medieval Period			Modern age	
Cereal replace plants	Early 6000 BC	Middle 5300 BC	Late 4700 BC	Early 4300 BC	Late 3500 BC	Early 3000 BC	Middle 1600 BC	Late 1300 BC	Hallstatt 900 BC	Skythian 700 BC	La Tène 500 BC	Pannonia 1st cent.	Barbaricum 1st cent.	Early 476 AD	Avar age 568 AD	Caroling age 811 AD	German, Slavonic 6–9th cent.	Hungarian conquest 895 AD	Early Árpád age 11th cent.	Late Árpád age 13th cent.	Hungarian Kingdom 1301–16th cent.	Cumanian 15th–16th cent.	Turkish occupation 16th–17th cent.	17th–18th cent.	19th century
Bromus secalinus rye brome																									
Chenopodium album white goosefoot/ fan-hen																									
Schoenoplectus lacustris common bulrush																									

Legend:

missing	0 piece
sporadic	1–10 pieces
little	10–100 pieces
much	100–1000 pieces
large	1000–10,000 pieces
extra large	10,000–100,000 pieces
extrem large	more than 100,000 pieces

Table 11. Frequency of gathered useful (wild) plants in Hungary based on macroremains, 1860–2006. Completed by F. Gyulai, 2006.

	Neolithic			Copper age		Bronze age			Iron age			Roman Period		Migration Period				Hungarian conquest time and Árpád age			Late Medieval Period			Modern age	
Gathered useful (wild) plants	**Early 6000 BC**	**Middle 5300 BC**	**Late 4700 BC**	**Early 4300 BC**	**Late 3500 BC**	**Early 3000 BC**	**Middle 1600 BC**	**Late 1300 BC**	**Hallstatt 900 BC**	**Skythian 700 BC**	**La Tène 500 BC**	**Pannonia 1st cent.**	**Barbaricum 1st cent.**	**Early 476 AD**	**Avar age 568 AD**	**Caroling age 811 AD**	**German, Slavonic 6–9th cent.**	**Hungarian conquest 895 AD**	**Early Árpád age 11th cent.**	**Late Árpád age 13th cent.**	**Hungarian Kingdom 1301–16th cent.**	**Cumanian 15th–16th cent.**	**Turkish occupation 16th–17th cent.**	**17th–18th cent.**	**19th century**
Glyceria maxima reed sweet-grass																									
Malva neglecta dwarf mallow																									
Rumex acetosa common sorrel																									
Rumex acetosella sheep's sorrel																									
Sinapis arvensis charlock																									
Thymus spec. thyme																									

Legend:

missing	0 piece
sporadic	1–10 pieces
little	10–100 pieces
much	100–1000 pieces
large	1000–10,000 pieces
extra large	10,000–100,000 pieces
extrem large	more than 100,000 pieces

Table 12. Frequency of gathered (wild) fruits in Hungary based on macroremains, 1860–2006. Completed by F. Gyulai, 2006.

	Neolithic			Copper age		Bronze age			Iron age			Roman Period		Migration Period				Hungarian conquest time and Árpád age			Late Medieval Period			Modern age	
Gathered (wild) fruits	Early 6000 BC	Middle 5300 BC	Late 4700 BC	Early 4300 BC	Late 3500 BC	Early 3000 BC	Middle 1600 BC	Late 1300 BC	Hallstatt 900 BC	Skythian 700 BC	La Tène 500 BC	Pannonia 1st cent.	Barbaricum 1st cent.	Early 476 AD	Avar age 568 AD	Caroling age 811 AD	German, Slavonic 6–9th cent.	Hungarian conquest 895 AD	Early Árpád age 11th cent.	Late Árpád age 13th cent.	Hungarian Kingdom 1301–16th cent.	Cumanian 15th–16th cent.	Turkish occupation 16th–17th cent.	17th–18th cent.	19th century
Cornus mas cornelian cherry/dogwood																									
Corylus avellana European filbert (haselnut)																									
Crataegus monogyna single-seed hawthorn																									
Fagus silvatica European beech																									
Fragaria vesca wild strawberry																									
Malus silvestris crab apple																									
Prunus avium L. subsp. silvestris wild cherry																									
Prunus cerasifera cherry plum																									
Prunus fruticosus ground cherry/dwarf cherry																									
Prunus mahaleb mahaleb cherry/St. Lucy cherry																									
Prunus padus European bird cherry																									
Prunus spinosa sloe/blackthorn																									
Pyrus achras wild pear																									

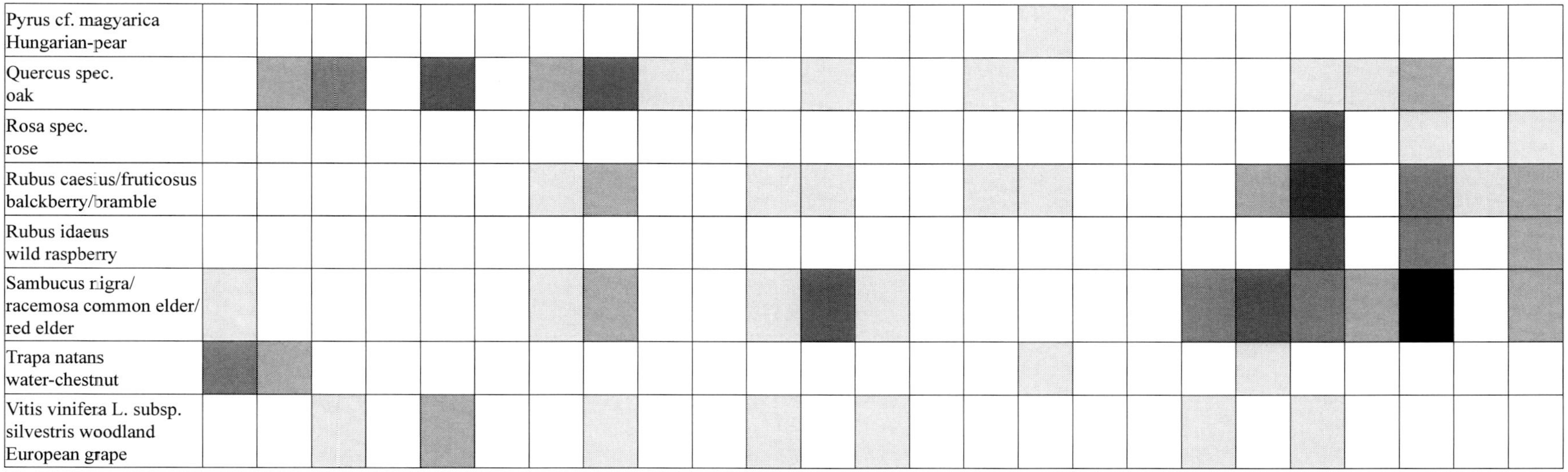

Legend:

missing	0 piece
sporadic	1–10 pieces
little	10–100 pieces
much	100–1000 pieces
large	1000–10,000 pieces
extra large	10,000–100,000 pieces
extrem large	more than 100,000 pieces

Table 13. Frequency of herbs in Hungary based on macroremains, 1860–2006. Completed by F. Gyulai, 2006.

	Neolithic			Copper age		Bronze age			Iron age			Roman Period		Migration Period				Hungarian conquest time and Árpád age			Late Medieval Period			Modern age	
Herbs	**Early 6000 BC**	**Middle 5300 BC**	**Late 4700 BC**	**Early 4300 BC**	**Late 3500 BC**	**Early 3000 BC**	**Middle 1600 BC**	**Late 1300 BC**	**Hallstatt 900 BC**	**Skythian 700 BC**	**La Tène 500 BC**	**Pannonia 1st cent.**	**Barbaricum 1st cent.**	**Early 476 AD**	**Avar age 568 AD**	**Caroling age 811 AD**	**German, Slavonic $6–9^{th}$ cent.**	**Hungarian conquest 895 AD**	**Early Árpád age 11th cent.**	**Late Árpád age 13th cent.**	**Hungarian Kingdom $1301–16^{th}$ cent.**	**Cumanian $15^{th}–16^{th}$ cent.**	**Turkish occupation $16^{th}–17^{th}$ cent.**	**$17^{th}–18^{th}$ cent.**	**19th century**
Anchusa arvensis L. alkanet																									
Aristolochia clematitis L. birthwort																									
Descurainia sophia (L.) Webb flixweed																									
Euphorbia lathyris L. caper spurge																									
Hypericum perforatum L. common Saint John's-wort																									
Marrubium vulgare L. white horehound																									
Matricaria cf. chamomilla wild chamomile																									
Melissa officinalis																									
Mentha spec. mint																									
Tilia spec. lime/linden																									

Legend:

missing	0 piece
sporadic	1–10 pieces
little	10–100 pieces
much	100–1000 pieces
large	1000–10,000 pieces
extra large	10,000–100,000 pieces
extrem large	more than 100,000 pieces

Table 14. Frequency of poisonous plants in Hungary based on macroremains, 1860–2006. Completed by F. Gyulai, 2006.

Poisonous plants	Neolithic: Early 6000 BC	Neolithic: Middle 5300 BC	Neolithic: Late 4700 BC	Copper age: Early 4300 BC	Copper age: Late 3500 BC	Bronze age: Early 3000 BC	Bronze age: Middle 1600 BC	Bronze age: Late 1300 BC	Iron age: Hallstatt 900 BC	Iron age: Skythian 700 BC	Iron age: La Tène 500 BC	Roman Period: Pannonia 1st cent.	Roman Period: Barbaricum 1st cent.	Migration Period: Early 476 AD	Migration Period: Avar age 568 AD	Migration Period: Caroling age 811 AD	Migration Period: German, Slavonic 6–9th cent.	Hungarian conquest time and Árpád age: Hungarian conquest 895 AD	Hungarian conquest time and Árpád age: Early Árpád age 11th cent.	Hungarian conquest time and Árpád age: Late Árpád age 13th cent.	Late Medieval Period: Hungarian Kingdom 1301–16th cent.	Late Medieval Period: Cumanian 15th–16th cent.	Late Medieval Period: Turkish occupation 16th–17th cent.	Modern age: 17th–18th cent.	Modern age: 19th century
Adonis aestivalis L. summer adonis	missing	missing	missing	missing	missing	missing	missing	missing	missing	missing	sporadic	sporadic	sporadic	missing	missing	missing	missing	missing	missing	sporadic	sporadic	sporadic	sporadic	missing	sporadic
Agrostemma githago L. korn-rade	missing	sporadic	sporadic	missing	little	little	much	little	sporadic	sporadic	little	little	little	little	missing	much	missing	sporadic	much	little	large	little	large	much	missing
Atropa bella-donna L. banewort/deadly nightshade	missing	missing	missing	missing	missing	missing	missing	missing	missing	missing	missing	missing	missing	missing	missing	missing	missing	missing	missing	missing	missing	missing	missing	missing	little
Datura stramonium L. thorn-apple	missing	missing	missing	missing	missing	sporadic	missing	missing	missing	missing	missing	missing	missing	missing	missing	missing	missing	missing	missing	missing	missing	missing	extrem large	missing	sporadic
Hyoscyamus niger L. common henbane	missing	sporadic	missing	missing	missing	sporadic	missing	sporadic	little	missing	sporadic	little	sporadic	missing	sporadic	missing	missing	missing	missing	much	much	sporadic	little	missing	large
Lolium temulentum L. darnel	sporadic	missing	missing	missing	missing	missing	missing	much	sporadic	sporadic	missing	missing	sporadic	missing	missing	missing	missing	missing	missing	missing	missing	missing	missing	sporadic	missing
Melampyrum spec. cow-wheat	missing	missing	sporadic	missing	missing	sporadic	sporadic	missing	missing	sporadic	sporadic	sporadic	sporadic	missing	missing	little	missing	missing	sporadic	missing	sporadic	missing	little	little	missing
Oenanthe aquatica water fennel	missing	missing	missing	missing	missing	missing	missing	much	missing	missing	missing	missing	little	missing	missing	missing	missing	missing	little	missing	missing	missing	missing	missing	sporadic
Sambucus ebulus L. danewort	little	little	sporadic	sporadic	little	sporadic	little	much	sporadic	missing	little	much	sporadic	missing	missing	little	missing	sporadic	missing	extra large	much	sporadic	extra large	missing	little

Legend:

missing	0 piece	(white)
sporadic	1–10 pieces	(very light grey)
little	10–100 pieces	(light grey)
much	100–1000 pieces	(medium grey)
large	1000–10,000 pieces	(dark grey)
extra large	10,000–100,000 pieces	(very dark grey)
extrem large	more than 100,000 pieces	(black)

Table 15. Frequency of religious plants in Hungary based on macroremains, 1860–2006. Completed by F. Gyulai, 2006.

	Neolithic			Copper age		Bronze age			Iron age			Roman Period		Migration Period				Hungarian conquest time and Árpád age			Late Medieval Period			Modern age	
Religious plants	Early 6000 BC	Middle 5300 BC	Late 4700 BC	Early 4300 BC	Late 3500 BC	Early 3000 BC	Middle 1600 BC	Late 1300 BC	Hallstatt 900 BC	Skythian 700 BC	La Tène 500 BC	Pannonia 1st cent.	Barbaricum 1st cent.	Early 476 AD	Avar age 568 AD	Caroling age 811 AD	German, Slavonic 6–9th cent.	Hungarian conquest 895 AD	Early Árpád age 11th cent.	Late Árpád age 13th cent.	Hungarian Kingdom 1301–16th cent.	Cumanian 15th–16th cent.	Turkish occupation 16th–17th cent.	17th–18th cent.	19th century
Convolvulus arvensis L. field bindweed																									
Convolvulus arvensis L. field bindweed																									

Legend:

missing	0 piece	
sporadic	1–10 pieces	
little	10–100 pieces	
much	100–1000 pieces	
large	1000–10,000 pieces	
extra large	10,000–100,000 pieces	
extrem large	more than 100,000 pieces	

Table 16. Frequency of possible animal fodder plants in Hungary based on macroremains, 1860–2006. Completed by F. Gyulai, 2006.

	Neolithic			Copper age		Bronze age			Iron age			Roman Period		Migration Period				Hungarian conquest time and Árpád age			Late Medieval Period			Modern age	
Possible animal fodder plants	**Early 6000 BC**	**Middle 5300 BC**	**Late 4700 BC**	**Early 4300 BC**	**Late 3500 BC**	**Early 3000 BC**	**Middle 1600 BC**	**Late 1300 BC**	**Hallstatt 900 BC**	**Skythian 700 BC**	**La Tène 500 BC**	**Pannonia 1 AD**	**Barbaricum 1st cent.**	**Early 476 AD**	**Avar age 568 AD**	**Caroling age 811 AD**	**German, Slavonic 9th cent.**	**Hungarian conquest 895 AD**	**Early Árpád age 11th cent.**	**Late Árpád age 13th cent.**	**Hungarian Kingdom 1301–16th century**	**Cumanian 15th–16th century**	**Turkish occupation 16th–17th century**	**17th–18th century**	**19th century**
Medicago cf. orbicularis button medic																				little					
Trifolium pratense red clover			sporadic					sporadic	sporadic	sporadic	sporadic	sporadic	sporadic			sporadic			sporadic	sporadic	sporadic	sporadic			sporadic

Legend:

missing	0 piece	
sporadic	1–10 pieces	
little	10–100 pieces	
much	100–1000 pieces	
large	1000–10,000 pieces	
extra large	10,000–100,000 pieces	
extrem large	more than 100,000 pieces	

Table 17. Frequency of building material plants in Hungary based on macroremains, 1860–2006. Completed by F. Gyulai, 2006.

	Neolithic			Copper age		Bronze age			Iron age			Roman Period		Migration Period				Hungarian conquest time and Árpád age			Late Medieval periode			Modern age	
Building material plants	**Early 6000 BC**	**Middle 5300 BC**	**Late 4700 BC**	**Early 4300 BC**	**Late 3500 BC**	**Early 3000 BC**	**Middle 1600 BC**	**Late 1300 BC**	**Hallstatt 900 BC**	**Skythian 700 BC**	**La Tène 500 BC**	**Pannonia 1st cent.**	**Barbaricum 1st cent.**	**Early 476 AD**	**Avar age 568 AD**	**Caroling age 811 AD**	**German, Slavonic 6–9th cent.**	**Hungarian conquest 895 AD**	**Early Árpád age 11th cent.**	**Late Árpád age 13th cent.**	**Hungarian Kingdom 1301–16th cent.**	**Cumanian 15th–16th cent.**	**Turkish occupation 16th–17th cent.**	**17th–18th cent.**	**19th century**
Phragmites australis common reed																									
Carex spec. Total sedge																									
Typha spec. Total cattail																									

Legend:
missing 0 piece
sporadic 1–10 pieces
little 10–100 pieces
much 100–1000 pieces
large 1000–10,000 pieces
extra large 10,000–100,000 pieces
extrem large more than 100,000 pieces

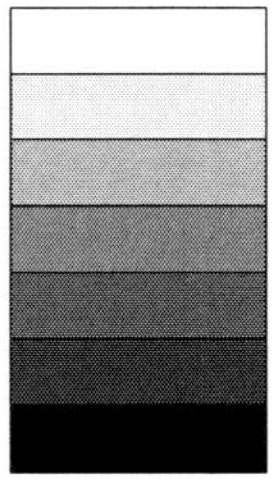

Table 18. Frequency of ornamental plants in Hungary based on macroremains, 1860–2006. Completed by F. Gyulai, 2006.

	Neolithic			Copper age		Bronze age			Iron age			Roman Period		Migration Period				Hungarian conquest time and Árpád age			Late Medieval Period			Modern age	
Spice plants	Early 6000 BC	Middle 5300 BC	Late 4700 BC	Early 4300 BC	Late 3500 BC	Early 3000 BC	Middle 1600 BC	Late 1300 BC	Hallstatt 900 BC	Skythian 700 BC	La Tène 500 BC	Pannonia 1st cent.	Barbaricum 1st cent.	Early 476 AD	Avar age 568 AD	Caroling age 811 AD	German, Slavonic 6–9th cent.	Hungarian conquest 895 AD	Early Árpád age 11th cent.	Late Árpád age 13th cent.	Hungarian Kingdom 1301–16th cent.	Cumanian 15th–16th cent.	Turkish occupation 16th–17th cent.	17th–18th cent.	19th century
Aesculus pavia horse chestnut																									
Chrysanthemum segetum corn marigold																									
Malva moschata musk mallow																									
Primula elatior oxlip																									
Silene armeria sweet-William catchfly																									
Viola odorata sweet violet																									

Legend:

missing	0 piece
sporadic	1–10 pieces
little	10–100 pieces
much	100–1000 pieces
large	1000–10,000 pieces
extra large	10,000–100,000 pieces
extrem large	more than 100,000 pieces

Table 19. Frequency of trace of plant protection remains in Hungary, 1860–2006. Completed by F. Gyulai, 2006.

	Neolithic			Copper age		Bronze age			Iron age			Roman Period		Migration Period				Hungarian conquest time and Árpád age			Late Medieval Period			Modern age	
Trace of plant protection	Early 6000 BC	Middle 5300 BC	Late 4700 BC	Early 4300 BC	Late 3500 BC	Early 3000 BC	Middle 1600 BC	Late 1300 BC	Hallstatt 900 BC	Skythian 700 BC	La Tène 500 BC	Pannonia 1st cent.	Barbaricum 1st cent.	Early 476 AD	Avar age 568 AD	Caroling age 811 AD	German, Slavonic 6–9th cent.	Hungarian conquest 895 AD	Early Árpád age 11th cent.	Late Árpád age 13th cent.	Hungarian Kingdom 1301–16th cent.	Cumanian 15th–16th cent.	Turkish occupation 16th–17th cent.	17th–18th cent.	19th century
Monilia on apricot (Prunus armeniaca) mumy																									
Mushrum screrotium																									
Ergot (Claviceps purpurea)																									
Bunt sporodochium																									
Mole/mildew (Tilletia spec.)																									
Nematode (Tilenchus tritici)																									
Nematode on wheat grain																									
Nematode on emmer																									
Nematode on barley																									
Carnivora excrement																									
Pestprint on sweet cherry drupa (stone)																									
Pestprint on sour cherry drupa (stone)																									
Pestprint on ground/ dwarf cherry drupa (stone)																									

Pestprint on charlock (Sinapis arvensis) seed																									
Weevil (Bruchus pisorum) prints on peas																									
Weevil (Calandra granaria) prints on grains																									

Legend:

missing	0 piece
sporadic	1–10 pieces
little	10–100 pieces
much	100–1000 pieces
large	1000–10,000 pieces
extra large	10,000–100,000 pieces
extrem large	more than 100,000 pieces

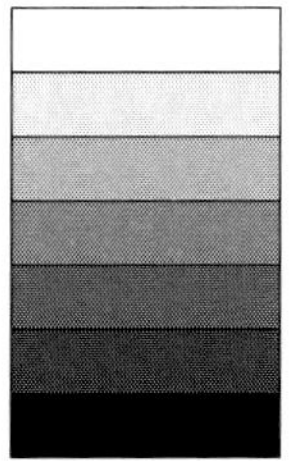

Table 20. Epoche of the Hungarian Archaeobotany.

	Number of scientist	**Identificated archaeological sites**
I. (1877–1917)		
Deininger, Lindau, Ascher, Wittmack	4	6
II. (1918–1954)		
Magyar Királyi Vetőmagvizsgáló Állomás	1	1
III. (1955–1964)		
Bogdán, Papp, Szabó, Hopf, Maác, Valkó, Wellmann, Mesch, Mándy, Tempír, Zsák	11	15
IV. (1965–1989)		
Hartyányi, Patay, Füzes, Facsar, Skoflek, Árendás, Hortobágyiné	7	163
V. (1990–2006)		
Jacomet, Dálnoki, Berzsényi, Kállay, Torma, Kovács, Gerócs, Verebes, Vályi, Bogard, Bending, Jones, Gyulai	13	229
Identification by non Archaeobotanist	–	40
Total	36	454

Table 21. Archaeobotanical Processing of Archaelogical sites in Hungary, 1877–2006.

Archaeobotanist	**Neolithic (6000–4300 BC)**	**Copper age (4300–3000 BC)**	**Bronze age (3000–900 BC)**	**Iron age (900–1 BC)**	**Roman age (1–middle 5th century AD)**	**Barbaricum (1st–5th century AD)**	**Migration period (476 AD–10th century)**	**Hungarian conquest time and Árpád age (895–1301 AD)**	**Late Medieval period (1301–1526 AD)**	**Modern age (1526 AD –**	**Uncertain age (prehistoric - historic)**	**Total**
Ascher – Wittmack (1877)				1								1
Berzsényi – Dálnoki (2005)	1											1
Bogard, Bending – Jones (2004)	1											1
Bogdán, Papp – Szabó (1963)										1		1
Dálnoki – Jacomet (2000)				2								2
Deininger (1877–1891)	1		2	1								4
Facsar (1984)				6								6
Füzes (1964–1990)	35	1	1		10		4	2	2	2		57
Gyulai (1983–2006)	34	24	33	12	17	16	16	26	21	5	10	214
Gyulai – Berzsényi (1998)			1									1
Gyulai – Dálnoki (1998)					1							1
Gyulai – Kállay (2002)					1							1
Gyulai – Torma (1999)			1									1
Gyulai, Torma, Kovács – Gerócs (1994)					1				1			2
Hartyányi (1967–1989)	12	1	12	3	16	3	6	13	13	7	7	93
Hopf (1962)			2									2
Lindau (1917)			1									1
Hung. Royal Seedstudy Station (Magy. Kir. Vetőmagvizsg. Áll.) (1940)								1				1
Maác (1955)	1											1
Skoflek – Árendás (1969–1971)						4						4
Skoflek (1984)									1			1
Skoflek, Hortobágyiné – Facsar (1973)									1			1
Torma (1996–1999)								1	2			3
Patay (1964)					1							1
Valkó (1960)			1									1
Vályi (2002)									1			1
Verebes (2000)			1									1
Wellmann, Mesch – Mándy (1963)										1		1
Tempír (1964)			3	1								4
Zsák – Tempír (1964)			1									1
Zsák (1960–1966)				1				2	1			4
Identification by non Archaeobotanist	5	1		1	11	1	4	1	6		10	40
Total	90	27	59	28	58	24	30	46	49	16	27	454

Table 22. Food, drink and other remains in Hungary, 1860–2006. Completed by F. Gyulai, 2006.

		Neolithic				**Copper age**			**Bronze age**			
Food and drink remains in English	**Remains in English**	2 sites	6 sites	4 sites	3 sites	7 sites	3 sites	2 sites	6 sites	7 sites	8 sites	1 site
		Early Neolithic (6000–5300 BC)	**Middle Neolithic (5300–4700 BC)**	**Late Neolithic (4700–4300 BC)**	**Neolithic without epoch**	**Early Copper age (4300–3500 BC)**	**Late Copper age (3500–3000 BC)**	**Copper age without epoch**	**Early Bronze age (3000–1600 BC)**	**Middle Bronze age (1600–1300 BC)**	**Late Bronze age (1300–900 BC)**	**Bronze age without epoch**
Bread	doughted fragment								26	2	6	
Breadcrust	doughted fragment											
Gruel/bread	fragment											
Gruel/bread	doughted fragment		9									
Gruel/bread	doughted fragment with fatty acid extract											
Cereal-gruel	non doughted fragment		5	42	11	10	15	10				
Cereal-gruel	non doughted fragment with fatty acid extract											
Cereal-gruel	non doughted fragment with imprint of the baked form on one side											
Cereal-gruel (clumsy milling product)	non doughted fragment	39	39			44			209	59	229	6
Cereal-gruel (fine milling product)	non doughted fragment		30	1		2	11					
Cereal-gruel (clumsy milling product) used fat by backing	non doughted fragment with fatty acid extract					29						
Cereal-gruel (medium clumsy milling product)	non doughted fragment								22	15		
Cereal-gruel (medium clumsy milling product)	non doughted fragment									3		
Cereal-gruel (fine milling product)	baked fragment with imprint of the baked form on one side											
Flour/cereal-gruel (fine milling product)	non doughted fragment								17			
Cereal-gruel (fine milling product)	non doughted fragment								68			
Cereal-gruel (fine milling product)	baked fragment with rectangular imprint of the baked form on one side								127			
Cereal-gruel (fine milling product)	baked fragment											
Cereal-gruel (fine milling product)/Cake	non doughted fragment								50			
Cereal-gruel (clumsy milling product) used fat by backing	non doughted fragment											
Cereal-gruel (fine milling product)/Cake	non doughted fragment with fatty acid extract		1									

Iron age			Roman age	Bar-baricum	Migration period				Hungarian conquest time and Árpád age				Late Medieval period			Modern age		Uncertain age	Total
4 sites	**1 site**	**5 sites**	**8 sites**	**8 sites**	**1 site**	**3 sites**	**1 site**	**1 site**	**2 sites**	**5 sites**	**6 sites**	**6 sites**	**9 sites**	**1 site**	**2 sites**	**0 site**	**2 sites**	**2 sites**	**118 sites**
Early Iron age (Hallstatt) (900–500 BC)	**Early Iron age (Skythian) (700–500 BC)**	**Late Iron age (La Tène) (5th century BC–1st century AD)**	**Roman age (1st–middle 5th century AD)**	**Barbaricum (1st–5th century AD)**	**Early Migration period (476–568 AD)**	**Late Migration period (Avar age) (568–895 AD)**	**Late Migration period (Karoling age) (811–895 AD)**	**Late Migration period (German, Slavonic) (6th–10th century AD)**	**Hungarian conquest time (895 AD)**	**Early Árpád age (1001–12th century AD)**	**Late Árpád age (13th century–1301 AD)**	**Árpád age without epoch**	**Hungarian Kingdom (13th–16th century AD)**	**Cumanian (15th–16th century AD)**	**Turkish occupation (16th–17th century AD)**	**17th–18th century–Total**	**19th century–Total**	**Uncertain age (prehistoric – historic)**	**Total**
			1	8						1	2	11	83		5		48	3	196
													1						1
				2													15	3	20
				1								4							14
												1							1
460	3	102		15		1831	8	1	5	x	35	5	151	28	32		1	6	2781
	1												13						14
				9							2								11
			148	xxxx								3							5776
			4	7									24					2	81
												2							31
										1	1	6							45
																			3
												7							7
																			17
1												9							78
																			127
				1								1							2
																			50
				9							23								32
												1						4	6

		Neolithic				Copper age			Bronze age			
Food and drink remains in English	**Remains in English**	2 sites	6 sites	4 sites	3 sites	7 sites	3 sites	2 sites	6 sites	7 sites	8 sites	1 site
		Early Neolithic (6000–5300 BC)	**Middle Neolithic (5300–4700 BC)**	**Late Neolithic (4700–4300 BC)**	**Neolithic without epoch**	**Early Copper age (4300–3500 BC)**	**Late Copper age (3500–3000 BC)**	**Copper age without epoch**	**Early Bronze age (3000–1600 BC)**	**Middle Bronze age (1600–1300 BC)**	**Late Bronze age (1300–900 BC)**	**Bronze age without epoch**
Cake/Dough	non doughted baked fragment used fine milling product											
Cereal-gruel/bread	fragment with trace of dough								405			
Cereal-gruel/bread	fragment with trace of dough											
Emmer (Triticum turgidum subsp. dicoccum) -gruel	Cereal-gruel with whole emmer wheat grains non doughted fragment									1		
Cereal-gruel/Bread/Cake	fragment											
Cereal-gruel/Bread/Cake with broken part of seeds	fragment with broken part of seeds											
One-course dish made with cereal milling produckt	fragment with fatty acid extract				12							
Foodremainss (cooked/ baked)	fragment											
White goosefoot (Chenopodium album) "cake" baked on leaves	non doughted fragment with rectangular imprint of the baked form on one side					20						
Bread made with gold-of-pleasure (Camelina sativa) seeds	fragment											
Bread made with millet (Panicum miliaceum) grains	fragment											
Millet-gruel	gruel-lumps	1	1	x					1	x	70	
Dish: in flour baked millet (Panicum miliaceum) used fat and blood	gruel-lumps											
Dish: in flour baked millet (Panicum miliaceum), field pea (Pisum sativum subsp. arvense) and on sun-dryed and grinded lamb												
Dish: in flour baked millet (Panicum miliaceum), foxtail millet (Setaria italica), field pea (Pisum sativum subsp. arvense) used fat by cooking												
Soup/vegetable-dish/one-course dish	burnt level an vessel and ceramicsurface				4					x		
Soup/vegetable-dish/one-course dish	independent fragment		1									

Iron age			Roman age	Bar-baricum	Migration period				Hungarian conquest time and Árpád age				Late Medieval period			Modern age		Uncertain age	Total
4 sites	1 site	5 sites	8 sites	8 sites	1 site	3 sites	1 site	1 site	2 sites	5 sites	6 sites	6 sites	9 sites	1 site	2 sites	0 site	2 sites	2 sites	118 sites
Early Iron age (Hallstatt) (900–500 BC)	Early Iron age (Skythian) (700–500 BC)	Late Iron age (La Tène) (5th century BC–1st century AD)	Roman age (1st–middle 5th century AD)	Barbaricum (1st–5th century AD)	Early Migration period (476–568 AD)	Late Migration period (Avar age) (568–895 AD)	Late Migration period (Karoling age) (811–895 AD)	Late Migration period (German, Slavonic) (6th–10th century AD)	Hungarian conquest time (895 AD)	Early Árpád age (1001–12th century AD)	Late Árpád age (13th century–1301 AD)	Árpád age without epoch	Hungarian Kingdom (13th–16th century AD)	Cumanian (15th–16th century AD)	Turkish occupation (16th–17th century AD)	17th–18th century–Total	19th century–Total	Uncertain age (prehistoric – historic)	Total
												2							2
		1																	406
												1							1
																			1
													32						32
																	18		18
																			12
													12					6	18
																			20
xxx																			500
xxx																			500
	1	50	1		12								xx						197
				xxx															500
									xxx										500
									xxx										500
													xxx						509
											21		17						39

		Neolithic				Copper age			Bronze age			
Food and drink remains in English	**Remains in English**	2 sites	6 sites	4 sites	3 sites	7 sites	3 sites	2 sites	6 sites	7 sites	8 sites	1 site
		Early Neolithic (6000–5300 BC)	**Middle Neolithic (5300–4700 BC)**	**Late Neolithic (4700–4300 BC)**	**Neolithic without epoch**	**Early Copper age (4300–3500 BC)**	**Late Copper age (3500–3000 BC)**	**Copper age without epoch**	**Early Bronze age (3000–1600 BC)**	**Middle Bronze age (1600–1300 BC)**	**Late Bronze age (1300–900 BC)**	**Bronze age without epoch**
Dish made with pea (Pisum sativum) seeds and emmer (Triticum turgidum subsp. dicoccum) grains	fragment											
Spice	leaves											
One-course dish used cereal milling produckt	burnt level an ceramicsurface										xx	
Meat-grual?	fragment											
Food remains	fragment			x								
Goulash soup mad from dough, vegetables, onion/ barlic and lamb	burnt level an ceramicsurface											
Lictarium (thin jam) made from sloe (Prunus spinosa), wild pear (Pyrus achras), crab apple (Malus silvestris), field rose (Rosa silvestris), cornelian cherry (Cornus mas)	burnt level an ceramicsurface											
Food remains with imprints of stem/leaves on the surface	fragment										234	
Caraway-seed soup	burnt level an ceramicsurface					5						
Fishsoup	burnt level an ceramicsurface											
Wild strawberry (Fragaria vesca) cake	fragment										164	
Drink remains (wine)	dryed material on ceramicsurface											
Drink remains (water mixed oil)	dryed material											
Drink remains (resin and oil)	dryed material											
Drinkholder vessel made from leather												
Total food an drink remains (calculated)		40	86	53	27	110	26	10	925	90	753	6

Legend (x=in case not mentioned number of seed):
x=sporadic (1–10 pieces, mean value 5),
xx=little (10–100 pieces, mean value 50),
xxx= much (100–1000 pieces, mean value 500),
xxxx= large (1000–10,000 pieces, mean value 5000),
xxxxx= extra large (10,000–100,000 pieces, mean value 50,000),
xxxxxx=extremely large (more than 100,000 pieces, mean value 500,000)

Iron age			Roman age	Bar-baricum	Migration period				Hungarian conquest time and Árpád age				Late Medieval period			Modern age		Uncertain age	Total
4 sites	1 site	5 sites	8 sites	8 sites	1 site	3 sites	1 site	1 site	2 sites	5 sites	6 sites	6 sites	9 sites	1 site	2 sites	0 site	2 sites	2 sites	118 sites
Early Iron age (Hallstatt) (900–500 BC)	Early Iron age (Skythian) (700–500 BC)	Late Iron age (La Tène) (5th century BC–1st century AD)	Roman age (1st–middle 5th century AD)	Barbaricum (1st–5th century AD)	Early Migration period (476–568 AD)	Late Migration period (Avar age) (568–895 AD)	Late Migration period (Karoling age) (811–895 AD)	Late Migration period (German, Slavonic) (6th–10th century AD)	Hungarian conquest time (895 AD)	Early Árpád age (1001–12th century AD)	Late Árpád age (13th century–1301 AD)	Árpád age without epoch	Hungarian Kingdom (13th–16th century AD)	Cumanian (15th–16th century AD)	Turkish occupation (16th–17th century AD)	17th–18th century–Total	19th century–Total	Uncertain age (prehistoric – historic)	Total
	1																		1
										57									57
																			50
																		34	34
																			5
									xxx										500
									xxx										500
																			234
																			5
		xxx																	500
																			164
x			x																10
			x																5
			x																5
						1													1
1466	6	653	169	5552	12	1832	8	1	2005	64	84	53	883	28	37	0	82	58	15119

15. Explanation of key terms in the text

Adult = fully developed individual plant.

Aerobe conditions = processes occurring in the presence of air (oxygen)

Agrobiodiversity = the multitude and abundance of plant varieties associated with or participating in crop production in agroecosystems, i.e. in the agricultural areas. Its role in the past is described by historical agrobiodiversity.

Anaerobe conditions = environment secluded from connection with air, contains no oxygen.

Anthropid = plant species introduced by humans or spread through human activities.

Anthropogenic category = classification group formed with regard to human use.

Archaeoadventive = plant species introduced in archaeological periods.

Archaeophyton = immigration of plant species that can be associated with well-specified periods.

Area = dispersion and spread of plant species.

Area type = geographical flora element types such as Balkanic, Carpathian, Mediterranean, Boreal, and so on.

Nutritional value = values suitable for the characterisation and comparison of specific species such as raw protein, hydrocarbon, raw fat, ash, raw fibre and dry matter content.

Biological species = a concept which considers species to be a constantly changing and forming biological organisation, being a "section", a "phase" in the permanent flow of evolution (evolutionary unit), as opposed to the phenotypical approach of species determination which can be characterised by the conventional morphological external (taxonomic) traits. In this context, the species is such a higher level of reproductive association where gene exchange might occur between members.

Boreal plant species = plant species of coniferous forests and marshes in the northern zone of the taiga. In Hungary they are relics, left over from a formerly existing cooler climate.

Convar. (convarietas) = (cultivated) subspecies, a group of cultivars consisting of variants being produced or grown in separate geographic regions.

Cytology = the science of cells.

Population boom = very rapid increase in the number of individuals in a population.

Diaspora = widely used term for joint indication of seeds and fruits.

Edaphic = different soil conditions formed on different basic rocks which affect the plants growing on them.

Composite growing (companion planting) = production of different plant species and cultivars at the same place at the same time (see "duplicate", composite sowing of rye and common wheat).

Original flora element = "spontaneous" species occurring in the former natural environment, which is not associated with human activities in the given area.

Erosion = deterioration of the soil surface. Soil erosion (runoff of fertile soil) can be caused by water or wind, and human activities might play a significant role.

Evolution = gradual development and transformation of the living world.

Extrazonal forest association = Due to special local climatic conditions such as valleys and slopes, forest associations belonging to a different zone appear (such as European beech associations – typical for higher elevations – in colder valleys within an oak-dominated association.

Variant = (cultivar) crop production unit created by targeted human activities such as crossing and selection, which differs from other cultivars at least in one economically significant trait (height, productability, resistance against diseases, and so on), and this trait is hereditary.

Phenotype = the form of appearance of individuals, the entirety of visible and measurable morphological, physiological and biochemical traits.

Fertile = fertile progeny of a crossing.

Gene centre = place of origin of a produced crop, where its wild ancestor is still found (primary gene centre) and having got into a new place and among new conditions, an even greater abundance of forms is developed (secondary gene centre).

Genetics = the science dealing with heredity.

Gene = unit responsible for a certain trait or property in heredity, located at a special locus on the chromosome. It is the smallest morphological unit of the chromosome, that base-sequence of the deoxyribonucleic acid chain that still contains information about protein amino acid composition.

Genotype = genetic structure of the organism, the entirety of its traits or genes.

Gradation of weeds = strong fluctuation and propagation in mass numbers of weed species.

Herbicide = chemical weed killer.

Herbology = knowledge of weeds.

Androgynous flower = a characteristic feature of the flower's sexual conditions, stamen (male) and pistil (female sexual organ) equally occur in such a flower.

Humification = humus generation, perishing and transformation into humus substances of plant residues in the soil.

Humus = slowly degrading organic matter in the soil made by microorganisms using plant and animal residues, consisting mainly of large molecules of dark coloured, colloid organic humic acids rich in carbon.

Intact layer = undisturbed layer in the soil.

Intensive weed = rapidly spreading, resistant weed, which is difficult to control.

Juvenile = young plant individual still in the process of development.

Calcinated seed = white coloured crystalline state of seeds after organic substances in them have been removed by heat. Morphological features can be more or less identified on such seeds.

Calibrated radiocarbon data = modern science provides the possibility for absolute dating of finds in calendar years as opposed to relative dating methods (stratigraphic, climatic, pollen, historical methods and the sequence created by archaeological artefacts). However, organic residue data obtained from radiocarbon (^{14}C) tests show a relatively wide range of variance. Using the independent method of absolute dating, which is customarily applied elsewhere but only currently being developed in Hungary by András Grynaeus, i.e. determination of growth rings of trees (dendrochronology), radiocarbon dates can be made more accurate and specified by years (BC, AD).

Carpology = knowledge of seeds and fruits.

Dioecious = (dioecious plant): separate flowers grow for stamina and pistils such as in common hop or nettle.

Continental elements = Plant species widespread in the inland, continental steppe and forested steppe regions of Eurasia.

Cosmopolitan = plant species found beyond the northern temperate zone, often with global spread.

Malacofauna = wildlife consisting of molluscs (Mollusca): bivalve molluscs ("shellfish") such as clams, scallops, and nucleobranches such as snails or slugs.

Meiotic cell division = nuclear division occurring in gametes (meiosis or miosis), resulting in only one set of chromosomes in the progeny cells, half of that in the somatic cells, where cell division is conservative (mitosis).

Biodegradation = deterioration of organic matter in soil with the help of living microorganisms (bacteria, fungi, algae).

Mycelium = the texture of fungi filaments (hyphens).

Nitrophil = plant species preferring nitrogen. They indicate richness of soil in nitrogen.

Plant macrofossils = plant residues, seeds, crops, bearings, fruits, twigs, mosses, as opposed to micro residues such as spores, pollen and tufts

Plant associations = phytocoenology.

Palynology = subject dealing with plant pollen, spores of fungi, in other words pollen analysis.

Phytolith = silicon-dioxide, calcium oxalate microscopic crystals (microfossils) found in the epidermis (epithelial tissue) of certain plant species and in the sclerenchyma (stereome) of others.

Polymerisation = chemical process where molecules of a substance are united in a web of cross-links, such as silicon rubber.

Population = the entirety of all the individuals of a species in any given area which are interconnected through mating.

Primary cultivar = improved varieties of grain crops selected in their respective gene centres which were further produced by humans at places more appropriate to their needs (primary cultivars). Some of the weeds occurring among them and found promising were selected for manipulation toward domestication (secondary cultivars): for instance fan-hen or white goosefoot (Chenopodium album).

Primary plant and animal biomass = total mass of all organisms found at a particular location at a particular time.

Relict species = leftover species, formerly widespread plants surviving in a more restricted area due to changing climatic conditions, where they still find suitable environment for their life conditions such as relict species from the Ice Age.

Ruderal plant associations = weed associations on downtrodden, degraded soils.

Secalietea species = grain weed in autumn sowing.

Subsp.= subspecies, a taxon within the species, plants belonging here might differ in the form they take, yet they are distinct in geographic distribution or ecological requirements; for instance, they may blossom in spring or in autumn.

Dry matter content = the substance of plant materials (for instance, grains) in a state free of water expressed in percentage value of their total mass. It is determined in a heating oven set at 105°C, with plant material dried up to the point where no more loss of mass occurs.

Segetal weeds = weed associations of sowing.

Secular development = a type of succession (climatic succession), temporal sequence of distinct plant associations in a specified area, where the changes take place over a very long period (for instance during an ice age) and they do not repeat themselves. As opposed to a biotic succession, where the very function of the plant association makes life conditions more and more unfavourable for itself leading to replacement by another association.

Selection = adaptation of living beings and associations to changing environmental conditions over a long period of time through the means of natural selection.

Subfossil seed = archaeological botanical macroscopic remains of plants preserved in a state suitable for identification due to continuous water cover.

Succession = temporal sequence of plant associations; for instance; changes in plant composition of silting-up dead-waters from aquatic associations to forest.

Characteristic species of an association = a species typical of a particular plant association as opposed to neutral species of an association.

Thermophylous plant species = plant species preferring warm environments such as a species originating in the Mediterranean.

Turfing = the creation of peat in bogs and marshes, where accumulated dead plant materials and humus are transformed by microorganisms into peat.

Var. (varietas) = intraspecific taxon, a series of plants having minor but hereditary characteristics in terms of form or quality, but not having distinct geographic distribution. For instance, there are 450 different known varieties of common wheat, and only one of these is erytrospermum ("red grain"), which all the Bánkúti breeds belong to.

Zonal forest association = the forest association most appropriate for the particular climatic conditions or altitude (e.g. European beech zone).

16. Archaeobotanical Database – supplement in CD

Compiled by Ferenc Gyulai 2006

A catalogue of the seed, fruit, food and beverage remains from Hungarian excavations, indicating species and number of items, classified on the basis of periods and sites, ranging from the Neolithic to the Modern Age, from the beginning of archeobotanical research up to the present in a tabular form.

About the author

Ferenc Gyulai (born in Debrecen/Hungary, 24th May 1955): agrochemist, environmental engineer, archeobotanist. – He graduated as an agricultural engineer from the Agrochemical Faculty of the University of Agriculture, Keszthely (1979), environmental engineer at the University of Petrochemical Industry, Veszprém (1986), University Doctor degree at University of Agriculture, Keszthely (1986), post-graduate (archaeobotanical) training at the University of Basel (1988–1990) by Prof. Stefanie Jacomet, Candidate of Science (CSc) degree in Biology (1994), PhD (1994), Széchenyi Professor Fellowship (1998), joint Fellowship of the American (National Science Foundation) and Hungarian Academy of Sciences (HAS) in North Carolina State University (Raleigh) (1999–2000) with Prof. Irwin Rovner, habilitation in Environmental Sciences (2003) and Doctor of HAS (DSc) degree in Agricultural Sciences (2005). – Curator for natural history at the Balaton Museum, Keszthely (1983–1988), Archeobotanist at the Hungarian Agricultural Museum (1990–1992), Senior member in HAS Institute of Archaeology (1992–1998), Biologist at the Institute for Agrobotany Tápiószele (1998–2005), Lecturer at Eötvös Lóránd University Budapest Institute for Archeological Sciences (1991–2000) and at the University of Pécs Department of Prehistoric Sciences (since 2004). Professor at Szent István University Gödöllő, Institute for Environmental Management, Department of Land and Landscape Management (since 2006), Leader of Nature Protection Engineering (MSc) Education (2006–) and Founder of Environmental Doctoral (PhD) School (2007–). Honorary Professor at Pannon University Georgicon Agricultural Faculty Keszthely (2007–). Founder of the biohistorical unit (historical agrobiodiversity) at Szarvasgede (1996). His special foci of interest are the study of plant macroremains (seeds and fruits) and food residues from archaeological excavations and the application of archaeometric methods. He has achieved significant results in the field of metric processing and statistical analysis of subfossil remains, investigation of native plant species and cultivars, conservation of agrobiodiversity.